KU-714-191

High Performance
Liquid
Chromatography

CHEMICAL ANALYSIS

A SERIES OF MONOGRAPHS ON ANALYTICAL CHEMISTRY AND ITS APPLICATIONS

Editor
J. D. WINEFORDNER
Editor Emeritus: **I. M. KOLTHOFF**

Advisory Board

Fred W. Billmeyer, Jr.
Eli Grushka
Barry L. Karger
Viliam Krivan

Victor G. Mossotti
A. Lee Smith
Bernard Tremillon
T. S. West

VOLUME 98

WILEY

A WILEY-INTERSCIENCE PUBLICATION

JOHN WILEY & SONS

New York / Chichester / Brisbane / Toronto / Singapore

High Performance Liquid Chromatography

Edited by

PHYLLIS R. BROWN

Department of Chemistry
University of Rhode Island
Kingston, Rhode Island

RICHARD A. HARTWICK

Department of Chemistry
Rutgers University
Piscataway, New Jersey

WILEY

A WILEY-INTERSCIENCE PUBLICATION

JOHN WILEY & SONS

New York / Chichester / Brisbane / Toronto / Singapore

Copyright © 1989 by John Wiley & Sons, Inc.

All rights reserved. Published simultaneously in Canada.

Reproduction or translation of any part of this work
beyond that permitted by Section 107 or 108 of the
1976 United States Copyright Act without the permission
of the copyright owner is unlawful. Requests for
permission or further information should be addressed to
the Permissions Department, John Wiley & Sons, Inc.

Library of Congress Cataloging in Publication Data:

High performance liquid chromatography / edited by Phyllis R. Brown.
　Richard A. Hartwick.
　　　p.　cm.—(Chemical analysis ; v. 98)
　　"A Wiley-Interscience publication."
　　Includes bibliographies and index.
　　ISBN 0-471-84506-X
　　1. High performance liquid chromatography.　I. Brown, Phyllis R.
II. Hartwick, Richard A.　III. Series.
QD79.C454H532　1988　　　　　　　　　　　　　88-726
543'.0894—dc19　　　　　　　　　　　　　　　　CIP

Printed in the United States of America

10　9　8　7　6　5　4　3　2　1

CONTRIBUTORS

D. D. Bly, E. I. duPont de Nemours & Company, Central Research and Development Department, Experimental Station, Wilmington, Delaware

Paolo Caliceti, National Institute of Diabetes and Digestive and Kidney Diseases, National Institutes of Health, Bethesda, Maryland

Peter W. Carr, Department of Chemistry, University of Minnesota, Minneapolis, Minnesota

Irwin M. Chaiken, National Institute of Diabetes and Digestive and Kidney Diseases, National Institutes of Health, Bethesda, Maryland

Stephen T. Colgan, Department of Chemistry and the Barnett Institute, Northeastern University, Boston, Massachusetts

Henri Colin, PROCHROM, Champigneulles, France

Giorgio Fassina, National Institute of Diabetes and Digestive and Kidney Diseases, National Institutes of Health, Bethesda, Maryland

J. Calvin Giddings, Department of Chemistry, University of Utah, Salt Lake City, Utah

Eli Grushka, Department of Inorganic and Analytical Chemistry, The Hebrew University, Jerusalem, Israel

Richard A. Hartwick, Department of Chemistry, Rutgers University, Piscataway, New Jersey

Jorgen Hermansson, Apoteksbolaget AB, Central Laboratory, Department of Biomedicine, Stockholm, Sweden

Roman Kaliszan, Medical Academy of Gdansk, Gdansk, Poland

B. Kaur, Wolfson Liquid Chromatography Unit, Department of Chemistry, University of Edinburgh, Edinburgh, United Kingdom

Laya F. Kesner, Department of Chemistry, University of Utah, Salt Lake City, Utah

J. J. Kirkland, E. I. duPont de Nemours & Company, Central Research and Development, Experimental Station, Wilmington, Delaware

J. H. Knox, Wolfson Liquid Chromatography Unit, Department of Chemistry, University of Edinburgh, Edinburgh, United Kingdom

Ira S. Krull, Department of Chemistry and the Barnett Institute, Northeastern University, Boston, Massachusetts

Donald J. Pietrzyk, Department of Chemistry, University of Iowa, Iowa City, Iowa

Thomas V. Raglione, Department of Chemistry, Rutgers University, Piscataway, New Jersey

Nicholas Sagliano, Jr. Department of Chemistry, Rutgers University, Piscataway, New Jersey

Göran Schill, Department of Analytical Pharmaceutical Chemistry, University of Uppsala, Uppsala, Sweden

Raymond P. W. Scott, The Perkin-Elmer Corporation, Norwalk, Connecticut

Carl M. Selavka, Department of Chemistry and the Barnett Institute, Northeastern University, Boston, Massachusetts

Richard C. Simpson, Department of Chemistry, University of Maryland, Baltimore County Campus, Catonsville, Maryland

William M. Skea, Millipore Corporation, Systems Division, Bedford, Massachusetts

U. Trüdinger, Institut für Anorganische Chemie und Analytische Chemie, Johannes-Gutenberg-Universität, Mainz, Federal Republic of Germany

K. K. Unger, Institut für Anorganische Chemie und Analytische Chemie, Johannes-Gutenberg-Universität, Mainz, Federal Republic of Germany

Stephen G. Weber, Department of Chemistry, University of Pittsburgh, Pittsburgh, Pennsylvania

W. W. Yau, E. I. duPont de Nemours & Company, Central Research and Development Department, Experimental Station, Wilmington, Delaware

Ilana Zamir, Department of Inorganic and Analytical Chemistry, The Hebrew University, Jerusalem, Israel

PREFACE

There has been a dramatic improvement in high-performance liquid chromatography (HPLC) since the first instruments were designed in the early 1960s. In fact, HPLC has come such a long way that it is almost a new separation technique. It has changed the information we get, revolutionized biomedical research, and contributed to the growth of the biotechnology industry.

Since the early 1970s there has been a tremendous surge not only in the development and refinement of HPLC instrumentation, but also in the applications of HPLC. In most biochemical and biomedical laboratories HPLC is indispensable for the exploration of biological systems, the determination of the composition of physiological fluids and cells, and the quantitation of minute amounts of biologically active constituents. The molecules can be very small (MW 100) or their molecular weights may range from thousands to hundreds of thousands. The analyses, which are rapid, can often be carried out without loss of the biological activity of the compound. HPLC can be readily automated; thus one technician can generate, in a relatively short time and with minimum effort, vast amounts of data which are highly reproducible. Quantitative results are produced automatically and tabulated clearly, with many routine calculations performed concomitantly. Since we can use HPLC to explore biochemical systems in a breadth and depth not possible previously, this technique has been instrumental in the development and explosive growth of biotechnology. Moreover, the use of HPLC has not been confined to analytical chemistry or the biological sciences; HPLC is now used as a powerful separation technique in all branches of science and industry that require the separation of components in a complex mixture or the isolation or purification of a substance.

In 1954 H. H. Strain et al. (1) wrote about chromatography:

> This increase in the number and applications of these basic analytical tools and a concomitant multiplication of the workers versed in their use have stimulated progress in all aspects of science concerned

with chemical substances and their reactions. The resultant expansion of knowledge has been so rapid, so great and so diverse that it cannot be cited here.

Thus what was written about chromatography in general over three decades ago is even more descriptive of HPLC today. HPLC has, indeed, stimulated progress in all areas of science and industry where good separations play a vital role.

In the first phase of the development of HPLC the emphasis, and rightly so, was on chromatographic theory. In the second phase, which was roughly between 1970 and 1980, the focus was on designing and refining instrumentation, columns, and column packings. In the third phase, the one we are now in, the direction is on applications and the practical use of the technique. Because of the rapid and extensive strides made recently, HPLC is not only an extremely important analytical tool but also vitally important in preparative and process-scale operations. Thus as a large-scale technique, HPLC is a powerful tool in industry for the isolation and purification of substances such as therapeutic drugs and biotechnology products for which a high degree of purity is required.

In 1978 Dr. Philip Elving and Dr. James Winefordner asked me to write a book on HPLC for the "Treatise on Analytical Chemistry" series. At that time I refused because I thought that there existed a plethora of books on HPLC. In addition I felt that the field had not stabilized; too many developments were taking place rapidly in both instrumentation and applications. However, I now feel that there are fewer real innovations and the field has stabilized somewhat. Also I have been grappling with questions on the future direction of separation science and, in particular, liquid chromatography. I decided that the time was right for a new book on HPLC, a book that would consolidate the basic theories of chromatography along with the more exciting technical developments in the field. In addition I wanted to address in this book some questions that concern all thoughtful researchers in separation science: What is the current state-of-the-art in liquid chromatography? Has the development of liquid chromatography plateaued? If so, what new methods will take its place or complement it? If not, where will the new frontiers be and what direction will liquid chromatography take?

Since the developments in the field of HPLC had come so rapidly and massively, I could not cover by myself in depth discussions of all these points. Therefore I agreed to edit a book on the condition that I could work with a coeditor and that we could have each chapter written by an investigator who either was a known authority or had done considerable work in the field. I was delighted that Dr. Richard Hartwick agreed to be my coeditor. I felt that the book needed the input of one of the leaders of the younger generation of chromatographers.

Together we outlined our goals for the book. First, we wanted to have the basic theoretical concepts available to those who will use and read this book. Second, we wanted the book to be a broad-based general book, a "classic" in the sense that it would be valuable to chromatographers, users of chromatography, and those who teach analytical chemistry (and especially separation science) 10 years from now as well as at present. Therefore we did not include any chapters on instrumentation per se, specific applications, or details that might be outmoded soon after the book is published. That material belongs in current analytical journals or journals devoted to specific topics. More importantly, we wanted knowledgeable and thoughtful researchers to discuss the "cutting edge" of HPLC in their areas of expertise.

I greatly appreciate the time, energy, and effort that went into writing each chapter and thank the authors for their cooperation and for their superb contributions. Their excellent manuscripts were so meticulously organized and written that they required almost no editing and no revisions. The expertise and talents of our authors will be evident to all who read this book. Finally, I thank Dr. Hartwick, who worked so closely with me and made my job an easy one.

PHYLLIS R. BROWN

Kingston, Rhode Island
August, 1988

CONTENTS

High Performance
Liquid
Chromatography

CHAPTER

1

THE THEORY OF THE DYNAMICS OF LIQUID CHROMATOGRAPHY

STEPHEN G. WEBER

Department of Chemistry
University of Pittsburgh
Pittsburgh, Pennsylvania

PETER W. CARR

Department of Chemistry
University of Minnesota
Minneapolis, Minnesota

LIST OF SYMBOLS

Roman Symbols

A_i	Weighting factor in optimization function
A_m	Area through which mobile phase flows (length^2)
A_s	Surface area of stationary zone (length^2)
B/A_f	Peak asymmetry
c	Concentration (mass·length^{-3})
$\bar{c}$	Solute concentration in specified volume (mass·length^{-3})
c^0	Injected concentration of solute
c^*	Equilibrium concentration [moles·(total column volume)$^{-1}$]
$K = \exp$	Laplace transform of mobile zone concentration (mass·length^{-3})
$c_{\max}$	Maximum concentration in peak (mass·length^{-3})
$\mathbf{D}$	Dispersion tensor (length^2·time^{-1})
D	Molecular diffusion coefficient (length^2·time^{-1})
D_H	Hydraulic diameter [4 × (cross-sectional area exposed to flow) (wetted perimeter)$^{-1}$]
$\mathscr{D}$	Dispersion coefficient (length^2·time^{-1})
d_p	Particle diameter
d_s	Diameter of micropores in a zeolite
$\mathbf{f}$	Function from which dispersion is calculated
f	Friction (mass·time^{-1})
f_a	Moles sorbed onto a unit of stationary phase over time d_t
f_d	Moles desorbed from a unit of stationary phase over time d_t
$f(c)$	Function of concentration
F	Volumetric flow rate (length^3·time^{-1})
h	$= H/d_p$
H	Height equivalent to theoretical plate, $= L/N$ (length)
K	Thermodynamic equilibrium constant
k	Rate constant (time^{-1}); Boltzmann's constant (energy·K^{-1})
k'	Phase capacity factor, $= (t_R - t_0)/t_0$
k''	Zone capacity factor, $= k_0 + (1 + k_0)k'$
k_0	$= \epsilon_p/\epsilon_e$
$\tilde{k}$	Ratio of k' values for the same solute on two stationary phases
k_{mt}	Mass transfer coefficient (length·time^{-1})
L	Column length
L_{ij}	Proportionality constant between flux i and gradient of chemical potential j
l	Jump length
m	Mean value of distribution

M_a	Molecular weight of solvent a (mass·mol^{-1})
N	Number of theoretical plates, $= L/H$
n	Peak capacity; moles (specified in context)
n	Number of steps taken in random walk
n_{eff}	Effective number of theoretical plates
n^0	Mass in pulse input (mass)
p	Laplace variable (time^{-1})
p	Exponent in empirical expression for diffusion coefficient
P	Probability
Pe	Peclet number, $= vl/\mathscr{D}$, where v is velocity, l characteristic length, and $\mathscr{D}$ dispersion coefficient; unless stated otherwise, $l = d_p$ and $\mathscr{D} = D$
p_i	Ratio of average peak height to trough height for pair of peaks
p_0	Reference value of p_i
q	Stationary phase concentration (mol·length^{-3} or mol·length^{-2}, sometimes mol·mass^{-1})
R	Solute zone velocity as a fraction of interstitial mobile phase velocity, $= (1 + k'')^{-1}$
R	Gas constant (energy·mol^{-1}·K^{-1})
r	Particle radius; variable in radial dimension (length)
r_0	Tube radius
Δr_o	Film thickness around a spherical particle
R_s	Resolution
Re	Reynolds number, $= u_e D_H \rho/\eta$
S_0	Concentration of adsorption sites (mol·length^{-3})
Sc	Schmidt number, $= \eta/\rho D$
Sh	Sherwood number, $= k_{mt} d_p/D$ for particles
T	Temperature (K)
t	Time
t_i	Retention time for solute i
t_L	Retention time of last component in mixture
t_M	Maximum retention time desired
t_R	Retention time
t_0	Retention time for unretained but permeating solute
u	Average mobile phase velocity in open conduit (length·time^{-1}); velocity of permeating but nonadsorbing solute, $= u_e/(1 + k_0)$
u_e	Interstitial velocity of mobile phase (length·time^{-1})
$\overline{v}$	Average zone velocity, $= Ru_e$ (length·time^{-1})
$\hat{v}$	Difference between average velocity and local velocity (length·time^{-1}), a vector

V_i	Molar volume of solute i at boiling point of i (length$^3 \cdot$mol^{-1})
w	Total mass of substance injected
W_i	Width at base of peak i
y	Void corrected dimensionless time, $= t/t_0 - 1$
z_i	Number of charges on species i

Greek Symbols

α	Separation factor, $= k_2'/k_1'$; selectivity
β	Weighting factor in optimization functions
γ	Tortuosity ($\gamma < 1$); dimensionless rate, $= k_d t_0$ (specified in context)
δ	Diffusion length; diffusion layer thickness (length)
$\delta(t)$	Delta function in time
ϵ	Nonequilibrium; porosity ratio, $= (1 - \epsilon_T)/\epsilon_T$ (specified in context)
$\bar{\epsilon}_{mob}$	Average value of nonequilibrium parameter
ϵ_p	Intraparticle porosity; volume of intraparticle space divided by total column volume, $= \epsilon_i(1 - \epsilon_e)$
ϵ_e	Excluded or interparticle porosity; volume of interparticle space divided by total column volume
ϵ_i	Particle porosity; volume of intraparticle space divided by particle volume
ϵ_T	Total column porosity, $= \epsilon_p + \epsilon_e$
ζ	Correction term for solvent association in Wilke–Chang expression
η	Viscosity (mass$\cdot$length$^{-1} \cdot$time^{-1})
κ	Rate constant (length$^3 \cdot$time^{-1})
λ	Constant in eddy dispersion, near 0.5 in the range $0.5 < \nu < 20$
μ_n'	nth moment about origin
μ_n	nth central moment
ν	Reduced velocity, $= u_e d_p/D$
χ	Dimensionless distance
ρ	Density (mass$\cdot$length^{-3})
σ	Standard deviation
σ^2	Variance
Φ	Phase ratio
Ψ^a	Electrostatic potential in phase a (energy$\cdot$charge^{-1})
ω	Constant in mobile phase mass transfer

1.1. INTRODUCTION

The theory of the dynamic processes involved in chromatography is a complex subject, the details of which interest few chromatogaphers. So why read on? Because an appreciation for the lessons learned by those who have studied the dynamics of liquid chromatography will lead to sound chromatographic procedures. Certainly separations are achieved based on thermodynamic differences in the properties of the eluates. But the possibilities allowed by thermodynamics are only manifested by following the principles that have evolved by studying accurate models of the dynamic events. Perhaps a small demonstration of the power of sound design with good dynamics is in order.

All chromatographers have been found at some time or other peering at a recorder tracing, rationalizing the relative retention of a set of peaks from a set of similar compounds. You have, haven't you? You probably talked about a hydrogen bond that is less likely to form here than there because of steric hindrance, or the different basicities in a pair of heterocycles, or some other equally interesting intermolecular interaction. Quickly, now, what is the free energy of formation of a hydrogen bond? Yes, it depends, but 2–10 kcal/mol is a reasonable answer. Now let us figure out what difference in free energy of transfer from the mobile phase to the stationary phase is required to separate a pair of components in a mixture. We shall base our example on the separation of a pair of peaks with a resolution of 1 (that is, almost baseline separated). From the resolution equation, and with the assumption that k' is large and the number of theoretical plates is 10,000, a separation factor α of 1.042 is calculated. (For details, see Appendix 1.1.) This translates into a difference between the transfer free energies of the two compounds of merely 0.025 kcal/mol. This is only a small fraction of RT (about 0.6 kcal/mol at room temperature) and is well inside the bounds of experimental uncertainty for typical thermochemical measurements. Thus while it remains an exciting and important area of research, the *a priori* prediction of retentions in liquid chromatography is not now possible. Dynamics to the rescue. If one cannot predict retention, the best way to obtain success without involved optimization is to use a system with as much efficiency, as much room for peaks as possible. In this same context Giddings has pointed out that in a complex (crowded) chromatogram the improvements in chemical selectivity in one pair of peaks often result in a decrease in other areas of the chromatogram. Thus the only real hope of improving the separation as a whole is to improve the efficiency of column performance. This means that the dynamic aspects of the process must be understood.

The future will bring new challenges to the separation scientist that will require a familiarity with the fundamentals of the theory of chromatography. We shall name just a few. Recent research has shown that the peak capacity of a column needs to be much greater than one might guess in order to have a reasonable probability of obtaining a complete separation of all the components in a mixture of moderate complexity (1–5). Thus improvement of the peak capacity of a column (see below) is of importance even for relatively "simple" mixtures containing 10–40 components. Postcolumn reaction detectors will become more important for selective detection. The design of such systems and the interface between them and chromatographic columns requires attention to dynamic theory. New detection techniques that are information rich, in combination with statistical operations such as factor analysis, are able to find two or three components in a single peak. Would it not be helpful if the chromatographic peak shape were accurately known? Then overlapping curve resolution (with or without the information provided by factor analysis) would be more certain. Improvements in the performance of separations of biomolecules will be of great importance in realizing the promise of biotechnology. Of course, the ability to improve old techniques and invent new ones is important, too. There are thus many reasons for understanding the theory of the dynamics of chromatography.

This chapter is divided into three parts following the Introduction. The first describes measures of column performance and their interrelationships. It also includes a section on the experimental determination of measures of column performance. The second part contains elements of the theory of chromatography. A description of the events that govern solute behavior is given first. This chapter is not meant to be a detailed exposition of the methodologies used to develop chromatographic theory; it is intended to be a discussion of the results of a large body of experimental and theoretical work, from both the chromatographic field and the engineering field, that is relevant to an understanding of the fundamental dynamic processes that take place in a liquid chromatographic column. The material is written for those who have been practicing chromatography and have had some exposure to theory at an advanced undergraduate level. An understanding of the essential features of retention and the meaning of parameters that describe retention (Chapter 2) is assumed. We strongly advise the interested reader to pursue the details by referring to the cited literature. In particular, we have attempted to bring the reader closer to the engineering literature, where much valuable work on separations has appeared. A brief discussion of the historical aspects of the development of the ideas that we now view as being established fact is worthwhile. Thus a historical discussion is presented. This is followed by a discussion of the (relatively) simple case of solute zone spreading in a straight open tube. This is not only of importance in flow injection techniques, but is a key factor in open tubular columns and in injection valves, detectors, and connecting tubing. The case of a bed packed with nonporous particles is taken up next. This is a complex system in its own right and deserves separate study. One can learn a great deal from the consideration of a column with one theoretical plate. A single-stage continuous-flow stirred tank reactor will be considered in this light. Finally, a discussion of packed beds for both linear and nonlinear chromatography concludes the chapter. The third part is a group of appendices in which more detailed treatments of a variety of topics can be found. References that have been marked by an asterisk are recommended reading.

1.2. MEASUREMENT OF SEPARATION EFFICIENCY

There are many measures of the "goodness" of a separation. They fall into three classes (see Table 1.1). One is the set of measures that can be made on a single peak, another is the set of measures that can be made on a pair of peaks, while the last is the set of measures that can be made on more than two peaks. This list is neither comprehensive nor are its

Table 1.1. Measures of "Goodness" of Separation

One Peak

μ_i	Statistical moments
N	Number of theoretical plates (column efficiency)
H	Height equivalent to a theoretical plate
h	Reduced plate height, $= H/d_p$
n_{eff}	Effective number of theoretical plates
B/A_f	Asymmetry (defined at some fraction of peak maximum)

Two Peaks

R_s	Resolution
n	Peak capacity
p_i/p_0	

More Than Two Peaks

map	Minimum alpha plot (window diagram)
CRF	Chromatographic resolution factor
COF	Chromatographic optimization function
ORM	Overlapping resolution map

members completely independent. However, it is a representative list that will give the reader the feeling for the information contained in such measures.

1.2.1.   Measurements Made on a Single Peak

Since a measurement made on a single peak cannot possibly be used to assess the quality of a separation, it is obvious that such measurements really say more about the column and its native abilities than about a particular separation. Before describing the measurements made on a peak, one must first consider the peak itself.

A chromatographic peak is best viewed as the distribution of transit times, that is, elution times, (6) of individual solute molecules (Fig. 1.1). In this sense, a chromatographic peak is similar to other distributions, such as the distribution of grades on an exam, of molecular weights of synthetic polymers, or of nuclear decay events. As is the case of all distributions, it is very useful to characterize chromatographic peaks by quantities termed moments. Moments are the parameters that characterize distributions. Since the concept is a statistical one, moments are commonly defined for distributions with unit area. A moment of any distribution can be calculated by dividing the relevant integral (see below) by the area of the distribution. This normalization process is equivalent to starting with a distribution that is unitless with an area of 1. Moments are really quite familiar in many areas of the physical sciences. For example,

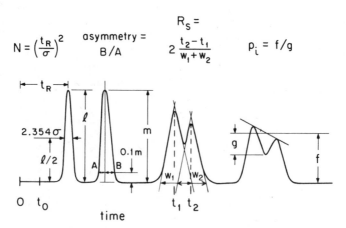

$$N = \left(\frac{t_R}{\sigma}\right)^2 \qquad \text{asymmetry} = \frac{B/A}{} \qquad R_s = 2\frac{t_2 - t_1}{w_1 + w_2} \qquad p_i = f/g$$

Figure 1.1. Liquid chromatogram showing calculations of various measures of the goodness of the separation. First peak—N; second peak—asymmetry; third and fourth peaks—resolution; fifth and sixth peaks—p.

the center of gravity corresponds to the first moment of a distribution of mass. Some definitions of moments of continuous distributions are shown in Table 1.2 (7). Thus the zeroth moment of a chromatographic peak is its area (total mass). The first moment (normalized) is the position of the center of mass, also called the peak centroid, and corresponds to the elution point of an average molecule. The peak centroid is the measure of retention that is independent of dynamics; it depends only on thermodynamics, mechanics, and physical dimensions.

Many of the processes that cause an initially narrow spike of concentration to broaden are formally similar to diffusion (8). For this reason, the Gaussian distribution of solute is often an appropriate, if idealized, model. A narrow concentration distribution of a solute in a stationary solvent will broaden with time as a Gaussian band due only to diffusion. In terms of moments one can write a Gaussian peak as

$$c(x) = c_{\max} \exp\left[-\frac{(x - \mu_1')^2}{2\mu_2} \right] \qquad (1)$$

(see Table 1.2 for a definition of the various terms), where $c_{\max}$ is the amplitude of the distribution and x is the dimension (time, volume, distance) over which the concentration varies. Notice that, unlike most real chromatographic peaks, a Gaussian distribution is perfectly symmetrical about its centroid. A narrow, or sharp, distribution has a relatively small μ_2. The quantity μ_2 is often denoted as the square of the standard deviation or variance σ^2. The normalized standard deviation σ/m is independent of the scale and units of the retention measurement, so it is a

Table 1.2. Moments of a Normalized (that is, Area = 1) Distribution $f(x)$

Moment	Mathematical Definition
nth moment about origin	$\mu_n' = \displaystyle\int_{-\infty}^{\infty} x^n f(x)\, dx$
nth central moment (moments about mean m)	$\mu_n = \displaystyle\int_{-\infty}^{\infty} (x - m)^n f(x)\, dx$
A primed moment is one taken about the origin.	
An unprimed moment is taken about the first moment about the origin μ_1'.	
μ_0' is the peak area	
μ_1' is the mean m	
μ_2 is the variance σ^2	

useful way to characterize peaks. Good chromatography corresponds to narrow peaks, a small σ/m. This must also be reflected in the normalized variance σ^2/m^2. The inverse of the latter quantity is the number of theoretical plates N. Large N means that the peaks will be relatively narrow. A column may be considered as being made up of a series of individual plates (but see Section 1.3.2.2). The number of plates in a column of length L is N, so the height equivalent to a theoretical plate, HETP or just H, is L/N. The ratio of H to the average diameter of the particles making up the packing d_p is called the reduced plate height h (9, 10). Although the theory from which this nomenclature is derived (11) is now considered highly oversimplified (12), this terminology is so deeply embedded that it would serve no good purpose to discard it. N, also frequently called the efficiency of the column, is a useful figure of merit. The higher N is, the better will be the column for doing any separation. The Gaussian peak shape equation can, by use of the above definitions, be rewritten in a very useful form,

$$c = w \sqrt{\frac{N}{2\pi}} \exp\left[-\frac{N(x - \mu'_1)^2}{2\mu'^2_1} \right] \tag{2}$$

where w is the mass of the solute injected onto the column. Inspection of Eq. 2 indicates that an increase in N, all else being constant, increases the peak amplitude (thereby improving detectability) and decreases the width of the peak.

There are other parameters or figures of merit that are closely related to the number of plates and the peak width. For example, the effective number of theoretical plates is derived by considering t_0 (the elution time of an unretained species) as the zero of the time axis rather than the point of injection. This results in the relationship (13)

$$n_{eff} = \left(\frac{k'}{1 + k'} \right)^2 N \tag{3}$$

Other figures of merit divide N or n_{eff} by the retention time to yield the rate of generation of theoretical plates.

Chromatographic peaks may spread asymmetrically. While such spreading influences the variance, the variance is not a proper measure of peak asymmetry ("tailing" or "fronting"). Tailing, or fronting, may be due to column overload (nonlinear chromatography). Tailing may also be due to a poorly packed column, a small number of theoretical plates, the existence of more than one type of adsorption site on the column, extracolumn band broadening (due to faulty technique or improperly designed or installed hardware), or excessive electronic filtering of the output signal. The latter effects can be estimated as the peak asymmetry. A rigorous statistic, the peak skew, is defined as μ_3/σ^3. It has the same qualitative information as the asymmetry but is more difficult to measure with precision and accuracy than is the asymmetry ratio B/A_f (see Table 1.1 and Fig. 1.1). Consequently, peak tailing is most easily and reproducibly determined using the asymmetry ratio.

1.2.2. Measurements Made on More Than One Peak

Measures that use information from two peaks are more informative about the quality of the separation than are measures made on a single peak. The most commonly used parameter is the resolution, given in two equivalent forms in Eqs. 4 and 5 (see Fig. 1.1),

$$R_s = 2\frac{t_2 - t_1}{w_1 + w_2} \tag{4}$$

$$= \frac{\sqrt{N}}{4}\left(\frac{\alpha - 1}{\alpha}\right)\left(\frac{k'}{1 + k'}\right) \tag{5}$$

In Eq. 5, k' refers to the second peak of the pair. The baseline width is w. It is equal to 4σ if the peak is Gaussian. Equation 4 is useful for laboratory calculations, while Eq. 5 is more useful for conceptual purposes. For peaks of equal size, an R_s of 0.5 will yield what looks like a broadened peak, while 0.7 yields a respectable trough. An R_s of 1.0 is sufficient for quantitative analysis if the determination is done by using peak height measurements, and an R_s of 1.25 yields baseline separation (14, 15). Of course, resolutions greater than this, for peaks of equal size, merely waste time. When the species of interest elutes on the tail of a solvent or major component peak, very high resolution may be required for adequate purification or quantitation.

Equation 5 is perhaps the most important and useful equation that results from all chromatographic theory. It should be thought about as expressing the following concept:

resolution = efficiency × selectivity × retention

The primary "laws of practical chromatography" are direct consequences of this formulation.

1. *To get any resolution the species must be retained ($k' > 0$).* The capacity factor of the later peak ought not be too close to zero. If it is near zero, the molecules of interest are interacting feebly with the stationary phase, or there is very little stationary phase. Note also that, for all practical purposes, once k' is greater than about 5 or so, there is no substantial reward for increasing it further, but the analysis time increases in proportion to the increase in k' of the last eluting species.

2. *If the selectivity α is close to unity, one must improve the selectivity to improve the separation of the desired pair of peaks.* This rule is based on the fact that separations are based on thermodynamics. There must be *some* difference in the distribution coefficients of the molecules in the mixture in order for a separation to occur. For $k' > 5$ or so, the efficiency N and the selectivity influence the separation. For a given desired resolution, a large α allows one to do fast separations since high-speed separations have low N. Thus increasing α can indirectly lead to a decreased analysis time.

3. *Make sure the column is as efficient as it can be.* The quality of the separation depends on the narrowness of the retention time distribution; narrower peaks allow a separation to occur with a smaller difference in the solute's distribution coefficients than that required for wider peaks.

Another derived parameter is the peak capacity n. This is the number of peaks that will just fit between two values of retention time t_1 and t_2 (1, 16, 17) with a given resolution. Equation 6 uses a resolution of 1,

$$n = 1 + \frac{\sqrt{N}}{4} \ln\left(\frac{t_2}{t_1}\right) \qquad (6)$$

If the chromatogram is sparsely populated with peaks, but the separation is incomplete (that is, there are more *components* in the mixture than *peaks* in the chromatogram), the chemistry should be altered in a first attempt to improve the separation. One way to keep track of the influence of the effect of alterations of the mobile phase, stationary phase, or temperature is by using a minimum alpha plot (18). With this method one plots α, the ratio of k' values for a pair of components, as a function of some change in the operating conditions (such as temperature, phase composition, ratio of lengths of two columns in series). This method was first applied in gas chromatography with two relatively weakly interacting stationary phases, so retention times of all the components in a given mixture are linear functions of the volume fraction of one of the stationary phases. Since the changes are linear, only a few experiments are required to characterize the system. When there are several components in the chromatogram, one can plot the α for all the pairs. In this case, one only cares about failure of the separation. If two or more pairs of components elute as single peaks, the separation is really no worse than if only one pair of components elutes as a single peak; neither separation works. It makes sense, then, to keep track of the minimum α. Of all the minimum α values, the conditions that produce the highest α are sought. This yields the conditions corresponding to the best separation.

In optimizing the separation of multicomponent mixtures, one finds that the resolution R_s is no longer a useful indicator of the goodness of the separation. A measure of the goodness of a multicomponent separation that obviates this shortcoming is the so-called chromatographic resolution factor CRF (19),

$$\text{CRF} = \sum \ln\left(\frac{p_i}{p_0}\right) + \beta(t_M - t_L) \tag{7}$$

Equation 7 shows the formula for it. For each pair of compounds a p_i is measured as shown in Fig. 1.1. The desired value of p is p_0. The second term accounts for the separation time t_L being longer than some desired maximum time t_M. β is a weighting factor that allows more or less importance to be given to the time of analysis. Glajch et al. (20) added a weighting factor for each pair of peaks, and used resolution in place of p, which resulted in the chromatographic optimization function COF shown in Eq. 8,

$$\text{COF} = \sum A_i \ln\left(\frac{R_i}{R_0}\right) + \beta(t_M - t_L) \tag{8}$$

In this equation R_i is the resolution between the two peaks and R_0 is a desired resolution. In this way, the overlap of two unimportant peaks can be made less important than the overlap of a pair of important peaks.

A drawback to the approaches discussed above is that the identity of the peaks is not necessarily known or important in calculating a CRF or a COF unless, of course, values of A_i vary widely. But it may be desirable to know the nature of the species that limit the goodness of the separation. Such considerations led Glajch et al. (20) to reconsider the use of R_s. They used a three-component mobile phase and obtained chromatograms for a mixture over a wide range of solvent compositions. The resolution between each pair of peaks can be calculated at each composition. When the resolutions corresponding to a particular pair of peaks are plotted on a triangular coordinate system of composition, one can visualize a pattern; there are regions in "mobile phase composition space" with high and low resolution. One may pick an acceptable resolution, say, $R_s = 1$, and color the region in composition space where $R_s < 1$ black. If this is done for each pair of peaks, and the resulting resolution maps are superimposed, the region in composition space left white yields acceptable resolution for all the species in the mixture.

1.2.3.   Making Measurements on Experimental Data

We now turn to the question of the best way to measure the parameters of the distribution, that is, the chromatographic peak, that have been discussed. Actually, the story is simple. At the base of all the measurements are the statistics that characterize the distribution. Generally, one is primarily concerned with the first moment, the mean or retention time, and the second central moment, the variance. We have already dispatched the skew in favor of the simpler to measure asymmetry. If the peak happens to be totally symmetrical, then the peak maximum is the mean value. However, most chromatographic peaks are tailed, so the mean occurs some time after the maximum. The direct calculation of the first moment (see Table 1.2) is possible by numerical integration. One must be aware that errors of several percent are likely if the baseline is poorly defined. With a flat baseline the percent standard deviation of the first moment is on the order of $100/N$.

The direct calculation of the variance is far less precise and is more error prone. Note in the formula for variance that it is a weighted average of the time of appearance in the detector, with the mean as the origin. The weights are the squared distances from the mean. This means that the data that have high signal to noise ratio, (snr), near the peak, are weighted lightly, while the data with the poorest snr, near the baseline, are weighted heavily.

In order to avoid the noise problem, curve fitting has been used. In this approach, the noisy chromatographic peak is approximated by any of several mathematical functions that have been shown to fit chromatographic peaks. The requirement is that the functions naturally be able to approximate a slightly skewed Gaussian-like peak to a high degree of accuracy. The exponentially modified Gaussian peak is one of these. It is the result of the convolution of the Gaussian peak with an exponential decay. Its use has recently been reviewed, and errors in the formulas used in the literature have been corrected (21). The beauty of this model is that the peak shape is only a function of the ratio of the standard deviation of the Gaussian peak (before it was convolved with the exponential) and the exponential decay time. It allows one, for example, to specify the amount of electronic filtering (which corresponds to convolution with an exponential if it is a simple RC low-pass filter) that can be tolerated given a certain peak's standard deviation (22). The difficulty with this approach is that it has only one asymmetry parameter, so the fit is not very flexible. Furthermore, there is no a priori reason to believe that it can fit peaks that are inherently tailed due to intracolumn processes.

The Chesler–Cram equation has been used by several workers (23–26). It has eight adjustable parameters, nine if one allows the baseline to be estimated as well. It can fit liquid chromatographic peaks quite accurately. But nine parameters? You could fit an elephant with nine parameters (27)! We have all learned that models with fewer parameters are better than those with many. That is true if the sole objective is the determination and interpretation of the parameters themselves. In this case, the parameters are not used. The best fitting equation, which is now noise-less, is well suited to the calculation of moments by the straightforward integrations suggested in Table 1.2. To some degree this procedure amounts to transferring one's lack of knowledge of the baseline to a distributed lack of knowledge of the shape of the peak. The latter is represented by the lack of perfect fit of the equation to the data, even when the data are noiseless. A seemingly better function than the Chesler–Cram equation is the Edgeworth–Cramer series (28, 29). It has a sound statistical basis and generally fits experimental data very well.

Less computationally intensive procedures are available and have distinct advantages (30–32). A very simple procedure, useful for a comparison of two columns, or for keeping track of column performance, is the height to area ratio. Noting that the area of a peak is proportional to the amount of substance injected w, one can see from Eq. 2 that the height to area ratio is proportional to $\sqrt{N}$.

Other simple procedures exist for obtaining more detailed information. They rely on the determination of the retention time, the asymmetry, and the width at a certain height above the baseline. When this height is between 0.1 and 0.5 of the peak maximum, reliable statistics are obtained almost trivially. The recommended height is 0.1 (31). It must be realized that the procedure is limited by the model used for determining the relationships between the statistics and the measured values. If the peak to which these equations are applied is not well described by the exponentially modified Gaussian, then the estimates of the parameters will be inaccurate. The original works (30–32) should be consulted for details.

However, one useful formula will be given as an example. An equation from which one can calculate the number of theoretical plates is given below. It is valid in the range of $1.09 < B/A_{0.1} < 2.76$ (31).

$$N_{sys} = \frac{41.7(t_R/W_{0.1})^2}{B/A_{0.1} + 1.25} \tag{9}$$

This can be compared to the formula one would use if the peak were Gaussian,

$$N = 16\left(\frac{t_R}{W_B}\right)^2 = \left(\frac{t_R}{\sigma}\right)^2 \tag{10}$$

For a Gaussian peak, $W_{0.1}$ corresponds to about $4.30\ \sigma$ units. Substituting this value into Eq. 9 and setting B/A to 1.0 (totally symmetric) yields Eq. 10 approximately. Equation 9 can now be seen to reveal the extent to which asymmetry decreases chromatographic efficiency. Further work along these lines has led to empirical equations that give accurate measures of the peak area from similar parameters (33).

1.3. BAND BROADENING IN ISOCRATIC ELUTION LIQUID CHROMATOGRAPHY

In this section band broadening is discussed from a fundamental point of view. In Section 1.2 of this chapter it was shown that the column performance and the adequacy of a separation are governed, at least in part, by the phenomena of both symmetrical and asymmetrical band spreading. The present discussion begins with a very general overview of the processes that cause this broadening. Following this, a brief look at the historical development of chromatographic dynamics will allow the reader to understand the important conceptual advances that have been made. Concise statements of these are highlighted in italic print. We then turn to detailed discussions of increasingly complex systems. The discussion begins with the apparently simple case of band spreading in open tubes. This is of relevance to open-column chromatography and to packed-bed chromatography. The relevance to the latter is twofold. As a practical matter, columns are connected to other system components by open tubes (connecting tubing), and those other system components are themselves quite often effectively open tubes (injectors, detectors). Also, the concepts for understanding the flow related phenomena in packed beds will be clearer after some exploration of the simpler flow processes in an open tube. The most difficult part of the theory, dispersion in the mobile phase, is presented in the next section. To introduce the influence of a stationary phase, and to give a feeling for one powerful theoretical method, we investigate the single theoretical plate column using the method of moments. After a brief look at capillary columns, we turn to linear chromatography in packed beds. Finally, a section on nonlinear, nonideal chromatography, especially important in affinity chromatography, concludes the chapter.

1.3.1. Overview

The following general overview is given with only one reference. We must acknowledge the leading role played by Prof. J. C. Giddings in the simplification and clarification of the thermodynamic and kinetic behavior of molecules in separation processes (34). Specific citations are given in the detailed discussions in the following sections.

1.3.1.1. Macroscopic Molecular Motion

The questions that the theorist addresses are related to the physics and chemistry of molecular motion. Solute molecules can move by two general phenomena. One is bulk flow. In this type of motion, some potential energy gradient exists that causes the bulk solution to move or flow. Molecules close to one another tend to move together because of the intermolecular forces that hold the molecules together. These forces evidence themselves as friction and viscosity. The second type of motion is specific motion. In this type of motion, molecules respond to a gradient of chemical potential, which may include a gradient of potential energy as well. However, different types of molecules in the same system can sense different potential energy gradients. Thus a solution of a salt in water that has a pair of electrodes immersed in it has a potential energy gradient for the anion that is the opposite of the potential energy gradient for the cation. Thus in the electrophoresis experiment, the anion and the cation move in opposite directions.

In chromatography both types of solute motion are important. The molecule-specific driving force that is of the most importance in chromatography is diffusion. Appendix 1.2 is an overview of diffusion that the reader may find helpful. The mechanics of bulk fluid flow in a packed bed are very complex, and an exact description is not possible. Even approximate treatments are difficult. One needs to appreciate this to understand the approaches taken. Approximations are routinely made.

1.3.1.2. The Chromatographic Band Shape

If one injected a rectangular zone of solute (that is, a zone of a finite width in distance, volume, or time that went from solute free to some concentration of solute back to solute free in a rectangular distribution) onto a column, one might expect zones of the identical rectangular shape to pass through the detector at some later time. Of course, this does not happen. The zones become mixed with solute-free mobile phase as they traverse the column leading to zones or bands that have been spread. The origins of the band spreading are in the dynamic events occurring in the column.

We are fortunate that chromatography yields average behavior that is predictable with thermodynamic, mechanical, and metric properties of the system (see Chapter 2). Thus the average of the exit times of molecules of solute is related to the equilibrium distribution constant of the solute. However, if we focus on a particular portion of the column, we will find that there is *local* departure from equilibrium. This should be expected; the chromatographic experiment bears a considerable resemblance to a perturbation experiment. To see this imagine a portion of a chromatographic column that is in equilibrium; the concentrations of solute in the two phases obey the appropriate equilibrium expression. We instantly replace the mobile phase outside of the porous particles with fresh solute-free mobile phase. The system is now not in equilibrium, there is an excess of solute in the mobile phase entrapped in the pores of the particles (the *stagnant mobile phase*). Molecules of solute begin to diffuse out of the pores in response to the chemical potential gradient. The rapidity with which the system reaches a new equilibrium depends on several *rates*. There is the rate of diffusive efflux from the particle, the rate of equilibration within the extraparticle space, and the rate of equilibration of the solute in the stationary phase with that in the stagnant mobile phase. The latter may be a chemical kinetic step, such as in affinity chromatography, or a diffusion step in a three-dimensional stationary phase. The rate of transfer of molecules in the interparticle space is a function of molecular diffusion as well as of bulk transport by the mobile fluid.

The distribution of exit times (but not the average) for a particular solute is influenced by the rates of the processes described. This can be appreciated by a simple statistical argument. We will consider a solute with a given retention time, so that the average of the distribution is set. Recall that the retention time is linearly related to the distribution coefficient. We can think of the experiment as being a measurement of the distribution coefficient. Our desire is for a narrow distribution. This can be said in statistically equivalent ways. Having a narrow distribution implies that the mean value of the distribution is relatively well known, if the spread of values about the mean is narrow it means that the "signal to noise ratio" of the determination of the mean is high. What is required, in general, to achieve a narrow distribution of the determination of the distribution coefficient? One must make a large number of measurements. But the total time that we have to make the measurements is fixed. Therefore to achieve a narrow distribution, each determination of the distribution coefficient must be made rapidly. The distribution coefficient is an equilibrium constant, so one must wait until equilibrium is attained before accepting the measurement as being valid. Given that we have a fixed time, the system in which equilibration is the most rapid will result in the narrowest distribution. Recalling the statistical arguments, if the deviations from the mean are normally distributed and uncorrelated, then a decrease in the equilibration time by a factor of 2 would double the number of "experiments" or equilibrium partitioning events, and that would result in a decrease in the width of the distribution of $2^{1/2}$.

Nonequilibrium is not the only cause of band broadening. Consider the case of a chromatographic column packed with solid, nonsorbing beads. Now there is no sorption of solute, so the considerations above are obviated. But still a narrow rectangular input band will elute as a broadened band. Extraparticulate processes occur to spread the band. One is simple molecular diffusion. Certainly during the time that a molecule is on the column it is diffusing. Another important process is called eddy dispersion. It results from the random trajectories taken by molecules as they travel down the tortuous paths provided by the interstitial, or interparticulate, space. The presence of the particles is not entirely deleterious. Indeed, if one attempted to do chromatography at ordinary flow rates with an open tube having the diameter of ordinary columns, the axial broadening due to flow dispersion would be intolerably large. The particles, in inducing secondary flow, cause radial mixing, decreasing the axial band broadening.

We have established that there are several steps in the equilibration of solute between phases. Each of these steps can be associated with a rate that is, in principle, calculable. It is likely that one or the other of the steps is rate limiting under different circumstances governed by such variables as flow rate, particle diameter and porosity, phase ratio and the distribution coefficient, particle shape, mobile phase viscosity, solute molecular diffusion coefficient, tortuosity of the inter- and intraparticle spaces, temperature, and pressure. There are additional processes, also influenced by these parameters, that are not dependent on the existence of an equilibrium between two phases for their expression as band broadening. The job of theory is to account accurately for all these processes and to determine the functional relationship of the distribution of solute at the exit of the column to the parameters listed above.

1.3.2. Historical Development of Theory

1.3.2.1. Early Investigations of Adsorbents by Chromatography

The term *chromatographic theory* can refer to the theory of adsorption or the theory of band broadening, that is, thermodynamics or kinetics. Today most general discussions of chromatography distinguish between the two quite clearly. However, in the first years of the theoretical description of chromatography this was not so. Much of the early work attempted to understand chromatographic retention, and attempted to relate it to adsorption isotherms. Wilson (35), using the Langmuir adsorption isotherm, presented the first equations for predicting the width and the velocity of a band. The mathematical model used was the mass balance model (see Appendix 1.3). In this model one focuses on a small disk of the column that is dx thick. It contains stationary and mobile phases. When a small volume of mobile phase moves through the slab (in some small time dt), there is convective flux in and out of the mobile phase,

$$\text{moles in} = c_x u_e A_m \, dt \tag{11}$$

$$\text{moles out} = c_{x+dx} u_e A_m \, dt \tag{12}$$

In these equations, the position is given by the subscript of c, the mobile phase velocity in the extraparticle space is u_e, and the area through which the convective transport occurs is A_m. The term u_e is also called the interstitial velocity. There will also be partitioning between mobile and stationary phases,

$$\text{moles sorbed} = f_a(c) \tag{13}$$

$$\text{moles desorbed} = f_d(q) \tag{14}$$

Together all of these changes result in a *net* change in the mobile phase concentration with time. This leads to a partial differential equation that can be solved in many cases, provided that Eqs. 13 and 14 are linear (see Appendix 1.3).

Early workers assumed that an instantaneous equilibrium between the phases was established and that there was no diffusion. Note that this eliminates all of the processes by which dynamic band broadening occurs. Primarily the objective was to relate retention to the adsorption isotherm.

In current terminology, Wilson suggested that for a single-component chromatogram Eq. 15 holds,

$$\frac{\bar{v}}{u_e} = \frac{1}{1 + \Phi[f(c^0)/c^0]} \tag{15}$$

In this expression $\bar{v}$ is the average solute band velocity, Φ is the phase ratio, and $f(c^0)$ is the relationship between moles adsorbed per gram of adsorbent q and the concentration of solute in the mobile phase c,

$$q = f(c) \tag{16}$$

Finally, c^0 is the injected concentration of the solute.

The mathematical difficulties increase when one considers the multiple-component chromatogram. In this case the adsorption isotherm for solute i is a function of all the concentrations of all the solutes in the system, as shown in Eq. 17,

$$f_i = f_i(c_i, c_j, \ldots) \tag{17}$$

Given its simplicity, Wilson's work was useful in providing a solid foundation for understanding retention. He also anticipated the importance of differences in adsorption energies in controlling the separation (that is, the $(\alpha - 1)/\alpha$ term in the resolution equation). As Giddings points out (36), Wilson also saw that an optimum velocity would exist for obtaining the best resolution. Although he did not derive equations, he realized the importance of both noninstantaneous achievement of equilibrium and molecular diffusion of solute along the chromatographic column (axial or longitudinal diffusion).

Later LeRosen (37) introduced the idea of referring the zone velocity to a standard, the mobile phase velocity. This yields R, the solute zone velocity as a fraction of the mobile phase velocity. He suggested an optimum of $0.2 < R < 0.3$ or, in modern terms, $2 \lesssim k' \lesssim 4$.

1.3.2.2. Importance of the Rate of Equilibration of the Solute and the Theoretical Plate Concept

Martin and Synge (11) described the effect of noninstantaneous equilibrium on the chromatographic zone shape. They used a linear isotherm $[f(c^0) = Kc^0]$. An essential concept of these workers is that since there is a moving phase, *the average time required for the system to equilibrate after a concentration perturbation defines a distance traveled by the mov-*

ing phase. This distance corresponds to a region in which local equilibrium exists. Theories of distillation (38) use this concept. The vapor, being of a composition different from the liquid below it, condenses in the column to become a new liquid. Above this newly condensed liquid there is a vapor phase of a different composition in equilibrium with it. Each condensation–evaporation step (in local equilibrium) corresponds to a theoretical plate. The number of such equilibrium steps, that is, theoretical plates, that the system undergoes establishes the purity of the distilled liquid. Thus the analogy between chromatography and distillation was developed. As Giddings has pointed out (12), the theoretical plate concept has many limitations. Two of major importance are its inability to incorporate actual chromatographic parameters, and the tendency of chromatographers to interpret it literally. One could not, at the time of Martin and Synge, predict the number of theoretical plates in a column; it was an input parameter. However, it was a tremendous advance at the time of its publication. Zone shapes at the time were considered to be rectangular (35) for a rectangular input, that is, a large injection volume. Having a series of zones in local equilibrium leads to the prediction of a *Poisson distribution of exit times.* At long times the Poisson distribution becomes adequately represented by the Gaussian distribution. The theory also predicts that the *zone width is related to* $\sqrt{HL}$. Note the simple square root relationship between the width of the distribution of exit times and the number of sorption–desorption events (proportional to L, all else being equal). This was also seen in the simple statistical argument used above.

1.3.2.3. *Other Influences on Band Broadening*

DeVault (39), and independently Weiss (40), using the mass balance approach, reinvestigated the problem considered by Wilson and predicted zone shapes for various isotherms. One can use the isotherm curvature to predict behavior as shown in Eqs. 18,

$$\frac{d^2 f(c)}{dc^2} < 0, \quad \text{tailing} \tag{18a}$$

$$\frac{d^2 f(c)}{dc^2} > 0, \quad \text{fronting} \tag{18b}$$

The work of Thomas (41) is noteworthy in that the expression for mass balance included a term for a *chemical reaction rate.* A second-order sorption rate constant and a first-order desorption rate constant, appropriate for ion exchange, were used and the equations were solved. The concentration profiles were predicted to be complicated functions of the reaction rates and the time on the column, as well as mobile phase linear velocity.

Taking a slightly different mathematical approach, Walter (42) attacked the same problem as Thomas. Consider the adsorption reaction to be irreversible. Then breakthrough curves will only be sharp if the "time of passage of the solution is of the order of 100 times the 'half-life' of the chemical reaction." The curve is less sharp for a reaction with an equilibrium constant of 1. For the experiment in which a strongly bound species is on the column initially, the breakthrough of a weakly bound species is not a sharp step. The thesis of the work is that, unless care is taken to use low flows, a correct equilibrium constant cannot be unambiguously determined from the data.

The details of the rate-limiting process involved in chromatography were described by Boyd et al. (43, 44). They considered the series of reactions: mass transport in the flowing mobile phase, mass transport within the stationary zone, and the kinetics of sorption (in this particular case, ion exchange). The result of their theoretical and experimental investigations was that *mass transfer in the mobile phase limits the rate of attainment of local equilibrium in systems with small particle radii and large partition coefficients*. This was an important result. First of all, the rate-limiting step stated above is a first-order process. The flux of solute through a film is proportional to the difference in the concentrations on the two sides of the film, as long as the solvent is the same on both sides. The first-order rate is given by

$$k = \frac{6D}{d_p \Delta r_0 K} \tag{19}$$

where D is the molecular diffusion coefficient, d_p is the particle diameter, Δr_0 is the film thickness, and K is the equilibrium constant for the adsorption. The thickness of the film surrounding a particle depends on the fluid velocity, so the rate-limiting step must also be velocity dependent. These authors carefully pointed out the physical picture leading to the "film" concept using the arguments of King (45). In a chromatographic system molecules are carried toward the sorbent particles by convection and diffusion. Far from the surface of a particle, solute motion toward the particle surface can occur more rapidly by convection than by diffusion under the typical conditions of flow rate and diffusion coefficients encountered. As the solute approaches the sorbent particle surface, its convective motion becomes increasingly tangential to the particle surface because this is how the fluid moves. At some distance from the surface, molecules are transported to the surface as rapidly by diffusion as by convection. Inside this distance diffusion plays the greater role. This is the hydrodynamically defined *diffusion layer thickness*. Diffusion through the film just described can limit the overall mass transport process. It is important to note that this film, sometimes called a *stagnant film*, need not be stagnant. To fit the simple model of diffusive flux through a film, the convective motion must be predominantly orthogonal to the flux in the film.

When the partition coefficient is smaller and the particle radius is larger, mass transfer within the particles limits the overall kinetics. Exact expressions for the system's progress toward equilibrium were derived, although no simple expression for a rate or time constant was given; this was done later by Glueckauf (46).

Boyd et al. derived an equation for a breakthrough curve based on the *linear driving force (film diffusion)* and a linear isotherm in conjunction with the requirements of mass balance. Results from the analytical solution, an infinite series of modified Bessel functions, were calculated using tables and were compared with experiment. The general agreement was good. Apparently unaware of the work of Martin and Synge, they described a peak resulting from an injection of a small quantity of analyte as being symmetrically broadened due to kinetic, not thermodynamic, processes.

Later Sillen (47) inveighed against the tendency to employ linear isotherms in theoretical work. He proceeded to show that breakthrough curves yield reliable thermodynamic data on adsorbents for a slow enough flow rate (48).

Beyond this point, increasing attention was paid to the great mathematical difficulties associated with the complex dynamic phenomena involved in chromatography. As mentioned, Glueckauf (46) determined the best rate constant to use in the case of diffusion within a spherical particle. This is one of the potential rate-limiting steps as determined by Boyd et al. The actual process that occurs within the particle of support is a complex one. The equation representing the concentration as a function of spatial coordinates and time is also complex. Glueckauf determined the best approximation to use for reasonably efficient chromatography ($N >$ 200): *the rate of change of the space-average concentration in the particle equals the product of a rate constant and the difference in the surface concentration and the internal space-average concentration,*

$$\frac{d\bar{c}}{dt} = \left(\frac{60D}{d_p^2}\right)(c_s - \bar{c}) \qquad (20)$$

Lapidus and Amundson (49) improved the mass balance model by incorporating longitudinal diffusion. Now the change in concentration in the mobile phase is influenced by partitioning, flow, and molecular diffusion along the separation axis.

1.3.2.4. Distribution of Exit Times Calculated from Adsorption and Desorption Probabilities

Giddings and Eyring (6) used a stochastic approach to describe chromatography. One considers a molecule to be in one of two states, adsorbed or desorbed. One cannot know whether or not a particular molecule will be adsorbed at a particular time; one can only describe a probability that the molecule is, or is not, adsorbed. As the molecule moves down the column, it undergoes a series of random events, moving between the stationary zone and the mobile zone. For a large number of steps the distribution of solute exit times is given by

$$p = \frac{(k_2 k_1 t_0)^{1/4}}{2\sqrt{\pi}} \frac{\exp[-(\sqrt{k_2 t} - \sqrt{k_1 t_0})]^2}{t^{3/4}} \tag{21}$$

where k_2 is the desorption probability and k_1 is the adsorption probability, both per unit time. The void time is t_0, and t is the time spent in the stationary phase.

1.3.2.5. Band Broadening Contributions from Independent Processes in Series Add

Boyd et al. (43, 44) described broadening due to two potential limiting steps and recognized that the *rates added inversely since the processes under consideration were serial,*

$$\frac{1}{k_T} = \frac{1}{k_a} + \frac{1}{k_b} \tag{22}$$

where k_T is the total rate constant, and k_a and k_b are the rate constants for the two serial processes of interest.

In 1956 Klinkenberg and Sjenitzer (50) derived contributions from the various processes that occur in chromatography with a linear isotherm, including longitudinal diffusion as well as mass transfer resistance (as discussed by Boyd et al.) and perhaps most importantly eddy dispersion. The term *eddy dispersion* arises from an analogy between this type of broadening and the "eddies" that exist in a turbulent flow system. It is evident to anyone who has ever taken a white-water canoe trip that the flow streams at any point are not time independent but are constantly changing—often quite suddenly. Instead of steady-state flow, eddies, or whirlpool-like pieces of fluid, randomly split off from the main flow direction. In turbulent flow there is a great deal of lateral and circular motion superimposed on the main downstream (longitudinal) movement of the fluid. Solute, moving with the fluid, experiences a random motion superimposed on the net downstream motion. This random motion has been viewed as being like diffusion, although it occurs on a much larger length scale. In a packed column, at velocities well below that yielding turbulent flow, the same type of random motion tends to occur, not due to eddies, but due to the fact that particles in the bed are arranged randomly. In chromatography the Reynolds number is usually below 1, so that the fluid is far from turbulent. In this hydrodynamic domain the influence of the particle surfaces is very large. Fluid flows more or less smoothly, but the stream is interrupted by the presence of particles. Klinkenberg and Sjenitzer reasoned that the continual splitting and confluence of streams would mix solute as do eddies in turbulent flow, hence the name. In the range of velocities and particle diameters for which the Reynolds number is less than 1 the viscous forces dominate inertial forces. In this case, the patterns of flow for two columns, one of which is an exact duplicate of the other except that all of its dimensions are increased or decreased by a factor, are geometrically similar. *The band spreading (expressed as a standard deviation of the distribution of solute along the separation axis) caused by eddy dispersion must be related to the particle diameter and the distance traveled, but not the velocity.*

The rate constants for the various processes were used by van Deemter, Zuiderweg, and Klinkenberg (51) to construct an equation giving *the variance of a chromatographic peak as the sum of the independent variances of the processes occurring in the column.* The van Deemter equation is

$$h = \frac{H}{d_p} = \frac{2\gamma D}{u_e d_p} + 2\lambda + \frac{C u_e d_p}{D} \tag{23}$$

In this equation (applicable to chromatography with a linear isotherm) the first term is due to molecular diffusion of solute in the mobile phase that is moving at velocity u_e in the interparticle space, the second term is due to eddy dispersion, and the third term is due to mass transfer within the chromatographic particles limiting the rate of attainment of equilibrium. The particle diameter is d_p. The tortuosity of the interparticle space is given by γ, and λ is an empirical constant that depends on the homogeneity of the packed bed. C is a collection of parameters, including the equilibrium constant, the phase ratio, the particle diameter, and molecular parameters that define the rate-limiting step in interphase mass transfer. The independence of the variances is an important concept. It allowed theoreticians to work on part of the problem, then add the parts together.

1.3.2.6. Random Walk, Stochastic Approach, and Band Broadening as an Effective Diffusion

Giddings (52) and Jones and Kieselbach (53) used the random walk approach. This is an appealingly simple approach that yields qualitatively correct results. The statistical approach deals quite naturally with the variances of distributions. Giddings (52) demonstrated the additivity of the variances of independent chromatographic processes.

The stochastic approach was again used by Giddings (8, 54) to arrive at an important relationship. Chromatography is modeled as being a simple first-order reaction characterized by a pair of first-order rate constants, the ratio of which gives the equilibrium constant for the reaction

$$A \underset{k_2}{\overset{k_1}{\rightleftharpoons}} B \tag{24}$$

where A represents the state of solute molecules in the mobile phase with velocity u_{e_1}, and B is in the stationary phase with velocity u_{e_2} (which, of course, is generally 0, but we shall retain a finite velocity in the expression below for generality). *The spreading of solute from its mean position due to mass transfer between relatively moving phases is analogous to diffusion* with an effective diffusion coefficient, given as

$$D = \frac{k_1 k_2 (u_{e1} - u_{e2})^2}{(k_1 + k_2)^3} \tag{25}$$

1.3.2.7. Brief Summary and Comment

The reader will recall that the initial investigations into the theory of chromatography focused on the influence of the adsorption isotherm on the retention process. The theory provided a set of criteria that, if met, would result in a valid determination of an adsorption equilibrium constant from a breakthrough curve. Gradually attention turned to the many rate processes that are involved in chromatography. Less attention was paid to the problem of nonlinear chromatography—a complete treatment of the spreading processes in the domain of linear chromatography was difficult enough. Several internally consistent formulations had arisen. The random walk treatment gave a band variance resulting from kinetics; the stochastic approach yielded a formal relationship between the "reaction rate constants" in chromatography and a diffusion coefficient. (We will shortly see that other processes are equivalent to diffusion. Coefficients with the dimensions length^2·time^{-1} will be referred to as dispersion coefficients $\mathscr{D}$, unless they are true molecular diffusion coefficients.) The height equivalent to a theoretical plate had been given as a sum of variances. Thus rate constants, dispersion coefficients, band variance, and the height equivalent to a theoretical plate are all interrelated (see Appendix 1.4).

A complete description of band shape, with explicit inclusion of rate constants for the various processes, has been very elusive. The mass balance, stochastic, and random walk approaches are either too difficult or insufficiently detailed to predict exact peak shapes given the parameters of a particular solute on a particular column.

1.3.2.8. Nonequilibrium Approach of Giddings

A very powerful approach, the nonequilibrium approach, was developed and applied to chromatography by Giddings (55–62). A general relationship between H and $\mathscr{D}$ can be given as (55)

$$H = \frac{2\mathscr{D}}{Ru_e} \qquad (26)$$

where Ru_e is the average solute velocity. According to the nonequilibrium approach H is directly proportional to the extent of disequilibrium. An expression for H is generated by relating the actual solute concentration c at any point in the column to the concentration that would exist if the column were at perfect equilibrium c^* by the following nonequilibrium expression:

$$c = c^*(1 + \epsilon) \qquad (27)$$

This approach is really a first-order perturbation method, and as such it really is only accurate when the displacement from equilibrium is small ($\epsilon < 0.01$). Each type of process that can cause disequilibrium (resistance to mass transfer in the moving zone, in the stagnant mobile phase, in the stationary phase, surface desorption kinetics, and so on) will generate a separate ϵ. For any individual mass transfer resistance, Eq. 28 may be used to determine the contribution to the plate height from that process,

$$H = \frac{-2\epsilon_{mob}}{\partial \ln c/\partial z} \tag{28}$$

where z is the spatial dimension along the column. The equation shows that a larger departure from equilibrium leads to a larger plate height. Of course, the extent of nonequilibrium is inversely related to the system's exchange rate. *High exchange rates lead to efficient chromatography.*

This formalism (55, 59) was used to derive expicit expressions for H caused by various kinetic processes (54), stationary zone diffusion (57, 58), and mobile phase mass transfer (57, 58), including the coupling theory (60–63). It was also used to describe band broadening due to the presence of several types of adsorption sites (55) (a problem that was the focus of many of Giddings's earlier works as well). The interested reader is referred to Appendix 1.5 for a synopsis of the details of the nonequilibrium approach.

The coupling theory was used to describe the relaxation of a concentration perturbation in the mobile phase. Giddings argued that mobile phase mass transfer and eddy dispersion are not independent serial processes, so their variances, and therefore their plate height contributions, do not add. Rather, since both processes act to reequilibrate a region of solution in which the concentration is not homogeneous, the processes are in parallel. These "reactions" are in parallel, so the overall corresponding reaction rate constant should be the sum of the eddy dispersion E and mobile zone mass transfer (MZMT) rate constants, as given by

$$k = k_E + k_{MZMT} \tag{29}$$

Now the rate constant k adds in series with rate constants for intraparticle (stationary zone) mass transfer (SZMT) and sorption–desorption kinetics K,

$$\frac{1}{k_T} = \frac{1}{k_{SZMT}} + \frac{1}{k_K} + \frac{1}{k_E + k_{MZMT}} \tag{30}$$

For velocities commonly used in liquid chromatography, the coupling concept leads to a convex H versus u_e curve. Giddings's research in this area gives a rather complete picture of the dynamics of chromatography, which is developed in his classic monograph (12).

1.3.2.9.   *Another Brief Comment*

At this point in the historical development of chromatographic theory, attention was firmly focused on linear chromatography. The works of Giddings and of van Deemter et al. led to the concept that most of the relevant processes have independent variances that are additive. This was a critical step in simplifying the problem of developing a complete quantitative model of chromatographic broadening. After investigating several theoretical approaches, Giddings found that the nonequilibrium approach was the most powerful and provided the most facile route to determining the contribution of a particular process to the total variance. Subsequently efforts aimed at describing the overall peak shape waned, while methods based on the computation of the peak moments became more common since this approach was simpler and contained most of the important information.

The van Deemter equation predicts that a plot of H versus u_e should have the shape shown in Fig. 1.2a. Note that at velocites well in excess of the minimum a linear increase in H with increasing velocity results. Thus an increase in the speed of analysis decreases the resolution. In contrast, the coupling theory predicts that the H versus u_e curve should

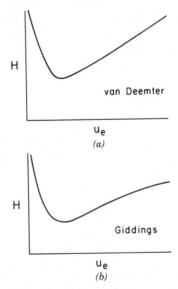

Figure 1.2. H versus u_e. (*a*) Van Deemter equation, which can be represented as $H = A + B/u_e + Cu_e$. (*b*) Giddings's equation, which can be represented as $H = (1/A + 1/C_{\text{MZMT}})^{-1} + B/v + C_{\text{SZMT}}u_e$.

roll off, as in Fig. 1.2b. Now an increase in u_e does not induce a proportional increase in H. High speeds are possible without a large sacrifice in efficiency. Therefore it makes sense to operate at high velocities. Operation in the range of velocities above the minimum in the curve puts one squarely in the domain in which mass transfer processes will be rate limiting. Since these processes often involve diffusion in and out of the particles, particle size must play a large role in defining the efficiency. Done and Knox (64), Giddings (65), and Huber (66) pointed out the importance of using small particle diameters to obtain increased efficiency. This brought with it the requirement for high-pressure pumps. HPLC was born.

1.3.2.10. Reduced Parameters

It is both intuitively appealing and simple to consider, rather than H and u_e, the reduced quantities h and v or Pe (10, 64, 67–69). The reduced plate height h is given by H/d_p, where d_p is the particle diameter. The reduced velocity v is given by $u_e d_p/D$. The reduced velocity may also be referred to as a *Peclet number*, Pe. The size scaling factor in Pe may be taken as another length, such as the tube radius, instead of the particle diameter. Also, the diffusion coefficient may be replaced by a dispersion coefficient $\mathcal{D}$. In the latter case it is often not the velocity that is the center of attention; rather the Peclet number is viewed as the inverse of the dimensionless dispersion coefficient. Care must be taken when reading the literature to determine whether the focus is on the velocity or on the dispersion coefficient in Pe. In this chapter we use the particle Peclet number, in which the characteristic length is d_p, unless otherwise noted. Furthermore we use it as a dimensionless velocity, unless otherwise noted. Knox and his coworkers have shown (10, 64, 68–70) that experiments on well-packed columns with the same mobile phase give superimposable h versus v plots for columns made with particles of different diameters. Thus the utility of the dimensionless coordinates is that they bring together apparently disparate data.

1.3.2.11. Attention to Detail

A conceptually appealing simplification, due to Huber (66, 71, 72), is the use of zone capacity factors. If the slow step in the mobile phase to stationary phase transfer process is a mass transfer event, then the important property of a molecule is whether or not it is mobile, not whether it is adsorbed. Thus k', the phase capacity factor, is not the appropriate property of the solute. The zone capacity factor k'' is more appropriate.

The two are related in the following way:

$$k' = \frac{n_s}{n_m} = \frac{n_s}{n_{fm} + n_{sm}} \qquad (31)$$

$$k'' = \frac{n_s + n_{sm}}{n_{fm}} \qquad (32)$$

$$k'' = \frac{\epsilon_p}{\epsilon_e} + k'\left(1 + \frac{\epsilon_p}{\epsilon_e}\right) = k_0 + k'(1 + k_0) \qquad (33)$$

In these equations n refers to the number (mass, moles) of the solute molecules in a phase indicated by the subscript; that is, s, stationary; m, mobile; sm, stagnant mobile; fm, flowing mobile. The porosity of the column (excluded or interparticle volume per total column volume) is ϵ_e, and ϵ_p is the porosity of the particles (pore or intraparticle volume per total column volume). The latter quantity is related to the porosity of the particles referred to the particle volume ϵ_i (included or pore volume per total volume of particles) by the relationship $\epsilon_p = \epsilon_i(1 - \epsilon_e)$. The total column porosity ϵ_T is just the sum of ϵ_e and ϵ_p. Note that the ratio ϵ_p/ϵ_e, called k_0, is the zone capacity factor k'' for a solute with no affinity for the stationary phase.

One of the most widely used expressions for h is the Knox equation (68), which is based on experimental data and correlations found to hold for gas chromatography,

$$h = Av^{1/3} + B/v + Cv \qquad (34)$$

The A term is empirical and accounts for eddy dispersion and mass transfer limitations occurring in the mobile phase. The B term represents the contribution of longitudinal diffusion, and the C term represents the contributions from slow steps inside the particles, namely, diffusion in the stagnant mobile phase and diffusion in the stationary phase. The relationship has been used successfully over a wide range of conditions, though it should be noted that Knox has cautioned that it is meant to work over a limited range of velocities.

Huber used a more detailed semiempirical approach (66, 71, 72). As was shown, general band broadening terms can be derived for complex processes if one hides one's lack of knowledge by specifying a "rate constant" or "mass transfer coefficient" to represent a process. (See Appendix 1.4 for a discussion of the relationship between them.) The theoretical challenge is to derive the rate constant as a function of measurable and controllable parameters. On the other hand, one can turn to the engineering literature and find correlations that describe the rate constants of complex processes over a wide range of conditions, usually in the form of a simple function of dimensionless variables. Huber did just that to arrive at an equation for h versus v.

More detail has recently been built into the functions A, B, and C. For example, by knowing the exact dependences of B and C on k' or k'', one can determine the diffusion coefficient of the solute in a C_{18} stationary phase (73–75). The effective diffusion coefficient of solute in the stationary zone is evidently identical in the B term and the C term (73, 76). The effective coefficient is the time-weighted average of the solute's diffusion coefficients in the various phases in which it may be found. Mathematically different, but conceptually similar approaches have been used to explore the rate-limiting step in intraparticle mass transfer (76, 77). The distribution of the number of sorption–desorption events plays a pivotal role in such an analysis. If the mean number of adsorption events that occurs once a solute is inside the porous particle is small (less than 1), then this process leads to band broadening. If the mean number of such events is larger, then the diffusion in the stationary zone is the rate-limiting step. Thus adsorption that occurs with a small adsorption rate constant, or has a small number of "sites" for adsorption, can cause band broadening.

With the advent of readily available computing power, numerical solutions of the differential equations, and numerical evaluation of the series analytical solutions, are becoming more attractive. A return to the computation of peak profiles for determining the rate-limiting processes has been strongly advocated by Lenhoff (78). In the same paper, Lenhoff has lucidly described the characteristic times of various potentially rate-limiting steps. A comparison of the characteristic times quite straightforwardly yields the conditions under which a given process becomes rate limiting.

1.3.3. Solute Dispersion in Open Tubes

The flow of a Newtonian fluid through a smooth straight open tube is well understood for Reynolds numbers from 0 to about 2000. The Reynolds number is one of several dimensionless numbers that are useful in describing momentum and mass transport in convective and diffusive processes. The Reynolds number is the ratio of inertial forces to viscous forces in a fluid and is defined as $Re = uD_H\rho/\eta$, where u is the average fluid velocity, D_H is the hydraulic diameter (4 × conduit cross-sectional area per conduit perimeter), ρ is the density of the fluid, and η is its viscosity. η/ρ is called the kinematic viscosity. It has the units length^2·time^{-1} and is often given by the symbol v. We will not use this symbol for kinematic viscosity; in this chapter, v is reserved to denote the reduced (dimensionless) velocity. For circular conduits, or tubes, the hydraulic diameter is the same as the diameter. For other shapes the hydraulic diameter seems to be the parameter of relevance to the hydrodynamics. At low Re (Re < 1) the force of the viscous drag due to the intermolecular forces in the fluid dominates the inertial forces. In this domain the flow is called viscous flow, and the shape of the space through which the fluid flows is of primary importance in governing the pattern of the flow. For Re > 1 [and in the limit of large Re corresponding to inviscid flow ($\eta = 0$)] inertial forces dominate. Other Re may be defined using different characteristic lengths. For example, in a packed bed the particle diameter d_p is more appropriate than the column's D_H. The volume of moving fluid to wetted surface area ratio has also been used. In the region defined (0 < Re < 2000) the flow in straight open tubes is laminar. A characteristic of such flows is that gradients of velocity are smooth. Fluid can be viewed as flowing in sheaths or layers that glide past one another with an exchange of momentum between layers. Part of the fluid's momentum is transferred between adjacent sheaths, and the pressure drop dissipated in transporting the fluid is relatively low. In pure viscous flow the sheath closest to the wall (at $r = r_0$) is essentially stationary. (This is called the no-slip boundary condition.) As one moves away from the wall toward the center of the tube (at $r = 0$), the velocity (momentum) increases to a maximum at the center of the tube. The local velocity $u(r)$ is therefore a function of the radial position in the tube, as expressed by the Hagen–Poiseuille equation,

$$u(r) = 2u\left[1 - \left(\frac{r}{r_0}\right)^2\right] \tag{35}$$

where u is the linear velocity averaged over the tube cross section. At very high Re (low viscosity) intermolecular forces are too weak to maintain a strong correlation between flow streams and turbulence results. Turbulence is characterized by a lack of correlation in the velocity of neighboring streams, the development of eddies, and a very high pressure drop per unit forward movement of the fluid. Furthermore, in turbulent flow the radial velocity profile flattens out. Note that turbulent flow in open cylinders begins to take place at Re > 2000 but is not fully established until considerably higher values. In a packed tube turbulent flow initiates at much lower Re (typically 20–100).

Imagine a long straight tube, with a perfect injector at the upstream

end and a perfect detector at the downstream end. Fluid is flowing through the system at an average velocity that yields laminar flow. Furthermore, imagine that you have "turned off" molecular diffusion. Inject into this system a perfectly narrow disk of solute at the inlet of the tube. What will the detector see? Certainly the molecules injected near the center of the tube will come screaming out at a velocity $2u$. But what about the molecules near the wall of the tube? Since they are prevented (in our minds) from diffusing to more rapidly moving laminas, they will move very slowly indeed. Clearly the initially narrow zone will elute as a broadened zone. Figure 1.3a shows the spatial distribution of the solute resulting from such a system. If one now allows diffusion to occur, but specifies that the average velocity be much greater than the effective diffusional velocity ($\approx D/r_0$, where D is the solute diffusion coefficient in the mobile fluid), then there will be a small effect only; see Fig. 1.3b (79–83). If we repeat the experiment for a slightly larger diffusional velocity, or at a slower fluid velocity, then the molecules that can diffuse away from the tube wall, increasing their velocity manyfold, will be the most affected. Differentiation of Eq. 35 shows that du/dr is greatest near the wall ($r \Rightarrow r_0$). Diffusion will bring about mixing over a distance δ given by $\approx \sqrt{2Dt}$. Then molecules that begin the transit near the center of the tube will experience a smaller range of velocities than those at the wall. The effect of radial diffusion of the solute is to bring the average velocity experienced by all solute molecules closer to the *mean* bulk fluid velocity. The resulting shape is shown in Fig. 1.3c. The formation of two overlapping peaks occurs with lower fluid velocity, as shown in Fig. 1.3d. At much lower fluid velocities, the peak appears symmetrical; Fig. 1.3e.

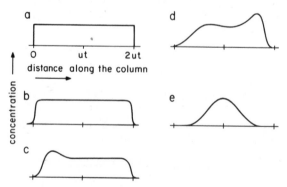

Figure 1.3. Distribution of solute in an open tubular column after injection of a spike of concentration that is radially homogeneous. (*a*) Very short times. (*b*), (*c*), (*d*) Longer times; (*e*) Long times. Taylor–Aris theory describes the case in (*e*).

Recent theoretical work has explored the set of circumstances leading to behavior in the range represented by Fig. 1.3*d* and *e* (84–86). This work follows earlier work by Taylor (87), Aris (88), and Gill (89–91). Among the most intriguing theoretical developments are the use of the random stumble (84) and the development of a general formalism for the computation of tortuosity and an effective dispersion coefficient in tubes and packed beds (86).

The random stumble is conceptually and computationally simple. Imagine a solute molecule moving in a zone of mobile phase of uniform velocity. Every now and then the molecule randomly stops for a while, then it starts up at its original velocity. The distribution of stopping times can be specified. This variation in the instantaneous axial velocity of a molecule bears some resemblance to what we observed in the case of flow through a tube, since in that case molecules were also in constant transit between laminas (zones) of high and low axial velocity. The peak shape that results at short times (Fig. 1.4) resembles those seen in Fig. 1.3 at analogous times. A simple explanation of the bimodal peak shapes also results from this simple two-velocity model. The molecules that find themselves initially in the moving phase have a finite probability of stopping. The molecules stop at random times. At times that are short with respect to the average exchange time a certain fraction of the injected molecules will not have stopped at all, another fraction only once, and so on. A similar state of affairs exists for the molecules that begin the experiment with zero velocity. In the case pictured in Fig. 1.4 the system has not had sufficient time for averaging to occur. Thus a bimodal (double-peaked) distribution results.

This situation in an open tube is exactly analogous to the formation of split peaks in a real chromatographic column. Split peaks are formed in a chromatographic column when the adsorption rate constant is so small

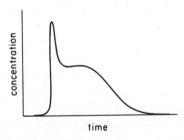

Figure 1.4. Exit time distribution for solute injected into a tube in which fluid is in the laminar flow regime. The fluid velocity is in the same range as that in Fig. 3*c* or *d*. Simulation was done using the random stumble model. [From Ref. (84) with permission.]

that a finite fraction of the solute can go through the column without being adsorbed even once. The phenomenon is evident in the theoretical treatment of Giddings and Eyring (6). The existence of bifurcated peaks is well known to all chromatographers who pack their own columns and is particularly common when very short (<5-cm) liquid chromatographic columns are studied. Part of the art of packing columns is devising methods to avoid channeling of the flow through low-density regions of the packed bed. What one observes with an initially badly packed column or a badly deteriorated column is a chromatogram in which the distribution of transit times (peak) for each component in the mixture is bifurcated or very badly shaped.

In a generalization of a previously enunciated concept (92), Paine and coworkers have developed the idea of a dispersion tensor (86) and have applied it to the problem under consideration. This formalism, although seemingly computationally difficult, appears to have some connection with the general nonequilibrium theory of Giddings (55). Since this topic is discussed below in conjunction with packed columns, it is now deferred.

Let us now consider the situation shown in Fig. 1.3e, the long time limit. The Taylor–Aris equation is used to define an "effective diffusion" coefficient or dispersion coefficient. This dispersion coefficient can be used in place of the diffusion coefficient in Fick's laws (see Appendix 1.2). The axial dispersion coefficient $\mathcal{D}_{ax}$ is the sum of the molecular diffusion coefficient and an axial dispersion coefficient caused by flow inhomogeneity,

$$\mathcal{D}_{ax} = D + \frac{r_0^2 u^2}{48D} \tag{36}$$

Several workers (79–91) have pointed out that this equation is strictly valid for long times only. Paine et al. point out the reason for this very well (86). They say: "The essential feature of this analysis is the recognition that the process is quasi-steady in a frame moving with the average velocity for times large compared to r_0^2/D."

If at first it seems contradictory that the solute's molecular diffusion should both help and hinder the overall dispersion, consider the thought experiment done above. When diffusion was not allowed, the spreading of the solute band was hydrodynamic in nature and severe. Diffusion is the means by which individual molecules experience an average velocity.

Several workers have given guidelines for the use of the Taylor–Aris equation. When the Taylor–Aris equation holds, one can inject a small plug of solute into a stream flowing in a straight tube and then record the solute passing through the detector as a Gaussian band. One can determine the molecular diffusion coefficient from the standard deviation of the Gaussian band. But for accurate data one must be sure that the conditions of the experiment satisfy the approximations made in the derivation of the Taylor–Aris equation. For straight tubes there are several answers that are in general agreement with one another. One is safe if the radial Peclet number Pe (Pe $= ur_0/D$) is greater than 10 and the dimensionless time Dt/r_0^2 is greater than 0.7 (81–86).

When these conditions are met and the Taylor–Aris theory applies, the injection of a concentrated spike of solute into the mobile phase can be used as the basis for an accurate method for the determination of molecular diffusion coefficients (82, 85). One must be careful to keep the tubing straight with a circular cross section. Looking at this in another way, using curved tubing with a noncircular cross section will induce secondary radial flow. Secondary flow will help mix the solute. Thus it assists diffusion in this job (81, 85, 93–95).

A statistical concept of use in chromatographic theory is that the variances of independent processes are directly additive (51, 52). This additivity is probably responsible for the popularity of the plate terminology since $H \propto \sigma^2$. Theorists have noted that the peak profile shortly after injection depends on the input profile, while for longer times this is not the case (84) (for similarly shaped input profiles). Golay and Atwood have pointed out an important consequence of this (80). Consider the case in which two short sections of tubing are connected in series. For large Pe and short dimensionless time the Taylor–Aris equation will not work. But the solute distribution traversing the system still has a variance, and both tubes contribute to the variance. But the variances are not additive. The reason is simply that the two systems are not independent. They are operating in a regime in which the input conditions are important. Thus the effect of the transit through the second tube on the peak shape depends on the shape of the profile at the output of the first tube. Be careful to note the distinction between the case of two independent linear processes adding in series and the case at hand. In the former case, the shape of the output profile is a function of the input profile going into the first process, the effect due to the first process, and the effect due to the second process. The effects of the processes are independent of the input shape. In the case of two short tubes in series, the shape of the input profile is one of the determining factors in the *effect* of the tube on the dispersion. The dispersions in the two tubes are not independent. As Golay and Atwood point out, the band broadening is actually less severe than one would calculate using Taylor–Aris theory.

Recently Wightman's group (96) has employed microelectrodes to determine dispersion and velocity as a function of radius in a tube. The velocity profile is parabolic, as expected. The dispersion is a function of radius; it is greater near the wall than in the center of the tube.

1.3.4.   Solute Dispersion in Beds Packed with Solid Particles, No Retention

This section deals with a cylindrical column of length L packed with solid particles of diameter d_p. It was demonstrated that the dispersion in an open tube is diminished by both molecular diffusion and secondary flow causing the solute molecules to sample the entire volume. Any mechanism that causes the solute to homogenize radially is beneficial (in the absence of wall effects), as long as the homogenization does not occur significantly in the axial direction. Packing the tube with particles increases dispersion. The crudest approximation to a packed bed is a bundle of parallel inter-connected capillaries. In the context of the above discussion, making many small capillaries with small r_0 allows diffusion to mix the solute among the laminas more rapidly. Another advantage to the packed bed are the random interconnections made by the various fluid streams.

1.3.4.1.   The Random Walk

The random interconnections of interparticle spaces suggest the use of random walk theory to model the dispersion due to this effect. As long as we let the fluid take enough steps [>10 (97)], the influence of starting conditions is forgotten, and the distribution of positions along the column is adequately represented as being Gaussian. The variance of a distribution generated from a random walk is given as $\sigma^2 = \eta l^2$, where the variance has units of length^2, η is the number of steps taken, and l is the length of a single step (97). The length of a step is certainly about d_p, the particle diameter. The number of steps taken during transit through the column must be related to the ratio L/d_p. Putting in a proportionality constant of 2λ, one arrives at

$$\sigma^2 = \eta l^2 = 2\lambda L d_p \qquad (37)$$

(Recall the discussion in Section 1.3.2.5.) By using the relationship between σ^2, H, and D (or $\mathcal{D}$) (8, 55) (see also Appendix 1.4) one arrives at an equation for the dispersion coefficient due to eddy dispersion (98, 99),

$$\mathcal{D}_{eddy} = \frac{u_e H}{2} = \frac{u_e \sigma^2}{2L} = \lambda d_p u_e \qquad (38)$$

This is not the only contributor to the dispersion of solute in the mobile phase. Molecular diffusion also carries molecules up- and downstream. Since eddy dispersion and molecular diffusion are independent (50, 51), their variances may be added. Equation 39 gives the axial dispersion coefficient $\mathscr{D}_{ax}$,

$$\mathscr{D}_{ax} = \gamma D + \lambda d_p u_e = D(\gamma + \lambda Pe) \tag{39}$$

Pe is defined implicitly by Eq. 39. In this expression the molecular diffusion coefficient has been multiplied by a tortuosity factor γ, which is always less than 1 and is typically 0.6–0.7 (100). It accounts for the fact that bumping into the particles slows the rate of progress of a molecule moving by molecular diffusion in a packed bed.

1.3.4.2.   Coupling of Axial Dispersion and Mobile Phase Mass Transfer

Giddings argued (60–63) that the dispersion due to the second term in Eq. 39 could be reduced by the relaxation of radial concentration gradients caused by eddy dispersion. Since the relaxation occurs in the mobile phase, it will be related to the rate of mass transfer through the mobile phase. This is entirely analogous to the open tube case considered above, in which molecular diffusion in the radial direction caused averaging to occur, limiting the spreading of a solute zone caused by radial velocity inhomogeneity. The form of the axial dispersion coefficient now becomes

$$\mathscr{D}_{ax} = D\left(\gamma + \frac{\lambda Pe}{1 + \omega/Pe}\right) \tag{40}$$

where ω is a numerical coefficient that depends on the factors that influence mobile phase mass transfer, such as the area to volume ratio. This theory has been very well documented in *gas* chromatography (99–102).

Expression 40 is obtained by considering the way parallel rates add. Recall Eq. 29; it showed the addition of parallel rates in an overall rate expression. Equation a4.12 (Appendix 1.4) shows the relationship between the plate height and a first-order interzone relaxation rate constant. Finally, with Eq. a4.14 one can convert the eddy dispersion term, Eq. 38, to an effective rate constant.

$$k_{MZMT} = \frac{2u_e}{H_{MZMT}} = \frac{2D}{\omega d_p^2} \tag{41}$$

$$k_E = \frac{u_e}{\lambda d_p} \tag{42}$$

$$k_{\text{coupled mob zone}} = k_E + k_{\text{MZMT}} \tag{43}$$

$$H_{\text{coupled mob zone}} = \frac{2u_e}{k_{\text{coupled mob zone}}} \tag{44}$$

$$H_{\text{coupled mob zone}} = \frac{1}{1/2\lambda d_p + D/\omega d_p u_e} \tag{45}$$

The same processes are expressed in purely plate height terms in

$$H \propto \frac{1}{1/H_E + 1/H_{\text{MZMT}}} \tag{46}$$

This is the term in Giddings's formulation resulting from the coupling of eddy dispersion and mobile phase mass transfer.

There is less experimental support for the importance of such a coupling process in liquids than there is for gases (99–102). The most recent data show that the axial dispersion in liquids is roughly described by Eq. 39 (see Fig. 1.5) (103). At large Pe, near 10^4 (recall that liquid chromatography is usually in the range of Pe $\approx$ 10), there is some flattening of the curve in Fig. 1.5. This is could be due to the use of columns that are too short so that the dispersion is influenced by the initial condition. This is very reminiscent of the situation encountered in open tubes when the tube is short or the velocity is high.

1.3.4.3.   The Same Physics Can Simultaneously Help and Hinder Separations

An important feature of the dispersion process in packed beds is the presence of both axial and radial dispersion. Figure 1.5 shows data for radial and axial dispersion in packed beds. Note that the two curves are similar, but that radial dispersion is lower by about an order of magnitude than longitudinal dispersion for Pe $\gg$ 1. In chromatography radial dispersion is largely beneficial. Rapid radial dispersion that brings solute to the wall of the column, where there may be channeling of flow, can be deleterious. But since there is no net bulk flow in the radial direction, dispersion represents the radial transport of solute entirely. The relaxation of radial concentration gradients and mass transport to particle surfaces are the benefits of radial dispersion. This was discussed in some detail. Thus one

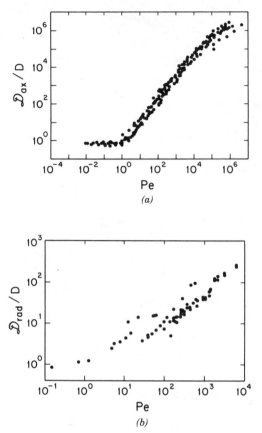

Figure 1.5. Experimentally determined (*a*) axial and (*b*) radial dispersion in packed beds. Note the greater axial than radial dispersion. [Adapted from Ref. (103) with permission of the American Institute of Chemical Engineers.]

wants the dispersion in the radial direction to be large, except in cases where wall effects are severe. On the other hand, dispersion in the axial direction is inimical to successful separations. Any theory of chromatography must take these two opposing influences of dispersion into account. In some cases the same physics is responsible for dispersion in both directions. Referring to the case of open tubes, we saw that molecular diffusion contributed directly to the dispersion coefficient because of axial diffusion, but also diminished the dispersion because of its ability to distribute solute throughout the entire radial dimension of the tube. Equations of the overall dispersion in a chromatographic column often have similar pairs of terms, one set of terms is in a denominator and one is in a numerator. This reflects the dual roles of convective mass transfer and diffusion.

1.3.4.4. The Use of Correlations

Huber developed a term for dispersion in the mobile zone that is similar in appearance to Eq. 40. In Eq. 41 (66) ω is a factor that depends on packing structure and geometry,

$$\mathscr{D} = D\left(\gamma + \frac{2\lambda \, Pe}{1 + \omega Pe^{-1/2}}\right) \tag{47}$$

Huber used a correlation of experimental data (104) for this term. Note the similarity to Eq. 40.

Kennedy and Knox (68) developed an empirical expression that included mobile phase mass transfer and eddy dispersion based on successful results from gas chromatography. The term for reduced plate height is given by Eq. 48. The value of the constant A and its relationship to the quality of packing has been discussed (70).

$$h_{\text{mobile phase dispersion}} = A\nu^{1/3} \tag{48}$$

1.3.4.5. The "Stagnant Film"—An Incorrect Application of the Diffusion Layer Concept

Horvath and Lin (105) argued that eddy dispersion occurs only in liquid that is actually moving between particles. They used the classical concept that under laminar flow conditions a stagnant boundary layer exists at the particle surface (106). The stagnant areas have been considered as a source of dispersion in packed beds because of their memory effect (102). The film traps solute at time t, and then later, at $t + \Delta t$, the solute concentration in the "film" is representative of the solute concentration that existed in the mobile phase near the film at time t. Horvath and Lin argued that an added effect of the film is to decrease the volume available to the flowing mobile phase. Furthermore, the particle boundary layer film thickness is velocity dependent and it must thin out at high velocity. They considered the stagnant fluid outside the particles as having a volume that was determined by the film thickness calculable from an empirically determined mass transfer coefficient. Thus as the velocity changes, the volume through which the fluid flows changes, and the eddy diffusion term changes. The effect is shown in

$$\mathscr{D}_{ax} = D\left(\gamma + \frac{\lambda Pe}{1 + \omega Pe^{-1/3}}\right) \tag{49}$$

Although in the final analysis, under typical chromatographic conditions, $\omega Pe^{-1/3}$ is much less than 1, so that Eq. 49 is indistiguishable from Eq. 39, there are reasons to doubt the validity of this treatment. For one thing, the film through which diffusive flux occurs is not necessarily stagnant. Although we must be quick to point out that stagnant zones may exist in the interparticle space, we also wish to point out the meaning of the film concept. As mentioned above, King (45) described this well. The thickness of this film is defined by the distance from the surface at which the rate of convective transport of solute toward the interface is equal to the rate of diffusive transport. Inside this distance the transport of solute normal to the interface occurs predominantly by diffusion. Because the flow is laminar, inside this distance the flow is predominantly parallel to the interface. Thus there *is* convective solute motion within the diffusion layer; however, it is parallel to the interface. The supposition that there is no convective solute motion within this film must be questioned.

1.3.4.6.   *New Theoretical Developments*

Recent theoretical developments yield insight into the relationships between the various broadening phenomena occurring in the column. Paine, Carbonell, Whitaker, and coworkers have developed a theory to describe dispersion in packed beds (86, 103, 107, 108). It is mathematically complicated, so it has only been applied to the problem of dispersion in open tubes (86). Complicated numerical computations for dispersion in packed beds have shown order of magnitude agreement with experiment for axial dispersion, but poor agreement with experiment for radial dispersion. In addition, in neither case was the Peclet number dependence in agreement with experiment.

In general they postulate that the system can be understood by considering the behavior of the gradient of the average concentration. For example, in the case of open tubes, the theory was developed in terms of the gradient (spatial derivative) in the axial direction of the concentration averaged over the cross-sectional area (r and Θ). A vector function $\mathbf{f(r)}$ is defined that relates the local deviation of the concentration from the average $\hat{c}$ to the gradient of the average concentration,

$$\hat{c} = \mathbf{f(r)} \cdot \nabla\langle c \rangle \tag{50}$$

Notice that $\hat{c}$ is a scalar function resulting from the operation of **f** on the gradient of the average. To solve for the peak shape, **f** must be determined, and this is mathematically difficult. It is, however, instructive to consider it qualitatively. The dispersion tensor (only the diagonal components enter into the equations) is given by Eq. 51. In this equation $\hat{v}$ is the difference between the average velocity and the local velocity. Let us walk through this qualitatively.

$$\mathbf{D} = -\frac{1}{V} \int_V \hat{v}\mathbf{f}\, dV = -\langle \hat{v}\mathbf{f} \rangle \tag{51}$$

Consider the case of a small eddy near a particle (see Fig. 1.6). (We have chosen an eddy to describe the application of the theory because it is a well-defined region of fluid. The treatment is perfectly general and does not depend on the presence of eddies for its applicability. We do not mean to suggest that eddies are predominant in chromatographic columns, nor do we mean to imply that eddies are an essential feature of hydrodynamic dispersion.) This section of the column is at the leading edge of the peak that is moving down the column. Then the z (axial) component of $\nabla \langle c \rangle$ is negative. Physically the eddy is not in the mainstream. Because of this, the concentration will be lower than average; $\hat{c} < 0$. This means that the z component of **f** will be positive. Considering Eq. 51 now, the z component of the vector $\hat{v}$ is negative (forward motion is lower here than on average), **f** is positive, and so the contribution of this section of the column to the integral in Eq. 51 is negative, and the

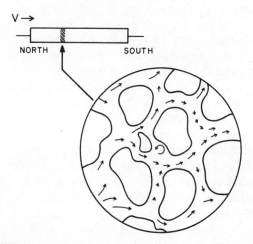

Figure 1.6. Possible flow in packed bed. See text for discussion.

contribution to **D** is positive. The "memory" of the relatively stagnant eddy causes axial dispersion.

What if, all of a sudden, the eddy were to break free from the particle against which it was nestled and be injected into a path flowing in the radial direction? According to the authors there is the same contribution to the axial dispersion, plus an additional component of radial dispersion. The vector **f** has remained essentially the same. However, the vector $\hat{v}$ now has a radial component. The radial component of $\hat{v}$ is reflected, through Eq. 51, in the dispersion tensor. The eddy could dissipate; but let us consider the case in which it now turns south (see Fig. 1.6), down the column (in the positive z direction). Repeating the preceding arguments, this time with $\hat{v}_z > 0$ (the eddy has caught a ride in a stream zipping between two particles), one finds that the contribution of this region of space to the integral is positive, so the net effect is to *decrease* the axial dispersion. This makes intuitive sense. Our eddy, with a concentration of solute more representative of solution commonly found downstream now (or commonly found at this postion some time ago), has been launched ahead, closer to its siblings.

The beauty of this treatment is that it has elements of both the coupling theory and the nonequilibrium theory of Giddings; it makes intuitive sense. Although not described here, the same ideas are used to develop a tortuosity tensor. The rather difficult to describe **f** is at the heart of axial dispersion.

The conservation of mass forms the basis for the recent theoretical achievements of Koch and Brady (109). These workers have taken a completely general approach to the problem of axial dispersion in packed beds of porous or nonporous particles. There are no assumptions made about the packing order, that is, an ordered array of particles *is not* assumed, although spherical, monodisperse particles *are* assumed. The use of a randomly packed bed is important. Both theory and experiment for ordered beds differ from the experimental results obtained on randomly packed beds. Koch and Brady point out that one should infer from this that there is a fundamental difference between these two types of packed beds. Chief among the differences is the role played by diffusion. In a randomly packed bed, but not in an ordered bed with flow parallel to a symmetry plane, mechanical dispersion occurs. Mechanical dispersion is defined as "true diffusive behaviour independent of molecular diffusion" (109). This is reflected by the term for eddy dispersion that is independent of the molecular diffusion coefficient. A purely mechanical dispersion will describe the events completely in a randomly packed bed (at large Pe) if the entire fluid space is hydrodynamically accessible. If the bed contains regions of zero velocity or of separated flow (flow in which there is an abrupt change in the velocity vector, such as the eddy in Fig. 1.6), then molecular diffusion is the only way for matter to enter and leave these regions. In such realistic systems the dispersion cannot be entirely mechanical. The results of the analysis of Koch and Brady are described in part below.

The detailed computations are carried out for a dilute bed ($\epsilon_e \approx 1$), but it is shown that the essential features of the results are useful for concentrated beds. Several important points are made in the paper; reading of the paper will be rewarded by a sound understanding of the events that control dispersion in the mobile phase despite fairly complex mathematics. It involves taking spatial averages of the mass conservation equations, and then casting those results in terms of conditional averages. The parameter in the conditional average is the distance from a particle surface. The results for $Pe \gg 1$ (of most relevance to chromatography) follow. Axial dispersion is approximately

$$\frac{\mathcal{D}_{ax}}{D_m} = 0.75Pe + \frac{\pi^2}{6}(1 - \epsilon_e)Pe \ln (Pe) \qquad (52)$$

while radial dispersion is approximately

$$\frac{\mathcal{D}_{rad}}{D_m} = 0.278(1 - \epsilon_e)^{1/2}Pe \qquad (53)$$

The first term in Eq. 52 is due to purely mechanical mixing of the fluid. The randomly packed bed of spherical particles causes the flow to split and recombine as discussed. The second term in Eq. 52 arises from the boundary layer around the particles. Because of the no-slip condition at interfaces, the velocity of the fluid becomes smaller as one gets closer to the surface of the particle. If there were no diffusion, then solute that had become trapped in a slowly moving stream around a particle surface would reside there for a long time, and the resulting dispersion would be large (recall the discussed open tube case). Koch and Brady made, and justified, the assumption that diffusion mixes solute over small distances, and that the average distance over which the mixing occurs is related to the diffusion layer thickness. Using this concept, a contribution to dispersion on the order of $Pe \ln (Pe)$ was found. This is an exciting development. It demonstrates that, even at large Pe, the axial dispersion is not entirely mechanical. The diffusional relaxation of concentration gradients caused by slowly moving (but not stagnant) regions of the mobile phase is similar to Giddings's coupling concept.

In the same paper, Koch and Brady evaluate dispersion at low Pe. As might be expected, the value of $\mathcal{D}_{ax}/D_m$ converges on the tortuosity factor, a result with which no one argues. The case of porous particles (and the related possibility of regions of mobile phase that are disconnected) was also considered. Terms due to the "holdup" of solute appear in the dispersion equation as Pe^2.

The data in Fig. 1.5 are predicted exceptionally well by the equations of Koch and Brady. Perhaps the most important contribution that these researchers have made is the development of a nonmechanical dispersion term at large Pe. They are currently working on related problems for short times and distances. As liquid chromatographic columns become smaller, their work will become even more relevant.

Another recent mathematical development that shows intuitive appeal, but is technically difficult, is the realization by Lee (110) that the theory for open tubes is simpler than that for packed beds because the tube case has periodic boundary conditions. He has modeled dispersion in periodic media with these ideas. This is neatly connected with the discussion above of Golay's work at short times in open tubes. The reader will recall that very short tubes, when connected, influence each other because the output profile of one is the input profile of the other, and the initial and boundary conditions govern output at short times. It certainly must be the same if one considers the interparticle space to be a series of short tubes. Therein lies the mathematical difficulty. If the output of one tube did not influence the effect of the next, the problem would be more straightforward; indeed, the series of stirred tanks model for chromatography is essentially the plate model, and it has been understood for decades.

Recently the shortcomings of several models of chromatographic broadening were considered (111). It was shown that the simple mass balance model predicts backmixing in the column and that a finite amount of solute will traverse the entire column in virtually no time. It was suggested that mixing cell models may be preferable, but even these models predict the "infinite speed of propagation." For practical chromatographic conditions, these issues are not of concern. However, for those wishing to have a good model for short times, there is still work to be done.

The fundamental bases for all of the theories of flow in porous media come down to three: mass balance, stochastic, and statistical mechanical. In a recent critical review, Sposito et al. (112) call into question the suitability of all of these models. It is suggested that one difficulty is that the equations are continuum equations, but events in porous packed beds may be governed by events on a molecular length scale. If the surfaces of the particles in the bed are rough on the length scale of a solvent molecular diameter, then it is possible that treating the fluid as a continuum will lead to errors in calculated behavior. A complete theoretical understanding of mass transport and dispersion in packed beds remains elusive.

1.3.4.7. *The Data Are in Agreement with Koch and Brady*

Although the theory remains incomplete, the data make a clear statement: the influence of mobile phase velocity on axial dispersion is nearly linear (99–102, 105, 113). Figure 1.5 shows many data taken over a wide range of Peclet numbers. Recall that for ordinary liquid chromatography Peclet numbers near 10 are typical. In this range casual inspection of Fig. 1.5 shows that $\mathscr{D}_{ax}/D$ is roughly proportional to Pe; more careful measurements show (103, 108, 113) that $\mathscr{D}_{ax}/D \propto Pe^{1.2}$. Note that the troublesome fractional, and not intuitively obvious, power of Pe (1.2) is completely explained by the Pe ln (Pe) term of Koch and Brady (109). [For $20 < Pe < 2000$, $0.57Pe \ln (Pe) \approx Pe^{1.2}$.] The most approximate form of the dependence of $\mathscr{D}_{ax}$ on Pe that will work for $Pe \gg 1$ and *a small range of Pe* is rearranged below,

$$\mathscr{D}_{ax}/D \approx Pe \qquad \therefore \qquad \mathscr{D}_{ax} \approx u_e d_p \qquad (54)$$

Surprised? The result obtained from simple random walk theory stands. This should not be interpreted as a justification for the abandonment of rigorously constructed mass, momentum, and energy conservation equations. It does point out the usefulness of the random walk treatment in getting a glimpse at the result before too much effort is expended.

For low flow rates, axial dispersion must become controlled by molecular diffusion. This can be seen in the asymptotic approach of the dispersion to the product of the molecular diffusion coefficient and a tortuosity factor (Fig. 1.5). When the dispersion coefficient approaches the molecular diffusion coefficient, the ordinate in Fig. 1.5 approaches log γ or log(0.65). Thus one has

$$\mathscr{D}_{ax} \cong \gamma D + \lambda u_e d_p \qquad (55)$$

Equation 55 will work satisfactorily for $Pe < 10$. For $Pe > 10$, and especially for correlating data over a large range of Pe, one should use the dependence given in Eq. 52. For a large range of Pe that encompasses $Pe \approx 1$, one must add at least a γ to Eq. 52 to account for molecular diffusion. The original reference should be consulted for further details (109).

Thus the theory is very difficult, but experimentally the answer is simple. One should not forget that the apparent simplicity in the oft used Eq. 55 hides quite a bit.

1.3.4.8. *Brief Discussion of the Influence of Retention or Adsorption on Dispersion*

It should be mentioned that most of this discussion has not admitted to the penetration of porous particles by solutes. This penetration causes terms on the order of Pe^2 to appear in the equation for axial dispersion (109). The retention of the species in the pores completely explains this aspect of dispersion. In certain cases, such as rapid and irreversible sorption of solute at short times, the porous particles influence the axial dispersion due to *mobile phase effects alone*. In other words, the effect is not one of mass transfer resistance (see below), but an effect mediated through the concentration changes that the adsorption brings about (68, 102, 110, 114). This can be understood as follows. As in an open tube, axial dispersion arises in large part from radial velocity gradients. If solute is strongly sorbed, then the solute in the region of the mobile zone near the interface will be depleted. But these are just the regions where one expects to find a lower velocity. Thus adsorption decreases the influence of the radial velocity inhomogeneity; the solute that remains in the mobile zone is distributed among regions more alike in velocity. The way in which the mobile zone dispersion depends on the solute's ability to permeate porous particles is currently unknown in general.

Finally it must be said that the flow in the column under conditions that are employed commonly could be in a transition region between flow dominated by viscous forces ($Re \ll 1$), in which case the mobile phase processes are geometrically similar for changes in the dimensions of the system (50), and flow dominated by inertial forces ($Re \gg 1$) (115). The uncertainty arises because it is difficult to be confident when applying dimensionless data taken from well-defined systems (such as open tubes) to packed beds because the appropriate length to use in the calculation of the Reynolds number is not obvious.

1.3.5. Retention and Dispersion in the Single Theoretical Plate Column

1.3.5.1. *Introduction*

As pointed out in other sections of this chapter, there are many approaches to the study of chromatographic systems, including the plate model (11), the random walk model (52, 53, 76), stochastic kinetic (6), general nonequilibrium (8) (see Appendix 1.5), and continuum mass balance models (see Appendix 1.3). Each of these models has its own strengths and limitations. Clearly the random walk model is physically very intuitive. In contrast the mass balance approach, of which the works of Kucera (116), Grushka (117), and Horvath and Lin (105, 118) are representative, involves quite complex mathematics. The thesis of this section is that the so-called continuously stirred tank reactor (CSTR), which has proven to be quite useful in developing simplified models of chemical reactors (119), is a convenient heuristic device for the investigation of chromatographic systems (see Fig. 1.7). Furthermore, it can give accurate estimates of the contribution of a number of processes to peak broadening in column chromatography. The chief virtue of the CSTR approach is that it does not involve the partial differential equations which arise in the continuum mass balance approach.

As shown in Fig 1.7, the basic system is very simple. The "reactor" is assumed to be well mixed, that is, each phase is taken to be homogeneous despite the existence of an interphase mass transfer resistance. The reactor is operated with matched inlet and outlet flows to simulate the balance of flow in a column. Only the so-called mobile zone, not to be confused with the mobile phase, some of which may be stagnant (trapped in pores), is allowed to leave the reactor. The stationary phase and any mobile phase which may be incorporated in the support particles are held within the reactor. This might be accomplished by use of a particle

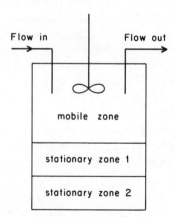

Figure 1.7. Continuously stirred tank reactor. The mobile zone is well stirred and inflow balances outflow. There are two stationary zones shown that are accessible in series. Fewer or more stationary zones may be added, and they may be accessible in series (as shown) or in parallel.

filter or a membrane that is not permeated by the stationary phase. The sample species is admitted to the reactor at some initial time designated as $t = 0$. For numerical simplicity we will initially take this to be an impulse input $\delta(t)$ of strength n^0. This model is obviously a very idealized one. A long series of such tanks will ultimately result in the production of a Gaussian elution profile at the exit of the last stage (119). As will be seen, a model based on the use of a single tank will generate equations that accurately predict the first moment and height equivalent to a theoretical plate (HETP) in agreement with other more complex approaches.

1.3.5.2. One Phase

The basic method is defined for all systems to be investigated by analysis of the simplest possible situation. Consider the case where initially there is no stationary or stagnant mobile phase within the tank. A mass balance on the material in the tank proceeds as follows:

$$\frac{dn}{dt} = Fc_{in} - Fc_{out} \tag{56}$$

where n is the number of moles of the sample in the reactor, F is the volumetric flow rate, c_{in} is the concentration of solute in the fluid flowing into the tank, and c_{out} is the solute concentration in the fluid leaving the tank. Since the tank is assumed to be well mixed, c_{out} is equal to the moving phase solute concentration in the tank. Because all of the sample is deposited in the tank as an impulse at $t = 0$, it is evident that there is no inflow of solute after $t = 0$; thus $c_{in} = 0$. We will designate the volume of moving phase within the tank as V_m. Consequently the rate of change of concentration inside the tank will be

$$\frac{dc_m}{dt} = \frac{1}{V_m}\frac{dn}{dt} \tag{57}$$

The initial solute concentration c^0 is evidently the strength of the impulse input divided by the volume of the moving phase n^0/V_m. The above equations and initial condition can be combined and solved. The solution is quite simple,

$$c_m = c^0 \exp\left(\frac{-Ft}{V_m}\right) \tag{58}$$

This is clearly an exponential (first-order) decay with time and the solute concentration–time profile is very asymmetric. The above result can be integrated to give the various peak moments (see Table 1.2). The total area under the curve (zeroth moment), the centroid (normalized first moment), and the second central moment (variance) are obtained from the following integrals:

$$\mu_0 = \int_0^\infty c_m(t)\, dt = \frac{n^0}{F} \tag{59}$$

$$\mu_1' = \frac{1}{\mu_0} \int_0^\infty c_m(t)t\, dt = \frac{V_m}{F} \tag{60}$$

$$\mu_2 = \frac{1}{\mu_0} \int_0^\infty c_m(t)(t - \mu_1')^2\, dt = \left(\frac{V_m}{F}\right)^2 \tag{61}$$

Certainly this type of device will not be able to perform the main task desired of a chromatographic column, that is, separate two or more species. It is nonetheless interesting to compute the number of theoretical plates using the conventional chromatographic definition as

$$N = \frac{(\mu_1')^2}{\mu_2} = 1 \tag{62}$$

We see then that this single tank very reasonably corresponds to a one-plate column. The height equivalent to a theoretical plate is conventionally written as the length of the column divided by N. Because there is no "length" associated with this system, we chose to establish the equivalent volume V_E of a theoretical plate as V_m/N.

The study of more complex systems will lead to the need to solve more difficult differential equations. Because the main concept to be explored here is whether the tank reactor model leads to accurate peak moments rather than accurate equations for the concentration–time dependence, it is a waste of effort to solve these equations.

At this point we introduce the well-known moments theorems from operational calculus. We transform the differential equation for c_m, solve for the transformed concentration variable in terms of the Laplace variable, and then apply the moments theorem.

We take the Laplace transform to be defined here as

$$c_m^* = \int_0^\infty \exp(-pt)c_m(t)\, dt \tag{63}$$

When applied to the differential equation resulting from the combination of Eqs. 56 and 57, this leads to

$$c_m^* = \frac{c^0}{p + FV_m} \tag{64}$$

Any desired noncentral nonnormalized moment can be obtained from the general relationship

$$m_n = (-1)^n \lim_{p \to 0} \frac{d^n c^*}{dp^n} \tag{65}$$

The normalized and centralized moments can be easily generated,

$$\mu_1' = \frac{m_1}{m_0} \tag{66}$$

$$\mu_2 = \frac{m_2}{m_0} - (\mu_1')^2 \tag{67}$$

1.3.5.3. Model I: Slow Interphase Transfer Kinetics (No Stagnant Mobile Phase)

The above system bears little relationship to what goes on in a chromatographic column since as presently constituted there is only a moving phase present. Suppose now that a second phase, which can be either a solid or an immiscible liquid, is added to the tank. Further we assume that the rate of transfer of solute between the two phases is subject to some impedance due to mass transfer or a chemical process as indicated below,

$$A_m \rightleftharpoons A_s \tag{68}$$

We now need to consider the mechanics of interphase transfer (see Fig. 1.8). Some thought will indicate that the rate of transport (moles per unit time) from left to right across the interface will be proportional to the concentration of solute in the mobile zone, and similarly the rate of transport from right to left will be proportional to the solute concentration in the stationary zone. At equilibrium the net rate of transport (the signed sum of left to right and right to left) across the interface will become zero. It should also be evident that the constants of proportionality κ_f and κ_b (see below) will vary in direct proportion to the interfacial area. The argument developed here is precisely that used in connection with the

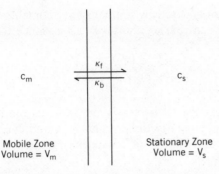

Figure 1.8. Interfacial region in single theoretical plate column.

discussion of Eq. a4.7. It should also be noted that these "rate constants," that is, κ, have dimensions of volume·time^{-1}. The differential equation describing the solute species A in the moving phase becomes

$$V_m \frac{dc_m}{dt} = -Fc_m - \kappa_f c_m + \kappa_b c_s \qquad (69)$$

$$V_s \frac{dc_s}{dt} = \kappa_f c_m - \kappa_b c_s \qquad (70)$$

Before solving these equations it is helpful to rewrite them in the following terms:

$$V_m \frac{dc_m}{dt} = -Fc_m - \kappa_b(Kc_m - c_s) \qquad (71)$$

$$V_s \frac{dc_s}{dt} = \kappa_b(Kc_m - c_s) \qquad (72)$$

with

$$K = \frac{\kappa_f}{\kappa_b} \qquad (73)$$

The term K is the equilibrium partition coefficient for the transfer of solute from the mobile phase to the stationary phase. Equations 71 and 72 can be solved subject to the impulse input of solute by taking the transform of both and substituting the transform of c_s into the equation for c_m. The resulting equation for the transform of the mobile phase solute concentration is then manipulated by applying the moments theorems to yield equations for the centroid μ_1' and the peak variance μ_2,

$$\mu_1' = \frac{V_m}{F}\left(1 + \frac{V_s}{V_m} K\right) \tag{74}$$

In Eq. 74 the ratio V_s/V_m is clearly the same as the usual definition of the phase ratio Φ. Similarly the product of K and Φ can be identified as the conventional chromatographic capacity factor,

$$\Phi = \frac{V_s}{V_m} \tag{75}$$

$$k' = \Phi K \tag{76}$$

Because there is no stagnant mobile phase present in the system, k' is equal to k''. The equation for the variance can be manipulated to yield an equation for the number of theoretical plates (see Eq. 62). Thus,

$$\frac{1}{N} = 1 + 2F\Phi \frac{k'}{(1 + k')^2} \frac{1}{\kappa_b} \tag{77}$$

This result is really very familiar. It predicts that the number of plates will decrease as the flow rate is increased; it predicts that a decrease in the interphase rate constant κ_b will broaden the concentration–time profile; and finally, the term $k'/(1 + k')^2$ is exactly what one gets from all approaches to broadening due to resistance to mass transfer in the stationary phase. We can use Eq. 77 to get a volume of mobile phase equivalent to one plate V_E by dividing V_m, which is a measure of the size of the "column," by N. Thus,

$$V_E = V_m + 2FV_s \frac{k'}{(1 + k')^2} \frac{1}{\kappa_b} \tag{78}$$

With the exception of the first term, this equation is identical to equations for the HETP where resistance to mass transfer in the stationary phase is the sole broadening process.

These results (Eqs. 74 and 77) are actually quite revealing. First note that the peak centroid μ_1' (Eq. 74) does not depend on the rate constants but only on their ratio K. This is in complete accord with the continuum mass balance approach when the boundary conditions are such that diffusion into and out of the column is prohibited (120). The centroid does depend on the volumetric flow rate, but this is a trivial consequence of the use of time rather than volume of eluent as a measure of retention. If μ_1' were expressed in volume units, then the dependence on flow rate would disappear.

The volume equivalent to a theoretical plate V_E is comprised of two terms (Eq. 78). The first term, V_m, which was also in the one-phase model discussed, reflects the geometry of the system. It represents the volume over which effective mixing occurs in the absence of retention. If one considered the more realistic model of many tanks in series, one would find that this volume-of-mixing term becomes smaller as the tank size becomes smaller. This is very similar to the eddy dispersion term. When axial dispersion is purely mechanical, the contribution to H is (Eq. 37) proportional to d_p. The dependence of the second term on the capacity factor and the interphase rate constant κ_b is exactly that obtained when other approaches are used. The flow rate dependence is also correct. The result is accurate even to the value of the numerical coefficient, namely, 2 in Eq. 78.

1.3.5.4. Model II: Stagnant Mobile Phase Present

As mentioned, a significant fraction of the mobile phase in a real packed column is stagnant because it is present in the pores of the support particle. We now consider this effect in the stirred tank reactor model. The process is formulated by assuming that material is transferred first from the moving mobile phase to the stagnant mobile phase and subsequently to the stationary phase. We now denote the volume of the moving or flowing mobile phase by V_{fm}. The total volume of mobile phase in the system V_m is clearly $V_{fm} + V_{sm}$. Because the moving and stagnant mobile phases are chemically identical, the effective chemical equilibrium constant for transfer from one to the other must be exactly unity. To be explicit, the equilibrium constant $K_{m,sm}$ defined in Eq. 80 will be taken as unity.

$$A_m \rightleftharpoons A_{sm} \tag{79}$$

$$K_{m,sm} = \frac{c_{sm}}{c_m} = 1 \tag{80}$$

$$V_{fm} \frac{dc_m}{dt} = -Fc_m - \kappa_{mt}(c_m - K_{m,sm}c_{sm}) \tag{81}$$

$$V_{sm} \frac{dc_{sm}}{dt} = \kappa_{mt}(c_m - K_{m,sm}c_{sm}) - \kappa_b(Kc_{sm} - c_s) \tag{82}$$

$$V_s \frac{dc_s}{dt} = \kappa_b (K c_{sm} - c_s) \tag{83}$$

The term κ_{mt} is the moving to stagnant mobile phase rate constant. This set of three differential equations must be solved subject to the usual initial condition. This is easily done by taking the Laplace transform and making substitutions to eliminate the stagnant mobile phase and the stationary phase solute concentrations. The final equation after the equilibrium constant $K_{m,sm}$ is set to unity is

$$\mu_1' = \frac{V_m}{F} (1 + k') \tag{84}$$

It is important to note that we define the k' based on the total amount of mobile phase whether or not it is in motion. Thus we have defined the phase ratio as

$$\Phi = \frac{V_s}{V_{fm} + V_{sm}} \tag{85}$$

This definition is identical to the one in Eq. 75. The second central moment is most compactly written in terms of the number of plates,

$$\frac{1}{N} = 1 + 2F\Phi \frac{k'}{(1 + k')^2} \frac{1}{\kappa_b} + 2F \frac{[k_0 + k'(1 + k_0)]^2}{(1 + k_0)^2 (1 + k')^2} \frac{1}{\kappa_{mt}}$$

$$= 1 + 2F\Phi \frac{k'}{(1 + k')^2} \frac{1}{\kappa_b} + 2F \frac{k''^2}{(1 + k'')^2} \frac{1}{\kappa_{mt}} \tag{86}$$

The term k'' is defined in Eq. 32. This equation clearly shows that the peak broadening is related to both the stagnant mobile phase and the stationary phase resistances. The two impedances add quite independently. The second term in Eq. 86 is identical to the stationary phase resistance to mass transfer in Eq. 77. The model was constructed so that solute can only enter the stationary phase after it passes through the stagnant mobile phase.

If κ_{mt} is allowed to become very large, then the second moment is controlled by the rate of desorption from the stationary phase κ_b as expected. Even in the absence of stationary phase the peak is broadened by the mass transfer resistance in the stagnant mobile phase.

It is well to observe that, although the preponderance of the net transfer process of the solute from the mobile to the stationary phase occurs through the stagnant mobile phase, the exterior surface of the support particle in a real chromatographic column is bathed in moving fluid. Thus a more realistic model would allow for some small rate of direct transfer from the moving zone to the stationary phase. In the present heuristic development of the single-plate column the treatment of this case is not warranted. It should be evident that the existence of a parallel path from the mobile zone to the stationary phase would prevent the total resistance to mass transfer from blowing up as the rate of transfer from the moving zone to the stagnant mobile phase κ_{mt} goes to zero.

1.3.5.5   *Concluding Remark*

The value of this approach is that although it does require knowledge of the theorems of operational calculus, one does not have to deal with partial differential equations to arrive at expressions for the peak moments. The only difference is that the steps needed to deal with any problem are purely algebraic. This approach should be compared with the approach using differential equations derived from mass balance alluded to previously in the historical development section (see also Appendix 1.3). This simplicity makes it far easier to deal with situations that are extremely tedious to handle by the continuum mass balance approach. For example, it is algebraically very easy to deal with a two-site adsorption model analog of Model II. One finds that each site contributes independent terms to both retention and broadening, which can be written by analogy to Eqs. 84 and 86.

1.3.6.   Dispersion in Open Tubular Chromatography

Dispersion in open tubular chromatography is similar to the Taylor–Aris problem, but with tubes whose walls are sticky. It was shown in the simple one-plate model that when molecules exist in volumes of the system that are not moving, dispersion results. The discussion of flow in open tubes pointed out that dispersion exists in the mobile phase because of velocity inhomogeneity. In open tubular chromatography both effects take place.

This problem was first solved by Golay (121) and was widely applied to capillary gas chromatography [see, for example, (122)]. The so-called Golay equation seems to describe capillary liquid chromatography as well (86, 123–126). Since peak volumes are so small in capillary chromatography, in order to obtain meaningful data on peak dispersion, careful attention must be paid to the band broadening caused by detectors and injectors (126–128). One form of the Golay equation is given below. Define

$h = H/r_0$; then

$$h = \frac{2}{v} + \frac{1 + 6k' + 11k'^2}{24(1 + k')^2} v + \frac{2}{3} \frac{k'}{(1 + k')^2} \left(\frac{d_f}{r_0}\right)^2 \frac{D_m}{D_s} v \qquad (87)$$

The first term is recognizable as an axial molecular diffusion term. In this equation we have used v, where v is ur_0/D, r_0 being the tube radius. The thickness of the stationary film, in which the molecular diffusion coefficient is D_s, is d_f. Note that the tortuosity constant in the open tube is unity, so it does not appear in the axial diffusion term. The second term is analogous to the second term in Eq. 36 from Taylor–Aris theory and becomes identical to it when $k' = 0$. It describes the influence of retention on the effect of radial mass transport in relaxing broadening caused by flow inhomogeneity. It is important to see that in the limit as $k' \to \infty$, the second term is only a factor of 11 larger than when $k' = 0$. The last term is due to slow mass transfer in the stationary phase. Because the film thickness d_f is usually small, at least when thick films are not used, the last term can usually be ignored. Note that this term, involving stationary phase mass transfer, approaches zero as $k' \to \infty$. Sketches of these two types of behavior are shown in Fig. 1.9.

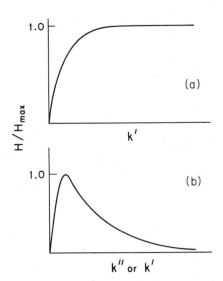

Figure 1.9. (a) For rate-limiting step of diffusion in stagnant mobile phase, H increases to a limiting value as k'' increases. (b) For rate-limiting step of diffusion in stationary phase, H reaches a maximum and decreases.

In certain cases the Golay equation underestimates h (126). An attempt was made to add a term for resistance to mass transfer at the mobile–stationary phase interface. Although the Golay equation has proven adequate for such things as optimization studies, the general problem is still under scrutiny. A general solution for the problems of diffusion, convection, and reaction in a multicomponent system is under active investigation (129–134). The Golay equation has also been successfully applied to small-diameter packed columns (128).

Knox and Gilbert (124), using the classical analysis of Knox and Saleem (10), have considered the optimization and outlook for the capillary column in liquid chromatography. They conclude that, for applications requiring on the order of 10^6 theoretical plates, capillary columns will be far better than packed columns. However, the tubing diameter required, near 10^{-5} m, will necessitate very small injector and detector volumes.

1.3.7. Dispersion in Packed Columns—Linear Isotherm

1.3.7.1. Overview

The title subject has been studied carefully, over four decades, by analytical chemists, physical chemists, and chemical engineers. Even so, a complete theory that can fully describe all of the processes that take place in a real column is not yet available. On the other hand, the ideas elaborated prior to 1970, for most purposes, suffice for a reasonable understanding of the theory.

By way of review, let us consider the primary processes leading to band broadening.

1. Axial molecular diffusion
2. Eddy dispersion
3. Mobile phase mass transfer resistance
4. Other mobile phase processes, such as streaming through bed cracks
5. Mass transfer at the mobile zone–stationary zone interface
6. Diffusion (that is, mass transfer) in the stagnant mobile phase
7. Chemical kinetics for the solute–stationary phase site reaction

In addition, other factors that can influence the band profile, and hence the separation, are

1. Temperature gradients

2. Multiple site adsorption
3. Nonlinear isotherms
4. Injector volume
5. Detector volume and response time

Order of magnitude estimates of the relative importance of these phenomena can often be obtained by judicious comparison of characteristic times or rates. For example, how much does axial diffusion in the flowing zone contribute to the spreading in comparison to eddy dispersion? We can compare the axial dispersion coefficient for eddy dispersion to the molecular diffusion coefficient. Equation 52 shows that for Pe around 10, eddy dispersion is significantly greater than axial molecular diffusion. Lenhoff (78) has given a rather complete set of dimensionless coefficients that facilitate a comparison of the time scales of the various processes. The reader can use the relationships in Appendix 1.4 to arrive at a set of parameters of the same units, such as rate constants in seconds^{-1}, for individual processes.

Another general comment is in order. In many textbooks and simpler treatments of the dynamic theory of chromatography there is a lack of attention to detail regarding the difference between mobile phase and stationary phase, and mobile zone and stationary zone. The phases are the materials (in the spirit of J. W. Gibbs), the zones are the parts of space with velocities implied by the adjectives. The mobile zone is that part of the column, about 40% of it, through which the flowing mobile phase moves. Stagnant mobile phase, mobile phase trapped in the pores of the stationary support, is part of the stationary zone, as is the stationary phase. Except in the case where the slow step in the chromatographic process is a chemical kinetic step at the interface, the zone formalism will make the resulting expressions simpler. In the same vein, care must be taken to use interstitial velocity u_e when using zones (and k'') to describe the dynamics. To obtain this velocity, one must determine the retention time of a solute that does not permeate the porous particles of the packing. In contrast, the average velocity u, determined from the retention of an unretained but permeating solute, combines naturally with a phase-based description of the dynamics (and k').

A general equation for band broadening is given in

$$h = Af(v) + \frac{B}{v} + \sum C_i v \qquad (88)$$

The second term, called the B term, is due to molecular diffusion in the axial direction. Of course molecular diffusion occurs in all directions, but only motion in the axial direction adds directly to axial dispersion. The third set of terms, the C term, is due to those processes in which the rate of relaxation of the nonequilibrium between the moving zone and the stationary zone is flow rate independent. The first term, known affectionately as the A term, is the least well understood. It contains flow induced dispersion as described in Section 1.3.4. It also contains terms due to mass transfer in which the mass transfer is flow rate dependent.

The formulation shown above is somewhat misleading. The van Deemter equation is the historical precedent for the dissection of the plate height expression into individual contributions due to independent processes. But the same process may contribute to axial dispersion and radial dispersion; the former is deleterious while the latter is helpful. Thus one should expect to see an expression for mobile zone mass transfer twice, and stationary zone mass transfer twice. In each case one of the pair would contribute directly to dispersion while the other would contribute inversely in its role as a means of relaxation of nonequilibrium. Because of this, the identification of a *single* term in the plate height equation with a *single* process is misleading. While we maintain the custom of A, B, and C terms, we encourage the reader to visualize the dual influences of mass transfer.

1.3.7.2. The B Term

The B term has been reinvestigated recently (26, 73, 74, 76). These studies all rely on the same concept; the effective diffusion coefficient in a heterogeneous medium is the time-weighted average of the diffusion coefficients for the solute in each individual phase (73). This will be true as long as the relaxation rate for mass transfer among the regions i in Eq. 89 is much faster than the rate of dispersion or the relaxation rate that incorporates the overall diffusion coefficient,

$$D_{\mathrm{eff}} = \sum D_i t_i \bigg/ \sum t_i \qquad (89)$$

Using this equation, and recalling that the ratio of the times a molecule spends in two zones is given by k'', one arrives at Eq. 90 for the contribution from axial molecular diffusion,

$$h_{\mathrm{ax,mol}} = \frac{2(D_{\mathrm{eff}}^{\mathrm{sz}} k'' + D_{\mathrm{eff}}^{\mathrm{mz}})}{Dv} \qquad (90)$$

where D is the solute diffusion coefficient in the mobile phase, k'' (Eqs. 32 and 33) is the zone capacity factor $t_{\mathrm{sz}}/t_{\mathrm{fz}}$, and k_0 is the ratio of porosities

ϵ_p/ϵ_e. The subscript eff indicates that there is a tortuosity factor incorporated in the D. $D_{\text{eff}}^{\text{sz}}$ is given by

$$D_{\text{eff}}^{\text{sz}} = \frac{D_{\text{eff}}^{\text{sm}} + k'(1 + k_0)D_{\text{eff}}^{s}}{1 + k'(1 + k_0)} \tag{91}$$

Since the B term is usually negligible, Knox and Scott (73) had to stop the flow of the eluent to allow molecular diffusion to occur to a significant degree. Having done this, they were able to determine the overall effective diffusion coefficient, the numerator in Eq. 90.

1.3.7.3. The A Term

The eddy dispersion term and the contribution of mobile phase mass transfer in relaxing the concentration gradients caused by eddy dispersion have been discussed above. Note that a portion of the dispersion that occurs in the mobile zone is due to molecular diffusion, as indicated in Eq. 39. However, this is ordinarily ascribed to the B term. Therefore in using expressions from Section 1.3.4 to construct an A term, one must be careful not to duplicate the molecular diffusion in the mobile zone that one might already have put in the B term.

The contribution of mass transfer in the mobile zone to the overall interzone mass transfer must also be considered. If one accepts a velocity independent mass tranfer coefficient, then the contribution to the plate height will be linear in velocity (Eq. a4.12) and it will appear in the C term, as in the van Deemter equation. As in the case of eddy dispersion, since this is a process taking place in a highly complex medium, exact analysis is precluded. Huber (66, 71, 72) and Horvath and Lin (105, 118) employed mass transfer coefficients obtained from engineering correlations.

Huber's analysis for mobile phase mass transfer is particularly straightforward. He inserts an expression for the mass transfer coefficient for a packed bed into Eq. a4.12. The mass transfer coefficient used was from work by Thoenes (135) and is

$$\text{Sh} = \frac{k_{\text{mt}}d_p}{D} = \frac{1}{\Psi_f}\,\text{Re}^{1/2}\text{Sc}^{1/3} \tag{92}$$

where Sh is the Sherwood number, a dimensionless mass transfer coefficient (Eq. a4.6). The factor Ψ_f is a geometrical factor near 1, Re is the Reynolds number, and Sc is the Schmidt number, the ratio of kinematic viscosity (viscosity/density) to diffusion coefficient. The Schmidt number is about 1000 for liquids, and the Reynolds number is usually around 0.01 in liquid chromatography. Equation 92 is used in Eq. a4.12 to arrive at the contribution to h from mobile zone mass transfer limits on interzone equilibrium,

$$h_{\text{MZMT}} = \frac{1}{3} \Psi_f \frac{\epsilon_e}{1 - \epsilon_e} \text{Sc}^{1/6} \left(\frac{k''}{1 + k''} \right)^2 \nu^{1/2} \tag{93}$$

The reader will recall that Huber also had a $\nu^{1/2}$ term in the contribution to mobile zone convective dispersion. In that term mobile zone mass transfer was viewed as relaxing concentration gradients caused by velocity inhomogeneity. In Eq. 93 the mobile zone mass transfer term plays a similar role in relaxing concentration gradients caused by sorption of solute.

Recall that Horvath and Lin (105, 118) used an analogous approach, but with a different mass transfer coefficient. Their analysis was explained above. When the mass balance equations were solved, the mobile zone mass transfer term became (for an unretained component)

$$h_{\text{MZMT}} = \frac{\kappa k_0^2 \nu^{2/3}}{(1 + k_0)^2} \tag{94}$$

where κ is a structure dependent factor.

For a retained solute one has

$$h_{\text{MZMT}} = \frac{\kappa (k_0 + k' + k_0 k')^2 \nu^{2/3}}{(1 + k_0)^2 (1 + k')^2} \tag{95}$$

Note that the term reaches a limiting value at large k'.

Snyder has used the empirical equation $H \propto \nu^n$ (136). This has made optimization strategies rather straightforward, but it suffers from a lack of theoretical underpinning. Nonetheless, such an approach may be quite useful for quickly determining, with a few tables (14, 15), the best conditions for carrying out a particular separation. Along similar lines, a global optimization routine has been developed that employs known physical characteristics of many column types (137). This software package is useful as a reference point to assess column performance, and to help achieve separations when chemical efforts (alteration in stationary and mobile phases) have been exhausted.

Following up on the successful treatment of gas chromatographic data, Knox and coworkers developed a combined mobile zone mass transfer–eddy dispersion term as $h = A\nu^{1/3}$. Although it is totally empirical, Knox and coworkers have shown that this equation works remarkably well over a wide range of conditions (64, 68, 70). Because of its simplicity and reliability, the Knox formulation is widely used in practical chromatographic work.

We should like to point out that the approach of using empirical formulas based on a wide range of data is a sound one. However, a difficulty exists with the correlations used by Horvath and by Huber. In the limit of low velocity each of the mass transfer coefficients is seen to go to zero. This is physically unrealistic; certainly as the velocity approaches zero, the rate of mobile zone mass transfer will not approach zero since molecular diffusion will occur even in the complete absence of convection. In recent years progress has been made in the correlation of mobile phase mass transport processes. It is worthwhile to bring up the latest correlation, and show how it should be used. Recently Wakao and Funazkri (114) arrived at an excellent correlation based on a large number of independently obtained data sets. All the data were critically evaluated and corrected for axial dispersion. The latter correction was almost nil for liquids, but was quite significant for gases. The correlation obtained is

$$Sh = 2 + 1.1Sc^{1/3}Re^{0.6} \tag{96}$$

This results in a mobile phase mass transfer term of

$$h_{MZMT} = \left(\frac{k''}{1 + k''}\right)^2 \frac{\epsilon_e}{1 - \epsilon_e} \frac{\nu}{6 + 3.3Sc^{-0.23}\nu^{0.6}} \tag{97}$$

The term $3.3Sc^{-0.23}$ is 0.67 for $Sc = 10^3$. Since ν is typically about 10, neither term in the denominator of Eq. 97 dominates. This means that the functional dependence of the mobile phase mass transfer term is probably not a simple power of velocity in the flow rate range of practical chromatography.

We must point out that the above correlation is well supported by data down to Re as low as 3. Typical chromatography occurs at Re below that. Caution must be used in applying the correlation to Re less than 3. In fact, it seems irresponsible. However, the limiting form of the equation at low Re yields a velocity independent mass transfer coefficient like that used by Giddings and van Deemter, as discussed. It should also be noted that the scatter in the data is more evident at low Re than at high Re.

One can now put an upper bound on the mobile zone mass transfer contribution to the reduced plate height. Note that the mass transfer coefficient becomes $2D/d_p$ as ν approaches zero. Taking a typical value of $\epsilon_e = 0.4$ and letting k'' be large, one arrives at the conclusion that h_{MZMT}

must be less than about $v/9$. In other words, the value of C for this process is never greater than 0.11. As will be shown, this contrasts dramatically with C for stationary zone processes that have, in principle, no upper bound.

In a recent experimental study Arnold et al. (138) used a correlation similar to that given above. It was shown that the mobile phase mass transfer term accounted for about 20% of the total plate height in a protein chromatography experiment. Thus it is not an inconsequential term, at least for this relatively low efficiency (d_p was 100 μm) experiment.

At this stage, the theory of mass transfer in packed beds is not sufficiently well advanced that the rationale for Eq. 96 will be forthcoming soon. While a molecular level theory is preferred (we are chemists, not engineers), for now it would seem that the semiempirical approach is best. Until further studies of mass transfer in packed beds have shown that the data review by Wakao and coworkers is in error, Eq. 97 is probably the best statement of the influence of mass transfer in the flowing zone in limiting the rate of equilibrium.

1.3.7.4. The C Term

The C terms are, in comparison, relatively simple. All the fluid and surfaces in the particles are stationary. While there may be several processes going on in the particle, one can often realistically model them as simple kinetic and diffusional events. Usually one treats the path from just inside the particle boundary to the adsorbed or partitioned state as being serial processes, and therefore the overall rate of the process is the inverse of the sum of the inverses of the rates (see Eq. 22). In other words, the plate heights for the processes add. The complexities begin to arise when one attempts to consider events on a molecular level in the stationary phase. Since we have little knowledge of the physical properties of the surfaces with which we work, and they can be almost arbitrarily complex, the experimental proof of theoretical assertions can be difficult.

Recent careful studies have begun to reveal something about dynamics at the molecular level in reversed-phase systems. The chemistry of the interfacial region is covered in Chapter 2. Only a few points of relevance to the current discussion will be made. The stationary phase in reversed-phase systems is influenced by the mobile phase composition. The work of Burke's group (139, 140) demonstrated partitioning of water and the organic component of the mobile phase into the stationary phase. Carr and Harris (141) have demonstrated the influence of such partitioning on the "polarity" of the stationary phase. Thus there is both direct and indirect evidence for a stationary phase that contains components of the mobile phase. The fluidity of the stationary phase can be inferred from measurements of ^{13}C nuclear longitudinal relaxation times in labeled stationary phases (142).

A general expression for the C term contribution to h from slow processes in the stationary zone is given by (26)

$$h_{SZMT} = 2 \frac{2k''}{(1 + k'')^2} \frac{Dt_{sz}}{d_p^2} v \qquad (98)$$

which was derived from the random walk model. t_{sz} is the average time a molecule of solute spends in the stationary zone.

Since t_{sz}/t_{fm} is just k'', one can also write

$$h_{MZMT} = \frac{2k''^2}{(1 + k'')^2} \frac{Dt_{fm}}{d_p^2} v \qquad (99)$$

Note the k'' dependences in the two equations. As was shown in the simple one theoretical plate column, when the mass transfer limitation is in the mobile phase, then the equation for h contains the ratio of two polynominals in which the numerator is second order in k'' and the denominator is second order in k''. It will make reading easier if we adopt the following shorthand for the ratio just described: $p(k''^2)/p(k''^2)$. Here p stands for polynomial, and the order is given as the exponent of k''. Table 1.3 shows the general form of the k'' dependence of the C term for various limiting mass transfer processes. These dependences come from a detailed consideration of the functional form of t_{sz} or t_{fm}. Separate discussions of each case will be undertaken below. We have included mobile phase mass transfer in this table since it may appear in the C term but, of course, it will not if the mass transfer rate is velocity dependent.

Table 1.3. Influence of k'' on h

Rate-Limiting Process	Order of $p(k'')$	
	Numerator	Denominator
1. Chemical rate of desorption	1	2
2. Diffusion through stationary phase	1	2
3. Diffusion within stationary zone	2	3
4. $'';D_s \to 0$	2	2
5. $'';D_s \to D_{sm}$	1	2
6. Mobile zone mass transfer	2	2

Entries 4 and 6 in Table 1.3 represent cases in which the rate-limiting step involves only the mobile phase. In entry 4, D_s is zero, as might be expected in adsorption chromatography, while in entry 6 the rate-limiting step is in the mobile zone. In these cases, and only these cases, the form of h is $p(k''^2)/p(k''^2)$. This is in agreement with Eq. 99. In discussions of the mobile zone processes above it was pointed out that the mobile zone mass transfer rate may depend on k'', as it does in the open tube (see the Golay equation, Eq. 87). If this were to be the case, then the dependence of t_{fm} on k'' would alter the conclusions just drawn and stated in Table 1.3.

In entries 1 and 2 in Table 1.3 the rate-limiting step involves only the stationary *phase*. The form of h resulting from this is $p(k'')/p(k''^2)$, and h approaches zero for large k''. In entry 3 we have the only case in which there is a well-understood influence of k'' on t_{sz}. When the rate-limiting step is intraparticulate diffusion, one must use the effective intraparticulate, or stationary zone, diffusion coefficient D_{sz}. As Eq. 91 shows, it is a function of k''. When D_s approaches zero, and rapid equilibration between stationary phase and stagnant mobile phase still exists, the rate-limiting process involves only mobile *phase*, and the dependence mimics that of entry 6, as described above. As D_s approaches D_{sm}, there is no k'' dependence on the intraparticle residence time, thus the dynamics mimic entries 1 and 2.

Stationary Zone Diffusion Limited Rate. Glueckauf (46) originally formulated the expression for the mass transfer coefficient for a spherical particle being filled or emptied diffusionally. In this case one has t_{sz} given by

$$t_{sz} = \frac{d_p^2}{60D_{eff}^{sz}} \tag{100}$$

Note carefully that this expression, while widely used, pertains only to cases in which the zones are nearly at equilibrium.

In Eq. 100 the expression D_{eff}^{sz} hides a host of unknown features. Recent work has followed the original detailed considerations of Giddings (143) in attempting to elucidate the influences of the various domains of the stationary phase on the band broadening (26, 73–76, 144, 145). The works of Knox and Scott (73) and of Weber (76) are in absolute agreement. The stationary zone diffusion coefficient is a time-weighted average of the diffusion coefficients of the various zones in which the solute resides (Eq. 89). This being the case, the intraparticle mass transfer term is given by Eq. 101. The effective diffusion coefficient in the stationary zone is given in Eq. 91.

$$h_{\text{SZMT}} = \frac{1}{30} \frac{k''}{(1 + k'')^2} \frac{D_m}{D_{\text{eff}}^{\text{sz}}} v \tag{101}$$

Note again that the same diffusional process acts in the B and C terms. Although it affects the value of h in opposite directions, the diffusion in the stationary zone has *exactly* the same mathematical form in the term for axial molecular diffusion (Eq. 90) and in Eq. 101. Effectively the same sort of result has been obtained by Kawazoe and Takeuchi (144) and by Raghavan and Ruthven (145). Ruthven (102) has shown that as long as the chromatography is fairly ideal (long time, large N, no point on the column very far from equilibrium), very complicated approaches to the general chromatographic problem reduce to the results obtained with the linear driving force approximation. This approximation postulates that the mass transfer processes can be described by means of a lumped mass transfer coefficient. This mass transfer coefficient results from the consideration of solute passing *serially* through each of three zones—flowing mobile phase, macropores, and micropores. This treatment is for gases. The general expression found by them for zeolites, which contain both macropores and micropores, is given in

$$\frac{1}{k_s K} = \frac{d_p}{6k_{\text{mt}}} + \frac{d_p}{60\epsilon_p D_1} + \frac{d_s}{60 K D_2} \tag{102}$$

The equation is for k_s, an effective first-order rate constant for mass exchange between the mobile and stationary zones. K is the equilibrium constant representing the distribution of solute between gas outside the particle and in the micropores. The first term on the right-hand side is the resistance to mass transfer term in the flowing mobile phase, the second is the Glueckauf term for resistance to mass transfer in the bulk of the stationary zone, and the third term is an analogous term for the microcrystalline stationary zone. The variable d_s is the diameter of the microporous regions.

Recently Stout et al. (74) developed a model in which the stationary zone is considered to consist of three regions, denoted by *i*, *ii*, and *iii*. Phase *i* is the stagnant mobile phase, while phases *ii* and *iii* are stationary phases. They postulated that phase *iii* had a lower diffusion coefficient than phases *i* and *ii*, and that this can lead to an unexpectedly large influence on the band broadening. This result, derived from the use of an unlikely model, is unrealistic. Investigation of this model in some detail will be beneficial since it involves multiphase partitioning, and such processes are becoming more important in separations.

There will be no real loss in generality if we simply consider two compartments in the stationary zone, say, a and b. The ordinary column is shown schematically in Fig. 1.10, top panel. Here the solute partitions into the mixed stationary phase. The phases a and b are intimately mixed, but retain their own distinct chemical properties. Thus they do not form a solution. We concern ourselves with the case in which diffusion in the stationary zone limits the band broadening. Since the domains of a and b are small in size with respect to the particle diameter, diffusion will certainly mix solute effectively between a and b in the time required to equilibrate the stationary zone with the mobile zone. (Recall Glueckauf's term for mass transfer with a spherical particle; the rate constant is proportional to D/r^2, where r is a domain radius and D is the diffusion coefficient for the solute in the phase.) In this case the intraparticle space will be in local equilibrium, even when there is nonequilibrium between zones. This is a reasonable assumption, and it can be relaxed later.

Since there are two phases in the stationary zone and a mobile phase, there are two independent equilibrium constants. (We will assume infinite dilution conditions.) One has

$$K_a = \frac{c_a}{c_m} \tag{103a}$$

$$K_b = \frac{c_b}{c_m} \tag{103b}$$

and the derived quantity K_{ab},

$$K_{ab} = \frac{K_a}{K_b} = \frac{c_a}{c_b} \tag{104}$$

Each phase has its own volume V. Thus one can write k' values for

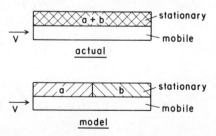

Figure 1.10. *Top.* Schematic diagram of liquid chromatographic column. Mobile phase enters from left and is exposed to mixed phase. *Bottom.* Model of Stout et al. Mobile phase first passes by phase a, then phase b.

each phase,

$$k'_a = \frac{K_a V_a}{V_m} = K_a \Phi_a \tag{105}$$

$$k'_b = \frac{K_b V_b}{V_m} = K_b \Phi_b \tag{106}$$

$$\tilde{k} = \frac{k'_a}{k'_b} \tag{107}$$

In this general model we are not defining the identity of the phases. That being the case, the use of k'' is inappropriate. We shall use k' exclusively. In the actual mixed-phase column, the measured k' is given by the number of stationary molecules of solute divided by the number of mobile molecules.

$$k' = \frac{n_a + n_b}{n_m} = \frac{c_a V_a + c_b V_b}{c_m V_m} \tag{108}$$

$$= k'_a + k'_b \tag{109}$$

The diffusion coefficient in the stationary zone will be the time-weighted average of the diffusion coefficients of the separate phases since local equilibrium is obtained inside the particle,

$$D_{sz}^{eff} = \frac{n_a D_a \gamma_a + n_b D_b \gamma_b}{n_a + n_b} \tag{110}$$

$$= \frac{\tilde{k} D_a \gamma_a + D_b \gamma_b}{1 + \tilde{k}} \tag{111}$$

In this equation the tortuosity factors γ have been included explicitly. Recall Eq. 101 for h_{SZMT}. Then one has for the stationary zone mass transfer

$$h_{SZMT} = \frac{1}{30} \frac{k'}{(1 + k')^2} \frac{(1 + \tilde{k}) D_m}{(\tilde{k} D_a \gamma_a + D_b \gamma_b)} v \tag{112}$$

Note that as either diffusion coefficient becomes small, h_{SZMT} increases to a limiting value defined by the other diffusion coefficient.

Let us now consider the model of Stout et al. shown in the lower panel of Fig. 1.10. Their assertion is that it is immaterial in what order the solute sees the two types of stationary zones. This is incontrovertibly true for the thermodynamic (retention) considerations, but only true under restricted circumstances for the dynamic (dispersion, H) considerations.

According to their model, the solute is only exposed to phase a in the first part of the column. The mobile phase volume exposed to phase a is V_{ma}; similarly the mobile phase volume exposed to phase b is V_{mb}. The phase ratio for the "partial" columns will be indicated by a prime. Thus Eq. 113 pertains to the region containing phase a; a similar relationship holds for the region containing b.

$$\Phi'_a = \frac{V_a}{V_{ma}} \tag{113}$$

The retention volume for the first part of the column must be added to the retention volume for the second part of the column to get the overall retention volume,

$$V = V_{ma}(1 + k'_{aa}) + V_{mb}(1 + k'_{bb}) \tag{114}$$

In this expression, the double letter subscript indicates k' calculated with the primed phase ratio. Since $V_{ma}k'_{aa}$ is equal to $V_m k'_a$ (and a similar relationship holds for b), it can easily be seen that Eq. 115 holds, and that the retention is indeed unaltered by segregation of the stationary zones,

$$V = V_m(1 + k'_a + k'_b) \tag{115}$$

Now let us consider the dynamics. Using Eq. 112 for each part of the column, one is led to

$$h_a = \frac{D_m k'_{aa}}{30(1 + k'_{aa})^2 D_a \gamma_a} v \tag{116}$$

$$h_b = \frac{D_m k'_{bb}}{30(1 + k'_{bb})^2 D_b \gamma_b} v \tag{117}$$

$$h_{\text{SZMT}} = h_a + h_b \tag{118}$$

Note that as either of the diffusion coefficients becomes small, the value of h_{SZMT} increases *without bound*. This should be compared with the behavior seen above for the mixed-phase column. Thus the two columns are not the same dynamically. There is a subtle point here that needs to be emphasized. The formula for h just given (Eqs. 116–118) is very close to the type of formula given for lumped mass transfer processes, such as the one shown in Eq. 102 without the mobile phase term. However, this is not the physical system that Stout et al. purport to be represented by their model. They begin their argument with our Eq. 101 and proceed to derive an expression for the diffusion coefficient and how it will influence C.

The difference in the dynamics exists because equilibrium between phases a and b is prevented in the segregated-phase model. To make the relation between the models clear, one can view these two chromatographic columns as chemical reactions. The first case considered was the mixed-phase column. It may be viewed as the chemical system shown by

$$\begin{array}{c} \text{fast} \\ A \underset{\text{slow}}{\overset{}{\rightleftharpoons}} B \\ M \text{ virtually negligible} \end{array} \qquad (119)$$

In the case under consideration we let the rate of conversion of A to M and of B to M be slow, while A to B was relatively more rapid. This is because we postulated stationary zone diffusion to be the rate-limiting step. If diffusion within B were slow, and if the mass transfer coefficient for phase B to M were very small (as shown in Eq. 119), then material would get from B to M by taking the path $B \rightarrow A \rightarrow M$.

In the segregated-phase column the analogous chemical reaction is given by

$$\begin{array}{c} \text{no} \\ \text{reaction} \\ A \qquad B \\ \text{fast} \diagdown M \diagup \text{slow} \end{array} \qquad (120)$$

In this case A and B do not interconvert at all. The only way for B to become M is by the direct route. Thus the set of conditions that would lead to the correct use of the model of Stout et al. is a very slow exchange between phases in the stationary zone, with a relatively rapid exchange between each phase and the mobile zone. This model is difficult to apply to porous particles.

For the sake of completeness we can relax the assertion of rapid $a \leftrightarrow b$ interconversion. The model that would apply when the rate-limiting step is stationary phase diffusion, or the desorption rate, is shown schematically by

$$A \underset{\text{fast}}{\overset{\text{slow}}{\rightleftharpoons}} B$$
$$\diagdown M$$ (121)

In this case, the processes are indeed in series.

The three equations shown for the A, B, and M reactions are really only simple limiting cases. In fact one may postulate more phases with a distribution of interphase transfer rate constants. These mass transfer rates would combine in series and in parallel to yield an overall transfer function that incorporates all the relevant processes.

Chromatographic evaluations of diffusion coefficients for solutes in reversed-phase liquid chromatography have been made. By careful determination of the dependence of the C term on k' for similar solutes one can measure the effective diffusion coefficient. Knox and Scott (73), using an octadecyl silane (ODS) phase with 60:40 methanol:water, found stationary phase diffusion coefficients to be around half of their value in the mobile phase. Nahar (75) found lower values, in the range of $3-5 \times 10^{-7}$ cm^{2}s^{-1}, for paraben esters in 55:45 methanol:water on ODS. The latter work included a temperature study in which it was found that the stationary phase diffusion had an activation energy of about 5.2 kcal/mol. Spectroscopic observations of the rate of formation of pyrene excimers indicate a value of 2.5×10^{-7} cm^{2}s^{-1} for the diffusion coefficient of pyrene. Since bonded phases tend to be very thin, the small diffusion coefficients generally do not have a large influence on the peak shape. The dynamic measurements that have been made support the static and spectroscopic measurements which suggest a stationary phase that is partially composed of mobile phase components, and consequently has appreciable fluidity.

Stationary Phase Egress Limiting the Rate. Another factor that will cause band broadening is the slow rate of solute leaving the stationary phase. By "rate" we mean either the rate of a chemical reaction or the diffusive transport rate in the stationary phase normal to the interface, as diagrammed in Eq. 121. In fact, simple chromatographic theory usually describes one of the contributors to band broadening as slow egress from the stationary phase characterized by a term k_d; Eq. 98 expresses this idea ($k_d \propto 1/t_{sz}$). This may be either a chemical rate or a diffusional rate proportional to d_f^2/D_s.

Recently the adsorption and desorption rates of 1-anilino-8-naphthalene sulfonate in the presence of cetyltrimethylammonium bromide were measured directly by a pressure jump relaxation technique (146). It is important to note that this method allows a direct measurement of the

rate of the chemical process free of macroscopic mass transport impedances. It was found that the adsorption rate of the ion pair onto a reversed-
phase column was nearly diffusion controlled. Desorption rates were on
the order of 500 s^{-1}. A comparison of this rate to the rate of mass transfer
within the stationary zone, which can be estimated to be $60D_{sz}/d_p^2$, will
indicate the relative importance of the two. Taking D_{sz} as $10^{-6} \text{ cm}^2\text{s}^{-1}$
and d_p as 10^{-3} cm leads to a value of about 60 s^{-1} for the latter rate. Of
course these are just estimated parameters, but it can be seen that the
stationary phase–solute dissociation rate can contribute to band broadening, albeit as a minor contribution, even for a reaction having a diffusion
controlled adsorption rate. For solutes with a similar k', but having a
smaller adsorption rate constant, the desorption rate will play an even
larger role.

The Role of Infrequent Adsorption. It is sometimes advantageous to consider the *adsorption rate* as the rate-limiting step. This is easily done using
a statistical approach that is related to the random walk. The use of this
statistical approach, the random fly, gives considerable insight into the
band broadening process (76). The random fly is more realistic than the
random walk in that it allows for a distribution of step lengths. This approach states that the distribution of step lengths that a solute takes in a
random fly can lead to band broadening. Consider the average time a
molecule exists in the stationary zone of a particle as our measurement
time. If the actual time that most individual molecules spend in the stationary zone is pretty close to the average, then the distribution of the
times spent in the stationary zone will be narrow. This will be the case
when a large number of adsorption–desorption events occur inside each
particle for each molecule. On the other hand, if the actual times spent
in the stationary zone vary from the average significantly, then the distribution will be wide, and this will be reflected in the breadth of the band.
This is a purely random process. The reason why the system would behave
as in the latter case is because it would only have a small probability of
adsorbing on the stationary phase. The distribution of mean values of
samples taken from a randomly distributed population is narrower the
larger the number of variates (measurements, determinations, observations, adsorptions) in the sample. In fact, the standard deviation of the
mean decreases as $\sqrt{n}$ increases. The argument here is very similar. The
result of this theoretical study is that the effect of chemical kinetics becomes dominant when the condition given in Eq. 122 is satisfied (76),

$$\frac{2}{q}\left(\frac{k'' - k_0}{k''}\right)^2 > 1 \tag{122}$$

where q is the average number of adsorptions per particle. It is important to realize that a low q may be due to a large activation barrier for adsorption, but may also be caused by a small number of adsorption sites.

With the exception of the difficult problem of axial dispersion in the mobile zone, and with the assumption of an interphase mass transfer rate that is a constant, the complete chromatographic problem for linear chromatography has been solved (147–152). The analytical expressions that result are so complex that it is difficult to gain any insight from inspecting them. Furthermore, the integrals in the final equations have oscillating kernels. It takes a great deal of care to integrate them properly. Nonetheless, it is quite a feat to have solved the problem for chromatography with an arbitrary number of types of particles in the bed, each of which has its own size distribution, shape, and chemical and physical properties (151).

1.3.7.5. Which A, B, and C Terms Are Correct?

The B term is reasonably straightforward; the formulation of Knox and Scott (73) (eqn. 90) is best and has been demonstrated to describe experiments well.

The A term involves axial dispersion in the mobile zone and mobile phase mass transfer resistance. Engineering correlations and the theory of Koch and Brady (109) show that the mobile zone axial dispersion coefficient is only approximately a linear function of the mobile phase velocity. It is more closely approximated by Eq. 52. Thus the mobile zone dispersion portion of the expression for h, valid only for Pe $\gg$ 1, would be given by $1.5 + \frac{1}{3}(\pi^2)(1 - \epsilon_e) \ln (\text{Pe})$.

Mobile zone mass transfer is described in another correlation as being related linearly to $v^{0.6}$ for large enough v. Equation 97 should be used in the final expression for h. We feel that "coupling" has been incorporated into the treatment of Koch and Brady. They arrived at the Pe ln (Pe) term from a consideration of how the diffusive motion of solute near the particle surface influences the dispersion. The diffusive motion of solute near the particle, in fact within the diffusion layer, is closely related to the mobile phase mass transfer coefficient. Thus there is an influence of mobile phase mass transfer on the axial dispersion.

One hopes that, in the future, theoretical advances will be able to explain these dependences. One should not forget that, without theoretical guidance and in the face of the difficulty of the experiments, the k'' dependence of these terms is simply not known. In any case, mobile zone mass transfer has to be recognized as playing a dual role, that of relaxing velocity inhomogeneities and as a mechanism for transporting the solute to the surface of the particles.

The C term is actually many terms. One may be for mobile phase mass transfer if its rate is velocity independent. A consideration of Eq. 97, which is part of the A term, shows that as $v \to 0$, h_{MZMT} becomes proportional to velocity. Under these conditions h_{MZMT} becomes part of the C term. This is needlessly confusing and arises because of the traditional dissection of h. There is another portion of the C term for stationary *phase* diffusion, one for stationary *zone* diffusion, and one for desorption from the stationary phase. In many cases one term will dominate and be the rate-limiting step at high velocities.

One can often adequately fit experimental data with the Knox equation over a wide range of velocities. Over a limited range of velocities the van Deemter equation has been shown by some to work quite well (26, 153), but not by others (73, 118). It is worth noting that the statistics of fitting complex curves are generally not adequate, in view of the precision in the measurement of HETP, to distinguish between the equations. Rather one must use the agreement of the computed values of A, B, or C with theory (and practical experience) to decide the adequacy of a particular model.

Recent experiments in Bloomington (J. E. Baur, E. W. Kristensen and R. Mark Wightman, submitted for publication June 1988) have shown that the retention time, and particularly the band broadening, varies significantly across the radius of a 10 cm. column packed with a 3 μm reversed phase. Correlations reported to date have necessarily used volume-averaged values of h. As detectors are developed that take advantage of the minimum h in the center of the column, numerical coefficients in these terms will change.

1.3.8. Dispersion in Packed Columns—Nonlinear Isotherms

This section deals with an advanced topic, namely, a chromatographic system that is both kinetically limited and overloaded. In this system so much sample has been injected that the sorption isotherm is no longer linear. This section assumes a familarity with the continuum mass balance model of chromatography. A brief review of this approach is given in Appendix 1.3.

A closed-form mathematical solution of the completely "general" nonlinear chromatographic problem is very likely impossible, and if such an equation were available, it would probably be computationally so difficult as to be almost useless. Thus different groups have devised a variety of approaches to obtain closed-form yet mathematically tractable solutions of the relevant nonlinear partial differential equation while preserving one or more important features of the problem. These approaches can be divided into three categories.

1. Simplification of the defining partial differential equation by neglect of one or more of the usual chromatographic mass transport related broadening factors. Specifically a closed-form solution can be obtained when axial dispersion is entirely neglected and one or more of the interphase mass transfer resistances are lumped together with the surface desorption rate constant. In this case the chromatographically realistic Langmuir nonlinear isotherm can be maintained. This approach is particularly relevant at flow rates which are significantly greater than the optimum since interphase transport and surface desorption kinetics are the dominant band broadening processes in this flow range.

2. The dispersive mass transport broadening term can be included in the partial differential equation, but it is necessary to assume perfect equilibrium between phases. Furthermore one can only approximate the isotherm curvature via the use of a physically unrealistic parabolic isotherm. This is the approach taken by Houghton and his coworkers (154) and exploited by Guiochon's group (155–157) for its use in both gas and liquid chromatography. This treatment is useful at flow rates below the optimum where the true dispersive term dominates band broadening.

3. The entire model can be preserved, but one can only obtain equations for the peak moments that correspond to a small deviation about the linear isotherm (158).

In one of the earliest theoretical studies of chromatography Thomas solved a very much simplified model of chromatography in which axial dispersion (eddy dispersion and longitudinal diffusion) and interphase resistance to mass transfer (in the moving phase, stagnant mobile phase, and diffusion in the stationary phase) were neglected. He focused his attention on the axial transport of solute by convective processes and the band broadening that results from slow chemical adsorption–desorption rates at the surface of the adsorbent. The so-called step input boundary condition was solved, and an equation that describes the breakthrough curve was obtained (41). The present importance of his work is twofold. First, he developed a mathematical transformation which changed a nonlinear problem into a difficult but still tractable linear problem. Second, as far as we are aware, his is the only solution in which a physically realistic isotherm, specifically a Langmuir isotherm, was preserved in toto. Thus it is possible to show what happens in the limit of a very high sample concentration.

Sometime after Thomas's work, Hiester and Vermeulen were able to show by use of the "linear driving force approach" that Thomas's mathematical solution could be generalized so as to allow for resistance to mass transfer between phases and slow surface adsorption–desorption kinetics (159). This comes about by introducing *effective* adsorption and desorption rate constants that incorporate a chemical desorption rate constant k_d and a mass transport resistance k_{mt}. As pointed out several times in this review, rate constants are analogs of conductances, and conductances in series add as do resistances in parallel. In the case of nonlinear adsorption chromatography it can be shown that the serial combination of mass transfer resistance and slow surface chemical kinetics leads to the following effective or net desorption rate constant:

$$\frac{1}{k_{d,\text{eff}}} = \frac{1}{k_d} + \frac{k'}{\epsilon(1 + Kc_0)k_{mt}} \tag{123}$$

Thus the only significant band broadening process that must still be left entirely unaccounted for is axial dispersion. Again Hiester and Vermeulen, following the method of Thomas, solved the breakthrough curve (step function input) problem. More recently several groups have examined various other types of inputs, including pulses (160, 161) and more general functions (162). Recently Wade and coworkers, at the University of Minnesota, were able to obtain a solution of the impulse input problem (163). This turned out to be a very challenging mathematical problem due to the extremely discontinuous nature of the impulse input. In linear chromatographic theory, or indeed in linear theory of any response system, it can be shown by means of the convolution theorem that the system response to an impulse input is simply the first time derivative of the system response to a step input. Furthermore one can show that the response of the column to any input function can be obtained by mathematical convolution of the impulse input response with the actual input function. This is absolutely not true in the general theory of nonlinear chromatography and is the source of tremendous mathematical complexity.

Based on Eq. a3.9 one can describe nonlinear chromatography upon neglect of the axial dispersion process by

$$\frac{\partial c}{\partial t} = -u\frac{\partial c}{\partial x} - \epsilon\frac{\partial q}{\partial t} \tag{124}$$

$$\frac{\partial q}{\partial t} = k_a c(S_0 - q) - k_d q \qquad (125)$$

where c is the local time dependent mobile phase solute concentration at axial position x (that is, c is a function of x and t), q is the corresponding solute concentration in the adsorbed state, and S_0 is the concentration of adsorbent sites, both filled and unfilled. Note that all concentrations are defined as per Appendix 1.3. The chromatographic velocity u is that of an unsorbed test species which is free to explore all the pores, and ϵ is the porosity ratio defined as $(1 - \epsilon_T)/\epsilon_T$, ϵ_T being the total porosity of the column.

In the case of a Langmuir isotherm where there is a fixed number of surface adsorption sites one can show that the sum of the concentration of adsorbed solute and the number of free adsortion sites is a constant,

$$q + S = S_0 \qquad (126)$$

At equilibrium it is easy to show that

$$q = \frac{KcS_0}{1 + Kc} \qquad (127)$$

where K is the equilibrium constant for adsorption ($= k_a/k_d$). Equation 127 describes the Langmuir adsorption isotherm.

The appropriate initial conditions for analytical zone elution chromatography are

$$c(x, 0) = 0 \qquad (128)$$
$$q(x, 0) = 0 \qquad (129)$$

For an impulse input (infinitesimally narrow pulse) the boundary condition at the entrance is

$$c(0, t) = c^0 \, \delta(t) \qquad (130)$$

where $\delta(t)$ is the unit strength impulse function defined as

$$\delta(t) = 0 \qquad \text{when } t > 0 \qquad (131)$$

$$\int_0^\infty \delta(t) \, dt = 1 \qquad (132)$$

The strength of the impulse c^0 is defined as the number of moles of solute injected divided by the volume of mobile phase in the column. The above problem can be solved for the mobile phase solute concentration as a function of time and axial position. In chromatography one observes the solute at the column exit ($x = L$ where L is the length of the column) as a function of time. The solution obtained by Wade et al. (163) using the transformation developed by Thomas is

$$\frac{c}{c^0} = \frac{1 - \exp(-\gamma Kc^0)}{\gamma Kc^0} \frac{[\gamma\sqrt{k'/y}\, I_1(2\gamma\sqrt{k'y}) + \delta(y)]\exp(-\gamma y - \gamma k')}{1 - T(\gamma k', \gamma y)[1 - \exp(-\gamma Kc^0)]}$$

$$(133)$$

where the dimensionless variable y and parameters γ, Kc^0, and k' have been introduced for notational simplicity. The accuracy of this solution has been checked by comparing the concentration–time curve so generated to that given by Goldstein's finite pulse input solution (160, 161). As the width of the finite pulse became small relative to the width due to the intracolumn broadening processes, the two equations produced identical numerical results.

$$y = \frac{t}{t_0} - 1 \tag{134}$$

$$\gamma = k_d t_0 \tag{135}$$

$$k' = \frac{k_a}{k_d} S^0 \epsilon \tag{136}$$

The term t_0 designates the transit time of an unsorbed molecule that explores all of the pores. The above equation is mathematically unfamiliar due to the presence of the function $T(\gamma y, \gamma k')$, which is defined below and referred to as the Thomas function. This function is not readily available nor are simple *generally* useful approximations. Goldstein has considered in detail the properties of related functions (160, 161). In practice the Thomas function is obtained from a numerical integration of its defining relationship (here u and v are general variables, not velocities),

$$T(u, v) = \exp(-v) \int_0^u \exp(-\lambda) I_0(2\sqrt{v\lambda})\, d\lambda \tag{137}$$

where λ is a dummy variable. I_0 and I are Bessel functions. Wade et al. (163) have developed a very fast algorithm for the evaluation of the integral based on rational approximations for the Bessel function and Gaussian quadrature numerical integration formulas. Despite its mathematical complexity, $T(u, v)$ is geometrically quite simple. It may be thought of as a switching function whose value changes from zero to unity in the vicinity of $y = k'$ (that is, near the peak maximum). In the limit as c^0 goes to zero the denominator approaches unity, and the classical linear isotherm result, which has been presented by Lapidus and Amundson (49) and Giddings and Eyring (6), is obtained.

As shown in Fig. 1.11, which was computed from Eq. 133, the qualitative effect of excess sample is to shift the peak maximum to shorter time, increase the peak width at the half-height, and generate a tail. Thus far we have not been able to find any closed-form equations for the various peak moments. Purely numerical results for the centroid, variance, and third central moment are shown in Figs. 1.12–1.14.

The most surprising result is that the peak variance is not a monotonic function of overload, that is, when the rate parameter γ is rather small, an increase in overload can decrease the second moment. In general one can neglect the effect of isotherm overload only when the parameter $\gamma K c^0$ is less than about 0.1.

Because the peak width at slight overload is dictated by γ, and γ establishes the number of plates on a column, one can make the following statement: *peak overload effects will be negligible when the amount of sample injected in a narrow pulse is less than the number of binding sites*

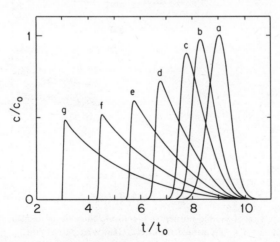

Figure 1.11. Theoretical peaks generated by Eq. 133. $k' = 8$; $\gamma = 100$; $Kc^0 = 0(a), 0.05(b)$, $0.10(c), 0.25(d), 0.50(e), 1.0(f), 2.0(g)$. [Reprinted from Ref. (163) with permission. Copyright 1987 American Chemical Society]

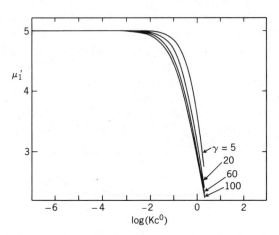

Figure 1.12. Behavior of normalized first peak moment μ_1' as a function of column overload. $k' = 4$. [Reprinted from Ref. (163) with permission.]

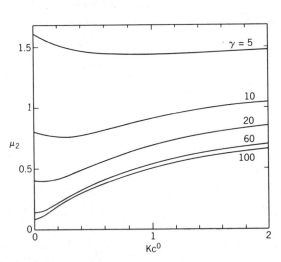

Figure 1.13. Behavior of second normalized, centralized peak moment μ_2 as a function of column overload. $k' = 4$. [Reprinted from Ref. (163) with permission.]

84

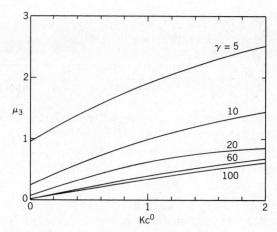

Figure 1.14. Behavior of third normalized, centralized peak moment μ_3 as a function of column overload. $k' = 4$. [Reprinted from Ref. (163) with permission.]

contained in a single plate. This is clearly in good agreement with chemical intuition.

ACKNOWLEDGMENTS

The authors would like to thank the National Institutes of Health (SGW) and the National Science Foundation (PWC) for support during the preparation of this chapter. They would also like to thank the members of their research groups, namely, Dr. Kamal Ismail, Lance Kuhn, John Long, Karen Sgroi, and Anne Warner (SGW), and Jim Wade (PWC), for valuable suggestions.

REFERENCES

1. J. M. Davis and J. C. Giddings, *Anal. Chem.*, *55*, 418 (1983).
2. D. P. Herman, M. F. Gonnord, and G. Guiochon, *Anal. Chem.*, *56*, 995 (1984).
3. M. Martin and G. Guiochon, *Anal. Chem.*, *57*, 289 (1985).
4. J. M. Davis and J. C. Giddings, *Anal. Chem.*, *57*, 2168 (1985).
5. J. M. Davis and J. C. Giddings, *Anal. Chem.*, *57*, 2178 (1985).
6. *J. C. Giddings and H. Eyring, *J. Phys. Chem.*, *59*, 416 (1955).
7. M. Abramovitz and J. A. Stegun, *Handbook of Mathematical Functions*, Dover, New York, 1965, p. 928.

8. J. C. Giddings, *J. Chem. Phys., 26,* 1755 (1957).

9. J. C. Giddings, *J. Chromatogr., 13,* 301 (1964).

10. *J. H. Knox and M. Saleem, *J. Chromatogr. Sci., 7,* 614 (1969).

11. *A. J. P. Martin and R. L. M. Synge, *Biochem. J., 35,* 1358 (1941).

12. J. C. Giddings, *Dynamics of Chromatography*, pt. 1, Dekker, New York, 1965, p. 20.

13. J. A. Perry, *Introduction to Analytical Chromatography*, Dekker, New York, 1981, p. 213.

14. *L. R. Snyder, *J. Chromatogr. Sci., 10,* 200 (1972).

15. *L. R. Snyder, *J. Chromatogr. Sci., 10,* 369 (1972).

16. J. C. Giddings, *Anal. Chem., 39,* 1027 (1967).

17. E. Grushka, *Anal. Chem., 42,* 1141 (1970).

18. R. J. Laub and J. H. Purnell, *J. Chromatogr., 112,* 71 (1975).

19. M. W. Watson and P. W. Carr, *Anal. Chem., 51,* 1835 (1979).

20. J. L. Glajch, J. J. Kirkland, K. M. Squire, and J. M. Minor, *J. Chromatogr., 199,* 57 (1980).

21. D. Hanggi and P. W. Carr, *Anal. Chem., 57,* 2394 (1985).

22. P. T. Kissinger, L. J. Felice, D. J. Miner, C. R. Preddy, and R. E. Shoup, in D. H. Hercules, G. M. Hieftje, L. R. Snyder, and M. A. Evenson, Eds., *Contemporary Topics in Analytical and Clinical Chemistry*, vol. 3, Plenum, New York, 1978, pp. 159–164.

23. S. N. Chesler and S. P. Cram, *Anal. Chem., 43,* 1922 (1971).

24. S. N. Chesler and S. P. Cram, *Anal. Chem., 44,* 2240 (1972).

25. S. D. Mott and E. Grushka, *J. Chromatogr., 148,* 305 (1978).

26. J. C. Chen and S. G. Weber, *Anal. Chem., 55,* 127 (1983).

27. D. H. Evans, personal communication.

28. F. Dondi, A. Betti, G. Blo, and C. Bighi, *Anal. Chem., 53,* 496 (1981).

29. F. Dondi, *Anal. Chem., 54,* 473 (1982).

30. D. J. Anderson and R. R. Walters, *J. Chromatogr. Sci., 22,* 353 (1984).

31. J. P. Foley and J. G. Dorsey, *Anal. Chem., 55,* 730 (1983).

32. W. E. Barber and P. W. Carr, *Anal. Chem., 53,* 1939 (1981).

33. J. P. Foley, *Anal. Chem., 59,* 1984 (1987).

34. *J. C. Giddings, in I. M. Kolthoff and P. J. Elving, Eds., *Treatise on Analytical Chemistry*, 2nd ed., pt. 1, vol. 5, Wiley, New York, 1981, sec. G, chap. 3.

35. J. N. Wilson, *J. Am. Chem. Soc., 62,* 1583 (1940).

36. J. C. Giddings, *Dynamics of Chromatography*, pt. 1, Dekker, New York, 1965, p. 15.

37. A. LeRosen, *J. Am. Chem. Soc., 64,* 1905 (1942).

38. R. H. McCormick, in B. L. Karger, L. R. Snyder, and C. Horvath, Eds., *An Introduction to Separation Science*, Wiley, New York, 1975.

39. D. DeVault, *J. Am. Chem. Soc.*, *65*, 532 (1943).

40. J. Weiss, *J. Chem. Soc.*, 297 (1943).

41. H. C. Thomas, *J. Am. Chem. Soc.*, *66*, 1664 (1944); *Ann. N.Y. Acad. Sci.*, *49*, 161 (1948).

42. J. E. Walter, *J. Chem. Phys.*, *13*, 332 (1945).

43. *G. E. Boyd, A. W. Adamson, and L. S. Meyers, Jr., *J. Am. Chem. Soc.*, *69*, 2836 (1947).

44. *G. E. Boyd, L. S. Meyers, Jr., and A. W. Adamson, *J. Am. Chem. Soc.*, *69*, 2849 (1947).

45. *C. V. King, *J. Am. Chem. Soc.*, *57*, 838 (1935).

46. *E. Glueckauf, *Trans. Faraday Soc.*, *51*, 34, 1540, (1955).

47. L. G. Sillen, *Nature, 166*, 722 (1950).

48. E. Ekedahl, E. Hogfelt, and L. G. Sillen, *Nature, 166*, 723 (1950).

49. *L. Lapidus and N. R. Amundson, *J. Phys. Chem.*, *56*, 984 (1952).

50. *A. Klinkenberg and F. Sjenitzer, *Chem. Eng. Sci.*, *5*, 258 (1956).

51. *J. J. van Deemter, F. J. Zuiderweg, and A. Klinkenberg, *Chem. Eng. Sci.*, *5*, 271 (1956).

52. *J. C. Giddings, *J. Chem. Ed.*, *35*, 58 (1958).

53. W. L. Jones and R. Kieselbach, *Anal. Chem.*, *30*, 1590 (1958).

54. J. C. Giddings, *J. Chem. Phys.*, *26*, 169 (1957).

55. J. C. Giddings, *J. Chem. Phys.*, *31*, 1462 (1959).

56. J. C. Giddings, *J. Chromatogr.*, *3*, 443 (1960).

57. J. C. Giddings, *J. Chromatogr.*, *5*, 46 (1961).

58. J. C. Giddings, *Anal. Chem.*, *33*, 962 (1961).

59. J. C. Giddings, *J. Phys. Chem.*, *68*, 184 (1964).

60. J. C. Giddings, *Anal. Chem.*, *34*, 1026 (1962).

61. J. C. Giddings, *J. Chromatogr.*, *5*, 61 (1961).

62. J. C. Giddings, *Nature, 184*, 357 (1959).

63. J. C. Giddings, *Anal. Chem.*, *35*, 1338 (1963).

64. *J. N. Done and J. H. Knox, *J. Chromatogr. Sci.*, *10*, 606 (1972).

65. J. C. Giddings, *Dynamics of Chromatography*, Pt. 1, Dekker, New York, 1965, ch. 7.

66. *J. F. K. Huber, *J. Chromatogr. Sci.*, *7*, 85 (1969).

67. J. C. Giddings, *J. Chromatogr.*, *13*, 301 (1964).

68. G. J. Kennedy and J. H. Knox, *J. Chromatogr. Sci.*, *10*, 549 (1972).

69. E. Grushka, L. R. Snyder, and J. H. Knox, *J. Chromatogr. Sci.*, *13*, 25 (1975).

70. J. H. Knox, *J. Chromatogr. Sci.*, *15*, 352 (1977).

71. J. F. K. Huber, *Ber. Bunsenges. Phys. Chem.*, *77*, 179 (1973).

72. J. F. K. Huber, *Fresenius Z. Anal. Chem.*, *277*, 341 (1975).

73. *J. H. Knox and R. P. W. Scott, *J. Chromatogr.*, *282*, 297 (1983).

74. R. W. Stout, J. J. DeStefano, and L. R. Snyder, *J. Chromatogr.*, *282*, 263 (1983).

75. M. Nahar, M.S. thesis, University of Pittsburgh, PA, 1983.

76. S. G. Weber, *Anal. Chem.*, *56*, 2104 (1984).

77. D. M. Scott and J. S. Fritz, *Anal. Chem.*, *56*, 1561 (1984).

78. *A. M. Lenhoff, *J. Chromatogr.*, *384*, 285 (1987).

79. M. J. Lighthill, *J. Inst. Math. Appl.*, *2*, 97 (1966).

80. M. J. E. Golay and J. G. Atwood, *J. Chromatogr.*, *186*, 353 (1979).

81. J. G. Atwood and M. J. E. Golay, *J. Chromatogr.*, *218*, 97 (1981).

82. J. T. Vanderslice, K. K. Stewart, A. G. Rosenfeld, and D. J. Higgs, *Talanta*, *28*, 11 (1981).

83. J. S. Yu, *J. Appl. Mech.*, *48*, 217 (1981).

84. E. B. Nauman and B. A. Buffham, *Mixing in Continuous Flow Systems*, Wiley, New York, 1983, p. 118.

85. B. Anderson and T. Berglin, *Proc. R. Soc. (London)*, *A377*, 251 (1981).

86. M. A. Paine, R. G. Carbonell, and S. Whitaker, *Chem. Eng. Sci.*, *38*, 1781 (1983).

87. *G. I. Taylor, *Proc. R. Soc. (London)*, *A219*, 196 (1953).

88. *R. Aris, *Proc. R. Soc. (London)*, *A235*, 67 (1956).

89. V. Ananthakrishnan, W. N. Gill, and A. J. Barduhn, *J. Am. Inst. Chem. Eng.*, *11*, 1063 (1965).

90. W. N. Gill and V. Ananthakrishnan, *J. Am. Inst. Chem. Eng.*, *13*, 801 (1967).

91. W. N. Gill, *Proc. R. Soc. (London)*, *A298*, 6 (1967).

92. A. Brenner, *Phil. Trans. R. Soc. (London)*, *297*, 81 (1980).

93. M. J. E. Golay, *J. Chromatogr.*, *186*, 341 (1979).

94. K. Hofmann and I. Halasz, *J. Chromatogr.*, *199*, 3 (1980).

95. R. Tijssen, *Anal. Chim. Acta*, *114*, 71 (1980).

96. E. W. Kristensen, R. L. Wilson, and R. M. Wightman, *Anal. Chem.*, *58*, 986 (1986).

97. *S. Chandrasekhar, *Rev. Mod. Phys.*, *15*, 1 (1943).

98. A. E. Sheidegger, *The Physics of Flow Through Porous Media*, 3rd ed., University of Toronto Press, Toronto, 1960, pp. 188–218.

99. O. Levenspiel and K. B. Bischoff, *Adv. Chem. Eng.*, *14*, 95 (1963).

100. C. L. DeLigny, *J. Chromatogr.*, *49*, 393 (1970).

101. E. Wicke, *Ber. Bunsenges. Phys. Chem.*, *77*, 160 (1973).

102. D. M. Ruthven, *Principles of Adsorption and Adsorption Processes*, Wiley, New York, 1982, p. 212.

103. H. W. Han, J. Bhakta, and R. G. Carbonell, *J. Am. Inst. Chem. Eng.*, *31*, 277 (1985).

104. J. W. Hiby, "Interactions between Fluids and Particles," *Proc. Symp. Inst. Chem. Eng.*, 1962, p. 312.

105. *C. Horvath and H.-J. Lin, *J. Chromatogr.*, *126*, 401 (1976).

106. C. F. Gottschlich, *J. Am. Inst. Chem. Eng.*, *9*, 89 (1963).

107. R. G. Carbonell and S. Whitaker, *Chem. Eng. Sci.*, *38*, 1795 (1983).

108. S. Eidsath, R. G. Carbonell, S. Whitaker, and L. R. Herrmann, *Chem. Eng. Sci.*, *38*, 1803 (1983).

109. *D. L. Koch and J. F. Brady, *J. Fluid Mech.*, *154*, 399 (1985).

110. H.-L. Lee, *Chem. Eng. Sci.*, *34*, 503 (1979).

111. S. Sundaresan, N. R. Amundson, and R. Aris, *J. Am. Inst. Chem. Eng.*, *26*, 529 (1980).

112. G. Sposito, V. K. Gupta, and R. N. Bhattacharya, in G. F. Pinder, Ed., *Flow Through Porous Media*, CML Publications, Southampton, 1983, p. 76.

113. H. Kaizuma, M. N. Meyers, and J. C. Giddings, *J. Chromatogr. Sci.*, *8*, 630 (1970).

114. N. Wakao and T. Funazkri, *Chem. Eng. Sci.*, *33*, 1375 (1978).

115. A. E. Sheidegger, (98), pp. 124–170.

116. E. Kucera, *J. Chromatogr.*, *19*, 237 (1965).

117. E. Grushka, *J. Phys. Chem.*, *76*, 2586 (1972).

118. *C. Horvath and H.-J. Lin, *J. Chromatogr.*, *149*, 43 (1978).

119. O. Levenspiel, *Chemical Reaction Engineering*, 2nd ed., Wiley, New York, 1972.

120. H. W. Hethcote and C. DeLisi, *J. Chromatogr.*, *240*, 269 (1982).

121. *M. J. E. Golay, in D. H. Desty, Ed., *Gas Chromatography 1958, Proc. 2nd Symp.* (Amsterdam), Academic, New York, 1958.

122. A. B. Littlewood, *Gas Chromatography*, 2nd ed., Academic, New York, 1970.

123. T. Tsuda and M. Novotny, *Anal. Chem.*, *50*, 632 (1978).

124. J. H. Knox and M. T. Gilbert, *J. Chromatogr.*, *186*, 405 (1979).

125. M. Krejci, K. Tesarik, M. Rusek, and J. Pajurek, *J. Chromatogr.*, *218*, 167 (1981).

126. R. Tijssen, J. P. A. Bleumer, A. L. C. Smit, and M. E. van Kreveld, *J. Chromatogr.*, *218*, 137 (1981).

127. V. L. McGuffin and M. Novotny, *J. Chromatogr.*, *218*, 179 (1981).

128. C. Z. Reese and R. P. W. Scott, *J. Chromatogr. Sci. 18*, 479 (1980).

129. A. Sahkarasubramanian and W. N. Gill, *Proc. R. Soc.* (*London*), *A333*, 115 (1978).

130. E. DeGance and L. E. Johns, *Appl. Sci. Res.*, *34*, 189 (1978).

131. E. DeGance and L. E. Johns, *Appl. Sci. Res.*, *34*, 227 (1978).

132. E. DeGance and L. E. Johns, *J. Am. Inst. Chem. Eng.*, *26*, 411 (1980).

133. E. DeGance and L. E. Johns, *Appl. Sci. Res.*, *42*, 55 (1985).

134. E. B. Nauman, *Chem. Eng. Sci., 36,* 957 (1981).

135. D. Thoenes, *Chem. Eng. Sci., 8,* 271 (1958).

136. L. R. Snyder, *J. Chromatogr., 7,* 352 (1969).

137. L. R. Snyder, Dry Lab, L. C. Resources, Inc.

138. *F. H. Arnold, H. W. Blanch and C. R. Wilke, *Chem. Eng. J., 30,* B25 (1985).

139. C. R. Yonker, T. A. Zwier and M. F. Burke, *J. Chromatogr., 241,* 257 (1982).

140. C. R. Yonker, T. A. Zwier, and M. F. Burke, *J. Chromatogr., 241,* 269 (1982).

141. J. W. Carr and J. M. Harris, *Anal. Chem., 58,* 626 (1986).

142. R. K. Gilpin and M. E. Gangoda, *Anal. Chem., 56,* 1470 (1984).

143. J. C. Giddings, (12), pp. 119–194.

144. K. Kawazoe and Y. Takeuchi, *J. Chem. Eng. Jpn., 1,* 431 (1974).

145. N. S. Raghavan and D. M. Ruthven, *Chem. Eng. Sci., 40,* 699 (1985).

146. D. B. Marshall, J. W. Burns, and D. E. Connolly, *J. Chromatogr., 360,* 13 (1986).

147. A. Rasmuson and I. Neretnieks, *J. Am. Inst. Chem. Eng., 26,* 680 (1980).

148. A. Rasmuson, *J. Am. Inst. Chem. Eng., 27,* 1032 (1981).

149. A. Rasmuson, *Chem. Eng. Sci., 37,* 787 (1982).

150. A. Rasmuson, *J. Am. Inst. Chem. Eng., 31,* 518 (1985).

151. A. Rasmuson, *Chem. Eng. Sci., 40,* 621 (1985).

152. A. Rasmuson, *Chem. Eng. Sci., 40,* 1115 (1985).

153. E. Katz, K. L. Ogan, and R. P. W. Scott, *J. Chromatogr., 270,* 51 (1983).

154. G. Houghton, *J. Phys. Chem., 67,* 84 (1963).

155. *A. Jaulmes, C. Vidal-Madjar, H. Colin, and G. Guiochon, *J. Phys. Chem., 90,* 207 (1986).

156. A. Jaulmes, M. J. Gonzales, C. Vidal-Madjar, and G. Guiochon, *J. Chromatogr., 387,* 41 (1987).

157. A. Jaulmes, C. Vidal-Madjar, H. Colin, and G. Guiochon, *J. Phys. Chem., 90,* 207 (1986).

158. K. Yamaoka and T. Nakagawa, *J. Phys. Chem., 79,* 522 (1975).

159. N. K. Hiester and T. Vermeulen, *Chem. Eng. Prog., 48,* 505 (1952).

160. S. Goldstein, *Proc. R. Soc. (London), A219,* 151 (1953).

161. S. Goldstein, *Proc. R. Soc. (London), A219,* 171 (1953).

162. R. Aris and N. R. Amundson, *Mathematical Methods in Chemical Engineering,* vol. 2, Prentice-Hall, Englewood Cliffs, NJ, 1973.

163. J. L. Wade, A. F. Bergold, and P. W. Carr, *Anal. Chem., 59,* 1286 (1987).

* Citations marked with an asterisk are recommended reading.

APPENDIX 1.1   THE RELATIONSHIP BETWEEN FREE ENERGY
DIFFERENCE AND RESOLUTION

By using the equation for resolution, one can arrive at the k' values for a pair of compounds whose peaks are just separated. The pair of k' values yields a ratio that is the same as the ratio of the associated equilibrium constants K. This is in turn related to the difference in the free energies of transfer of the species.

The equation for resolution is

$$R_s = \frac{\sqrt{N}}{4} \frac{\alpha - 1}{\alpha} \frac{k'}{1 + k'} \qquad \text{(a1.1)}$$

where N is the number of theoretical plates, α is the separation factor, the ratio of k' values, and the k' in the last term is the k' of the second eluting component. A resolution of 1 leads to almost baseline resolution for peaks of equal size. We can allow the k' term to be 1, since it is close to 1 for $k' > 5$. For 10,000 theoretical plates one then has

$$\alpha = 1.042 \qquad \text{(a1.2)}$$

Since α is the ratio of k' values, it is also the ratio of K values and one has

$$\alpha = \frac{K_2}{K_1} = \exp\left(\frac{\Delta G_1^0 - \Delta G_2^0}{RT}\right) \qquad \text{(a1.3)}$$

Finally, rearranging Eq. a1.3, one has the value of $\Delta\Delta G^0$,

$$\Delta G_1^0 - \Delta G_2^0 = RT \ln(1.042) \qquad \text{(a1.4)}$$

which, at room temperature, is about 25 cal/mol. Here the ΔG^0 values are the transfer free energies for solute going from mobile phase to stationary phase at infinite dilution. This difference in energy is exceedingly small. It would correspond to a difference of 0.018 in pK_a for the ionization of two acids, or to 1.08 mV in the formal potential of two species. It corresponds to a shift of about eight wave numbers in the infrared.

APPENDIX 1.2   AN OVERVIEW OF THEORIES OF DIFFUSION

Chromatography works because many microscopic molecular "extractions" occur in a typical chromatographic experiment. In the macroscopic world, a series of extractions is performed by shaking the extraction vessel to increase the interphase mass transfer rate. In a chromatographic column the same opportunity does not exist for shaking, but the rate of interphase transport is clearly still important; molecules must be mixed by diffusion. On the other hand, the reader with some exposure to chromatographic theory will recognize that the process of molecular diffusion can degrade chromatographic performance. Thus an understanding of diffusion is essential for a complete understanding of chromatography.

The following discussion will only cover diffusion in systems having diffusion coefficients that are not functions of the solute concentration. This will be true for dilute solutions such as those found in analytical chromatography.

One of the goals of modern theory is to predict diffusion coefficients from the intermolecular forces between solutes and solvents. This was also a goal of Fick's (1). Fick's laws of diffusion are analogous to the laws of Fourier for the diffusion of heat in a gradient of temperature, and of Ohm for the diffusion of charge in a gradient of electrical potential. Fick's first law is

$$\mathbf{J} = -D\nabla c \qquad (a2.1)$$

$$\mathbf{J} = -D\left(\frac{dc}{dx}\right) \qquad (a2.2)$$

where $\mathbf{J}$ is the molecular flux, D is the diffusion coefficient, and ∇c is the spatial gradient of the concentration. The equation is also given for one dimension, a form of some utility in certain circumstances. Fick's second law, which can be derived from the consideration of a mass balance in a very small volume of fluid, is given as

$$\frac{\partial c}{\partial t} = D\nabla^2 c \qquad (a2.3)$$

$$\frac{\partial c}{\partial t} = D\left(\frac{\partial^2 c}{\partial x^2}\right) \qquad (a2.4)$$

Again, the law has been given in a one-dimensional form as well as in the full three-dimensional form.

Thermodynamic reasoning shows (2, 3) that the flux of matter is proportional to the gradient of the chemical potential. In fact, Onsager has shown that all fluxes are driven by gradients of all potentials. This is expressed as

$$\mathbf{J}_i = -\sum_j L_{ij} \nabla \mu_j \qquad \text{(a2.5)}$$

In this equation $L_{ij} = L_{ji}$ (4). The use of the coupling of one potential gradient to the flux of a species not ordinarily associated with the potential has been discussed in the context of separations (5). For the most part in chromatography one deals with the cases in which one chemical potential gradient is dominant and $(L_{ij})_{i \neq j}$ are small. An exception to this is in the diffusion of charged species in an electrolyte in which the coupling through the electrostatic potential is strong. Consider the single driving force case, then, when

$$\mathbf{J}_i = -L_{ii} \nabla \mu_i \qquad \text{(a2.6)}$$

which, for a one-dimensional system, becomes

$$\mathbf{J}_i = -L_{ii} \left(\frac{\partial \mu_i}{\partial x} \right) \qquad \text{(a2.7)}$$

The chemical potential of a species i at infinite dilution in a two-phase system that may contain charged species is given by

$$\mu_i^a = \mu_i^{0,a} + RT \ln(c_i^a) + z_i F \Psi^a \qquad \text{(a2.8)}$$

where μ_i is the chemical potential of solute species i. Superscript a means in phase a and superscript 0 indicates the standard state chemical potential. The concentration is given by c and the charge on the species by z. The potential is given by Ψ, F is the Faraday, R is the gas constant, and T is the temperature. The standard state is taken as the solute acting as if it were at infinite dilution, that is, only solute–solvent effects are important. The concentration of the solute in the standard state is taken as 1 molar. Note that this is a fictitious system. It is commonly called extrapolated infinite dilution. Consider the case of a single phase with no electrolyte and no externally applied potential, and with the solute at infinite dilution. Then one has

$$\mu_i = \mu_i^0 + RT \ln(c_i) \qquad \text{(a2.9)}$$

$$\frac{d\mu_i}{dx} = \frac{RT/c_i}{dc_i/dx} \tag{a2.10}$$

where the superscript a has been dropped because this is a homogeneous system.

From Eqs. a2.2, a2.7, and a2.10 one can deduce that

$$L_{ii} = \frac{D_{a,i}^{\infty} c_i}{RT} \tag{a2.11}$$

holds. In Eq. a2.11 the superscript ∞ is used to indicate infinite dilution conditions, and the subscripts a for the phase and i for the solute have been included in the diffusion coefficient $D_{a,i}^{\infty}$ to remind the reader that the solute's diffusion coefficient depends on the medium in which the solute is dissolved.

The proportionality constant between the chemical potential and the flux of matter is L_{ii} in the simple system described. Note that this coefficient is proportional to the diffusion coefficient. The phenomenological picture of a concentration gradient driving a flux is a correct one; it is observed experimentally.

The following question arises. Why, if diffusion moves molecules randomly, can a collection of molecules move concertedly to effect a net transport from one point to another? The answer lies in the existence of a concentration gradient. Consider a plane that divides a volume element. Let us take the concentration to the left of the plane to be higher than that to the right. While all molecules on either side move randomly, it is evident that there are more molecules on the left available to move to the right than vice versa. This must result in net transport from left to right.

Consider once again the expression a2.8 for the chemical potential. One can see that, in the general case, flux is driven by gradients in solvent composition, concentration, and electrostatic potential. All three certainly can occur in liquid chromatography.

1.2.1. Molecular Friction

Simple intuitive models die slowly. This is so with the notion of diffusion as a body being moved through a continuous viscous fluid. The gradient of a potential is a force $\mathbf{F}$, and it has been shown that the diffusive flux is driven by a gradient of chemical potential. In a classical system, a force acting on a body in a viscous medium results in a steady velocity $\mathbf{v}$ being obtained such that Eq. a2.12 holds,

$$\mathbf{F} = f\mathbf{v} \tag{a2.12}$$

In this expression f is a frictional coefficient with units of mass per time. One can imagine a solute molecule moving through a solvent and interacting with neighboring molecules (causing friction) as it is hammered by other molecules (giving it energy for motion). The diffusion coefficient can be thought of (6) as the ratio of the energy of the diffusing particle (proportional to kT, k being Boltzmann's constant) to the friction,

$$D = \frac{kT}{f} \qquad (\text{mass·length}^2\text{·time}^{-2}/\text{mass·time}^{-1}) \qquad (a2.13)$$

If one assumes that molecular diffusion can be modeled as a wettable sphere moving through a viscous continuum, then the frictional coefficient is given by (7)

$$f = 6\pi\eta_a a_i \qquad [\text{mass·time}^{-1}$$

$$= (\text{mass·length}^{-1}\text{·time}^{-1})\text{·length}] \qquad (a2.14)$$

where η_a is the viscosity of phase a and a_i is the radius of sphere i. The frictional factor is made up of two terms—$4\pi\eta a$ is from the pressure caused by the viscous medium on the "front" of the moving sphere; $2\pi\eta a$ is caused by viscous drag on the "sides" of the sphere. If the sphere does not wet, then the drag is nil—the sphere slips through the medium—and the frictional factor contains only the pressure contribution as shown in

$$f = 4\pi\eta_a a_i \qquad (a2.15)$$

If n is used in place of the numerical value, then the expression for the frictional factor may be generalized to

$$f = n\pi\eta_a a_i \qquad (a2.16)$$

The parameter n may be calculated for shapes that differ from spherical, for example, for ellipsoids of revolution (8). One may now write an expression for the diffusion coefficient as

$$D_{a,i}^{\infty} = \frac{kT}{n\pi\eta_a a_i} \qquad (a2.17)$$

where $D_{a,i}^{\infty}$ is the diffusion of solute i in phase a at infinite dilution. One can rearrange Eq. a2.17 to obtain a relationship that would appear to depend on solute properties only,

$$D_{a,i}^{\infty}\eta_a = \frac{kT}{n\pi a_i} \tag{a2.18}$$

The left-hand side of Eq. a2.18 is often called the Walden product, although Walden originally formulated the idea for limiting ionic conductance and viscosity, that is, $\Lambda^{\infty}\eta$ = constant. The right-hand side of the expression includes terms that depend only on the solute. This is useful, since one can in principle calculate the quantity on the right-hand side. Solvent viscosities are tabulated, so that the diffusion coefficient can be predicted. The form of Eq. a2.18 also suggests an experimental test of the Stokes–Einstein equation, Eq. a2.17.

For large entities, such as latex spheres and colloidal particles, the equation works well (9). However, as might be expected, for cases in which the solute and solvent molecules are of similar size, the relationship fails. Empirically it has been found that

$$D_{a,i}^{\infty}\eta_a^p = \text{constant} \tag{a2.19}$$

Furthermore the value of the constant depends on the size and shape of the molecules in the system. In a detailed study of approximately spherical molecules, Evans et al. (10) found that the exponent p depends on the solute size. Smaller solutes yielded smaller exponents; for example, for argon $p = 0.44$, while for tetradodecyltin $p = 0.94$. For a perfect large sphere the exponent p would be 1, so from this analysis it appears that tetradodecyltin acts almost as a sphere. An extrapolation leads to the prediction that solutes with radii greater than about 8 Å have $p = 1$, at least in the solvents studied. The solvents used were C_6, C_{10}, and C_{14} linear alkanes, and some lower alcohols. Charged species deviate less from the Stokes law than do neutrals. This does not mean that they are better approximations to spheres; rather it indicates the large role played by dielectric friction, particularly for small species. Charged molecules suffer an extra friction, dielectric friction, caused by the dielectric relaxation of the solvent as the ion passes through it (11). This is a complex phenomenon; approximate theories due to Zwanzig (11) and Hubbard and Onsager (12) have been used.

The usefulness of the Stokes–Einstein relationship lies principally in its simplicity; both the solvent and the solute are characterized by a single parameter. However, for all but spherical solutes at infinite dilution in a medium made of molecules that are small with respect to the solute, the theory fails. Theories that account for molecular interactions are required, and indeed are under development (9).

1.2.2. Diffusion as an Activated Process

Imagine a solute molecule, comfortably solvated, that is being buffeted about by neighboring solvent molecules. A fluctuation causes the formation of an incipient solvation environment nearby, on the order of a solute or solvent molecular diameter away. The molecule becomes desolvated, at least partially, in a jump from its present site to the contiguous newly formed site. The point when the molecule is between solvated equilibrium states is a transition state. Thus rate theory can be used to describe the process (13).

The development of this idea (13) leads to an equation for molecular flux that can be compared to Fick's first law to yield a relationship between the derived quantities and the macroscopically observed diffusion coefficient, Eq. a2.20,

$$D = l^2 k \qquad \text{(a2.20)}$$

where l is the jump length and k is the jump rate, given by a rate expression. A major experimental observable, the exponential dependence of the diffusion coefficient on temperature, is predicted by this expression. Using a derived expression for viscosity, one can replace $l^2 k$ with terms that are reminiscent of the Stokes–Einstein equation, as

$$D = \frac{kT}{n\pi\eta a} \qquad \text{(a2.21)}$$

Glasstone et al. derived the value of n from a consideration of the various intermolecular distances in a flowing liquid; its value will be near 1. This contradicts the prediction of the Stokes–Einstein expression, but this might be expected since the treatment of Glasstone et al. explicitly includes the discreteness of the solvent, while the hydrodynamic theory of Stokes explicitly excludes it. The comparison of the rate theory with experiment leads to an overestimation of diffusion coefficients by Eq. a2.21 (with n near 1). On the other hand, the treatment of temperature dependent diffusion coefficient data using Arrhenius plots leads to acceptable correlations of the energy required to reach the transition state and the energy required to vaporize a mole of solvent. If the energy of creating a hole into which a solute molecule can go is important to the rate of the process, then there should be a correlation between the heat of vaporization for the solvent and the activation barrier for the diffusion of a solute. Such relationships were used to support the theory (13).

Recent experimental results indicate that the jump distance is on the order of 0.1 Å, clearly inconsistent with molecular dimensions. This study

(14) used Mössbauer spectroscopy of an iron complex in an aqueous phosphoric acid glass. The study calls into question the validity of the Glasstone et al. model, but exponential dependence of the rate of diffusion on temperature and the relationship between activation barriers for diffusion and heat of vaporization are supportive of the theory.

It is perhaps best to view the diffusion process as consisting of steps that are smaller than a solvent diameter. It is intuitively appealing to relate the process of diffusion to the creation and destruction of voids in the solvent. Local fluctuations in free volume allow solute motion in small steps, and the size and frequency of the fluctuations are undoubtedly related to the energetics of solvent–solvent interactions. Thus the correlations used to support the Glasstone et al. model would also support this model.

1.2.3.   Hindered Diffusion

In chromatography molecules are required to diffuse in spaces that may approach the size of the molecules themselves. When the radius of a pore r through which a solute is diffusing approaches the radius of the solute a, the diffusion coefficient is decreased. A brief and informative review of the theory of this phenomenon has recently appeared (14). Several formulas are given, and the statement is made that they are essentially indistinguishable within the typical experimental error associated with such measurements. The simplest one is given by

$$D_{\text{pore}} = D_{a,i} \exp(-4.6\chi) \qquad (a2.22)$$

where χ is a/r. Thus mass transfer within a porous support particle will be slowed as the size of the solute approaches the size of the pore.

1.2.4.   Empirical Relationships

Tyrrell and Harris (9) have reviewed empirical relationships for diffusion coefficients. One that has been widely used is the Wilke–Chang expression

$$\frac{D_{a,i}\eta_{a,i}}{T} = \frac{7.4 \times 10^{-8}(\zeta M_a)^{1/2}}{\overline{V}_i^{0.6}} \qquad (a2.23)$$

where M_a is the solvent molecular weight, $\overline{V}_i$ is the solute molecular volume (in milliliters per mole at the boiling point), and ζ is a correction term for solvent association. This has a value of 1.0 for unassociated solvents and values of 1.5 for ethanol, 1.9 for methanol, and 2.6 for water (15).

Edward (16) has provided a critical review of the Stokes–Einstein equation and, in particular, the relationship between n in Eq. a2.21 and the solute size. The question of the best method to measure the solute size is a difficult one. One must not only consider the basis of the primary calculation, such as whether it should be based on molar volume or calculated van der Waals radii, but also one must be careful to account for important secondary influences. These include the molecular shape and the extent of hydration of the solute in water. Edward suggests the use of van der Waals radii, and he provides a table of atomic and group increments. Note that the table is more current and different than the table provided in the work of Wilke and Chang (15).

Recently Katz and Scott (17) have reexamined an empirical relationship due originally to Arnold (18),

$$D_{a,i} = \frac{F(1/M_i + 1/M_a)^{1/2}}{(\overline{V}_i^{1/3} + \overline{V}_a^{1/3})^2} \qquad (a2.24)$$

As before, a represents the solvent and i the solute, M is molecular weight, and $\overline{V}$ is a molar volume. Katz and Scott found that, for a single solvent system at a given temperature and pressure, the diffusion coefficients of a large number of solutes fit the following relationship:

$$\frac{1}{D} = F_1 + F_2 \overline{V}^{1/3} M^{1/2} = F_1 + \frac{F_2 M^{0.833}}{\rho^{1/3}} \qquad (a2.25)$$

All of the variables refer to the solute. F_1 and F_2 are constants. The correlation obtained with Eq. a2.25 was good. The precise data obtained were used in determining the dependence of various band broadening terms on molecular diffusion. Finally, the authors were able to determine a solute's molecular weight from the band broadening in its chromatographic peak.

References to Appendix 1.2

1. H. J. V. Tyrrell, *J. Chem. Ed., 41,* 397 (1964).
2. G. S. Hartley, *Phil. Mag., 12,* 473 (1931).
3. L. Onsager and R. Fuoss, *J. Phys. Chem., 36,* 2689 (1932).
4. L. Onsager, *Phys. Rev., 37,* 405 (1931).
5. J. C. Giddings, in I. M. Kolthoff and P. J. Elving, Eds., *Treatise on Analytical Chemistry,* 2nd ed., pt. 1, vol. 5, Wiley, New York, 1981, sec. G, chap. 3.

6. A. Einstein, *Ann. Phy.*, *17*, 549 (1905).

7. M. Lauffer, *J. Chem. Ed.*, *58*, 250 (1981).

8. K. E. Van Holde, *Physical Biochemistry*, Prentice Hall, Englewood Cliffs, NJ, 1971, p. 81.

9. H. J. V. Tyrrell and K. R. Harris, *Diffusion in Liquids*, Butterworths, London, 1984, chap. 7.

10. D. F. Evans, T. Tominaga, and C. Chan, *J. Solut. Chem.*, *8*, 461 (1979).

11. R. Zwanzig, *J. Chem. Phys.*, *52*, 3625 (1970).

12. J. B. Hubbard and L. Onsager, *J. Chem. Phys.*, *67*, 4850 (1977).

13. S. Glasstone, K. J. Laidler, and H. Eyring, *The Theory of Rate Processes*, McGraw-Hill, New York, 1941, pp 516 ff.

14. R. E. Baltus and J. L. Anderson, *Chem. Eng. Sci.*, *38*, 1959 (1983).

15. C. R. Wilke and P. Chang, *J. Am. Inst. Chem. Eng.*, *1*, 264 (1955).

16. J. T. Edward, *J. Chem. Ed.*, *47*, 261 (1970).

17. E. D. Katz and R. P. W. Scott, *J. Chromatogr.*, *29*, 270 (1983).

18. J. H. Arnold, *J. Am. Chem. Soc.*, *52*, 3937 (1930).

APPENDIX 1.3   CONTINUUM MASS BALANCE MODEL OF CHROMATOGRAPHIC COLUMNS

In order to comprehend this approach, the reader should examine Fig. a3.1 in detail. The mathematical technique used is to carry out a detailed mass balance on the finite element of length Δz. This finite element is also termed the control volume. Such an analysis leads to the following equation for the rate of accumulation of solute in Δz:

rate of accumulation in Δz

$$= u_e A_e [c(z - \Delta z, t) - c(z, t)] + A_e [J(z - \Delta z, t) - J(z, t)] \quad \text{(a3.1)}$$
$$\underbrace{}_{\text{accumulation by convection}} \underbrace{}_{\text{accumulation by diffusion}}$$

where u_e is the linear velocity in the intersitial volume around the par-

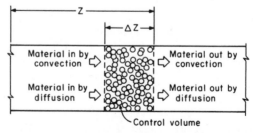

Figure a3.1. Control volume used for establishing conditions of mass balance. [Reprinted from Ref. (1) with permission. Copyright 1973 Elsevier Science Publishers]

ticles, A_e is the free cross-sectional area of the column, and $c(z, t)$ refers
to the solute concentration in the flowing mobile phase (moles per liter
of mobile phase). For simplicity in this derivation we assume that there
is no stagnant mobile phase anywhere in the column, that is, the support
particle is nonporous. Furthermore we assume that diffusive or dispersive
transport along the column axis takes place only in the mobile phase. This
is evidently a very reasonable approximation to make for gas chromatography. It is also reasonable in liquid chromatography when mechanical
dispersion is large with respect to molecular diffusion. Finally the term
$J(z, t)$ represents the flux of solute due to the axial diffusive–dispersive
processes.

An expression must now be written for the rate of change of the mass
of solute inside the control volume. There are necessarily two contributions, the mobile and the stationary compartments. The net rate of
change of solute in the mobile phase is written as

$$A_e\Delta z \frac{\partial c}{\partial t} \tag{a3.2}$$

Extreme care is called for in representing the stationary phase solute
concentration. The reader is warned that different authors define the
amount of stationary phase in quite different ways, which leads to subtle
differences in the results. We first define the volume of stationary phase
as the column volume that is not occupied by pure mobile phase. In the
case of a porous particulate bonded phase, the stationary phase volume
includes the support particle volume, the bonded phase volume (as it
exists in equilibrium with a particular mobile phase), but not the stagnant
mobile phase volume. We here define the volume described above per
unit length of column as $\overline{V}_s$. Thus the rate of change of solute in the
stationary phase in the finite element is

$$\overline{V}_s\Delta z \frac{\partial q}{\partial t} \tag{a3.3}$$

where q is the number of moles of solute in the stationary phase in Δz
divided by the volume of stationary phase in Δz. Clearly q is the stationary
phase solute concentration. The term $\overline{V}_s$ has units of area. Combining all
of the above expressions leads to

$$A_e\Delta z \frac{\partial c}{\partial t} + \overline{V}_s\Delta z \frac{\partial q}{\partial t}$$
$$= u_eA_e[c(z - \Delta z, t) - c(z, t)] + A_e[J(z - \Delta z, t) - J(z, t)] \tag{a3.4}$$

We now divide through both sides by Δz, take the limit as Δz goes to zero, and ignore the finite size of the particles. The effect of finite particle size has been discussed in Chilcote and Scott (1).

$$A_e \frac{\partial c}{\partial t} + \overline{V}_s \frac{\partial q}{\partial t} + u_e A_e \frac{\partial c}{\partial z} = -A_e \frac{\partial J}{\partial z} \tag{a3.5}$$

The flux $J(z, t)$ due to the axial dispersive processes (eddy dispersion and molecular diffusion) can be related by Fick's laws to the gradient of concentration in the axial direction (see Eq. a2.2),

$$J(z, t) = -\mathscr{D}_{ax} \frac{\partial c}{\partial z} \tag{a3.6}$$

Thus the equation can be written as

$$\frac{\partial c}{\partial t} + \frac{\overline{V}_s}{A_e} \frac{\partial q}{\partial t} + u_e \frac{\partial c}{\partial z} = \mathscr{D}_{ax} \frac{\partial^2 c}{\partial z^2} \tag{a3.7}$$

For a nonporous support particle it is easily shown that the ratio $\overline{V}_s / A_e$ is equal to $(1 - \epsilon_e)/\epsilon_e$, where ϵ_e is the external void fraction, that is, the fraction of the cross-sectional area that is available for fluid flow. In the case of nonporous particles ϵ_e is identical to the total porosity of the column ϵ_T. Thus in final form we have

$$\frac{\partial c}{\partial t} + \epsilon \frac{\partial q}{\partial t} + u_e \frac{\partial c}{\partial z} - \mathscr{D}_{ax} \frac{\partial^2 c}{\partial z^2} = 0 \tag{a3.8}$$

where ϵ is the porosity ratio [in this case $(1 - \epsilon_e)/\epsilon_e$].

When the support is porous and a large fraction of the mobile phase is contained in the pores, the preceding derivation must be altered to allow for these changes. The mobile phase solute concentration must be defined relative to the total amount of mobile phase in the column. The definition of q remains as above. The final result is

$$\frac{\partial c}{\partial t} + \epsilon \frac{\partial q}{\partial t} + u \frac{\partial c}{\partial z} - \mathscr{D}_{ax} \frac{\partial^2 c}{\partial z^2} = 0 \tag{a3.9}$$

where u is the chromatographic velocity, that is, the velocity of an unsorbed solute which explores all of the pores, and ϵ must now be taken as the ratio $(1 - \epsilon_T)/\epsilon_T$, with ϵ_T being the total porosity fraction of the column. Note that in this treatment of a porous column the rate of exchange between the moving zone and the stagnant mobile phase is taken as being instantaneous.

From this point it is evident that some relationship between c and q must be found to allow one to eliminate q as a dependent variable. The nature of this relationship defines the various forms of chromatographic theory. For example, if one assumes that c and q are in equilibrium, then this defines what is called "ideal" chromatography. If a first-order kinetic relationship between c and q is used, then this defines linear nonideal chromatography. Finally if a nonlinear kinetic relationship is applied, then we have nonlinear, nonideal chromatography. This is clearly the most complex situation. The results for this case are presented in Section 1.3.8 without derivation.

Case 1.   Ideal Chromatography

When dispersion is neglected, interphase mass transfer is assumed to be so fast that perfect equilibrium exists between phases and the equilibrium isotherm is perfectly linear ($q = Kc$). It can be shown that for an impulse input function $(n^0/V_m)\delta(t)$ the solute concentration at the column exit is given as

$$c(t) = \frac{n^0}{V_m} \delta(t - t_0(1 + k')) \tag{a3.10}$$

This result simply says that the solute will appear at the exit as a perfectly narrow pulse when $t = t_0(1 + k')$. All the solute elutes at exactly the retention time t_R. It should be evident that as the variance of a Gaussian peak approaches zero, the peak shape will approach Eq. a3.10.

Case 2.   Linear Chromatography with Slow Interphase Kinetics

In this case the dispersion term $\mathcal{D}_{ax} (\partial^2 c/\partial z^2)$ is set to zero. The solution to this situation is the limit of Eq. 133 as c^0 goes to zero. Thus we obtain the far simpler equation

$$\frac{c}{c^0} = \left[\gamma\sqrt{\frac{k'}{y}} I_1(2\gamma\sqrt{k'y}) + \delta(y) \right] \exp(-\gamma y - \gamma k') \tag{a3.11}$$

All of the terms used here are defined in the main text in the discussion of Eq. 133. The term $\delta(y)$ is an impulse peak, which elutes at the column dead volume. It is only significant when the kinetics of adsorption and desorption are so slow that a solute molecule can migrate through the whole column without ever being adsorbed. In the limit of fast kinetics (large γ) it can be shown that the peak becomes Gaussian.

Case 3.   Linear Chromatography with Axial Dispersion

When interphase transfer kinetics are fast and the sorption isotherm is linear, Eq. a3.9 can be solved. It is first necessary to specify a second boundary condition at the column exit. This is still a highly debated issue, so we will not go into it. A useful solution to the problem is

$$c(t) = A \frac{\bar{v}}{2\sqrt{\pi D^* t}} \exp\left[-\frac{(L - \bar{v}t)^2}{4D^* t} \right]$$ (a3.12)

where L and $\bar{v}$ have been defined previously as the column length and the zone velocity, while D^* is defined as

$$D^* = \frac{\mathcal{D}_{ax}}{1 + k'}$$ (a3.13)

In this expression there is an implicit assumption that all the processes leading to axial dispersion occur while the solute is in the mobile *phase*.

The term A denotes the integral under the curve of $c(t)$ versus time. Equation a3.12 is really not a Gaussian despite the square in the exponential term. When D^* is very small, the concentration will only be finite in the immediate vicinity of $t = t_R$. Thus replacing t with t_R in the preexponential term and in the denominator of the exponential term, leads to the perfect Gaussian form

$$c(t) = A \frac{\bar{v}}{2\sqrt{\pi D^* t_R}} \exp\left[-\frac{(L - \bar{v}t)^2}{4D^* t_R} \right]$$ (a3.14)

Case 4.   Axial Dispersion with a Parabolic Isotherm

As discussed in Section 1.3.8, Guiochon and coworkers (2, 3) have used the approach of Houghton to derive an equation for an impulse input into a column that has dispersion, but where the phase transfer kinetics are so fast that perfect equilibrium is obtained. The system of equations can be solved mathematically when the curvature in the isotherm is approximated as being parabolic,

$$q = Kc + \beta c^2 \tag{a3.15}$$

where q and c are the stationary phase and the mobile phase solute concentrations. It is obvious that as c gets small, the second term becomes negligible with respect to the first. It is important to note that no physical isotherm can be described by a parabola. Thus Eq. a3.15 can only be used to simulate a small departure from linearity in a real isotherm, for example, of the Langmuir or Freundlich type. A second consequence is that the solution to this problem does not exhibit mass conservation except in the limit as the amount of sample injected approaches zero. This is not the case for the solution of the kinetic Langmuir isotherm described in Section 1.3.8. According to the work of Guiochon the solution is

$$c(t) = \frac{2}{\lambda \bar{v}} \left(\frac{D^*}{\pi t} \right)^{1/2} \frac{\exp[-(L - \bar{v}t)^2/4D^*t]}{\coth(\mu/2) + \exp[(L - \bar{v}t)/2\sqrt{D^*t})]} \tag{a3.16}$$

The new terms are the "leaning" coefficient λ, which is related to the curvature in the isotherm, and μ, which is a function of λ. As λ goes to zero, the hyperbolic cotangent dominates the denominator and Eq. a3.16 becomes identical to Eq. a3.12. When λ is positive, the isotherm is convex (Langmuirian) and the peak develops a tail. Conversely a negative value of λ causes the peak to front. Because λ, at least for small departures from a real isotherm, can be related to the curvature in the isotherm, one can use this peak shape equation to fit overloaded peaks. However, because the basic model is physically unrealistic and does not conserve mass, it must be used with caution. Finally this approach has the advantage, in contrast to the exact solution to the kinetic Langmuir problem, that it can be used to fit fronted peaks.

References to Appendix 1.3

1. D. D. Chilcote and C. D. Scott, *J. Chromatogr.*, *87*, 315 (1973).
2. A. Jaulmes, M. J. Gonzales, C. Vidal-Madjar, and G. Guiochon, *J. Chromatogr.*, *387*, 41 (1987).
3. A. Jaulmes, C. Vidal-Madjar, H. Colin, and G. Guiochon, *J. Phys. Chem.*, *90*, 207 (1986).

APPENDIX 1.4   THE RELATIONSHIP BETWEEN H , $\mathcal{D}$, k, k_{mt}, and σ^2

It is appropriate to discuss the relationship between various parameters of the dynamic events under scrutiny, specifically, the interrelationships between the variance of the exit time distribution, the height equivalent to a theoretical plate H, the dispersion coefficient $\mathcal{D}$, the first-order rate constant k, and the mass transfer coefficient k_{mt} are derived in an approximate fashion (1, 2).

1.4.1. H and $\mathcal{D}$

The variance, H, and $\mathcal{D}$ are related in a simple fashion. By definition one has

$$H = \frac{d\sigma^2}{dx} = \frac{\sigma^2}{L} \qquad (a4.1)$$

and the Einstein equation states that

$$\sigma^2 = 2\mathcal{D}t \qquad (a4.2)$$

Of course, the Einstein equation was derived for molecular diffusion, but the same relationship is obtained from a consideration of any random walk observed over a long enough time (3). Note that Eq. a4.2 is valid for one dimension. If the medium in which diffusion is occurring is isotropic, then the diffusion coefficients in each direction are the same. Since diffusion in each of the Cartesian directions x, y, and z is independent of the others, and variances of independent processes add, the numerical coefficient in Eq. a4.2 for three dimensions is 6. Also note that the variance is the length variance. From Eqs. a4.1 and a4.2, and using Eq. a4.3 for the time,

$$t = \frac{L}{\bar{v}} \qquad (a4.3)$$

one arrives at Eq. a4.4, which relates the dispersion coefficient and the height equivalent to a theoretical plate,

$$H = \frac{2\mathcal{D}}{\bar{v}} \qquad (a4.4)$$

where $\bar{v}$ represents the average zone velocity Ru_e.

1.4.2. k and k_{mt}

A first-order rate constant may naturally arise, such as in the case of limiting slow chemical kinetics. Often, however, the limitation in the overall rate is a mass transfer limitation. A mass transfer rate constant (units time^{-1}) and a mass transfer coefficient (units length·time^{-1}) are related through the area to volume ratio. The inverse of the first-order rate constant is a characteristic time. Consider a volume of fluid V in contact with a surface area A_s. The volume is emptied of solute by a mass transfer process with an associated mass transfer coefficient k_{mt}. The flux of the solute through the interface is governed by the mass transfer coefficient and the difference in concentration Δc between the two phases,

$$J = k_{mt}\Delta c \tag{a4.5}$$

In the engineering literature mass transfer coefficients are compiled and correlated by using the dimensionless Sherwood number,

$$\text{Sh} = \frac{k_{mt}d_p}{D} \tag{a4.6}$$

The rate at which concentration changes in the volume V will be greater the greater the surface area through which the flux occurs and the smaller the volume from which the solute is being transferred. Thus Eq. a4.7 can be seen to relate k and k_{mt},

$$\frac{dc}{dt} = k\Delta c = \frac{k_{mt}A_s}{V}\Delta c \tag{a4.7}$$

For example, when considering the flux of matter across spherical particle boundaries in liquid chromatography, one must use the particle surface area and the interstitial volume. The easiest way to do that is to determine the surface area to volume ratio for a single particle, then multiply this by the particle volume divided by the interstitial volume to arrive at

$$\frac{A_s}{V_{mz}} = \frac{6(1 - \epsilon_e)}{\epsilon_e d_p} \tag{a4.8}$$

1.4.3. k and H

The variance resulting from a random walk can be used to find the relationship between k and H. Recall that the variance due to a random walk is given as nl^2, where n is the number of steps and l is the length of the equidistant steps. The number of forward steps taken is certainly the total time the molecule is in the flowing zone divided by the time a molecule is in the flowing zone during one equilibration step t_{fm}. Thus one has

$$n_f = \frac{L}{u_e t_{fm}} \tag{a4.9}$$

At any point along the column the number of forward steps equals the number of backward steps, so one arrives at Eq. a4.10 for the number of steps,

$$n = \frac{2L}{u_e t_{fm}} \tag{a4.10}$$

The forward step length is the distance a molecule gains over the population average by being in the flowing zone. This is $(1 - R)u_e t_{fm}$. Equation a4.11 is the resulting variance,

$$\sigma^2 = 2L(1 - R)^2 u_e t_{fm} \tag{a4.11}$$

One can consider the time in Eq. a4.11 as being the inverse of a first-order rate of mass transfer from the mobile zone to the stationary zone. Thus one has for H due to this rate,

$$H = \frac{\sigma^2}{L} = \frac{2(1 - R)^2 u_e}{k_{fm}} \tag{a4.12}$$

Noting that k'' is just t_{sz}/t_{fm}, one can write Eq. a4.13 for a stationary zone mass transfer rate controlled process,

$$H = \frac{2R(1 - R)u_e}{k_{sz}} \tag{a4.13}$$

1.4.4. $\mathscr{D}$ and k

One can now derive the effective dispersion coefficient due to a relaxation between phases that is characterized by the first-order rate constant k,

$$\mathscr{D} = \frac{f(R)u_e^2}{k} \tag{a4.14}$$

Equation a4.14 may also be used for algebraic convenience to derive an equivalent relaxation rate constant k from a natural dispersion process, such as eddy dispersion, characterized by $\mathscr{D}$.

References to Appendix 1.4

1. J. C. Giddings, *J. Chem. Phys.*, *26*, 1755 (1957).
2. J. C. Giddings, *J. Chem. Phys.*, *31*, 1462 (1959).
3. S. Chandrasekhar, *Rev. Mod. Phys.*, *15*, 1 (1943).

APPENDIX 1.5   AN INTRODUCTION TO THE GIDDINGS GENERAL NONEQUILIBRIUM THEORY OF BAND BROADENING IN CHROMATOGRAPHY

1.5.1. Theoretical Background

Subsequent to developing the random walk model of chromatography Giddings investigated a distinctly different approach, which is really the most general and powerful method yet devised for estimating the HETP for many different types of zone broadening processes. The approach was developed in a series of papers starting in 1959 [see Refs (55–63) in the main part of this chapter]. The reader should see Giddings's definitive monograph [chaps. 3 and 4 in Ref. (12) in the main section] for a detailed summary of the method and results. The following introductory overview is provided for those who only want the underpinnings of the treatment. The authors have used the sequence of arguments and much of the terminology that is found in Ref. (12), chap. 3.

The method has been applied to the calculation of the HETP from resistance to mass transfer in the mobile phase, the stationary phase, and preliminarily to eddy dispersion processes. It can handle adsorption kinetics, diffusion limitations, multisite adsorption, and combinations of all of the preceding broadening processes.

At heart the nonequilibrium method is a long-time or near equilibrium approach. As such it cannot be used to compute the exact peak shape. In essence the peak is assumed to be Gaussian and the method generates an estimate of the peak variance.

A peak is broadened when the mobile and stationary zones are not in full equilibrium because at the front of the zone there is an excess of solute in the mobile phase relative to that which would be present if the system were at complete equilibrium. Similarly there is a deficit in the mobile phase concentration at the rear of the zone. Consequently the front of the band moves too rapidly and the rear too slowly relative to the center of the zone, and the peak is therefore broadened. The nonequilibrium approach formulates these excesses and deficits as fluxes relative to the flux that would be observed if the system were at equilibrium. This "excess" flux is then shown to be mathematically analogous to a diffusion process and a dispersion coefficient is computed.

The details of the nonequilibrium approach are best illustrated by the stepwise computation of the plate height for a one-site system limited by slow adsorption. For simplicity the stagnant mobile phase is ignored, as are slow transport through the mobile phase and slow diffusion within the particle. The only significant band broadening process is assumed to be the adsorption and desorption of the solute from a fixed site on the surface of the stationary phase.

We first consider the concentrations of solute in the stationary and mobile phases c_s and c_m. These are the actual concentrations defined per unit volume of column in the stationary and mobile phases. (Note the difference in the definition of concentration from that given in Appendix 1.3.) The corresponding concentrations, if complete equilibrium were to be obtained, are c_s^* and c_m^*. The actual and equilibrium concentrations are related by an implicit equation that defines the extent of departure from equilibrium,

$$c_s = c_s^*(1 + \epsilon_s) \qquad (a5.1)$$

$$c_m = c_m^*(1 + \epsilon_m) \qquad (a5.2)$$

In order for the peaks to be approximately Gaussian, the departure from equilibrium must be small. Thus in general the ϵ values will typically be less than about 0.01. It should be evident, based on a mass balance, that the two nonequilibrium parameters are related. It is easily shown that

$$\frac{\epsilon_m}{\epsilon_s} = -k' \qquad (a5.3)$$

The rate of exchange of solute s_m between phases is related to the nonequilibrium parameters as

$$s_m = \left(\frac{dc_m}{dt}\right)_{\text{mass transfer}} = k_d c_s - k_a c_m \qquad (a5.4)$$

where k_d and k_a are the desorption and adsorption rate constants, respectively. In view of the fact that at equilibrium the forward and backward rates are equal,

$$k_d c_s^* = k_a c_m^* \tag{a5.5}$$

Equation a5.4 can be written as

$$s_m = k_d c_s^* \epsilon_s - k_a c_m^* \epsilon_m = -c_m^* \epsilon_m (k_a + k_d) \tag{a5.6}$$

The next step is to relate the rate of exchange between phases to the axial transport of solute by convection. We refer the reader to Appendix 1.3 at this point. Since we are neglecting axial dispersion effects, that is, molecular diffusion and eddy dispersion, and are considering only the mobile phase, it is evident that

$$\frac{\partial c_m}{\partial t} = \left(\frac{dc_m}{dt}\right)_{\text{mass transfer}} + \left(\frac{dc_m}{dt}\right)_{\text{flow}} = s_m + \left(\frac{dc_m}{dt}\right)_{\text{flow}} \tag{a5.7}$$

The first term on the right in Eq. a5.7 is, as indicated, given by s_m. The second term can be obtained from the considerations given in Appendix 1.3 as

$$\left(\frac{dc_m}{dt}\right)_{\text{flow}} = -u\frac{\partial c_m}{\partial z} \tag{a5.8}$$

One of the major simplifications of the nonequilibrium theory is now used to replace the true concentrations with the almost identical equilibrium concentrations. Thus Eq. a5.7 when combined with Eq. a5.8 can be restated as

$$s_m = \frac{\partial c_m^*}{\partial t} + u\frac{\partial c_m^*}{\partial z} \tag{a5.9}$$

The next step is to obtain an equation for $\partial c_m^*/\partial t$. This is done by understanding that c_m^* is equal to $c/(1 + k')$, where c is the sum of the mobile and stationary phase concentrations. Further the rate of change of c is much easier to compute than either c_m or c_s since the only contributor to it is the flow term. Thus,

$$\frac{\partial c}{\partial t} = \left(\frac{dc_m}{dt}\right)_{\text{flow}} = -u\frac{\partial c_m^*}{\partial z} \tag{a5.10}$$

The relationship between c and c_m^* is now used to get an equation for the rate of change of the equilibrium mobile phase concentration,

$$\frac{\partial c_m^*}{\partial t} = \frac{1}{1 + k'}\frac{\partial c}{\partial t} = -\frac{1}{1 + k'}u\frac{\partial c_m^*}{\partial z} \qquad \text{(a5.11)}$$

This equation is substituted for the first term on the right in Eq. a5.9 to get

$$s_m = \frac{k'}{1 + k'}u\frac{\partial c_m^*}{\partial z} \qquad \text{(a5.12)}$$

Equation a5.6 is now substituted in Eq. a5.12 to obtain an explicit relationship for ϵ_m,

$$\epsilon_m = -\frac{1}{k_a + k_d}\frac{k'}{1 + k'}\frac{1}{c_m^*}\frac{\partial c_m^*}{\partial z}u = -\frac{1}{k_a + k_d}\frac{k'}{1 + k'}\frac{\partial \ln c_m^*}{\partial z}u \qquad \text{(a5.13)}$$

Equation a5.13 can be reformulated by dividing the numerator and the denominator by k_d. Recognizing that the ratio k_a/k_d is the capacity factor k', we obtain

$$\epsilon_m = -\frac{k'}{(1 + k')^2}\frac{1}{k_d}\frac{\partial \ln c_m^*}{\partial z}u \qquad \text{(a5.14)}$$

The next step in the derivation is to relate the nonequilibrium parameter to the excess flux. Subsequently that flux will be related to the HETP. The total flux of solute along the column axis, ignoring diffusive processes, is

$$J = uc_m = uc_m^* + uc_m^*\epsilon_m = J^* + \Delta J \qquad \text{(a5.15)}$$

The first term, represented as J^*, is clearly the equilibrium flux and therefore does not cause any broadening. It is only the excess flux ΔJ that broadens the peak. The excess flux is evidently related to ϵ_m. Thus,

$$\Delta J = -\frac{k'}{(1 + k')^2}\frac{u^2}{k_d}\frac{\partial c_m^*}{\partial z} \qquad \text{(a5.16)}$$

This can be related to the gradient of the total concentration c by the previously stated relationship $c_m^* = c/(1 + k')$.

$$\Delta J = -\frac{k'}{(1 + k')^3} \frac{u^2}{k_d} \frac{\partial c}{\partial z} \tag{a5.17}$$

Equation a5.17 is the basis for the analogy between diffusion and kinetically induced peak broadening. This lies in the observation that the excess flux is proportional to the gradient of the concentration, as is the case with a diffusionally induced flux, when the diffusion coefficient is concentration independent and the concentration is a good measure of the solute activity, that is,

$$J_{\text{diff}} = -D \frac{\partial c}{\partial z} \tag{a5.18}$$

Consequently the axial dispersion coefficient $\mathcal{D}_{\text{ax}}$, which one could use to generate broadening that is completely equivalent to that resulting from kinetically limited adsorption–desorption, would be

$$\mathcal{D}_{\text{ax}} = \frac{k'}{(1 + k')^3} \frac{u^2}{k_d} \tag{a5.19}$$

(Compare Eqs. 25 and a4.14 to Eq. a5.19.) As Giddings pointed out, the fact that the kinetic band broadening can be written as a diffusion process means that in the long-time limit the peak shape will be Gaussian. Thus we expect that for relatively fast kinetics where ϵ_m is small the peak shape will be very nearly symmetric.

We now employ the well-known relationship between the longitudinal diffusion coefficient and H (Eq. a4.4),

$$H = \frac{2\mathcal{D}(1 + k')}{u} \tag{a5.20}$$

Substitution of the effective dispersion coefficient in Eq. a5.20 leads to the now very familiar result

$$H = 2 \frac{k'}{(1 + k')^2} \frac{u}{k_d} \tag{a5.21}$$

While it must be admitted that the overall procedure is a great deal more complex and much more tedious than that involved in the random walk model or the CSTR approach, the nonequilibrium model is very general and can be used to study a very wide variety of broadening processes.

1.5.2. Resistance to Mass Transfer in the Stationary Phase

The nonequilibrium method has been applied to the computation of the C term for resistance to mass transfer in the stationary phase. This process is purely diffusional in nature. For a homogeneous material of any geometry it can be shown that

$$C = q \frac{k'}{(1 + k')^2} \frac{d^2}{D_s} \tag{a5.22}$$

where q is a numerical factor that depends on the geometry of the diffusional field and d is the appropriate length. For diffusion in a uniform thin film of thickness d, as in a capillary column, q is equal to $\frac{2}{3}$. For diffusion in a uniform sphere of diameter d_p the q factor is equal to $\frac{1}{30}$. This factor would apply with good accuracy to ion exchange chromatography or to a bulk stationary phase diffusion system. In the case of paper chromatography d is the diameter of the fiber and q is equal to $\frac{1}{16}$. When the stationary phase has a composite geometry, the C term can be written as

$$C = \sum q_i \left(\frac{V_i}{V_s} \right) \frac{d_i^2}{D_s} \tag{a5.23}$$

where V_i is the volume of the element, d_i is the depth or height of the diffusion path, and q_i is the relevant numerical factor. V_s is the total volume of the stationary phase.

1.5.3. Resistance to Mass Transfer in the Mobile Phase

The simplest case assumes only diffusional mass transfer. We know of course that convective contributions are very important (see Eq. 97). In this case,

$$C = \frac{\omega d_p^2}{D_m} \tag{a5.24}$$

The result for a parabolic velocity distribution in a capillary tube has been presented elsewhere. For the hypothetical case of a capillary of radius r_c in which the velocity is uniform across the tube, the nonequilibrium approach gives the result

$$C = \frac{1}{4} \frac{k'^2}{(1 + k')^2} \frac{r_c^2}{D_m} \tag{a5.25}$$

For a packed column in which the velocity varies parabolically from the wall to the center,

$$C = \left(\frac{\zeta - 1}{\zeta + 1}\right)^2 \frac{r_c^2}{24D_m\gamma} \tag{a5.26}$$

where ζ is the ratio of velocities, r_c is the tube radius, and γ is the tortuosity factor.

1.5.4. Slow Surface Desorption Kinetics

This was the case examined in detail in Section a1.5.1, that is, slow desorption from a single type of site,

$$C = 2 \frac{k'}{(1 + k')^2} \frac{1}{k_d} \tag{a5.27}$$

If there are many different types of sites on the surface so that the desorption kinetics differ from site to site, then it can be shown quite rigorously by the nonequilibrium approach that C will be related to the average desorption time $\bar{t}_d$

$$C = 2 \frac{k'}{(1 + k')^2} \bar{t}_d \tag{a5.28}$$

Based on this it is easy to see that a small number of strong adsorption sites can have a significant effect on the peak width.

CHAPTER

2

MECHANISM OF SOLUTE RETENTION IN CHROMATOGRAPHY

RAYMOND P. W. SCOTT

The Perkin-Elmer Corporation
Norwalk, Connecticut

2.1. INTRODUCTION

Any discussion on the mechanism of solute retention must start with the classical definition of a chromatographic separation. A chromatographic separation is achieved by the distribution of a solute mixture between two immiscible phases, a mobile phase and a stationary phase. Those solutes distributed preferentially in the mobile phase will pass through or from the system more rapidly than those substances that are distributed preferentially in the stationary phase. As a consequence the solutes will be eluted in the order of the magnitude of their distribution coefficients with respect to the stationary phase. This definition, although perfectly correct, is somewhat trite since although it introduces the essential requisites of a mobile and a stationary phase, it obscures the basic process of selective retention in the term distribution.

The relative distribution of a solute between two immiscible phases can be described by the magnitude of the distribution coefficient which, in turn, can be defined mathematically by the equation

$$K_A = \frac{c_{S(A)}}{c_{M(A)}} \tag{1}$$

where K_A = distribution coefficient of solute A between stationary phase and mobile phase

$c_{S(A)}$ = concentration of solute A in stationary phase under equilibrium conditions

$c_{M(A)}$ = concentration of solute A in mobile phase under equilibrium conditions

It follows from the definition of a chromatographic separation that solutes will be eluted from the chromatographic system (column) in the order of increasing magnitude of K. The retention volume V_R of a solute A can be shown (1, 2) to be given by

$$V_{R(A)} = V_M + K_A V_L \tag{2}$$

where V_L is the total volume of stationary phase in the chromatographic system and V_M is the total volume of mobile phase in the system.

In a liquid–solid system V_L can be replaced by the total mass of *stationary* phase and V_M by the total mass of *mobile* phase in the chromatographic system. Under these circumstances the distribution coefficient would be defined as the ratio of the concentrations of the solute in the two phases under conditions of equilibrium, where concentrations are defined as mass of solute per unit mass of phase. It should also be noted that V_M includes all volumes of mobile phase from the point of injection to the detector cell and thus includes all extracolumn volumes.

It follows that to control $V_{R(A)}$ it is necessary to control the value of K_A, and there are two basic theories that attempt to aid in the prediction of K_A and thus the chromatographic selectivity. These two theories are the thermodynamic theory and the molecular interaction theory. However, before discussing these theories, the dependence of chromatographic selectivity on solute retention needs to be discussed, and to do this, the processes that take place in a chromatographic column must be understood.

During the development of a separation, two processes progress continuously and simultaneously in the chromatographic column. Primarily the solutes are moved apart as a result of the difference in the magnitudes of the distribution coefficient of each respective solute. Second, as a result of the design of the column, the dispersion of the individual solute bands is constrained such that the individual solutes are eluted discretely. The mechanism of solute band dispersion is not germane to this chapter and is dealt with elsewhere in this book. The relative movement of the peaks apart, however, constitutes the column selectivity, and it is the control of the relative retention of a given pair of peaks that is pertinent to this discussion.

The degree of separation can therefore constitute a means of measuring the selectivity of the phase system. The selectivity for two specific solutes could be determined from the difference in retention volume of the two solutes. For example, for two solutes A and B,

$$V_{R(A)} - V_{R(B)} = (V_M + K_A V_L) - (V_M + K_B V_L) = V'_{R(A)} - V'_{R(B)}$$

or

$$V'_{R(A)} - V'_{R(B)} = (K_A - K_B)V_L \tag{3}$$

where $V'_{R(A)}$ is the corrected retention volume of solute A and $V'_{R(B)}$ is the corrected retention volume of solute B.

Obviously Eq. 3 gives a measure of how far the peaks are moved apart, but unfortuantely it includes the volume of stationary phase in the column and thus pertains only to the specific column that was used to measure the corrected retention volume. A more appropriate measure would be one that applied to the phase system only and consequently could be used for *any* column that employed the *same* phase system. The measure used is the retention ratio α of the two solutes, which is taken as the ratio of their corrected retention volumes,

$$\alpha = \frac{V'_{R(A)}}{V'_{R(B)}}$$

or

$$\alpha = \frac{K_A V_L}{K_B V_L} = \frac{K_A}{K_B}$$

It is seen that the separation ratio α depends only on the distribution coefficient of the two solutes, which in turn depends only on the phase system selected and is *independent of the properties of the column*.

Furthermore, the selectivity of a phase system for a given pair of solutes increases as the magnitude of their separation ratio α. Consequently, for any given complex mixture, the phase system must be chosen to maximize α (the separation ratio of the pair of solutes eluted closest together) to obtain the separation of the total mixture in the minimum time (3).

2.2. THE THERMODYNAMIC THEORY OF RETENTION

Solute retention and selectivity are directly related to the distribution coefficient. Thus the classical thermodynamic expression relating the distribution coefficient with the excess free energy can be employed to describe the control of retention in thermodynamic terms,

$$RT \ln K = - \Delta G_0 \tag{4}$$

where R = gas constant
T = absolute temperature
ΔG_0 = excess free energy

bearing in mind that

$$\Delta G_0 = \Delta H_0 - T \Delta S_0 \tag{5}$$

where ΔH_0 is the excess free enthalpy and ΔS_0 the excess free entropy. Substituting for ΔG_0 from Eq. 5 into Eq. 4,

$$\ln K = - \left(\frac{\Delta H_0}{RT} - \frac{\Delta S_0}{R} \right) \tag{6}$$

or

$$K = \exp - \left(\frac{\Delta H_0}{RT} - \frac{\Delta S_0}{R} \right) \tag{7}$$

Equation 6 gives a direct relationship between the distribution coefficient K (and thus retention) and the thermodynamic functions *excess free enthalpy* and *excess free entropy*. In fact by determining the corrected retention volume of a solute in a given phase system over a range of temperatures the magnitudes of ΔH and ΔS can be determined. The respective values of the distribution coefficient at each temperature can be calculated from

$$K_A = \frac{V'_{R(A)}}{V_L}$$

If a liquid–solid system is employed, then the total volume of stationary phase in the column must be replaced by the total mass of stationary phase. It should be noted that, in practice, a precise value for the proportion of the total stationary phase present in the system that is *chromatographically available* can be extremely difficult to identify. Nevertheless, if the logarithm of the distribution coefficient is plotted against the reciprocal of the absolute temperature, then a Vant Hoff plot is produced, an example of which is shown in Fig. 2.1. From the slope of the curve the excess free enthalpy can be calculated and the intercept will give a value for the excess free entropy,

$$\frac{\partial \ln(K)}{\partial(1/T)} = \frac{\Delta H_0}{R}$$

Although such information can give some insight into the nature of the distribution, it is unfortunately knowledge acquired after the fact. At present it is not possible to predict values for the excess free enthalpy or excess free entropy of a given distribution system. Consequently, the thermodynamic theory cannot help in predicting retention or selectivity, or even give guidance on the choice of an appropriate phase system to employ. It can, however, as already mentioned, help explain the nature of a given distribution system once it has been achieved.

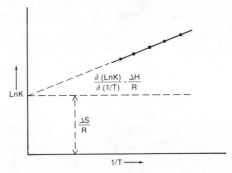

Figure 2.1. Graph relating logarithm of distribution coefficient to reciprocal of absolute temperature.

2.3. THE MOLECULAR INTERACTION MODEL FOR SOLUTE RETENTION

The thermodynamic properties of any physicochemical system are, in fact, bulk properties of the system. Their magnitude represents the combination of a number of different effects that may take place on the molecular scale. Since the excess free enthalpy or excess free entropy of a distribution system cannot be allotted to any specific stationary or mobile phase effect, this renders the use of thermodynamic data useless for the prediction of retention or selectivity.

To avoid the ambiguity of bulk property data, attempts have been made to relate the molecular properties of the solute and phase system to retention and selectivity. The two molecular properties that have been considered are the molecular diameter of the solute and the type of molecular interactions that could occur between the solute molecule and the two phases. Molecular diameter can be used to explain separations that are achieved by exclusion chromatography, and intermolecular forces between solute and mobile and stationary phases employed to describe the selectivity that is obtained in normal partition chromatography.

The molecular interaction model suggests that distribution occurs between two phases as a result of the different molecular forces that exist between the solute molecule and the molecules of the two phases. Those solute molecules that experience stronger forces between them and the stationary phase will be retained longer in the chromatographic system than those molecules that experience stronger interactions with the mobile phase. Consequently retention can be enhanced by choosing a stationary phase that is likely to exhibit stronger interactions with the solute than the mobile phase. Now there are a number of different types of intermolecular forces that can be utilized in controlling retention and selectivity, ionic forces, dispersive forces, and polar forces. Hydrogen bonding could be considered a fourth type of molecular interaction, but in the context of this discussion, it is better considered as a very strong polar force.

2.3.1. Ionic Interactions

Ionic interactions result from permanent charges residing on the molecule and are exploited in ion exchange chromatography. If it is required to retain cations in the chromatographic system, then the stationary phase must contain anions for them to interact with. Conversely, if the chromatographic column is to be selective toward anions, then the stationary phase must contain cations. In general the mobile phase in ion chromatography normally carries the counter ion to the stationary phase.

2.3.2. Polar Interactions

Polar forces result from molecular interactions between dipoles or induced dipoles in the solute and phase molecules. A polar molecule carries an electric charge, but no *net* charge. The charge can arise from a permanent dipole such as that caused by the carbonyl group in an aliphatic ketone or by the induction polarization of a molecule or group such as an aromatic nucleus. Polar interactions can arise between a molecule with a permanent dipole and a polarizable molecule. For example, when a hydroxyl group comes in close proximity to an aromatic nucleus, the electric charge from the dipole of the alcohol induces an opposite charge on the polarizable aromatic nucleus with resulting molecular interaction. It follows that a stationary phase containing hydroxyl groups could be employed effectively to retain aromatic compounds, but the mobile phase would have to be devoid (or nearly so) of polar substances to ensure that the major retentive force resided in the stationary phase.

2.3.3. Dispersive Interactions

Dispersive interactions are more difficult to describe. Although electric in nature, they result from charge fluctuations rather than from permanent electric charges on the molecule. Examples of purely dispersive interactions are the molecular forces that exist between hydrocarbon molecules. *n*-Heptane is *not* a gas due to the collective effort of all the dispersive interactions that hold the molecules together as a liquid. To retain solutes selectively, solely on a basis of dispersive interactions, the stationary phase must not contain polar substances but hydrocarbon-type materials such as the reverse-bonded phases now so popular in liquid chromatography. It also follows that to allow the dispersive selectivity to predominate in the stationary phase, the mobile phase must be polar and significantly less dispersive—hence the use of methanol–water and acetonitrile–water mixtures as mobile phases in reversed-phase chromatographic systems.

In any distribution system it is extremely rare that only one type of interaction occurs in a given phase and if it does, it will almost certainly be dispersive in nature. Polar interactions are always accompanied by dispersive interactions and ionic interactions are almost always accompanied by both polar and dispersive interactions. Nevertheless it is possible to choose phases in which one particular type of interaction can dominate.

2.4. SELECTIVITY BY MOLECULAR EXCLUSION

Exclusion liquid chromatography is now a well-established separation technique used largely in the separation of high molecular weight substances, such as polymer fractions and materials of biological origin. The separation depends on the use of a stationary phase that is contained in an inert porous support, the pores of which cover a relatively wide range of diameters. The support can be silica gel of appropriate porosity or an organic gel.

The retention volume of a given solute can be given by Eq. 2,

$$V_R = V_M + KV_L = V_M + K'As$$

where As is the surface of the support and K' is the distribution coefficient of the solute between the mobile phase and the surface.

However, if the support is inert, or rendered inert by the choice of mobile phase (such as when silica gel employed with an aqueous mobile phase), then K is zero and

$$V_R = V_M \tag{8}$$

Now V_M can be divided into two parts, the interstitial volume V_i and the pore volume V_p. Thus Eq. 8 can be put in the form

$$V_R = V_M = V_i + V_p \tag{9}$$

The contents of pore volume V_p is, in effect, a stationary phase itself, and thus Eq. 9 can be put in the form

$$V_M = V_i + KV_p \tag{10}$$

where K is now the ratio of the concentrations of the solute in the "moving"-mobile phase to that in the "stationary"-mobile phase under conditions of equilibrium.

However, since the content of the pores is, in most instances (but not all), the same as that in the mobile phase, $K = 1$, and thus Eq. 9 is the pertinent equation that describes retention in exclusion chromatography. If the pores have a diversity of size, then some molecules may be too large to enter some pores and thus Eq. 9 has to be further modified to take the pore size into account,

$$V_{R(A)} = V_i + V_{p(A)} \tag{11}$$

where $V_{p(A)}$ is the pore volume that is available to solute A.

It is seen that the retention volume of the solute depends not merely on the pore volume, but on the volume of those pores that have diameters equal to or greater than that of the solute molecule. Thus the range of pore sizes and the pore volume associated with them in a given exclusion medium control both the retention and the selectivity.

Exclusion chromatography is a very effective technique for the separation of solutes on the basis of molecular size but suffers from one serious drawback. The exclusion chromatogram has a very limited peak capacity since the complete sample is eluted between the interstitial volume and the thermodynamic dead volume V_M. In normal elution chromatography peaks can be eluted between the dead volume and the retention volume of the last eluted peak, which may be equivalent to 20 or even 50 dead volumes. Thus as the pore volume of a column is approximately half the dead volume, this means that the exclusion chromatogram has a peak capacity of two to three orders of magnitude less than the elution chromatogram. Consequently, the exclusion chromatogram can often give only the molecular weight distribution of the sample in the form of a broad peak unless the sample has very few components. An alternative would be to employ very high column efficiencies (4), an example of which is shown in Fig. 2.2. The chromatogram shows the separation of benzene and the C_2 to C_8 alkyl benzenes. All the solutes are separated and they individually only differ by two methylene groups. There are, however, some serious disadvantages since although the column has an efficiency of over 600,000 theoretical plates, it needs to have a length of 14 meters, made up in 1-meter sections, which involves somewhat difficult fabrication. Furthermore, the solutes take nearly 10 hours to elute and thus can hardly be used for routine analyses. Nevertheless, the use of such columns could make exclusion chromatography practical for very special separation needs.

Curves relating pore volume and pore size for a number of different silica gels are shown in Fig. 2.3. It is seen that the pore size varies widely between different silica gels. Partisil 10, for example, has a relatively small pore volume of about 0.6 mL/g and a pore diameter range between 10 and 200 Å. In contrast, Porasil C with a pore volume of 1 mL/g has a pore diameter range between 80 and 8000 Å. It is obvious that for optimal separations the silica gel has to be carefully chosen to suit the molecular

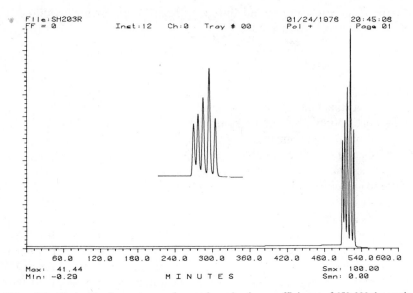

File:SH203R
FF = 0   Inst:12   Ch:0   Tray # 00   01/24/1978   20:45:08
                                                    Pol +   Page 01

60.0   120.0   180.0   240.0   300.0   360.0   420.0   480.0   540.0 600.0

Max: 41.44   Smx: 100.00
Min: -0.29   M I N U T E S   Smn: 0.00

Figure 2.2. Exclusion chromatogram from column having an efficiency of 650,000 theoretical plates. Column length 14 m; I.D. 1 mm; Spherisorb 5 μm; flow rate 25 μL/min; Sample 0.5 μL of a 10% v/v solution of C_2 to C_8 alkyl benzenes in THF.

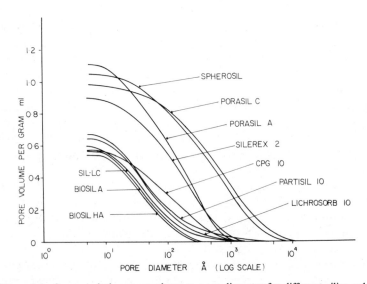

Figure 2.3. Curves relating pore volume to pore diameter for different silica gels.

126

EXCLUSION CHROMATOGRAMS FROM DIFFERENT SILICA GELS

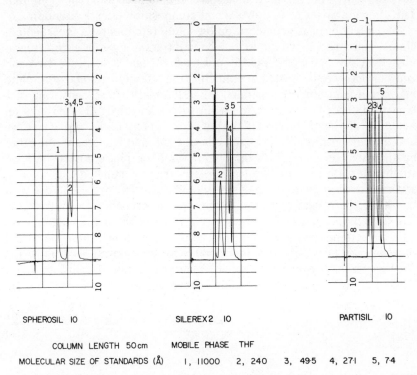

Figure 2.4. Exclusion chromatograms from different silica gels. Column length 50 cm; mobile phase THF; molecular size of standards: (Å) 1—11,000 Å, 2—240 Å, 3—49.5 Å, 4—27.1 Å, 5—7.4 Å.

size of the solutes to be separated. The use of three of the silica gels for the separation of four solutes having mean molecular diameters of 7.4, 27.1, 49.5, 240, and 11,000 Å, respectively, is shown in Fig. 2.4. The advantages of employing the silica gel with the correct range of pore sizes for the sample concerned is abundantly clear.

2.5. ELUTION CHROMATOGRAPHY: SOLVENT COMPOSITION AS A MEANS OF CONTROLLING RETENTION

Having chosen the solvent with the necessary functional groups to interact with the solute molecules, its effect can be modified by mixing with another solvent that interacts with the solutes to a markedly less extent. Appropriate mixtures of the two solvents can then be used for isocratic development or the two individual solvents employed for gradient elution. The question arises as to how solvent composition effects retention.

The effect of phase composition on solute retention was investigated, almost simultaneously, and in two quite different areas of chromatography, by Laub and Purnell (5, 6) and by Scott and Kucera (7, 8). Laub and Purnell were interested in predicting the retention volume of a solute in *gas chromatography* separated on a stationary phase consisting of a binary mixture. They wished to do this from a knowledge of the retention volume of the solute on each of the pure components of the binary mixture and the composition of the mixed phase. Scott and Kucera, on the other hand, were concerned with the effect of mobile phase composition on solute retention in *liquid chromatography*. They postulated that, even in a binary mixture, a solute molecule could only effectively interact with one molecule of phase at any instant and the probability of interaction was directly proportional to the *concentration* of the interacting moiety. This concept has a parallel in the kinetic theory of gases where the partial pressure of a gas (which is proportional to the concentration of molecules per unit volume) is linearly related to the probability of collision. Katz, Ogan, and Scott (9) enunciated this concept in the form of the equation

$$K = \frac{\sum_{r=1}^{r=n} \phi_r v_r}{\sum_{p=1}^{p=m} \phi_p' v_p} \tag{12}$$

where K is the overall distribution coefficient of a solute between two immiscible solvents, one solvent (phase 1) containing n components and the other (phase 2) m components, $v_r + v_p$ are the volume fraction of components r and p in phases 1 and 2, respectively, and ϕ_r and ϕ_p' are constants characteristic of the interactions of the solute with components r and p, respectively.

For two pure immiscible solvents $v_r = v_p = 1$ and $n = m = 1$. Thus,

$$K = \frac{\phi_r}{\phi_p'}$$

Therefore ϕ_r/ϕ_p' is the distribution coefficient of the solute between components r and p. Thus Eq. 12 can be put in the form

$$K = \sum_{r=1}^{r=n} \left(\frac{v_r}{\sum_{p=1}^{p=m} K_r^p v_p} \right) \tag{13}$$

where K_r^p is the distribution coefficient of the solute between components p and r.

Consider a particular application of Eq. 12 where one phase is a binary mixture of n-heptane and chloroform and the other pure water. In Eq. 12, $n = 2$ and $m = 1$, and thus the equation reduces to

$$K = \frac{\phi_1}{\phi_1'} v + \frac{\phi_2(1 - v)}{\phi_1'}$$

$$= K_1 v + K_2(1 - v)$$

$$= K_2 + (K_1 - K_2)v \qquad (14)$$

where v is the volume fraction of component 1, $\phi_1/\phi_1' = K_1$ is the distribution coefficient of the solute between component 1 and water, and $\phi_2/\phi_1' = K_2$ is the distribution coefficient of the solute between component 2 and water.

Equation 14 indicates that there should be a linear relationship between the volume fraction of one component of the solvent mixture and the distribution coefficient of the solute between the two phases, the distribution coefficient being measured with respect to the binary mixture.

In Fig. 2.5 the results obtained by Scott et al. (9) are shown as curves relating the distribution coefficient of n-pentanol to the volume fraction of one component of a binary mixture. Curves for three different binary mixtures, n-heptane/n-heptyl chloride, n-heptane/toluene, and n-heptane/n-heptyl acetate, are included. In Fig. 2.6 similar curves are shown for three different *solutes*, but one binary mixture of n-heptane and n-butyl chloride. It is seen that the effect of solvent composition on the distribution coefficient is relatively simple and predictable, but unfortunately is subject to certain caveats. First, if there is an association between the solvent components, for example, in methanol–water mixtures, then the solvent is no longer a simple binary mixture. Consequently, the linear relationship between the distribution coefficient and the volume fraction of methanol no longer exists. Second, if the two phases are not in equilibrium, such as in a reversed-phase column where the amount of solvent absorbed on the stationary phase is still dependent on the mobile phase composition, then the simple relationship will again not be realized. Both these exceptions are discussed later.

Equation 14 also applies to gas chromatography. The corrected retention volume of a solute is proportional to the distribution coefficient of the solute with respect to the stationary phase. Consequently, the retention volume of a solute should also be linearly related to the volume

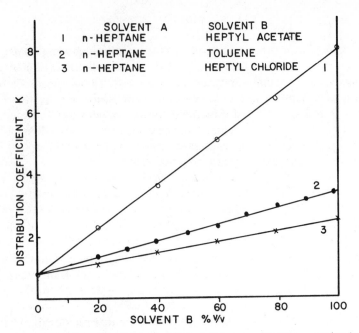

Figure 2.5. Graphs showing distribution coefficient of *n*-pentanol between water and a binary solvent mixture plotted against solvent composition.

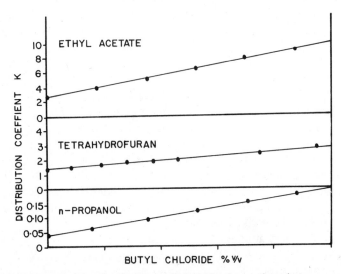

Figure 2.6. Graphs showing distribution coefficients for three solutes between water and a mixture of *n*-heptane and *n*-butyl chloride plotted against solvent composition.

130

fraction of one component of the binary mixture employed as the stationary phase. This linear relationship was confirmed many times by Laub (10), who again supported the basic concept that concentration controls the probability of molecular interaction. The same caveats apply, however, in gas chromatography. When association between the components of the stationary phase occurs, then the stationary phase is no longer a binary mixture and consequently the linear relationship between phase composition and retention volume breaks down.

2.5.1. Liquid Chromatography Systems

In practical liquid chromatography using, for example, silica gel as the stationary phase, changes in solvent composition can affect solute retention in two ways. First, the surface of the stationary phase can be modified by the adsorption of a layer of solvent molecules on the surface and thus change the nature of the solute interactions with the silica. Second, changes in solvent composition will affect retention by changing the interactions in the mobile phase in the manner previously discussed.

Solvent molecules are adsorbed on the surface of silica gel to provide either a monolayer or a bilayer, depending on the polarity of the solvent. Mixtures of n-heptane with chloroform, butyl chloride, or benzene form monolayers according to the Langmuir isotherm, examples of which are shown in Fig. 2.7. These isotherms are typical for relatively nonpolar solvents. It is seen that the layer is virtually complete at solvent concentrations of about 20% w/v, and consequently the nature of the stationary

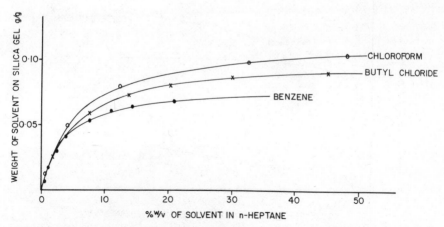

Figure 2.7. Langmuir adsorption isotherms for three nonpolar solvents. Data curve fitted to Langmuir function $Y = x(A + Bx)$.

phase surface is now constant. Above this solvent concentration, solute retention will be controlled solely by the interactions taking place in the mobile phase. Thus the simple linear relationship between distribution coefficient and solvent composition will hold. It must be emphasized, however, that *the distribution coefficient is defined with respect to the mobile phase*, which is the *inverse* of that normally employed in chromatography retention. In chromatography the distribution coefficient is normally referred to the stationary phase.

Polar solvents in *n*-heptane normally form double layers of solvent on the silica gel surface due to the very strong interaction between the silanol groups of the silica and the polar groups of the solvent. An example of the bilayer absorption of ethyl acetate on the silica gel is shown in Fig. 2.8. It is seen that the monolayer is formed very quickly and is virtually complete at an ethyl acetate concentration of only 1% w/v. The second layer is very weakly held and forms much more slowly.

Monolayers of adsorbed solvent are also formed on the surface of reverse phases. This solvent layer also modifies the nature of the interaction of the solute with the stationary phase and reduces retention. Examples of the adsorption isotherms for the lower aliphatic alcohols in water are shown in Fig. 2.9; these isotherms are also Langmuir in type. It is seen that methanol, the commonly used solvent with water, forms a monolayer relatively slowly and thus even at a concentration of 20% w/v the monolayer has only just completely formed. It will be seen later that the

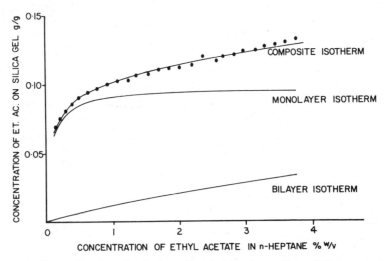

Figure 2.8. Composite adsorption isotherm for ethyl acetate on silica gel. Data curve fitted to bilayer Langmuir-type function $Y = A - (A + ABx/2)/(1 + Bx + Cx^2)$.

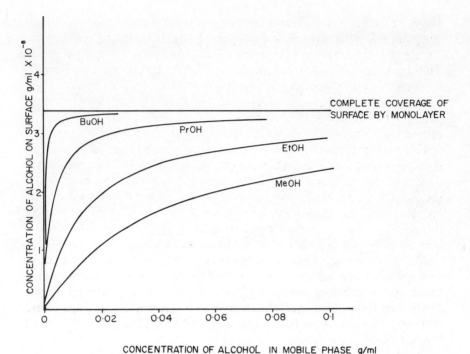

CONCENTRATION OF ALCOHOL IN MOBILE PHASE g/ml

Figure 2.9. Adsorption isotherms for C_1 to C_4 alcohols on ODS2 reverse phase.

simple Eq. 14 cannot be applied to reverse-phase systems due to the association of the methanol with water.

Consider the application of Eq. 14 to a liquid chromatography system employing silica gel as the stationary phase and n-heptane–ethyl acetate mixtures as the mobile phase. Reiterating Eq. 14,

$$K = K_2 + (K_1 - K_2)v \tag{14}$$

Now K is defined as C_m/C_s, that is, the distribution coefficient is referred to the mobile phase. In liquid chromatography K_c is defined with respect to the stationary phase, that is,

$$K_c = \frac{C_s}{C_m} = \frac{1}{K}$$

Thus Eq. 14 becomes

$$\frac{1}{K_c} = K_2 + (K_1 - K_2)v \tag{15}$$

Now

$$V_R' = K_c V_L$$

Thus

$$\frac{V_L}{V_R'} = K_2 + (K_1 - K_2)v$$

or

$$\frac{1}{V_R'} = a + bv \qquad\qquad (16)$$

where

$$a = \frac{K_2}{V_L} \quad \text{and} \quad b = \frac{K_1 - K_V}{V_L}$$

Experimental evidence supporting the validity of Eq. 16 has been provided by a number of workers (11–14). A set of curves relating the reciprocal of the corrected retention volume to solvent composition, for a number of solutes, is shown in Fig. 2.10. It is seen that the retention of

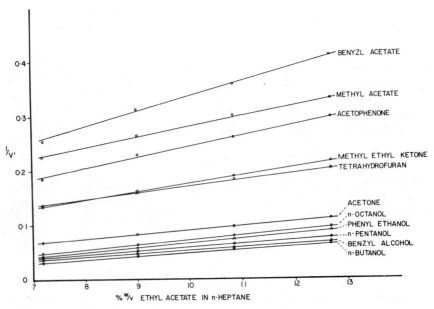

Figure 2.10. Graphs relating reciprocal of corrected retention volume to % w/v of ethyl acetate in *n*-heptane for different solutes.

a solute could be easily predicted by Eq. 16, provided that the stationary phase surface was constant in nature (any solvent layer likely to form on the stationary phase surface was complete) and the retention volume of the solute was known at two different solvent concentrations. Prediction of retention data from reversed-phase systems is, however, far more complex.

2.5.2. Aqueous Solvent Systems

When methanol and water are mixed, association takes place between the methanol and the water to form a methanol–water complex. Any mobile phase consisting of these two solvents is, in fact, ternary in nature, consisting of free or unassociated water, free or unassociated methanol, and methanol associated with water (9). The term "free water" or "free methanol" assumes that the solvent molecules are still associating with themselves but not with each other. In fact, acetonitrile and tetrahydrofuran also associate with water, but to a much lesser extent. Lochmuller et al. have suggested that a fourth component may also be present, resulting from the association of two methanol molecules with one water molecule (15). The quantity of the trimer present, however, appears to be very small and the free water, methanol, and singly associated methanol are the components that have, by far, the major effect on solute retention. One of the several manifestations of this association is the change in volume on mixing. This change is very significant for methanol–water mixtures and has been used (14) to calculate the equilibrium constant for the association of methanol with water. From the equilibrium constant the curves relating the concentration of free water, free methanol, and methanol associated with water to the initial composition of the solvent mixture can be constructed. A set of such curves is shown in Fig. 2.11.

It is clear that the apparently simple mixture of methanol and water is, in fact, very complex. Solute retention would not be described by a simple linear relationship between retention volume and methanol concentration. The curves indicate that up to a nominal volume fraction of 0.4 methanol, the solvent actually consists of a binary mixture of free water and associated methanol. Similarly, between a nominal volume fraction of 0.8 and 1.0 methanol, a different binary mixture exists, consisting of associated methanol and free methanol.

Between nominal volume fractions of 0.4 and 0.8 of methanol, an even more complex situation arises, since there is a ternary mixture consisting of free water, free methanol, and associated methanol. The real nature of the solvent systems is depicted in Fig. 2.11 and explains a number of

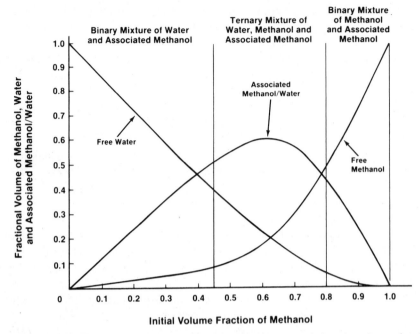

Figure 2.11. Graphs showing relative volume fraction of water, methanol, and associated methanol against original volume fractions of methanol in mixture.

anomalous results from methanol–water mixtures that have been noted in the past. Scott and Simpson (16) measured the adsorption isotherms for methanol on a reversed-phase matrix and demonstrated that a Langmuir-type isotherm was involved. However, Lochmuller and Wilder (17) claimed that at the low nominal concentrations of methanol over which the measurements were taken, the reversed-phase material contains a considerable amount of water on the surface as well as the methanol. From the curves shown in Fig. 2.11 it is apparent that the adsorption isotherms measured by Scott and Simpson were probably for associated methanol–water, not methanol, and, furthermore, that the water identified by Lochmuller was not free water but the water contained in the associated methanol–water complex. Scott and Simpson also calculated the surface area of some reversed-phase matrices from the maximum amount of methanol adsorbed and the area of the methanol molecule. The values they obtained were rather low, and it is now obvious that this error arose, at least in part, as a result of employing the area of the methanol molecule in their calculations and not that of associated methanol.

A linear gradient of nominal volume fractions of methanol with water, in fact, now can be seen to be a very complex gradient system, if the nature and the concentration of the individual interacting solvent species are taken into account. Initially there is an approximately linear change from free water to methanol associated with water until the nominal volume fraction of water is about 0.4. Subsequently, from a nominal volume fraction of methanol of 0.4 to 0.8, the volume fraction of free water falls to almost zero, the volume fraction of associated methanol goes through a maximum, and the free volume of methanol begins to rise very rapidly. Finally, the volume fraction of free methanol increases almost linearly, while the volume fraction of associated methanol decreases linearly. As, in general, the strength of the interactions of methanol with a solute is likely to be much greater than that between the solute and free water or water associated with methanol (that is, the eluting strength of free methanol is much greater than that of free water or associated methanol), then solutes would be eluted very rapidly toward the end of the linear program. The rapid manner in which the free methanol volume fraction changes toward the end of a linear program, as deduced from the curves in Fig. 2.11, explains the need for using convex program gradient profiles with methanol–water mixtures.

It would appear that despite the claims by some that methanol is the ideal solvent to mix with water,* the converse is the case. Figure 2.11 shows solvent changes in three distinct stages across the entire concentration range. Furthermore, and as a result, there are at least two distinct changes in the nature of the stationary phase surface.

The fact that there is little free methanol in the solvent mixture until the nominal volume fraction of methanol is in excess of about 0.5 also accounts for the fact that proteins and other biological materials that are readily denatured in the presence of free methanol can be separated by reversed-phase chromatography, provided the nominal methanol volume fraction is less than about 0.6. The surprising evidence that proteins are not denatured when methanol–water mixtures are used as the mobile phase is explained by the fact that virtually all of the methanol is associated with water and thus is sufficiently less active than free methanol in causing denaturation.

---

* Such claims are based on the observation that plots of ln k' versus volume fraction of methanol in water are linear "over practical ranges of retention." Such ranges are those over which "free" methanol is present according to current theory. The linearity of ln k' versus volume fraction of methanol is accidental therefore and one does not need to invent reasons for a linear free energy relationship between distribution coefficient and composition (see below).

2.5.3. Effect of Methanol Concentration on Distribution Coefficient

Consider the distribution of a solute between n-hexadecane and methanol–water mixtures. The liquid–liquid distribution system eliminates the complexity of surface changes in reverse-phase chromatography and thus isolates the effect of solvent–solute interactions.

Due to the presence of associated methanol, there are three components in the methanol–water phase, consequently returning to Eq. 12, $n = 3$. Furthermore, since there is a single solvent component present in the hydrocarbon phase, $m = 1$, Eq. 12 thus becomes

$$K = \frac{\phi_1 v_1 + \phi_2 v_2 + \phi_3 v_3}{\phi'} \tag{17}$$

where v_1 = volume fraction of free methanol
  v_2 = volume fraction of associated methanol
  v_3 = volume fraction of free water

Now $\phi_1 \phi' = K_M$, which is the distribution coefficient of the solute between free methanol and n-hexadecane; $\phi_2/\phi' = K_{MW}$, which is the distribution coefficient of the solute between associated methanol and n-hexadecane; and $\phi_3/\phi' = K_W$, which is the distribution coefficient of the solute between free water and n-hexadecane. Thus Eq. 17 becomes

$$K = K_M v_1 + K_{MW} v_2 + K_W v_3 \tag{18}$$

Scott et al. (9) determined the distribution coefficients of n-pentanol and vinyl acetate between n-hexadecane and a range of methanol–water mixtures. Their results are shown in Fig. 2.12. Taking the values of the distribution coefficient of each solute at the center and the extremes of the concentration range examined, the three distribution coefficients for methanol, associated methanol, and water were calculated using Eq. 18. This was achieved by solving the three simultaneous equations for each solvent composition and for each solute. Values for v_1, v_2, and v_3, the volume fractions of free water, associated methanol, and free methanol, were taken from the data shown in Fig. 2.11. The values of the different distribution coefficients are included in Fig. 2.12. As would be expected, the dominant distribution coefficient for both solutes is that with respect to methanol, indicating that the strongest solute–solvent interactions are with the free methanol. The weakest interaction for n-pentanol is with the associated methanol, whereas there is relatively weak, but nevertheless significant, interaction between the hydroxyl group and the free

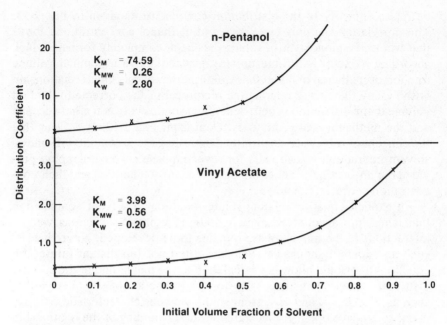

Figure 2.12. Graphs relating solute distribution coefficient between methanol–water mixtures and *n*-hexadecane against initial volume fraction of methanol. Lines—theoretical; points—experimental.

water. This is indicated by the magnitude of the distribution coefficient of the *n*-pentanol between the free water and the *n*-hexadecane, approximately 2.8. In contrast the vinyl acetate appears to interact more strongly with the associated methanol ($K_{MW} = 0.56$) than the free water ($K_W = 0.2$). Both interactions are, however, very weak compared with the interaction of the vinyl acetate with the free methanol ($K_M = 3.98$).

Unfortunately, in reversed-phase chromatography the relationship between retention and solvent composition is more complicated. Changes in solvent composition not only affect the interactions in the mobile phase by changing the fractions of free methanol, associated methanol, and free water present, but also cause complicated changes in the nature of the stationary phase surface by creating a complex adsorption isotherm.

As already discussed, the retention volume of a solute is directly proportional to the distribution coefficient with respect to the stationary phase but inversely proportional to the distribution coefficient with respect to the mobile phase. As a consequence, Eq. 16 applies and the reciprocal of the corrected retention volume, measured on a reversed-phase column over a series of different methanol–water mixtures, should be related linearly to the distribution coefficient, as given in Fig. 2.13. This correlation can only be expected at methanol concentrations above that where the monolayer of solvent has been completely formed, which according to Scott and Simpson (16) must be above a nominal volume fraction of methanol of about 0.4. Confirmation of this linear relationship is shown by the curves relating the reciprocal of the corrected retention volume to the distribution coefficient of the solutes between n-hexadecane and the methanol–water mixtures taken from Fig. 2.12. It is clear that the concept of solvent concentration controlling the probability of solute–solvent interaction is also valid for reverse-phase chromatography, provided the interacting moieties are unambiguously identified and their concentrations accurately known.

An explicit equation relating solute retention to solvent composition that takes into account changes in stationary phase surface as well as interactions in the mobile phase remains to be developed. Nevertheless, such an equation should be available in the not too distant future. An equation that would allow the calculation of the magnitude of the specific distribution coefficients necessary for the prediction of solute retention, such as K_W, K_{MW}, and K_M, from solute–solvent characteristics still remains to be developed. A much greater understanding of the relationship between the bulk electrical characteristics of solute and solvent and the magnitude of the molecular interaction needs to be attained before a com-

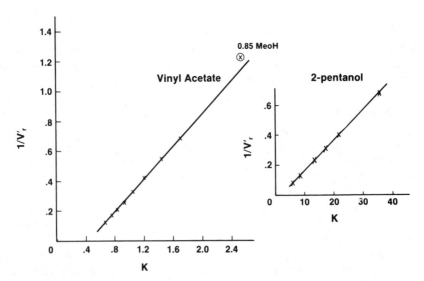

Figure 2.13. Graphs of $1/V_r'$ for two solutes against their distribution coefficients.

plete understanding of solute retention in liquid chromatography will be achieved.

2.6. ALTERNATIVE EMPIRICAL RELATIONSHIPS BETWEEN RETENTION AND SOLVENT COMPOSITION

In the early days of liquid chromatography the dual effect of solvent composition on solute retention was not fully understood. It was thought that solvent composition only affected the retentive capacity of the stationary phase and that interactions in the mobile phase were, at best, insignificant, if not nonexistent. As a consequence, no attempt was made to treat the two effects individually, and a simple expression was sought that would describe solute retention as a function of solvent concentration over the entire range of mobile phase composition. The function, suggested somewhat arbitrarily, was one in which the logarithmic relationship between corrected retention volume or capacity factor and solvent composition was proposed. This relationship took the form

$$\log k' = A + Bc \tag{19}$$

where k' = capacity factor of solute
c = concentration of one mobile phase component
A, B = constants

This relationship, with the exception of the extremes of concentration range, gave reasonable correlation with a significant fraction of the retention data available. It failed, however, in many instances, particularly with reverse-phase systems and in certain ternary systems involving water as one of the solvents. As we have seen, this is not surprising, but even if solvent composition did not have a dual effect on solute retention as described, the logarithmic relationship would still not be theoretically reasonable. A logarithmic relationship between capacity factor and solvent composition presumes that the excess free energy of the solute between the two phases is linearly related to the solvent composition. There is no reasonable theoretical basis for this. Nevertheless, attempts were made to expand the simple relationship given in Eq. 14 to take into account different types of stationary phase interactions and improve the correlation with experimental data.

Indeed the logarithmic relationship has been employed in a number of computer programs purporting to result in the optimization of solvent composition to provide maximum resolution. Such programs have met with varying degrees of success, but for the want of better alternatives are used fairly extensively to give approximate guidance to optimum solvent composition. Such programs must be used, however, with extreme caution as the recommended optimum solvent composition can be misleading.

The use of a simple arbitrary relationship to describe retention behavior that is not based on a sound physicochemical approach has another disadvantage. There is a temptation to improve the correlation between the function and experimental data by expanding the equation to include further terms on quasi-theoretical reasoning. There are a number of good reasons to wish for an interpolative function to predict retention as a function of solvent composition. In such cases, a simple curve fitting routine could be used. However, it should be understood from the start that while, for example, a polynomial function could be employed, this should be done without attempting to justify each term on physicochemical grounds.

The effect of a given solvent and its concentration on solute retention is complex but is beginning to be understood. The role of concentration is now fairly clear. However, considerably more basic research is required in this field before a complete understanding of the mechanism of retention is achieved. Progress will be gradual, but the prospect in the near term for a practical and useful function, based on sound reasoning, that will predict retention in a precise manner, is good.

REFERENCES

1. A. S. Said, *Theory and Mathematics of Chromatography*, Huthig, Heildberg, Basle, New York, 1981, p. 121.
2. R. P. W. Scott, *Contemporary Liquid Chromatography*, Wiley, New York, 1976, p. 25.
3. E. Katz, K. L. Ogan, and R. P. W. Scott, *J. Chromatogr., 289,* 65 (1984).
4. R. P. W. Scott and P. Kucera, *J. Chromatogr., 125,* 251 (1976).
5. R. J. Laub and J. H. Purnell, *J. Chromatogr., 112,* 71 (1975).
6. R. J. Laub, in J. C. MacDonald, Ed., *Inorganic Chromatographic Analysis*, Wiley, New York, 1985, chap. 2.
7. R. P. W. Scott, *J. Chromatogr., 122,* 35 (1976).
8. R. P. W. Scott and P. Kucera, *J. Chromatogr., 112,* 425 (1975).
9. E. D. Katz, K. L. Ogan, and R. P. W. Scott, *J. Chromatogr., 352,* 67 (1986).
10. R. J. Laub, in P. Kuwana, Ed., *Physical Methods in Modern Chemical Analysis*, Academic, New York, 1983, chap. 4.
11. R. P. W. Scott and P. Kucera, *J. Chromatogr., 149,* 93 (1978).

12. W. K. Robbins and S. C. McElroy, *Liquid Fuel Technol., 2,* 113 (1984).

13. R. J. Hurtubise, A. Hussain, and H. F. Silver, *Anal. Chem., 53,* 1993 (1981).

14. M. McCann, S. Madden, J. H. Purnell, and C. A. Wellington, *J. Chromatogr., 294,* 349 (1984).

15. C. H. Lochmuller, M. A. Humzavi-Abedi, and Chu-Xiang Ou, *J. Chromatogr.,* in press.

16. R. P. W. Scott and C. F. Simpson, *Faraday Symp., 15,* 69 (1980).

17. C. Lochmuller and M. Wilder, *Anal. Chim. Acta, 130,* 31 (1981).

CHAPTER

3

OXIDE STATIONARY PHASES

K. K. UNGER and U. TRÜDINGER

Institut für Anorganische Chemie und Analytische Chemie
Johannes Gutenberg-Universität
Mainz, FRG

3.1. INTRODUCTION

A large number of metals and semimetals form solid hydrous oxides. Silica, alumina, titania, magnesia, ceria, thoria, and urania, for example, are produced in large tonnages and have gained widespread technical application as pigments, fillers, thickeners, desiccants, adsorbents, and as basic materials in ceramic and nuclear industries. Thus it is not surprising that silica and alumina were among the first adsorbents employed in classical column liquid chromatography (1). Unlike the low-cost technical grade oxides, chromatographic materials are high-cost products due to specific manufacturing processes by which the physical structure and the surface chemistry are controlled and tailored. The outstanding chromatographic separation capability of oxides results from a combination of several favorable features: the rigid solid structure, the high surface area to volume ratio, the hydrophilic and polar nature of the surface, and their acid–base properties.

The major breakthrough in the manufacture of defined oxide adsorbents (mainly silica and alumina) occurred between 1960 and 1970, accompanied by extensive research on the surface structure of these materials. The rapid expansion of HPLC to the analysis of biological samples, as well as the high demands on the purity of pharmaceutical products and industrial chemicals, had a considerable impact on the further improvement of the quality of oxides as packings and stationary phases.

This chapter addresses the most relevant aspects of the physical and chemical structures of oxides and discusses their chromatographic behavior in column liquid chromatography.

3.2. THE CHEMISTRY OF OXIDES

The term *hydrous oxide* includes solid substances ranging from hydroxides to oxide-hydroxides to hydrated oxides. The metal atoms in the compounds are linked to oxygen, hydroxide, and water as ligands with coordination numbers between 4 and 8. The oxides form crystalline layerlike and three-dimensional structures or behave X-ray amorphous. There are several routes for processing oxides. The most common method is the hydrolysis of respective metal compounds, for example, a salt or an alkoxide accompanied by polycondensation and polymerization. In this way nuclei are formed which grow to crystallites or colloidal particles. The crystallites precipitate as well as the aggregated colloids. Usually hydrogels are formed as coherent systems of three-dimensionally linked particles. The hydrogels undergo aging; that is, further condensation, polymerization, dissolution, redeposition, aggregation, and recrystallization occurs. Upon aging and drying, the hydrogels shrink and yield xerogels with a dense porous structure. Substitution of the intermicellar water of the hydrogel prior to aging by organic solvents of low surface tension and drying under gentle conditions avoid the collapse of the gel structure and yield highly porous aerogels (2). Finely divided oxides with particles of colloidal size are produced by flame hydrolysis of volatile halo compounds such as silicon tetrachloride and aluminum chloride. These so-called pyrogenic products form powders of low density (3–5). Porous oxides are also processed by thermal decomposition of solid precursors, as for carbonates

$$MCO_3 \rightleftharpoons MO + CO_2 \tag{1}$$

where M denotes a bivalent metal.

The manufacture of porous alumina accomplished by thermal treatment of gibbsite, bayerite, or boehmite may serve as another example (6). It is widely acknowledged that the nature of the parent material, the route of manufacture, the processing and aftertreatment conditions all together have a significant influence on the physical and chemical properties of hydrous oxides. Relevant properties are the particle size and size distribution, the particle morphology, the surface area, the porosity, the pore size distribution, the phase composition, the degree of crystallinity, and hydration. For chromatographic application of oxides the stationary phase composition is of more fundamental importance than the bulk structure.

In oxides the surface atoms are terminated by hydroxyl groups which provide the hydrophilic character and cause the uptake of water. Hydroxyl groups have been identified by physicochemical methods, such as infrared (IR) spectroscopy and solid-state nuclear magnetic resonance cross-polarization magic angle spinning (NMR CP MAS) spectroscopy (7). Most of these methods are bulk methods; they detect both internal and surface hydroxyl groups. In order to assess surface hydroxyl groups separately, combined techniques must be applied, such as infrared spectroscopy with isotopic exchange employing D_2O.

Methods to discriminate between surface hydroxyl groups and physisorbed water have been described. This can be accomplished by IR spectroscopy because the absorption bands attributed to the deformation vibrations of hydroxyl groups and water differ greatly. For silica, isotopic exchange with deuterated trifluoroacetic acid in combination with proton nuclear magnetic resonance spectroscopy has been applied (8).

In principle, the hydroxyl group can be linked to one, two, or even three surface metal atoms as exemplified for γ- and η-alumina (Table 3.1). The differently linked hydroxyl groups of alumina were identified by their absorption bands between 3000 and 4000 cm^{-1}. Contrary to aluminum with the coordination numbers 4 and 6, silica possesses only tetrahedrally coordinated silicon atoms. On silica, three different surface hydroxyl groups were identified by IR and ^{29}Si NMR CP MAS spectroscopy as single, geminal, and vicinal hydroxyl groups (Fig. 3.1). The total concentration of surface hydroxyl groups of silica in the completely hydroxylated state was assessed at 8–9 μmol OH/m^2 or 4.8–5.4 OH/nm^2 (10). Thermal treatment of silica above 273 K leads to a release of physisorbed water and water from the condensation of hydroxyl groups. At higher temperatures, water from the solid bulk phase may be liberated. Hence measurement of the surface hydroxyl group content by the weight loss on ignition at 1400 K may lead to false results, that is, to higher values. It should be emphasized that the dehydration and dehydroxylation pattern is different from oxide to oxide and even shows deviations within a given oxide.

Table 3.1. Types of Hydroxyl Groups at Surface of γ- and η-Alumina and Corresponding Infrared Absorption Bands

Designation	Structure	Wave Number (cm^{-1})
Type Ia	OH | Al	3760–3780
Type Ib	OH | Al	3785–3800
Type IIa	H O Al Al	3730–3735
Type IIb	H O Al Al	3740–3745
Type III	H O Al Al Al	3700–3710

SOURCE: From Ref. (9), reprinted by permission of the publisher.

In the case of silica the vicinal hydroxyl groups condense first, up to about 800 K. Below this temperature the dehydroxylation is reversible when the annealed silica is exposed to water vapor or immersed in water. Dehydroxylation in the range between 400 and 800 K forms reactive strained siloxane groups. Above 800 K the hydroxyl group concentration

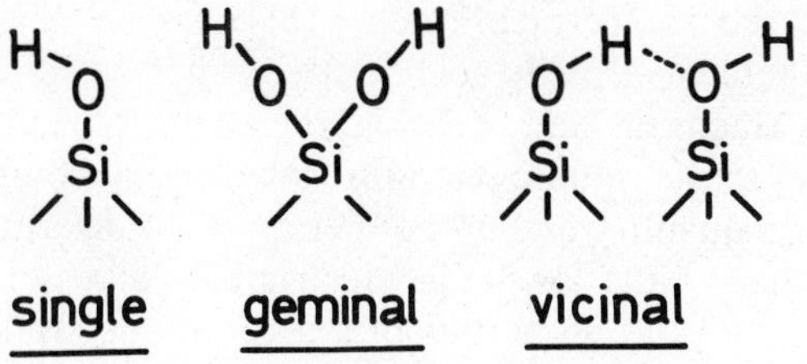

Figure 3.1. Bronsted surface sites at silica.

further declines through condensation of geminal and free hydroxyl groups. In addition, the strained siloxane groups convert into stable ones that render the silica hydrophobic and largely inhibit rehydroxylation. Expressed simply, dehydroxylation of hydroxyl groups on silica can be written as follows:

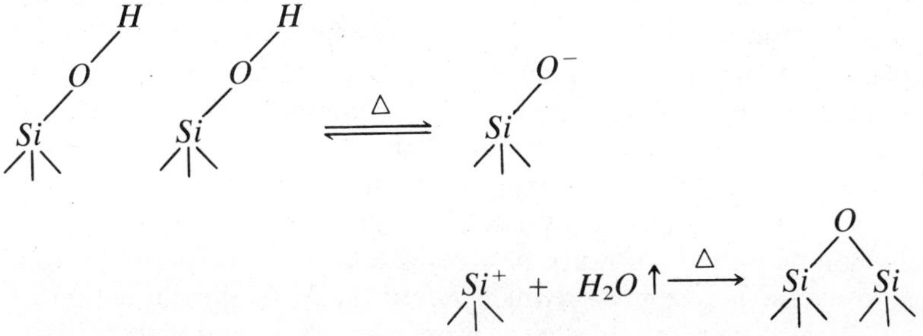

$$(2)$$

Transitional aluminas have Bronsted acid and basic as well as Lewis acid and Lewis basic surface sites. The concentration of Lewis sites increases with the calcination temperature of the transitional alumina at the expense of Bronsted sites. Bronsted and Lewis acid sites are analysed by adsorption of pyridine and by monitoring the absorption bands by infrared spectroscopy (7).

Oxides such as zirconia and titania, provided they are amorphous, have the tendency to crystallize at annealing above 500 K. Concurrently the pore structure collapses, that is, surface area and porosity decrease drastically. In conclusion, heat treatment remains a sensitive method to alter the surface chemistry and pore structure of hydrous oxides.

A common feature of oxides, with a noticeable effect on their chromatographic use, is the acidic and basic character of the surface. Subjected to electrolyte solutions, these acidic and basic groups are capable of ion exchange. In order to compare these properties of oxides, the pka is a helpful parameter. The pka values of oxides, derived from the wave number shifts produced by benzene adsorption (12), are listed in Table

3.2 together with the isoelectric pH values (11). It is seen that some oxides, such as anatas, behave strongly acidic, others such as silica and alumina, weakly acidic. Magnesia and zirconia represent typical basic oxides. When discussing these values, one must take into account that the data reflect an average quantity and do not permit us to discriminate fine gradations. The same is valid for the isoelectric pH values of oxides, which scatter depending on the properties of a given oxide. As a general rule, oxide particles exhibit zero charge at the isoelectric pH. Above the isoelectric pH they possess cation, below anion exchange properties (13). The ion exchange capacity is not only affected by the pH but also determined by the valence of charged ions and the charge distribution.

In addition to their ion exchange properties, oxides e.g. such as zirconia, ceria, and thoria that contain highly coordinated metal atoms, exhibit a pronounced complexation ability toward chelating ligands. These properties can be utilized for selective binding of analytes. Apart from ion exchange and complexation, the hydroxyl groups of oxides are known to undergo a variety of reactions, such as esterification with alcohols or condensation reactions (7), which often cause irreversible adsorption when analyzing polar and active solutes by column liquid chromatography. The solubility of oxides in neat aqueous media causes a severe stability problem when chromatographic separations are carried out in hydroorganic or buffered eluents. While alumina, zirconia, and titania are sparingly soluble in the pH range between 2 and 10, silica dissolves to a noticeable extent above pH 9, which is attributed to the formation of soluble silicates. The manifold surface properties of oxides and their variations make the oxides into versatile stationary phases.

Table 3.2. Isoelectric pH and pka Values of Hydrous
Oxides

Oxide	Isoelectric pH*	pka†
SiO_2	2.2	7.0
γ-Al_2O_3	7.5	8.5
TiO_2	4.7	—
Anatas	—	0.5–2.0
Rutile	—	0.5
ZrO_2	6.7	10.5
MgO	—	18.5

* From Ref. (11).
† From Ref. (12).

3.3. INDIVIDUAL OXIDES AS ADSORBENTS

3.3.1. Porous Silica

Silica occupies a leading position among oxide-type packings. This development came about because of two significant features. First, porous silica can be reproducibly made with a controlled particle shape and size and with a graduated specific surface area, pore size, and porosity. Second, the extensive knowledge accumulated in silica surface chemistry and organosilicon chemistry has greatly facilitated the rapid development of bonded silica phases. There is now broad agreement that the quality of bonded phases in terms of batch-to-batch and column-to-column reproducibility, defined properties, and enhanced stability is to a very large extent determined by the parent silica and its silanization.

The term *porous silica* calls for a more precise definition. Synthetic amorphous silica is divided into three types according to the manufacturing process (4). Pyrogenic silica, made by flame hydrolysis, is a low-density powder composed of loosely aggregated colloidal silica particles and thus is mainly employed as a filler. Precipitated silica is manufactured from sodium silicate and sulfuric acid under specific conditions. It is a porous powder with a density of about 0.2 g/mL and is used as thickener and filler. Silica xerogels are made by mixing sodium silicate and sulfuric acid, maintaining an acidic or neutral pH, and yielding a jelly named *silica hydrogel*. Washing and drying of the hydrogel gives lumps of hard porous xerogel particles with a density of 0.5 g/mL. Of the above three, the xerogel type silica is employed for chromatographic purposes. The desired particle size is achieved by consecutive milling and sizing. In the beginning of HPLC, pellicular packings, which have a thin porous layer on an impervious glass core, were preferred to totally porous particles (14), but they are now displaced completely by the latter. As a speciality, monodisperse nonporous 1.5-μm silica particles were recently introduced as parent packings for bonded phases for protein and peptide separation by HPLC (15).

The porous silica particles made by the conventional gel process are usually angular in shape. Spherical particles require specific beading processes. The most common are:

1. Small silica sol droplets are generated, which are solidified to hydrogel beads in an immiscible two-phase system.
2. A water-soluble organic monomer is added to a silica sol, forming a liquid complex. The resulting coacervate is subjected to poly-

merization, yielding beads. The polymer must be burnt out to obtain porous particles.

3. A silica sol is spray-dried through a nozzle at high temperatures, whereby the dispersing liquid evaporates and spherical xerogel particles are formed.

The process conditions allow an adjustment not only of the particle size, but also simultaneously of the surface area, pore size, and porosity. In order to manipulate the physical and chemical properties further, the particles may be subjected to various thermal and chemical treatments. The different starting materials, ranging from water glass to sols, the manifold steps of the process, and the conditions are responsible for the variety of silica products. Table 3.3 lists the trade names and the suppliers of commercial HPLC silica packings. Silica and alumina adsorbents of particle diameters $d_p > 20$ μm used for preparative and upscale chromatography were recently collated by Unger and Janzen (16).

Clearly the tendency is to spherical rather than to angular packings in HPLC. Beads of uniform size and shape permit generating a more homogeneous, dense, and stable column compared to angular packings. The superior properties, however, have to be paid for with a 5–10 times higher price. The particle size encompasses a range from 3 to 100 μm of mean particle diameter. Analytical columns are packed with particles of $3 < d_p < 10$ μm by applying the slurry technique, while larger particles are common for preparative columns. Particles of $d_p < 3$ μm are produced in small quantities mainly for research pruposes. The sizing as well as the size analysis of these small particles open up a number of problems. Furthermore, due to the inverse proportionality of the column pressure drop and the square of the mean particle diameter of the packing, columns with 1 to 2 μm particles generate high back pressures so that rather short columns must be used. Further, frits have to be placed at the column outlet with < 0.2-μm porosity.

Size analysis of 3 to 100 μm particles is performed by the Coulter counter, the laser photosedimentometer, and other techniques (17). Although theory and practice of particle size analysis, including computation, have achieved a high standard, the majority of chromatographic packings remain poorly characterized; for example, only a size range or an undefined mean value of d_p is given. The lack of precise size distribution data together with the method employed for size analysis make a reliable comparison of the performance of HPLC columns almost impossible. The size distribution of packings for analytical columns should be narrow, that is, the ratio of d_{p90} to d_{p10} value of the cumulative size distribution should be 1.5–2.0. The d_{p10} (d_{p90}) value corresponds to the

Table 3.3. Survey of Commercial Silicas by Manufacturer

Trade Name	Form†	d_p (μm)	a_s (m^2/g)	p_d (nm)
Alltech Associates, Inc.				
Applied Science Labs.				
2051 Wankegan Road				
Deerfield, IL 60015, USA				
Adsorbosphere silica	b	3, 5, 10	220	10
Econosphere silica	b	3, 5	220	10
Versapack silica	b	10	300	15
Amicon Europe				
Grace Industrial Chemicals, Inc.				
Av. de Montchoisi 35				
CH-1001 Lausanne, Switzerland				
Matrex	*, b	5, 10, 15, 20, 30	540	6
Silica		50, 105	320	10
Media			320	25
XWP 500 Å	*, b	30–100	60	50
XWP 1000 Å	*, b	30–100	40	100
XWP 1500 Å	*, b	30–100	28	150
J. T. Baker Chemical Co.				
222 Red School Lane				
Phillipsburgh, NJ 08865, USA				
Silica gel	o, c	3, 5	170	12
Silica gel	*, b	10	300	15
Silica gel	*, b	40	480	6
Flash silica gel	*, b	63–200, 100–425	—	6
Beckman Instruments, Inc.				
2500 Harbor Blvd.				
Fullerton, CA 92634, USA				
Ultrasil-Si	*, b	10	—	—
Ultrasphere-Si	o	5	180	8
Bio-Rad				
Chemical Division				
1414 Harbour Way South				
Richmond, CA 94804, USA				
Bio-Sil HA	*, b	<100	—	—
Bio-Sil A	*, b	20–44, 80–150	—	—
Chrompack				
P.O. Box 3				
4330 AA, Middleburg, Netherlands				

Table 3.3. (*continued*)

Trade Name	Form†	d_p (μm)	a_s (m^2/g)	p_d (nm)
ChromSpher Si	o, b, c	5	160	12
CP-Spher Si	o, b, c	8	—	8
	Du Pont de Nemours Corp.			
	Bioresearch Systems			
	Concord Plaza			
	Wilmington, DE 19898, USA			
Zorbax SIL	o, c	3, 5–6	350	7–8
	ICN Biochemicals GmbH			
	Postfach 369			
	D-3400 Eschwege, FRG			
ICN Silica	*, b	3–6, 7–12, 10–18, 18– 32, 32–63, 0–63, 63– 100, 63–200, 100–200, 200–600	550	6
	Kali Chemie AG			
	Hans-Böckler-Allee 20			
	D-3000 Hannover 1, FRG			
KC-Mikroperl M	o, b	30–60	210	10
			130	18
			76	27
			65	34
			55	43
			47	54
			38	68
KC-Mikroperl L	o, b	100–200	—	—
	Macherey-Nagel & Co.			
	Postfach 307			
	D-5160 Düren, FRG			
Nucleosil 50	o, b	5, 7, 10	450	5
Nucleosil 100	o, b	5, 7, 10, 15– 25, 25–40, 40–63	350	10
Nucleosil 120	o, b	3, 5, 7, 10	200	12
Nucleosil 300	o, b	5, 7, 10, 15– 25, 25–40, 40–63	100	30

Table 3.3. (*continued*)				
Trade Name	Form†	d_p (μm)	a_s (m²/g)	p_d (nm)
Nucleosil 500	o, b	5, 7, 10, 15–25, 25–40, 40–63	35	50
Nucleosil 1000	o, b	5, 7, 10, 15–25, 25–40, 40–63	25	100
Nucleosil 4000	o, b	5, 7, 10, 15–25, 25–40	10	400
Polygosil 60	*, b	5, 7, 10, 15–25, 25–40, 40–63, 63–100	450	6
Polygosil 100	*, b	5, 7, 10, 25–40, 40–63	300	10
Polygosil 300	*, b	7, 15–25, 25–40, 40–63	100	30
Polygosil 500	*, b	7, 15–25, 25–40, 40–63	35	50
E. Merck Frankfurter Landstr. 250 D-6100 Darmstadt 1, FRG				
LiChrosorb Si 60	*, c	5, 7, 10	500	6
LiChrosorb Si 100	*, c	5, 7, 10	300	10
LiChrospher Si 100	o, c	5, 10, 20	250	10
LiChrospher Si 300	o, c	10	250	25
LiChrospher Si 500	o, c	10	60	50
LiChrospher Si 1000	o, c	10	30	100
LiChrospher Si 4000	o, c	10	10	400
Fractosil 200	*, b	40–63, 63–125	150	18
Fractosil 500	*, b	40–63, 63–125	50	50
Fractosil 1000	*, b	40–63, 63–125	20	100
Fractosil 2500	*, b	63–125	8	250
Fractosil 5000	*, b	63–125	3	500
Fractosil 10000	*, b	63–125	1.5	1000
Fractosil 25000	*, b	63–125	0.6	25000
Superspher Si 60	o, c	4	650	6
Superspher Si 100	o, c	4	350	10
Kieselgel 40	*, b	63–200, 200–500	>500	4
Kieselgel 60	*, b	<63	500	6
		63–200	500	6
		200–500	500	6

155

Table 3.3. (*continued*)

Trade Name	Form†	d_p (μm)	a_s (m²/g)	p_d (nm)
Kieselgel 100	*, b	63–200	250	10
		200–500	250	10
Kieselgel 60	*, b	15–40, 40–63, 63–100	500	6
Kieselgel 60 reinst	*, b	63–200	500	6
LiChroprep Si 40	*, b	15–25, 40–63	—	—
LiChroprep Si 60	*, b, c	15–25, 40–63	—	—
LiChroprep Si 100	*, b	15–25, 40–63	—	—

Millipore
Waters Chromatography Division
34 Maple Street
Milford, MA 01757, USA

Trade Name	Form†	d_p (μm)	a_s (m²/g)	p_d (nm)
μ Porasil	*, c	10	300	6
Resolve silica	o, c	5, 10	—	9
Porasil A	o, b	37–75, 75–125	300–500	—
Porasil B	o, b	37–75, 75–125	140–230	—
Prep-PAK 500 silica		55–105	—	—

Orpegen GmbH
Czerny Ring 22
D-6900 Heidelberg, FRG

Trade Name	Form†	d_p (μm)	a_s (m²/g)	p_d (nm)
HD-Sil-60	*, b	10, 20, 30, 60	—	6
HD-Sil-100	*, b	10, 20, 30, 60	—	10
HD-Sil-SS-80	o, b	5	—	8
HD-Sil-SS-100	o, b	5	—	10
HD-Gel-WP	o, b	10, 30, 60	—	25
HD-Gel-300	o, b	7	—	30

Phase Separations Ltd.
Deeside Industrial State
Queensferry, Clwyd CH5, 2LR, UK

Trade Name	Form†	d_p (μm)	a_s (m²/g)	p_d (nm)
Spherisorb SW	o, b	3, 5, 10	220	8
Spherisorb SX	o, b	5, 10	190	30
Spherisorb VLS	o, b	20	190	30

Reichelt Chemietechnik GmbH & Co.
Englerstr. 18
D-6900 Heidelberg, FRG

Trade Name	Form†	d_p (μm)	a_s (m²/g)	p_d (nm)
Thomasorb Si 60	x, b	5, 7.5, 10	500	6
Thomaspher Si 50	o, b	5, 7.5, 10	500	5
Thomaspher Si 500	o, b	5, 7.5, 10	300	10

Table 3.3. (*continued*)

Trade Name	Form†	d_p (μm)	a_s (m²/g)	p_d (nm)
	Separations Group			
	17434 Mojave Street			
	P.O. Box 867			
	Hesperia, CA 92345, USA			
Vydac TP silica	*, b, c	5, 10, 20–30	80	30
Vydac HS silica	o, b, c	5, 10, 20–30	500	8
	Serva Feinbiochemica GmbH & Co.			
	Postfach 105260			
	D-6900 Heidelberg, FRG			
Daltosil 75	o, b	40–10, 100–200	—	7.5
Daltosil 100	o, b	4	—	10
Daltosil 150	o, b	40–100, 100–150, 150–200, 100–200	—	15
Daltosil 300	o, b	40–100, 100–150, 150–200, 100–200	—	30
Daltosil 500	o, b	40–100, 100–200	—	50
Daltosil 1200	o, b	40–100, 100–200	—	120
Daltosil 3000	o, b	40–100, 100–200	—	300
Si 60	*, b	3, 5, 10, 15, 20–40, 50–100, 125–150	—	6
Si 100	*, b	3, 5, 10	—	10
Si 300	*, b	3, 5, 10, 20–40	—	30
	Shandon Southern Products Ltd.			
	Chadwick Rd			
	Astmoor, Runcorn, Cheshire WA 7 1PR, UK			
Hypersil	o, b, c	3, 5, 10	170	12
Hypersil WP-300	o, b, c	5, 10	60	30
	Showa Denko K.K.			
	13-9 Shiba Daimon i Chome			
	Minato-Ku, Tokyo 105, Japan			
Silicapak E-411	o, c	5	—	—

157

Table 3.3. (*continued*)

Trade Name	Form†	d_p (µm)	a_s (m²/g)	p_d (nm)
Supelco, Inc.				
Supelco Park				
Bellefonte, PA 16823-0048, USA				
LC-Si	o, c	3, 5	170	10
LC-3-Si	o, c	5	—	30
LC-5-Si	o, c	5	—	50
PLC-Si	o, c	15	170	10
Tosoh Corporation				
Toso Building, 1-7-7 Akasaka				
Minato-Ku, Tokyo 107, Japan				
TSK-Gel LS-310 SIL	o, b, c	5, 10, 15	—	—
Whatman, Inc.				
Bridewell Place				
Clifton, NJ 07014, USA				
Partisil	*, b, c	5, 10	350	9
LPS-1	*, b	13–24	250	—
LPS-2	*, b	37–53	450	—
YMC, Inc.				
P.O. Box 492				
Mt. Freedom, NJ 07090, USA				
100 A Silica	o, c	3, 5, 15, 25–44	—	—

† Abbreviations: o—spherical; *—irregular; c—column; b—bulk material.

mean particle diameter at 10% (90%) of the cumulative size distribution. In the case of a broad distribution the chromatographer has to pay twice. The fines enhance the pressure drop, and the larger particles reduce the plate number of the column since the plate number is inversely proportional to d_p at constant column length.

According to their pore size, adsorbents and packings have been classified into three ranges (18):

Microporous	$p_d < 2$ nm
Mesoporous	$2 < p_d < 50$ nm
Macroporous	$p_d > 50$ nm

where p_d is the mean pore diameter. Chromatographic silicas have been produced with pore sizes covering the mesoporous and macroporous ranges. For HPLC of low molecular weight compounds, silicas with mean pore diameters between 6 and 15 nm are the packings of choice, corresponding to specific surface areas of 500–200 m^2/g (Table 3.4). The separation of high molecular weight compounds such as synthetic polymers and biopolymers by means of size exclusion and interaction chromatography calls for pore sizes in the range of 15–100 nm. Packings containing micropores of $p_d < 2$ nm give rise to tailed peaks in chromatography, which are attributed to the slow kinetics of solutes and/or column overload effects.

As illustrated in Table 3.4, the mean pore diameter determines the value of the specific surface a_s: the larger the pores, the lower is the specific surface area. a_s is assessed by nitrogen sorption at 77 K, applying the BET method, and represents the total internal surface area accessible to nitrogen as adsorptive. The external surface area of 5 to 10 μm particles accounts for less than 1 m^2/g and hence is negligible. It is essential in column chromatography to express a_s in square meters per milliliter of column volume because the silicas vary in their packing density ρ_p. In other words, a_s in square meters per gram must be multiplied by the packing density ρ_p in grams per milliliter of column. The packing density of silicas is an inverse function of the specific pore volume and ranges from 0.2 to 0.8 g/mL, the latter value resulting for low-porosity silicas (19). Silicas with a pore volume of 1 mL/g have a packing density of 0.5 g/mL. While a large internal surface area or stationary phase is needed in the chromatography of low molecular weight compounds, biopolymers have been observed to be sufficiently retained on bonded silica packings of low specific surface area (on the order of a few square meters per

Table 3.4. Relationship between Average Pore Diameter
p_d and Specific Surface a_s of Porous Silicas

p_d (nm)	a_s (m^2/g)	a_s (m^2/mL of column volume)
6	~500	~250*
10	~250	~125*
30	~100	~50*
50	~50	~25*
100	~20	~10*
400	~5–10	~3–5*
Nonporous, $d_p \sim 1$ μm	~3	~5

* Assuming a packing density of 0.5 g/mL.

milliliter of column volume) (20). Retention in this case responds to the multisite attachment of the large molecules at the stationary surface.

Analogous to the particle size distribution, the pore size distribution of porous silica assessed by nitrogen sorption or mercury porosimetry measurements also shows a spread. Usually computed pore size distributions cover from one to three decades of the mean pore diameter, depending on the origin of the silica (19). The pore size distribution of silicas with a uniform pore size spans less than one decade of pore diameter. Except in the size exclusion mode, the width of the pore size distribution is known to have a minor effect on chromatographic properties. It should be emphasized that the computed pore size and its distribution are based on simplified pore shape models. In reality, the pore structure of silicas is much more complicated, since it represents a system of interconnected voids and constrictions of various shapes.

Silicas are manufactured with graduated specific pore volumes (v_p) of 0.3–2.0 mL/g. The latter value corresponds to about 80% of the particle porosity. Silicas with larger particle porosities have been synthesized, but these become highly fragile when high pressures are applied to pack the materials into analytical HPLC columns. Typically, chromatographic silicas posses v_p values of 0.5–1.2 mL/g when v_p is multiplied by ρ_p, we obtain the internal column volume. The sum of the internal and the interstitial column volumes gives the total column porosity, which is about 0.8 for most of the silica columns. The interstitial column porosity has been reported to vary from 0.35 to 0.5 for analytical columns (21). For spherical particles of uniform shape and size the interstitial porosity is calculated to be 0.26, which concurrently provides the most stable packing due to the high contact number of particles, namely, 12. The relatively high interstitial porosity reported for HPLC columns clearly indicates how far the packing density is from the optimum.

In common with other hydrous oxides, the surface activity of silica is the fundamentally important parameter controlling retention and selectivity in column chromatography. The surface activity is equal to the specific surface area that the oxide provides for interaction. In addition, the surface activity of a given oxide can be varied over a wide range through preadsorption or equilibration of polar compounds such as water. Silica shows its highest activity in the absence of physically adsorbed water, whereby the solute molecules interact directly with the surface hydroxyl groups. This situation is hard to achieve and maintain in chromatographic practice because waterfree eluents must be employed and humidity totally excluded. When exposed to humidity, a dried silica adsorbs water, thereby deactivating the surface and reducing the surface activity. Adsorption of water vapor through the gas phase yields the typical isotherms illustrated in Fig. 3.2 for three commercial silicas. Ad-

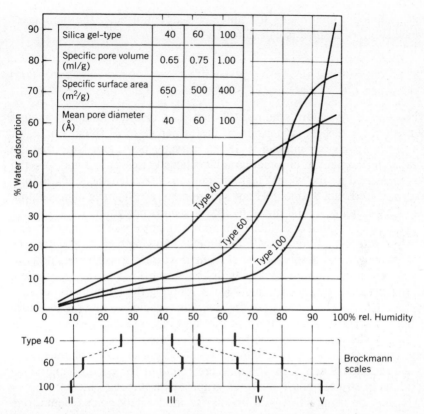

Figure 3.2. Water vapor adsorption isotherms of three commercial silica packings. [From Ref. (22), reprinted by permission of the publisher.]

sorption occurs by multilayer adsorption followed by capillary conden-sation and ultimately filling the whole pore volume. The control of water adsorption of silica through the gas phase is applied in thin-layer chro-matography (TLC) (22). Historically, the first attempt to standardize the surface activity of oxides in chromatography was made by Brockmann and Schodder (23, 24), advocating the so-called Brockmann activity scale. This scale is based on the elution pattern of selected dyes on the respective oxide. Silica spans four of the five Brockmann scales established for alumina.

In column liquid chromatography employing silica and apolar eluents the control of surface activity is equally relevant but somewhat differently approached than in TLC.

In liquid adsorption the uptake of water or polar organic compounds from an apolar solvent is confined to a monolayer. Assuming a molecular cross-sectional area $a_m = 0.2$ nm^2 per water molecule, the monolayer capacity x_m of a silica is calculated by

$$x_m = \frac{a_s}{a_m N_L} \qquad (3)$$

where N_L is Avogadro's number. When a_s is inserted in square meters per gram, x_m has the dimension of moles per gram of silica. For $a_s = 300$ m^2/g, x_m gives 1 mmol/g = 18 mg of water per gram. Note that the total water content of chromatographic silicas varies from 3 and 15% (w/w), measured as the weight loss on ignition. Provided the silica is fully hydroxylated with 8 μmol OH per square meter, a 300-m^2/g material contains 2.4 mmol/g = 40.8 mg of hydroxyl groups per gram.

The water content of a silica column in straight-phase chromatography is adjusted by conditioning with the eluent, applying blends of water-free and water-saturated solvents or solvent mixtures. Another means is to use a saturator column, packed with coarse particles of the same silica as suggested by Engelhardt and Elgass applying the controlled-moisture system (CMS) (25). In this case the eluent is recycled so that the moisture is held at a constant level in the entire chromatographic system. Equilibration of the silica column with water-modulated solvents is a slow process compared to the conditioning of reversed-phase silica columns. Several hundred column void volumes are required for conditioning silica columns in straight-phase chromatography with unpolar eluents.

Hence, on using silica with unpolar to moderately polar eluents, one is faced with the problem of control and adjusting the surface activity and—when working in aqueous or hydroorganic eluents—of considering the ion exchange properties. Although the surface acidity of pure silica is assessed at a pKa of 7.0 ± 0.5, aqueous suspensions of chromatographic grade silica often deviate markedly from this value, showing a pH between 3 and 9 (26, 27). Clearly this deviation is caused by basic or acidic impurities of the silica surface originating from the manufacturing process and does not reflect differently acidic surface sites of the silica. Hansen et al. (27) convincingly demonstrated this finding by titrating silica suspensions with 0.01N HCl and 0.01N NaOH (Table 3.5). In order to overcome this problem the surface pH of silicas was adjusted either by treatment of the bulk material by an appropriate buffer (28–30) or in situ by purging the column with the buffer or aqueous solution (27). Such standardized silica columns were found to be highly suitable for the separation of polar solutes and served as supports for liquid–liquid partitioning systems.

Ternary solvent systems with partial miscibility, such as water,

Table 3.5. pH of 1% Suspension of Silicas

Trade Name	Batch				True pH
	A^*	B	$C^\dagger$	D	
LiChrosorb Si 40		6.1			
LiChrosorb Si 60	7.8	7.4	7.4	7.4	5.7
LiChrosorb Si 100	7.0	7.5	6.8	7.4	
LiChrospher Si 60		5.1			
LiChrospher Si 100	5.3	5.8	5.7		
LiChrospher Si 300		5.9			
LiChrospher Si 500	9.9	8.3			
LiChrospher Si 1000	9.2	8.6			
Hypersil	9.0	8.5	8.4		6.3
Nucleosil 50		8.1	6.7		
Nucleosil 100	5.7	6.0	6.0		5.9
Nucleosil 100V		8.1			
Polygosil 60	8.0	7.1			
Partisil	7.5	5.6	5.3	6.3	5.9
Sperisorb S5 W	9.5	9.0	9.5		6.2
Spherosil XOA 600		3.8			
Spherosil XOA 800		5.7			
Zorbac SIL	3.9	4.1	5.6		
Sepralyte SI		8.2			
Vydac 90		6.1			
Serva Si 600		7.1			5.9
Serva Si 100		7.1			
Serva Si 200		6.7			
Chromosorb LC 6		7.1			

Source: From Ref. (27), reprinted by permission of the publisher.

* Batch series A are values measured by Engelhardt and Müller (26).

† Column materials used for further investigations and for titration curve experiments.

ethanol, 2,2,4-trimethylpentane or acetonitrile, ethanol, and 2,2,4-trime-thylpentane in combination with silica or reversed-phase silicas, were chosen to carry out separations according either to an adsorption or to a liquid–liquid partitioning mechanism (Fig. 3.3). On adjusting a solvent composition within the regime of partial miscibility, the mixture splits into two phases. One is held stationary by the support, the other serves as a mobile phase. Under these circumstances, partitioning of the solutes between the two phases occurs. When moving beyond the equilibrium line into the regime of total miscibility, a phase system is created which resolves the solutes according to an adsorption mechanism (32, 33).

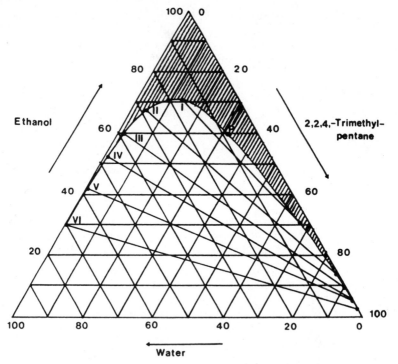

Figure 3.3. Triangular phase diagram of ternary solvent system composed of 2,2,4-trimethylpentane, ethanol, and water. [From Ref. (31), reprinted by permission of the publisher.]

In order to circumvent the problems encountered with the elution of polar ionic solutes, such as strong acids, buffers were added to the eluent that shift the pH to a range where the ionization of the solutes is suppressed (34, 35). Vice versa, the ionization of surface hydroxyl groups of the silica is drastically reduced by adding small amounts of acids, such as acetic acid, to the eluent.

A great deal of work has been directed at the elimination of the residual hydroxyl group activity of reversed-phase silicas, particularly when basic compounds are chromatographed. There are several alternatives to reduce this undesired phenomenon: the parent silica was aftertreated by acids before modification (36); amines, such as triethylamine, were added as modifiers in low concentrations to the hydroorganic eluent (37) to block the acidic surface centers; a polymer coating was applied to completely cover the surface (38). The magnitude of the residual polarity of reversed-phase silica was monitored by chromatographing basic test solutes and measuring the change of the solute capacity factor and the peak shape (37–39).

3.3.2. Porous Alumina

Alumina is manufactured by refinement of natural bauxite through the Bayer process and is mainly used for the production of aluminum metal. Minor quantities are produced for adsorbents, catalysts, ceramics, refractories, and for the synthesis of aluminum compounds (40). The term hydrous alumina stands for a variety of crystalline and amorphous products of the composition $Al_2O_3 \cdot n\text{-}H_2O$, where n varies between 3 and 0 (Table 3.6). Nonporous crystalline trihydroxides exist as hydrargillite or gibbsite, bayerite, and nordstrandite. Amorphous alumina gels are produced from aluminum salts, alkaline aluminates, and aluminum alkoxides. $AlO(OH)$, an aluminum oxide hydroxide, exists in two crystalline forms

Table 3.6. Decomposition Sequence of Alumina Hydroxides

| | Conditions Favoring Transformations | |
	Path a	Path b
Pressure	>1 atm	1 atm
Atmosphere	Moist air	Dry air
Heating rate	>1°C/min	<1°C/min
Particle size	>100 μm	<10 μm

Source: From Ref. (40), reprinted by permission of the publisher.

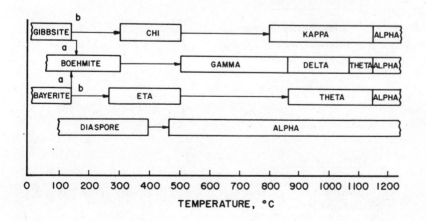

as boehmite and diaspor. Porous or active aluminas comprise the family of so-called transition aluminas known by the greek letters γ, δ, η, χ, κ, and ι. These oxides are formed by controlled thermal decomposition of the water-rich precursors according to the routes indicated in Table 3.6. The major component of the transition aluminas is γ-alumina, which has a lattice structure close to the spinal type $MgAl_2O_4$ with tetragonal deformation. γ-Alumina is made by thermal decomposition of gibbsite, boehmite, or pseudoboehmite. The two latter develop specific surface areas of up to 500 m^2/g with a pronounced micropore structure. While boehmites and pseudoboehmites made from bauxite contain sodium as impurity, those manufactured via hydrolysis, condensation, and dehydration of aluminum alkoxides are free of inorganic impurities. Boehmites or pseudoboehmites convert into γ-alumina by heat treatment of 750 and 1000 K. Above 1000 K the crystallization of κ, δ, and θ-alumina is favored. The end product of calcination at 1000 K is crystalline nonporous α-alumina called corundum.

Along with dehydroxylation the products develop slit-shaped pores between the microcrystallites. γ-Aluminas are mesoporous materials with some microporosity. The a_s values of about 200 m^2/g are highest at low calcination temperatures and drop to about 50 m^2/g at 1000 K. Similarly, the specific pore volume decreases from 0.6 to 0.2 mL/g. Attempts have been made to synthesize large pore size aluminas with a higher porosity.

The surface structure of aluminas, in particular of transition aluminas, has been the subject of thorough investigations. A model of the alumina surface was proposed by Peri (41), which was later refined by Közinger and Ratnasamy (9). On account of the high calcination temperature the surface bears Bronsted as well as Lewis sites. Five distinct acid–base

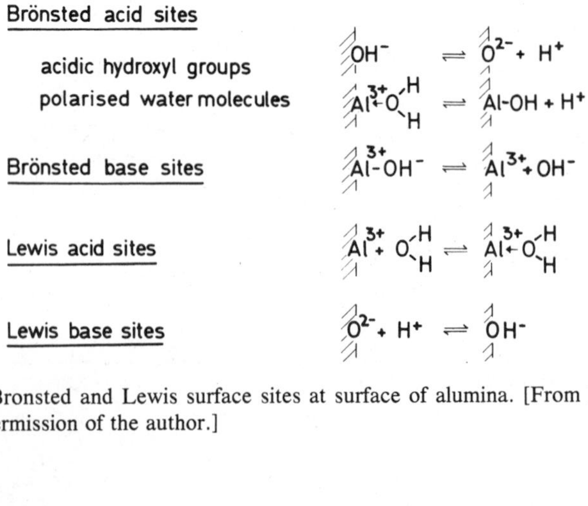

Figure 3.4. Bronsted and Lewis surface sites at surface of alumina. [From Ref. (42), reprinted by permission of the author.]

sites were observed, as shown in Fig. 3.4 (42). These were identified by IR spectroscopy and by adsorption measurements employing acidic and basic test compounds. For instance, the total surface hydroxyl group concentration of an γ-alumina activated at 470 K was 10 μmol/m^2; the amount of Lewis acid sites was assessed to 0.6 μmol/m^2 by means of pyridin adsorption.

Chromatographic grade aluminas are fewer in number than silicas and are listed in Table 3.7.

They are manufactured by the thermal decomposition route as dis-

Table 3.7. Survey of Commercial Aluminas by Manufacturer

Trade Name	Form†	d_p (μm)	a_s (m^2/g)	p_d (nm)	pH
J. T. Baker Chemical Co. 222 Red School Lane Phillipsburgh, NJ 08865, USA					
Aluminum oxide					
Acid washed	*, b	50–200	—	6	4.5
Neutral	*, b	50–200	—	6	7
Acid, activity 1	*, b	50–200	—	6	4.5
Basic, activity 1	*, b	50–200	—	6	10.0
Neutral, activity 1	*, b	50–200	—	6	7.5
Bio-Rad Chemical Division 1414 Harbour Way South Richmond, CA 94804, USA					
Neutral alumina AG 7	*, b	—	—	—	6.9–7.1
Basic alumina AG 10	*, b	—	—	—	10.0–10.5
Acid alumina	*, b	—	—	—	3.5–4.5
ICN Biochemicals GmbH Postfach 369 D-3400 Eschwege, FRG					
ICN alumina N	*, b	18–32, 32–63	200	6	7.5
ICN alumina A	*, b	18–32	200	6	4.5
ICN alumina B	*, b	18–32	200	6	10
Macherey-Nagel & Co. Postfach 307 D-5160 Düren, FRG					
Alox 60—5, 10	*, b	5, 10	155	6	Basic alumina 9.5

Table 3.7. (*continued*)

Trade Name	Form†	d_p (μm)	a_s (m²/g)	p_d (nm)	pH
E. Merck					
Frankfurter Landstr. 250					
D-6100 Darmstadt 1, FRG					
Aluminiumoxid 60, aktiv basisch	*, b	63–200	—	6	9.0 ± 0.5
Aluminiumoxid 90, aktiv basisch	*, b	63–200	—	9	9.0 ± 0.5
Aluminiumoxid 90, aktiv neutral	*, b	63–200	—	9	7.3 ± 0.5
Aluminiumoxid 90, aktiv sauer	*, b	63–200	—	9	4.0 ± 0.5
Aluminiumoxid 150, basisch	*, b	63–200	—	15	9.0 ± 0.5
LiChroprep Alox T	*, b	25–40	—	—	Basic alumina
LiChrosorb Alox T	*, b	5, 10	—	15	Basic alumina
Phase Separations Ltd.					
Deeside Industrial State					
Queensferry, Clwyd CH5 2LR, UK					
Spherisorb AY	o	5, 10	90	13	—

† Abbreviations: o—spherical; *—irregular; c—column; b—bulk material.

cussed or by spray-drying of alumina sols, whereby the size of the colloidal alumina particles determines the pore size. The first products were already produced around 1930 (1) and then standardized for chromatographic use by Brockmann (23, 24). Five activity grades were distinguished according to the amount of adsorbed water. Alumina made from hydrargillite contains sodium as inorganic impurity, which gives rise to a pH of 9 of the suspension in water. The apparent surface pH can be shifted to the neutral and acidic range by treatment with acids. In this way basic, neutral, and acidic aluminas are obtained. The isoelectric pH value of alumina is about 7.5. Above pH 8 alumina behaves as a cation-, below as an anion-exchanger (13). γ-Alumina is pH stable over a pH range of 2–12, which allows to perform separations at high pH values (43, 44).

3.3.3. Porous Titania, Zirconia, and Ceria

Compared to silica and alumina, very few applications are reported on the use of titania, zirconia, and ceria in column liquid chromatography.

Spherical oxide particles are obtained by the following procedures:

1. Controlled hydrolysis of the corresponding metal compounds (45)
2. Spray-drying of colloidal suspension (46)
3. Sol-gel processes in a two-phase system (47–50)
4. Coating of silica with the metal halides followed by hydrolysis (51).

The formation of the pore structure is largely dependent on the conditions of the processes employed (52). The determining parameters are the type of starting materials, the pH during formation and aging of the gels, the temperature and the kinetics of dehydration, and the kinetics of crystallization. The amorphous gels of zirconia, titania, and ceria share the common feature that crystallization occurs at temperatures of about 550 K (53). Combined with crystallization is a loss in porosity and surface area (54). Amorphous zirconia forming a lamellar structure (55) crystallizes at about 650 K into a cubic phase and forms a monocline crystalline modification at about 1000 K. Amorphous titania converts into anatas (550 K) and starts to form rutil at 850 K.

The amorphous gels are microporous products with little mesoporosity. Typical a_s values exceed 400 m^2/g, the specific pore volume remains low (<0.2 mL/g), and the mean pore diameter d_p is smaller than 3 nm. Several attempts have been made to avoid or eliminate the microporosity. Procedures applied were enlargement of pores by thermal and hydrothermal treatment (56, 57), use of pore fillers with subsequent burning out (58, 59) precipitation at constant pH (45), and spray drying of sols of colloidal particles of appropriate size (46).

Apart from the acidic and basic character of the surface sites, the oxides have a tendency toward complexation, which leads to strong or even irreversible binding of solutes.

3.4. CONTROL OF RETENTION AND SELECTIVITY OF OXIDE STATIONARY PHASE COLUMNS

3.4.1. Retention Models

Retention in liquid adsorption chromatography arises from the reversible interaction of solute molecules with the active sites of the stationary surface. Since the surface is preadsorbed with solvent molecules M, the solute molecules X compete for the adsorption sites during elution. Hence the basic mechanism is a displacement process and the adsorption–desorption equilibrium is expressed by the equation

$$X_n + n\text{-}M_a \rightleftharpoons X_a + n\text{-}M_n \tag{4}$$

where the subscripts n and a denote the nonadsorbed and adsorbed states of X and M. As a first approximation, the retention is proportional to the difference between the free energy of interaction of the solute and the surface, and the solvent and the surface. Equation 4 serves as the basis of the displacement model in liquid adsorption chromatography pioneered by Snyder (60) and Soczewinski (61), which was later refined (62) and today is called the Snyder–Soczewinski model. It is anticipated that the displacement model is generally applicable to polar hydrophilic adsorbents such as silicas, aluminas, and polar nonionic bonded silicas with eluents of nonpolar, moderately polar, and nonamphoteric solvent mixtures (63). The term nonamphoteric denotes solvents capable of self-hydrogen bonding (64). In particular it allows to predict the retention of solutes as a function of the stationary and mobile phase characteristics and to optimize the resolution of solutes of different structure on polar adsorbents (65). According to Snyder (60) the solute capacity factor k' is related to the properties of the phase system and the solute as follows:

$$\log_{10} k' = \log_{10}\left(\frac{V_a W}{V_m}\right) + \alpha'(S^0 - A_s\epsilon^0) + \Delta_{\text{eas}} \qquad (5)$$

where V_a is the volume of the monolayer of adsorbed solvent per unit mass of adsorbent, W the amount of adsorbent in grams, V_m the volume of eluent in milliliters corresponding to the column dead volume, α' an adsorbent activity parameter proportional to the adsorbent surface energy ($\alpha' = 1$ for a standard adsorbent), S^0 the dimensionless free energy of adsorption of the solute on a standard adsorbent of $\alpha' = 1$, A_s the molecular cross-sectional area of the adsorbed solute, and ϵ^0 the solvent strength parameter. Δ_{eas} is a second-order term for corrections in case the system deviates from regular behavior.

Equation 5 states that the logarithmic capacity factor of different solutes at a given adsorbent yields linear plots against the solvent strength ϵ^0 for different solvents, the slopes of the plots being proportional to A_s. The solvent strength parameter ϵ^0 is a dimensionless quantity varying from zero to unity with n-pentane as standard ($\epsilon^0 = 0$). Solvents are grouped into eluotropic series in the sequence of increasing ϵ^0 values. Table 3.8. lists the eluotropic series for silica and alumina (66, 67).

The refined displacement model (62, 65) corrects for specific solute–surface and solute–solvent localization effects such as the following.

1. Restricted access delocalization of the solvent. This effect accounts for the adsorption of solvents of different polarities.
2. Site competition delocalization of the solute and the solvent. This

Table 3.8. Elutropic Series of Solvents for Silica and Alumina

Solvent	Solvent Strength ϵ^0		Selectivity Group
	Silica	Alumina	
3-M fluorochemical FC-78	-0.2	-0.25	-
n-[Hexane	0.01	0.01	-
n-Heptane	0.01	0.01	-
Isooctane	0.01	0.01	-
1-chlorobutane	0.20	0.26	V
Chloroform	0.26	0.40	VIII
Methylene chloride	0.32	0.42	V
Isopropyl ether*	0.34	0.28	I
Ethyl acetate	0.38	0.58	VI
Tetrahydrofuran	0.44	0.57	III
Propyl amine	0.5	—	I
Acetonitrile	0.50	0.65	VI
Methanol	0.7	0.95	II

Source: From Ref. (82), reprinted by permission of the publisher.

* Methyl *t*-butyl ether has been recommended as a preferred alternative to either isopropyl or ethyl in liquid–solid chromatography and other liquid chromatography procedures. Its advantages include much reduced susceptibility to peroxide formation (so much so that antioxidants need not be added during use), a solvent strength ϵ^0 similar to that of ethyl ether (and somewhat greater than for isopropyl ether), and less potential hazard.

refers to cases when localizing and nonlocalizing solvent and solute molecules compete in interaction with the surface sites.

3. Intramolecular delocalization of the solute or solvent. It refers to situations where solvent or solute molecules possess more than one adsorbing functional group.

Solute–solvent interactions can no longer be neglected when hydrogen bonding occurs between the solvent and the solute in the mobile phase and the adsorbed phase. As a result, Eq. 5 must be corrected by adding the term Δ_{eas}.

The effects mentioned greatly contribute to solvent selectivity and provide a fundamental basis to optimize the retention of solutes on oxide stationary phases.

In contrast to Snyder and Soczewinski, Scott and Kucera (68–71) postulated a solvent interaction model for straight-phase chromatography on silica, which was critically examined by Snyder and Poppe (64). An extensive thermodynamic treatment of liquid adsorption processes was described by Jaroniec and coworkers (72–77), with particular emphasis on the surface heterogeneity of adsorbents. The data obtained were correlated to the retention of solutes in liquid adsorption chromatography.

3.4.2.   Control of Retention and Selectivity

The effect of the adsorbent on retention and selectivity is manifested by the quantities V_a and α' of the Snyder equation, Eq. 5. V_a is equal to the average thickness of the adsorbed monolayer of the eluent multiplied by the specific surface area a_s of the adsorbent. At a given eluent composition and at a constant surface activity parameter α' the solute capacity factor increases linearly with a_s, normalized to unit column volume. This dependence was observed for superficially porous silicas with a_s values between 5 and 35 m^2/g and the same packing density (78), and for the LiChrospher products Si 100, Si 500, Si 1000, and Si 4000 at identical mobile phase composition (79). The addition of a polar modifier, such as water, methanol, and acetonitril, which are preferentially adsorbed at the most active surface sites, reduces both V_a and α'. In other words, with increasing amounts of polar modifier in the eluent, the surface becomes more and more deactivated and more homogeneous. The plot of the logarithmic solute capacity factor k' against the relative water content of the eluent gives a straight line with a negative slope. Deactivation of the surface not only decreases the retention but also improves the peak shape due to the loss of highly active and heterogeneous surface sites. Solute retention on polar nonionic bonded silicas, e.g. as of the diol type, was observed to be scarcely affected by the relative water content of the eluent. It is conceivable that through surface modification the most active sites react preferentially, and thus the surface is already deactivated to a large extent. Even a slight increase of the slope of the function $\log_{10} k'$ against the relative water content is observed to reflect the weak reversed-phase character of the diol-bonded silica (80). The corresponding graphs are shown in Fig. 3.5. The role of water and other modulators in column liquid chromatography on silica and alumina has been discussed in depth by Engelhardt and Elgass (25).

In a given phase system the retention of solutes is dominated by the S^0 and A_s values in the Snyder equation. Typical S^0 values are R—CH=CH$_2$ (0.5), C$_6$H$_5$—Cl (1.3), C$_6$H$_6$ (1.5), R—Cl (1.7), R—J (1.9), naphthalene (2.0), C$_6$H$_5$—R (2.8) C$_6$H$_5$—OR (3.3), C$_6$H$_5$—COOR (5.0), C$_6$H$_5$—OH (5.7), C$_6$H$_5$—NH$_2$ (6.6.), and C$_6$H$_5$—COOH (7.6) (60). Distinct retention differences arise with multifunctional solutes. These are associated with the subtle steric orientation of the adsorbed molecules at the stationary surface. Some data are collected in Table 3.9. Even diastereomers were observed to be resolved on straight-phase systems. The

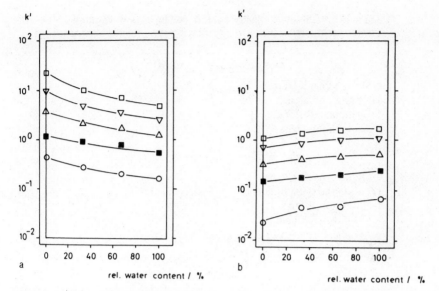

Figure 3.5. Logarithmic solute capacity factor versus relative water content of eluent. Solutes—oligophenylenes; eluent—n-heptane. (a) LiChrosorb Si 100 column. (b) Column with diol modified LiChrosorb Si 100. [From Ref. (80), reprinted by permission of the publisher.]

elution pattern of solutes with complex structures is difficult to predict by equations, since the net interactions are governed by many parameters. Often a retention reversal occurs for solutes when transferring a separation from one brand of silica to another brand of different origin.

The most powerful means for modulating retention consists of changing the mobile phase composition (82). As stated, retention is controlled by the solvent strength parameter ϵ^0. The eluotropic series list single solvents with increasing ϵ^0. Increasing ϵ^0 means a stronger solvent and a lower capicity factor at otherwise constant conditions. The solvent strength parameters ϵ^0 of different individual adsorbents are related to each other. For instance, between silica and alumina and between diol-bonded silica and silica the following relationships were found:

$$\epsilon^0_{\text{silica}} \sim 0.7\epsilon^0_{\text{alumina}} \qquad (60)$$

$$\epsilon^0_{\text{diol silica}} \sim 0.3\epsilon^0_{\text{silica}} \qquad (80)$$

For binary mixtures of A/B, where B is the stronger solvent, ϵ^0 does not vary linearly with the content of B. At low concentrations of B a steep rise in ϵ^0_{AB} is observed, which levels off toward higher B contents. The

Table 3.9. Solute Capacity Factors of Silicas and Aluminas

	k'
Compound Type[a] (Silica)	
2-Methoxynaphthalene	0.6
1-Nitronaphthalene	1.8
1-Cyanonaphthalene	2.7
1-Acetonaphthalene	5.5
Number of Functional Groups[a] (Silica)	
2-Methoxynaphthalene	0.6
1,7-Dimethoxynaphthalene	1.4
1-Nitronaphthalene	1.8
1,5-Dinitronaphthalene	6.1
Acetophenone[b]	1.1
3-Nitroacetophenone[b]	1.6
Isomers (Alumina)	
m-Dibromobenzene[c]	3.8
p-Dibromobenzene[c]	6.9
Quinoline[d]	5.4
Isoquinoline[d]	18.6
1,2,3,4-Dibenzanthracene $(C_{22}H_{14})$[e]	0.6
Picene $(C_{22}H_{14})$[e]	12.0

Source: From Ref. (83), reprinted by permission of the publisher.
[a] 23% v CH_2Cl_2–pentane; [b] 60% v CH_2Cl_2–pentane; [c] pentane; [d] benzene; [e] CH_2Cl_2.

solvent strength of binary, ternary, and quaternary solvent mixtures can be calculated (60, 62, 84).

There are two additional concepts for classifying solvents according to their retention power. One is based on the solubility parameter of Hildebrand δ, another on the solvent polarity index P' (85). P' expresses the relative ability of a solvent to interact as proton acceptor, as proton donator, or as dipole. The single contributions are measured by weighting the factors x_a, x_d, and x_n, where $x_a + x_d + x_n = 1.0$.

Similar to the eluotropic series with ϵ^0 as decisive parameter, solvents are classified by the solvent polarity P', ranging from -2 to 10. The three parameters ϵ^0, δ', and P' are interrelated. The polarity P' of a solvent mixture is the arithmetic average of the P' values of the pure solvents in the mixture, weighted according to the volume fraction of each solvent. Solvents in column liquid chromatography are classified into eight groups based on their proton donor (x_d, proton acceptor x_a, and strong dipole

x_n contributions, as listed in Table 3.10 and arranged in a so-called solvent selectivity triangle (Fig. 3.6).

The general approach in modulating retention is to adjust ϵ^0 or P'. Holding ϵ^0 and P' constant and varying the mobile phase composition leads to a fine tuning of the polarity differences of the eluent and results in a pronounced improvement of selectivity. In detail, a practical optimization strategy to gain high resolution has the following steps (86, 87).

1. Adjustment of the best solvent strength of the eluent, where the capacity factors are in the range of $0.5 \le k' \le 20$. This is achieved by varying the composition of a binary solvent mixture, such as n-hexane and methylene chloride.

2. At constant ϵ^0, further spacing of overlapping bands is achieved by employing ternary or quaternary solvent mixtures, such as n-hexane, methylene chloride, methyl tert-butyl ether, and acetonitrile. If this is not effective, solvent–solute hydrogen bonding in the stationary phase or solvent–solute hydrogen bonding in the mobile phase is generated. For instance, a proton-acceptor solvent such as triethylamine or a proton-donor solvent such as methanol can be added to the eluent.

3. When silica does not provide the desired selectivity, alumina or polar nonionic bonded silica can be used as adsorbent.

4. In some cases the change of the column temperature improves the resolution.

Table 3.10. Classification of Solvent Selectivity

Group	Solvents
I	Aliphatic ethers, tetramethylguanidine, hexamethyl phosphoric acid amide, (trialkyl amines)
II	Aliphatic alcohols
III	Pyridine derivatives, tetrahydrofuran, amides (except formamide), glycol ethers, sulfoxides
IV	Glycols, benzyl alcohol, acetic acid, formamide
V	Methylene chloride, ethylene chloride
VI	a. Tricresyl phosphate, aliphatic ketones and esters, polyethers, dioxane b. Sulfones, nitriles, propylene carbonate
VII	Aromatic hydrocarbons, halo-substituted aromatic hydrocarbons, nitro compounds, aromatic ethers
VIII	Fluoroalkanols, m-cresol, water, (chloroform)

Source: From Ref. (88), reprinted by permission of the publisher.

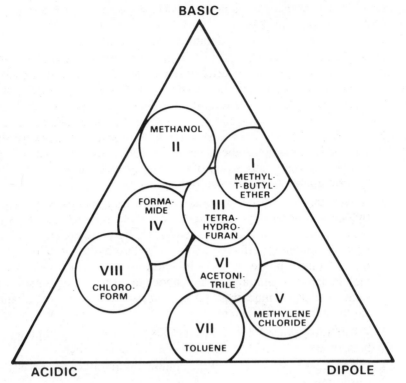

Figure 3.6. Selectivity triangle for solvents. [From Ref. (88), reprinted by permission of the publisher.]

5. Finally, an adjustment of the pH and the addition of a complexing agent can lead to selectivity changes of the unresolved solute pairs.

Although silica adsorbents, together with a variation of the mobile phase composition, bear a high resolution power, alumina offers a specific selectivity toward isomers of conjugated aromatics and acids (basic aluminas). A further use for oxide adsorbents—after the addition of ion-pair reagents—is the separation of polar, nonionic, and ionic solutes by ion-pair chromatography. Applications of this kind were reported on silica, alumina (see Section 3.4.4), ceria, and zirconia (88, 89).

3.4.3. Optimization of Resolution

Chromatographic resolution R_{ji} between a pair of adjacent peaks is related to the parameters r_{ji}, k_i', and N_i,

$$R_{ji} = (r_{ji} - 1) \left(\frac{k_i'}{1 + k_i'} \right) N^{1/2} \tag{6}$$

where r_{ji} is the selectivity coefficient equal to the capacity factor ratio k_j'/k_i'; k_i' is the capacity factor of the less retained and k_j' the capacity factor of the more retained compounds, and N is the average plate number.

It is evident from Eq. 6 that R_{ji} is dominated by the selectivity coefficient r_{ji} and to a lesser extent dependent on N. In general, the selectivity of the phase system is adjusted by the following means:

1. Choice of optimum stationary phase
2. Choice of optimum mobile phase and mobile phase composition utilizing binary, ternary, and quaternary solvent mixtures
3. Coupling of columns that contain different stationary phases
4. Use of columns with blended stationary phases
5. Column switching.

The optimization strategies applied fall into three categories:

1. Empirical approaches by trial and error based on the experience of the chromatographer
2. Statistical design and computation based on selected rules, guidelines, and systems
3. Straightforward application of a profound theory and optimization, including all chromatographic parameters.

Glajch et al. (86, 87) developed a method of data analysis called *overlapping resolution mapping* (ORM), where they combined the Snyder solvent selectivity triangle concept with a mixture-design statistical technique. Three solvents, A, B, and C, were chosen offering different interaction capabilities through proton donor, proton acceptor, and dipole–dipole properties. A fourth solvent, D, was employed that functioned simply to control retention, that is to ensure that the sample mixture is eluted within an adequate period of time. The apices of the triangle then corresponded to mixtures of A/D, B/D, and C/D of equal solvent strength ϵ^0. Ten runs, that is, chromatographic experiments, were then carried out and using ORM the optimum solvent composition was predicted. Figure 3.7 shows an overlapping resolution map for a straight-phase steroid separation on a Zorbax NH_2 column. The corresponding chromatograms are reproduced in Fig. 3.8.

Overlapping Resolution Map for Normal Phase Steriod Separation

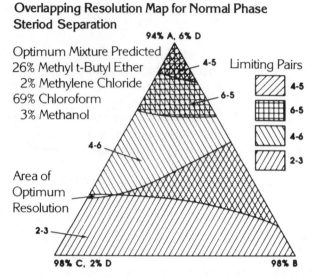

Figure 3.7. Overlapping resolution map for normal phase steroid separation. [From Ref. (81), reprinted by permission of the publisher.]

Another interesting approach to optimizing resolution via the stationary phase composition attempts the use of segmented columns and columns with blended stationary phases (90). The total capacity factor of the solute was shown to be a linear combination of the k' value assessed from the single columns multiplied by the volume fraction of the stationary phase

Table 3.11. Capacity Factors k' of Pesticide Carbamates on Single Columns of LiChrosorb Si 100, LiChrosorb Diol I, and LiChrosorb CN and on a Column Containing 10% (v/v) LiChrosorb Si 100, 66% (v/v) of LiChrosorb Diol I, and 24% (v/v) of LiChrosorb CN at Constant Eluant Composition

	Single Column			Blended Column	
Solute	Si 100	Diol I	CN	Measured	Calculated
CIPC	0.55	0.92	0.62	0.68	0.69
Benomyl	1.25	1.78	0.88	1.15	1.13
Barban	1.70	3.21	1.76	2.13	2.10
Promecarb	1.79	1.80	1.42	1.53	1.55
Baygon	2.48	3.48	2.90	2.94	3.00
Carbaryl	3.91	4.98	3.55	3.80	3.93
Phenmedipham	6.90	17.04	7.40	9.43	9.63
Methomyl	17.46	19.63	44.7	35.4	36.1

Source: From Ref. (90), reprinted by permission of the publisher.

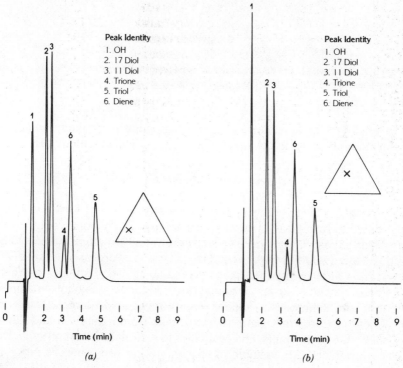

Operating Conditions:
Instrument: Du Pont HPLC
Column: Zorbax NH$_2$, 25 cm × 4.6 mm i.d.
Carrier: 26% A. 2% B. 69% C. 3% D
A. Methyl t-Butyl Ether
B. Methylene Chloride
C. Chloroform
D. Methanol
Flow Rate: 3.0 cm^3 min
Temperature: 35 C
Detector: UV Absorbance (254 nm)

Operating Conditions:
Instrument: Du Pont HPLC
Column: Zorbax ˜ NH$_2$. 25 cm · 4.6 mm i.d.
Mobile Phase: 40% A. 15% B. 42% C. 3% D
A. Methyl t-Butyl Ether
B. Methylene Chloride
C. Chloroform
D. Methanol
Flow Rate: 3.0 cm^3 min
Temperature: 35 C
Detector: UV Absorbance (254 nm)

Peak Identity
1. OH
2. 17 Diol
3. 11 Diol
4. Trione
5. Triol
6. Diene

Peak Identity
1. OH
2. 17 Diol
3. 11 Diol
4. Trione
5. Triol
6. Diene

Time (min)

(a)

Time (min)

(b)

Figure 3.8. Steroid separation. (*a*) Normal phase with computer. (*b*) Normal phase with visual optimization. [From Ref. (81), reprinted by permission of the publisher.]

in the blended mode. The measured capacity factors of a series of carbamate pesticides on blended stationary phases (LiChrosorb Si 100, LiChrosorb Diol I, LiChrosorb CN) were in close agreement with the computed values taken from the single stationary phases (Table 3.11).

Analogous to the solvent selectivity triangle, a stationary phase selectivity triangle can be constructed, resulting in domains where the selectivity coefficient for a pair of solutes is highest (Fig. 3.9).

The most sophisticated approach to optimize resolution, analysis time,

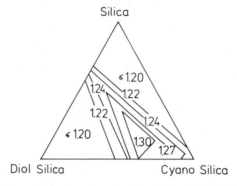

Figure 3.9. Selectivity coefficient r_{ii} of a pair of pesticide carbamates as a function of stationary phase blend. [From Ref. (90), reprinted by permission of the publisher.]

and sensitivity of detection consists of applying the column switching technique and carrying out multidimensional chromatography (91).

3.4.4. Applications (92, 93)

Traditionally, oxide stationary phases such as silicas and aluminas are best suited to resolve nonionic, unpolar to moderately polar compounds, particularly those with multifunctional groups and structural isomers. The elution order (94) follows the sequence saturated hydrocarbons (small k') < olefins < aromatic hydrocarbons ≈ organic halides < sulfides < ethers < nitrocompounds < esters ≈ aldehydes ≈ ketones < alcohols ≈ amines < sulfones < sulfoxides < amides < carboxylic acids (large k').

Sample mixtures that were assayed range from natural oil and flavor extracts, plant protectants, and pesticides to pharmaceutical and synthetic organic products. An example of the separation of isomers is given in Fig. 3.10, where retinyl esters were resolved at trace level (95). For method development, thin-layer chromatography (TLC) is frequently employed as a pilot technique (96, 97). Retention data can be transferred from TLC to column chromatography when standardized conditions are chosen. Several attempts have been undertaken to extend the application range of oxide columns to the resolution of polar ionic compounds, such as bases and acids. Aromatic acids, nucleosides, and nucleobases have been chromatographed on bare silicas employing polar mobile phases and buffers as additives that suppress the ionization of polar solutes (33, 34).

Loading the silica with buffer salts that are insoluble in the mobile phase was advocated by Schwarzenbach (28–30). This permitted the assay of strongly polar sample mixtures without tailing. The use of volatile

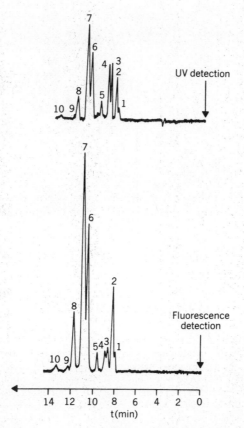

Figure 3.10. Separation of retinyl esters. Column—125 × 4 mm. Spherisorb 3 μm. Eluent—
n-hexane/diisopropylether (98.5/1.5 v/v). Detectors—UV 0.04 AUFS, fluorescence detector.
Flow—1.5 mL/min. Elution sequence: 1—13 *cis* stearate; 2—13 *cis* palmitate; 3—11 *cis*-
stearate; 4—11 *cis* palmitate; 5—9 *cis* palmitate; 6—all *trans* stearate; 7—all *trans* palmitate;
8—all *trans* oleate; 9—all *trans* palmioleate; 10—all *trans* linolate. [From Ref. (95), reprinted
by permission of the publisher.]

eluents facilitates semipreparative isolation and allows coupling of HPLC
with mass-spectrometric detection. An enormous flexibility in gaining selectivity is achieved by using silica or alumina columns in the ion-pair or
liquid–liquid partitioning mode.

   In the first-mentioned mode an ion-pair reagent, such as a tetraalkylammonium salt, is added to resolve basic solutes. Retention is modulated
by the buffer ions, the pH of the buffer, the type of counterion, and the
cosolvent of the hydroorganic eluent. Figure 3.11 provides an example
for barbiturates. Alumina was found to provide a much wider pH range

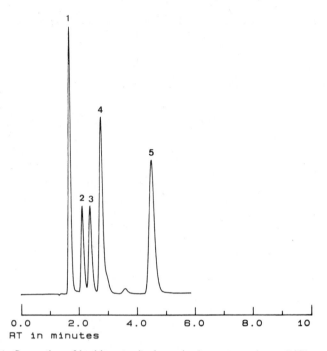

Figure 3.11. Separation of barbiturates by ion-pair chromatography on LiChrosorb Si 60, 5 μm, 125 × 4-mm column. Eluent—methanol/water/phosphate buffer, 0.2M, pH 7.3 (80/15/25 v/v/v) + 0.006M CTMABr detection—UV 254 nm; F_v = 1.0 mL/min; elution sequence—barbital, revonal, luminal, prominal, thiogenal. (Courtesy of Dr. W. Jost, E. Merck, Darmstadt, FRG.)

for applications than silica (43, 44). Lingemann et al. (98) have demonstrated the utility of such systems in the assay of pharmaceutical formulations, such as tetracyclines.

Converting a liquid adsorption to a liquid–liquid partitioning system simply by changing the mobile phase composition or the column temperature was accompanied by a marked improvement in selectivity, as evidenced by Huber et al. for carbamate pesticides (31, 32). Buffer-treated silicas with adjusted apparent surface pH applied in the liquid–liquid partitioning mode serve as an additional example (27).

The alternatives discussed clearly demonstrate that the drawbacks of silica and alumina associated with the varying surface chemistry and sensitivity toward polar solutes can be eliminated or circumvented.

Apart from analytical applications straight-phase chromatography attracts increasing attention in the preparative mode. Purity control of industrial products and structural elucidation of constituents from synthetic mixtures call for selective isolation procedures. The strategy applied in preparation work is dictated by the resolution, the purity, and the throughput of the products. One of the decisive system parameters is the column loadability, which corresponds to the maximum sample input that causes a certain loss of resolution as achieved under analytical conditions (99–102). The choice in carrying out a preparative isolation is either to maintain the nonoverload conditions by increasing the column bore under otherwise constant conditions, or to overload the column and cut fractions. Overload can be accomplished by injecting a highly concentrated sample mixture of small volume or a diluted sample of high volume. Concentration and volume overload have different effects on the peak shape (103). There are some aspects that favor the use of silica rather than reversed phase silica in the preparative mode: bare silicas are much cheaper than bonded silicas. With silicas a broad spectrum of volatile solvents is available; this facilitates the removal of solvents better than hydroorganic eluents. Furthermore, the solubility of isolates is higher in organic solvents than in water-rich mobile phases.

3.5. SUMMARY AND FUTURE PROSPECTS

In this chapter the various aspects of oxide stationary phases in column liquid chromatography have been critically examined. Among the oxides, silica will maintain its premier position as hydrophilic inorganic adsorbent. Substantial efforts are being put into the synthesis of better defined and more reproducible silicas. Nevertheless, subtle differences in the chromatographic properties of silicas remain, arising from the different manufacturing processes. Adjustment of the surface chemistry by employing buffers helps to standardize the properties. Highly flexible phase systems are generated by applying bare silica and alumina in the ion-pair and liquid–liquid partitioning mode simply by adjusting and modulating the mobile phase composition. Surprisingly, these versatile techniques have not yet found widespread recognition and application in practice.

Undoubledly there is potential for the application of low-price spherical silicas of an average particle diameter $d_p > 20$ μm and graduated pore sizes for preparative purposes. However, such silica materials, with average particle diameters in the range of $\geqslant 100$ μm, are already manufactured for catalytic purposes as supported liquid phase catalysts (104). It is thus of great interest to have such materials available on the liquid chromatography market with adapted particle sizes.

REFERENCES

1. U. Wintermeyer, in K. K. Unger, Ed., *Packings and Stationary Phases for Chromatographic Techniques*, Dekker, New York, in print.
2. J. Fricke, Ed., *Aerogels*, Springer Verlag, Heidelberg, 1986, pp. 2–19.
3. F. Hund and K. H. Schulz, in K. Winnacker and L. Küchler, Eds., *Chemische Technologie*, vol. 2, *Anorganische Technologie II*, Carl Hanser, Munich, 1970, pp. 141–194.
4. E. Ferch, in H. Harnisch, R. Steiner, and K. Winnacker, Eds., *Chemische Technologie*, vol. 3: *Anorganische Technologie II*, Carl Hanser, Munich, 1983, pp. 75–90.
5. G. D. Parfitt and K. S. W. Sing, *Characterization of Powder Surfaces*, Academic, New York, 1976.
6. K. Wefers and G. M. Bell, Tech. Paper 19, Alcoa Research Lab., Alcoa Comp., East St. Louis, IL, 1972.
7. H. P. Boehm and H. Knoezinger, in J. R. Anderson and M. Boudart, Eds., *Catalysis*, vol. 4, Springer Verlag, Heidelberg, 1983, pp. 40–189.
8. H. Holik and B. Matejkova, *J. Chromatogr., 213,* 33 (1981).
9. H. Knötzinger and P. Ratnasamy, *Catal. Rev. Sci. Eng., 17,* 31 (1978).
10. K. K. Unger, *Porous Silica, J. Chromatogr. Libr.*, vol. 16, Elsevier, Amsterdam, 1979, pp. 57–72.
11. G. A. Parks, *Chem. Rev., 65,* 177 (1965).
12. N. E. Tretyakov and V. N. Filimonov, *Kinet. Katal., 13,* 815 (1972).
13. D. J. Pietrzyk, in K. K. Unger, Ed., *Packings and Stationary Phases for Chromatographic Techniques*, Dekker, New York, in print.
14. L. R. Snyder and J. J. Kirkland, *Introduction to Modern Liquid Chromatography*, Wiley, New York, 1979, pp. 168–202.
15. K. K. Unger, G. Jilge, J. N. Kinkel, and M. T. W. Hearn, *J. Chromatogr., 359,* 61–72 (1986).
16. K. K. Unger and R. Janzen, *J. Chromatogr., 373,* 227–264 (1986).
17. T. Allen, *Particle Size Measurement*, Chapman and Hall, London, 1981.
18. K. S. W. Sing, *Pure Appl. Chem., 51,* 1 (1979).
19. K. K. Unger, B. Anspach, and H. Giesche, *J. Pharm. Biomed. Anal., 2,* 139–151 (1984).
20. K. K. Unger, B. Anspach, R. Janzen, G. Jilge, and K. D. Lork, in C. Horvath, Ed., *HPLC: Advances and Perspectives*, vol. 5, Academic, New York, in print.
21. R. Ohnmacht and I. Halasz, *Chromatograhia, 14,* 155–162 (1981).
22. F. Geiss, *Die Parameter der Dünnschicht-Chromatographie*, Vieweg, Braunschweig, 1972, p. 131.
23. H. Brockmann, *Angew. Chem., 53,* 384 (1940).

24. H. Brockmann and H. Schodder, *Ber. dtsch. chem. Ges.*, *74B*, 73 (1941).
25. H. Engelhardt and H. Elgass, in C. Horvath, Ed., *HPLC: Advances and Perspectives*, vol. 2, Academic, London, 1980, pp. 57–108.
26. H. Engelhardt and H. Müller, *J. Chromatogr.*, *218*, 395–407 (1981).
27. S. Hansen, P. Helboe, and M. Thomsen, *J. Chromatogr.*, *368*, 39–47 (1986).
28. R. Schwarzenbach, *J. Liquid Chrom.*, *2*, 205 (1979).
29. R. Schwarzenbach, *J. Chromatogr.*, *202*, 397 (1980).
30. R. Schwarzenbach, *J. Chromatogr.*, *334*, 35 (1985).
31. J. Kraak and J. P. Crombeen, in H. Engelhardt, Ed., *Practice of High Performance Liquid Chromatography*, Springer Verlag, Heidelberg, 1986, pp. 180–198.
32. J. F. K. Huber, M. Pawlowska, and P. Markl, *Chromatographia*, *17*, 653–663 (1983).
33. J. F. K. Huber, M. Pawlowska, and P. Markl, *Chromatographia*, *19*, 19–28 (1984).
34. W. T. Brugman and J. C. Kraak, *J. Chromatogr.*, *205*, 170 (1981).
35. W. T. Brugman and J. C. Kraak, *Chromatographia*, *15*, 282 (1982).
36. J. Köhler, D. B. Chase, R. D. Farlee, A. J. Vega, and J. J. Kirkland, *J. Chromatogr.*, *352*, 275 (1986).
37. A. Paulus, Ph. D. dissertations, Universität Tübingen, Tübingen, FRG, 1986.
38. G. Schomburg, J. Köhler, H. Figge, A. Deege, and U. Bien-Vogelsang, *Chromatographia*, *18*, 119 (1984).
39. T. Daldrup and B. Kardel, *Chromatographia*, *18*, 81 (1984).
40. G. Mac Zura, K. P. Goodboy, and J. J. Koenig, in *Kirk-Othmer Encyclopedia of Chemical Technology*, vol. 2, Wiley, New York, (1962).
41. J. B. Peri, *J. Phys. Chem.*, *69*, 220 (1965).
42. R. Steyrer, Ph. D. dissertation, Universität München, Munich, FRG, 1978.
43. C. Laurent, H. A. H. Billet, and L. de Galaan, *Chromatographia*, *17*, 253 (1983).
44. H. Billiet, C. Laurent, and L. Galan, *Trends Anal. Chem.*, *4*, 100–103 (1985).
45. German patent DOS 2, 409, 407. C. H. Marsh (1973).
46. U.S. patent 4, 010, 242. R. K. Iler et al. (1973).
47. J. L. Woodhead, *Sci. Ceram.*, *9*, 29 (1977).
48. V. Baran, R. Caletka, M. Tympl, and V. Urbanek, *J. Radioanal. Chem.*, *24*, 353 (1975).
49. R. Caletka and M. Tympl, *J. Inorg. Nucl. Chem.*, *39*, 669 (1977).
50. J. L. Woodhead, *Sci. Ceram.*, *10*, 169 (1980).
51. R. Aigner-Held, W. A. Aue, and E. E. Pickett, *J. Chromatogr.*, *189*, 139 (1980).
52. J. P. Bonsack, *J. Colloid Interface Sci.*, *44*, 430 (1973).

53. G. Gimblett, A. A. Rahman, and K. S. W. Sing, *J. Chem. Tech. Biotechnol.*, *30*, 51 (1980).

54. R. C. Asher and S. J. Gregg, *J. Chem. Soc.*, 5057 (1960).

55. M. J. Torralvo-Fernandez and M. A. Alario-Franco, *J. Colloid Interface Sci.*, *77*, 29 (1980).

56. V. M. Chertov and N. T. Okopnaya, *Izv. Akad. Nauk Mold. SSR, Ser. Biol. Khim. Nauk, 6*, 63 (1974).

57. V. M. Chertov, N. T. Okopnaya, and I. E. Neimark, *Dokl. Akad. Nauk SSR, 209*, 876 (1973).

58. T. M. El-Akkad, *J. Tech. Biotechnol.*, *30*, 497 (1980).

59. T. M. El-Akkad, *J. Colloid Interface Sci.*, *78*, 100 (1980).

60. L. R. Snyder, *Principles of Adsorption Chromatography*, Dekker, New York, 1968.

61. E. Soczewinski, *Anal. Chem.*, *41*, 179 (1969).

62. L. R. Snyder, in C. Horvath, Ed., *HPLC: Advances and Perspectives*, vol. 3, Academic, New York, 1983, pp. 157–222.

63. L. R. Snyder and T. C. Schunk, *Anal. Chem.*, *54*, 1764–1772 (1982).

64. L. R. Snyder and H. Poppe, *J. Chromatogr.*, *184*, 363 (1980).

65. L. R. Snyder, J. L. Glajch, and J. J. Kirkland, *J. Chromatogr.*, *218*, 299–326 (1981).

66. L. R. Snyder, *Principles of Adsorption Chromatography*, Dekker, New York, 1968, pp. 192–199.

67. L. R. Snyder and J. J. Kirkland, *Introduction to Modern Liquid Chromatography*, Wiley, New York, 1979, p. 366.

68. R. P. W. Scott and P. Kucera, *Anal. Chem.*, *45*, 749 (1973).

69. R. P. W. Scott and P. Kucera, *J. Chromatogr.*, *149*, 93 (1978).

70. R. P. W. Scott and P. Kucera, *J. Chromatogr.*, *171*, 37 (1979).

71. R. P. W. Scott, *J. Chromatogr. Sci.*, *18*, 297 (1980).

72. M. Jaroniec and A. Patrykiejew, *J. Chem. Soc. Faraday Trans. I, 76*, 2486 (1980).

73. M. Jaroniec and J. Piotrowska, *J. High Resolut. Chromatogr., Chromatogr. Commun., 3*, 257 (1980).

74. J. K. Rozylo, J. Oscik, B. Oscik-Mendyk, and M. Jaroniec, *J. High Resolut. Chromatogr., Chromatogr. Commun., 4*, 17 (1981).

75. M. Jaroniec, J. A. Jaroniec, and W. Golkiewicz, *J. High Resolut. Chromatogr., Chromatogr. Commun., 4*, 89 (1981).

76. M. Jaroniec and J. Oscik, *J. High Resolut. Chromatogr., Chromatogr. Commun., 5*, 3 (1982).

77. J. Ośćik, *Adsorption*, Ellis Horwood, Chichester, West Sussex, UK, 1982.

78. K. K. Unger, *Porous Silica, J. Chromatogr. Libr.*, vol. 16, Elsevier, Amsterdam, 1979, pp. 198–199.

79. J. F. K. Huber and F. Eisenbeiß, *J. Chromatogr., 149,* 127–141 (1978).

80. N. Becker and K. K. Unger, *Fresenius Z. Anal. Chem., 304,* 374–381 (1980).

81. *Techscan, 1*(4), 6 (Sept. 1981), E. I. du Pont de Nemours, Analytical Instruments Div., Wilmington, DE.

82. L. R. Snyder and J. J. Kirkland, *Introduction to Modern Liquid Chromatography,* Wiley, New York, 1979, pp. 365–389.

83. L. R. Snyder and J. J. Kirkland, *Introduction to Modern Liquid Chromatography,* Wiley, New York, 1979, p. 356.

84. L. R. Snyder and J. L. Glajch, *J. Chromatogr., 214,* 1 (1981).

85. L. R. Snyder and J. J. Kirkland, *Introduction to Modern Liquid Chromatography,* Wiley, New York, 1979, pp. 246–267.

86. J. L. Glajch, J. J. Kirkland, K. M. Squire, and J. M. Minor, *J. Chromatogr., 199,* 57–79 (1980).

87. J. L. Glajch and J. J. Kirkland, *Anal. Chem., 55,* 319–336A (1983).

88. M. T. Gilbert and R. A. Wall, *J. Chromatogr., 149,* 341 (1978).

89. Y. Ghawmi and R. A. Wall, *J. Chromatogr., 174,* 51 (1979).

90. R. Hirz, Ph. D. dissertation, Universität Wien, Vienna, Austria, 1982.

91. H. Hulpke and U. Werthmann, in H. Engelhardt, Ed., *Practice of High Performance Liquid Chromatography,* Springer Verlag, Heidelberg, 1986, pp. 143–157.

92. J. N. Done, J. H. Knox, and J. Loheac, *Applications in HPLC,* Wiley, London, 1974.

93. A. Pryde and M. T. Gilbert, *Applications in HPLC,* Chapman and Hall, London, 1979.

94. L. R. Snyder and J. J. Kirkland, *Introduction to Modern Liquid Chromatography,* Wiley, New York, 1979, p. 361.

95. K. H. Biesalski, Habilitationsschrift, Johannes Gutenberg-Universität Mainz, Mainz, FRG, 1986.

96. L. R. Snyder and J. J. Kirkland, *Introduction to Modern Liquid Chromatography,* Wiley, New York, 1979, pp. 383–389.

97. W. Jost, H. E. Hauck, and F. Eisenbeiß, *Kontakte (E. Merck), 3,* 45–51 (1984).

98. H. Lingemann, H. A. van Munster, J. H. Beynen, W. J. H. Underberg, and A. Hulshoff, *J. Chromatogr., 352,* 261–274 (1986).

99. A. W. J. de Jong, J. C. Kraak, H. Poppe, and F. Nooitgedacht, *J. Chromatogr., 193,* 181 (1980); A. W. J. de Jong, H. Poppe, and J. C. Kraak, *J. Chromatogr., 209,* 432 (1981).

100. A. Wehrli, in H. Engelhardt, Ed., *Practice of High Performance Liquid Chromatography,* Springer Verlag, Heidelberg, 1986, pp. 109–142.

101. A. Wehrli, U. Hermann, and J. F. K. Huber, *J. Chromatogr., 125,* 59 (1976).

102. F. Eisenbeiß, S. Ehlerding, A. Wehrli, and J. F. K. Huber, *Chromatographia, 20,* 657 (1985).

103. J. H. Knox and H. M. Pyper, *J. Chromatogr., 363,* 1 (1986).

104. Shell International, Chemical Company, London, UK SE1 7PG, Shell Centre.

BIBLIOGRAPHY

H. Engelhardt, *High Performance Liquid Chromatography*, Springer Verlag, Berlin, 1979.

H. Engelhardt and H. Elgass, in C. Horvath, Ed., *HPLC: Advances and Perspectives*, vol. 2, Academic, New York, 1980, pp. 57–108.

F. Geiss, *Die Parameter der Dünnschicht-Chromatographie*, Vieweg, Braunschweig, 1972.

D. L. Saunders, *J. Chromatogr. Sci., 15,* 372 (1977).

L. R. Snyder, *Principles of Adsorption Chromatography*, Dekker, New York, 1968.

L. R. Snyder, in C. Horvath, Ed., *HPLC: Advances and Perspectives*, vol. 3, Academic, New York, 1983, pp. 157–222.

L. R. Snyder and J. J. Kirkland, *Introduction to Modern Liquid Chromatography*, Wiley, New York, 1979.

K. K. Unger, *Porous Silica, J. Chromatogr. Libr.*, vol. 16, Elsevier, Amsterdam, 1979.

CHAPTER

4

CARBON IN LIQUID CHROMATOGRAPHY

J. H. KNOX and B. KAUR

Wolfson Liquid Chromatography Unit
Department of Chemistry
University of Edinburgh
Edinburgh, UK

4.1. INTRODUCTION

Carbon was used in the early days of liquid chromatography as an adsorbent, but later fell into disuse mainly because of the emergence of

189

materials such as the silica gels, which had substantially superior chromatographic properties, and because of the development and wide popularity of thin-layer chromatography, which required a translucent and colorless partitioning material. The recent resurgence of interest in carbon, and more particularly graphite, has arisen for reasons which mirror those for its original demise, namely, the recognition that existing packing materials, and in particular silica gels, possess several undesirable features, including solubility in certain eluents and unwanted interactions with sensitive analytes. Advances in surface science and in chromatography since the 1950s have enabled us to formulate the requirements for good chromatographic packing materials and to appreciate the causes of the poor performance and undesirable features of the early materials. This in its turn has led us to more effective control of the properties of adsorbents for chromatography and of the methods for their production. The improvements in structure and performance have been particularly pronounced with respect to silica gels and their bonded forms, which are now the basis of modern high performance liquid chromatography (HPLC) methodology. We believe that this understanding can now be applied to carbon materials and will lead to a revolution in the development and use of nonpolar adsorbents.

In this chapter we discuss first the basic methods of manufacture of carbon, the requirements for HPLC packing materials, previous uses of carbon in liquid chromatography, the physical and adsorptive properties of graphites, the chromatographic properties of uncoated carbons and their applications, and finally the chromatographic properties of coated carbons.

4.2. PREPARATION AND DIFFERENT FORMS OF CARBON

Carbon exists in a bewildering variety of configurations, ranging from the crystalline forms—graphite and diamond—via numerous intermediate forms to amorphous carbons. These materials differ in their degree of crystallinity, in their porosity and pore dimensions, in the accessibility of their pores, and in the proportion and nature of the functional groups attached to carbon atoms. Carbon is widely used on a large scale in industry (1). The main industrial carbons are diamonds, synthetic graphites, active carbons, carbon blacks, glassy carbons, and pyrolytic carbons. Since bulk industrial products have nearly always been the historic starting points for developing chromatographic packings, it is relevant to ask whether any of these forms can be used directly or with minor modification for liquid chromatography purposes.

Diamonds are the most expensive form of carbon, but being nonporous and of low specific surface area, they are unlikely ever to be useful in chromatography. When examined by Telepchak (2), their performance was disappointing: they remain a curiosity.

Synthetic electrographites (1) are made by mixing granulated coke with about 30% of its weight of pitch, extruding to the shape required, and heating in stages to 2600–3000°C. This gives a so-called graphitized carbon, which contains about 30% void space in closed, inaccessible pores. Although such materials can be ground to small particles, their lack of significant mesoporosity makes them unsuitable for chromatography.

Glassy carbons are made by programmed pyrolysis of organic polymers (1, 3), such as phenolformaldehyde resin or polyacrylonitrile, up to about 1000°C. The final glassy carbon, with a density of about 1.6 g/mL, has 30–40% of the original polymer volume and contains micropores which are largely closed and inaccessible. On heating to between 2500 and 3000°C, glassy carbons partially reorganize into graphitic ribbons with two-dimensional order (see later). These high-temperature glassy carbons contain about 30% of closed pores in an otherwise dense matrix. They contain no mesopores and, like electrographites, appear to be unsuitable for chromatography.

Active carbons are made by oxidative pyrolysis of natural materials, such as wood, vegetable shells, coal, lignite, and bone (1). They always contain significant quantities of mineral ash, which can sometimes be removed by chemical treatment, and normally a high proportion of oxygen. Their densities are between 0.3 and 0.7 g/mL, and they have enormous BET surface areas of up to 2000 m^2/g, comparable to the computed surface area of a single carbon atom (2500 m^2/g). They are microporous and are used extensively to remove organic materials from air and aqueous solutions. Although they were used in the early days of chromatography (4), the problems arising from the severe nonlinearity of their adsorption isotherms render them unsuitable for modern elution chromatography.

Carbon blacks, commonly known as soots, are formed at around 1000°C by combustion of carbonaceous materials in a deficiency of oxygen (1). Carbon black particles are fragile aggregates of colloidal units of between 100 and 1000 Å in diameter, loosely held together by van der Waals forces. They have specific surface areas of 5–100 m^2/g. When formed at 1000°C, the colloidal carbon particles are spherical and amorphous, but on heating to 2500–3000°C they assume a polyhedral form characteristic of the so-called turbostratic carbons (5) (see later). While graphitized carbon black (GCB) has been extensively used in gas chromatography (6), it is too fragile to be useful for modern liquid chromatography.

Pyrolytic carbons are formed by high-temperature deposition (above

900°C) from organic vapors [see (7)]. They are dense nonporous materials and are again unsuitable for chromatography.

Contrasting with industrial silica gels, none of the widely available industrial carbons, with the possible exception of carbon black, can be used directly as the starting point for producing a carbon adsorbent for modern liquid chromatography. This crucial difference is the main reason for carbon's absence from the range of popular support materials for liquid chromatography. If carbon is to be used in modern liquid chromatography, new forms must be developed specifically for the purpose. The substantial progress that has now been made toward this objective (6–9) is the subject of this chapter, but first we consider the requirements for a satisfactory HPLC packing material and the extent to which graphitic materials can meet these requirements.

4.3. REQUIREMENTS FOR AN HPLC COLUMN PACKING MATERIAL

4.3.1. Bed and Pore Structure

In the broadest sense a chromatographic column is a bed possessing two ranks of porosity. The first rank of porosity, closely related to the particle size, determines the flow characteristics of the bed. The second rank of porosity, associated with the pores inside the particles, determines the mass transfer rates of the solute molecules within the particles, and to some extent their partition ratios. The pore diameter must be at least equal to the diameter of the largest molecules to be chromatographed, preferably three times as large, and they must not be so small that their walls appear to be curved at the molecular level, for then nonlinear adsorption isotherms and tailed peaks result. The ideal adsorbent particles for HPLC have diameters of between 3 and 10 μm and contain only mesopores, that is, pores having diameters within the range of 60–1000 Å. Preferably the pore size distribution should be narrow, unless there are special reasons against this. Specific surface areas should not exceed about 200 m²/g.

4.3.2. Particle Porosity

The optimal particle porosity is a compromise. High porosity is desirable to provide good mass transfer and accessibility of the adsorbent surface, but too high porosity makes particles fragile. In practice, a particle porosity of 50–70% is suitable for retentive chromatography, although a porosity of up to 80% may be acceptable for size exclusion chromatography (SEC), where a high ratio of pore volume to extraparticle volume is desirable.

4.3.3. Surface Homogeneity

For a material to act as a good adsorbent for chromatography it must have an energetically homogeneous surface to provide linear adsorption isotherms at low coverage. Most adsorbents used in chromatography, being amorphous, do not have homogeneous surfaces. The chromatographer's cure for this deficiency is to deactivate the most energetic surface sites by adsorption of a partial monolayer of a strongly adsorbed modifier. For silica gel, when used as an adsorbent in liquid chromatography, water is widely used at 30–50% of a monolayer coverage. If the material is not to be used as an adsorbent, the undesirable features can be virtually eliminated by chemical reaction, as is done in producing chemically bonded silica gels. The original features of the adsorbent are then lost, and the adsorbent is used, not as the stationary phase itself, but as a structure upon which to carry a stationary phase.

The perfect adsorbent surface would be totally homogeneous with no adventitious functional groups, and could be totally masked by adsorption or chemical reaction to give entirely different functionalities. In spite of its great versatility, silica gel is defective in this respect since it contains surface silanol groups with a wide spectrum of adsorptive strengths, and since, for steric reasons, only about 50% of them can be derivatized by silane reagents. Even the most exhaustively silanized silica gels show residual silanol activity.

4.3.4. Mechanical Strength and Rigidity

Any HPLC adsorbent should possess sufficient mechanical strength to withstand a dynamic compressive pressure of up to 3000 bar/m. A strong silica gel, for example, will withstand at least 10,000 bar/m. Adequate mechanical strength for use in HPLC requires that the fundamental colloidal units in the packing material be firmly welded together in a continuous three-dimensional structure. It is equally important that the particles neither swell nor shrink when the eluent composition is changed. This has always been a problem with polymeric supports in liquid chromatography.

4.3.5. Particle Shape

There has been much discussion as to whether particles of packing for HPLC should be spherical or angular. The advantages claimed for spheres are the following. (1) Spheres are manufactured specifically for HPLC and are made from higher quality starting materials than chips. Their properties should therefore be more reproducible. (2) With spheres fines, which produce high column back pressures, are more easily avoided and do not form during storage and transportation. (3) Spheres can better withstand compressive stress and erode less at their points of contact under HPLC conditions. (4) Columns packed with spheres have a better shelf life.

4.3.6. Chemical Stability

Chemical stability in eluents is most important. Packings must not dissolve in eluent under normal operating conditions. Silica dissolves significantly in aqueous solutions at pH > 7, and even in pure water or methanol it has a finite solubility. The well-known bed settling which occurs with silica gel columns is almost certainly due to stress-induced solution of the silica gel at the points of contact of the particles. Organic moieties bonded to the silica gel surface are not necessarily affected, but are cleaved by hydrolysis at pH below 2 or above 7. The ideal adsorbent and bonded adsorbent would be stable to all eluents and at all pH. Carbon undoubtedly meets this criterion.

4.3.7. Reproducibility

One of the most common criticisms of silica gel, both bare and in bonded forms, is its irreproducibility from batch to batch and from manufacturer to manufacturer. This is probably unavoidable and arises from the nature of the silica gel surface itself, which is made up of silanol groups with a wide and uncontrolled range of adsorptive activity. The mix of activity depends critically on how the silica gel is manufactured and on the impurities it contains. The best way to avoid such problems would undoubtedly be to use a reproducible crystalline surface.

4.3.8. Summary of Requirements for an Ideal HPLC Packing Material

1. Particles should be spherical and available in particle diameters of 3–10 μm.
2. Particles should be strong enough to withstand pressure gradients up to 3000 bar/m. They should not swell or shrink.
3. Particles should have a porosity in the range of 50–70%, extending to 80% for size exclusion chromatography.

4. Particles should contain no pores smaller than ≃60 Å in diameter and should have a uniform pore size distribution.

5. Particles should be available with a range of mean pore diameters of 60–1000 Å.

6. The internal surface of the material should be energetically homogeneous.

7. The internal surface should be capable of modification by bonding, adsorption, or coating (or by a combination of them) to provide a range of surface functionalities. It should be possible to mask the original character of the surface completely.

8. Packing materials should be chemically stable under all conditions of pH and eluent composition.

9. The physicochemical characteristics of the material should be reproducible from batch to batch and from manufacturer to manufacturer.

Silica gel fails on counts 6, 8, and 9, and partially on count 7. GCB fails on count 2 and partially on count 6; it meets count 8 and probably 9. Recent work by the authors (8, 9) shows that porous graphite can now be made with adequate mechanical strength for HPLC and with a surface homogeneity superior to that of GCB. Thus counts 2 and 6 are now met. The new carbon adsorbents appear, in principle, to meet all key requirements for an ideal HPLC adsorbent.

4.4. PREVIOUS USE OF CARBON IN LIQUID CHROMATOGRAPHY

Following the revival of the chromatographic method in the early 1930s, a wide range of adsorbents was employed in column liquid chromatography, as can be seen from the comprehensive survey published by Lederer and Lederer in 1953 (10). In the 1940s charcoal was used extensively by Tiselius and coworkers (4, 11–14) for frontal and displacement chromatography of amino acids (12), peptides (13), and acids (14). In 1946 Claesson (15) obtained excellent separations by frontal chromatography on 5–40-μm activated carbon detecting eluted solutes by a refractometer. In this work, a forerunner of HPLC, he applied a pressure of 3 bar to the head of the 100 × 10-mm column. The nonlinear adsorption isotherms exhibited by charcoal were ideal for these chromatograhic techniques, but totally unsuitable for elution chromatography as used today. Accordingly, charcoals soon gave way to silica gels and aluminas. Their eclipse was complete with the advent of thin-layer chromatography, which required colorless adsorbents to enable separated bands to be visualized by color reagent sprays.

Following the elegant work of Isirikyan and Kiselev (16), who showed that the surfaces of GCB were highly reproducible even when coming from different sources, GCB became widely used as a nonpolar adsorbent for gas chromatography (6). Kiselev argued that GCB with its ordered atomic surface provided a unique surface for fundamental study (17) and therefore the ultimate stationary phase for gas chromatography (18), which would have the added advantage of possessing unique stereoselectivity (19). Under the trade name Carbopack, GCB has proved itself as a unique gas chromatographic adsorbent in a wide variety of specialist applications (6, 20). Kiselev's work did not, however, lead to the use of GCB in liquid chromatography, and it was only in 1981 that it appeared in a successful liquid chromatography experiment (21).

Five years before this, interest in carbon for liquid chromatography was revived by Guiochon and coworkers, who recognized the potential of carbon black and attempted to overcome its basic physical weakness. Colin, Eon, and Guiochon (7, 22, 23) strengthened carbon black by deposition of pyrolytic carbon from benzene vapor at 900°C. The composite material could be graphitized, which brought about some improvement in peak shape and HETP, but its overall performance was not sufficiently good to make pyrocarbon on graphite a viable alternative to reversed-phase bonded silica gels. Its surface was still energetically heterogeneous, and there was the further problem that the initial 15–20-μm GCB particles were extremely fragile and difficult to handle. Figure 4.1 shows a typical chromatogram obtained with pyrocarbon on GCB. It is notable that peak shapes are poor whenever solutes are significantly retained (k' values greater than 1 or 2).

Unger et al. (24) attempted in 1980 to produce HPLC graphites from purified active carbons and cokes. Their starting materials had surface areas ranging from 20 to 900 m^2/g. On calcination at 1800°C and above, these fell to between 1 and 5 m^2/g. Although these carbons were very strong and readily withstood HPLC pressures, they gave disappointing chromatographic performance and had small adsorptive capacities.

Another attempt to produce carbons for HPLC was made by Plzak and coworkers (25–27). Their basic material was produced by electrochemical reduction of polytetrafluoroethylene (PTFE) at an alkali-metal/mercury amalgam electrode. The initial carbon had an extremely high surface area of about 2000 m^2/g, indicating a microporous active carbon, but on heating to graphitization temperatures this was reduced to about 20 m^2/g. This material showed good mechanical strength but poor chromatographic properties.

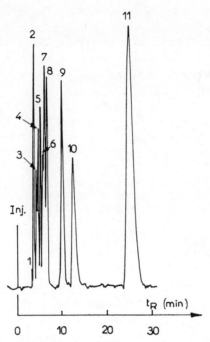

Figure 4.1. Separation of methyl benzenes on Black Pearls L coated with 44% pyrocarbon (22). Eluent: acetonitrile. Solutes: 1) unretained; 2) benzene; 3) 3-xylene; 4) 2-xylene; 5) 1,3,5-trimethylbenzene (TMB); 6) impurity; 7) 1,2,4-TMB, 8) 1,2,3-TMB; 9) 1,2,4,5-tetramethylbenzene (TeMB); 10) 1,2,3,4-TeMB; 11) pentamethylbenzene. [From Ref. (22). Reproduced by permission of the *Journal of Chromatography*.]

Following these somewhat disappointing results, it was encouraging when Liberti and coworkers (21) confirmed in 1981 that GCB did indeed provide satisfactory peaks not only in gas chromatography but also in liquid chromatography, and that GCB could become an ideal liquid chromatography adsorbent if only it was less fragile. One of their chromatograms using 20-μm Carbopack B is shown in Fig. 4.2, where the excellent peak symmetry at high k' values is notable.

Two years earlier Knox and Gilbert (28) had invented a novel method for making a mesoporous glassy carbon, using silica gel as a template to provide the desired pore size. They impregnated a silica gel template with phenol-formaldehyde resin, carbonized the silica–resin composite at 1000°C to give a carbon–silica composite, and finally dissolved out the silica to leave a porous glass carbon. This material became useful as a chromatographic adsorbent when heated to above 2000°C to induce graphitization. This eliminated microporosity but left the mesoporous

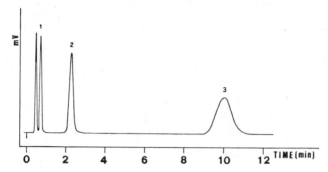

Figure 4.2. Separation of analgesics on 30 μm Carbopack B (a GCB). Eluent: methanol. Solutes: 1) phenylacetamide; 2) phenacetin; 3) caffeine. [From Ref. (21). Reproduced by permission of the *Journal of Chromatography*.]

structure of the original silica gel intact. The final porous graphitic carbon (PGC) had a specific surface area of 100–300 m^2/g. In the first experiments (8) this new material performed well in gas chromatography but disappointingly in liquid chromatography, where it was only marginally superior to Guiochon's pyrocarbon on carbon black (7). However, improved procedures for making PGC have now provided materials (9) which give an HPLC performance equivalent to that of octadecyl (ODS) bonded silica gels. A typical chromatogram is shown in Fig. 4.3.

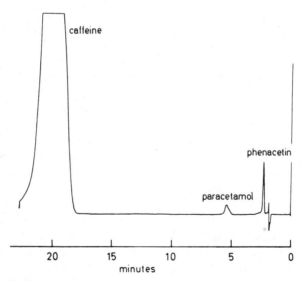

Figure 4.3. Separation of analgesics on 7-μm PGC. Eluent—methanol:water (95:5). Solutes as indicated. [From Ref. (9). Reproduced by permission of the *Journal of Chromatography*.]

In 1982 a similar method of making porous carbon was published by Novak and Berek (29). Limited chromatographic data on this material have been reported by Skutchanova et al. (30).

Current results with mesoporous graphitic carbons, when coupled with previous data on the adsorptive and chromatographic properties of GCB, suggest that the problem of providing a porous graphite with the correct physical structure for HPLC has now been essentially solved.

4.5. PHYSICAL AND ADSORPTIVE PROPERTIES OF GRAPHITES

The work just cited shows that only materials loosely referred to as "graphitic" are likely to be useful in modern liquid chromatography. Accordingly we concentrate mainly on such carbons in the rest of this chapter.

In the industrial context "graphitization" is loosely used to cover any process whereby a carbon is heated to temperatures above about 2000°C. In most cases this does not bring about true graphitization, although it certainly causes fundamental structural changes in an initially amorphous material. Strictly the term graphitization should be reserved for those processes which produce true three-dimensional graphites with the Bernal structure (31). This consists of sheets of hexagonally arranged carbon atoms stacked in the order $ABABA$. . . . The sheets are 3.354 Å apart, and the carbon atoms in each sheet are 1.42 Å apart. The X-ray diffraction pattern shows $\{h, k, l\}$ reflections, where h, k, and l can simultaneously be nonzero.

Many carbons are not graphitizable in this sense, but on heating to high temperatures, they produce two-dimensional graphites in which the graphitic sheets are still parallel but are no longer properly registered with respect to the sheets above and below. Two-dimensional graphites show X-ray diffraction patterns with only $\{0, 0, l\}$ and $\{h, k, 0\}$ reflections. The atomic spacing within the sheets is slightly less than in true graphite at 1.40 Å, while the spacing of the sheets (given by the diffraction angle of the 002 reflection) is slightly greater at 3.44 Å. Two-dimensional graphites are sometimes called turbostratic carbons, following Biscoe and Warren (5), who coined the term to describe GCB. Carbons which have not been heated above 1000°C are amorphous and show almost no structure under X-ray analysis.

Figure 4.4 shows X-ray diffractograms of porous spheres of PGC and of a mixture of spheres and graphitic needles with which they are sometimes contaminated. The PGC spheres consist of two-dimensional graphite, while the needles consist of three-dimensional graphite. The distinc-

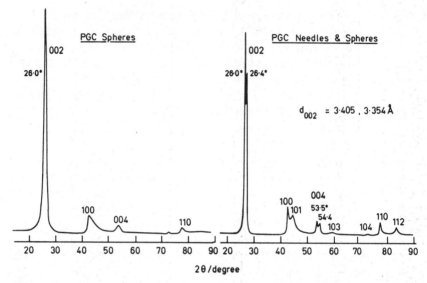

Figure 4.4. X-ray powder diffractograms of PGC spheres (left) and PGC spheres contaminated with graphitic needles (right), contrasting the patterns obtained from two- and three-dimensional graphites. [From Ref. (9). Reproduced by permission of the *Journal of Chromatography*.]

tion between the diffraction patterns of two- and three-dimensional graphites is now clearly seen: the 002 and 004 peaks for the mixture are doublets indicating the two distinct phases with their distinct layer spacings. The asymmetric 100 peak is typical of two-dimensional graphite spheres, while the oblique reflections (labelled 101, 103, 104, 112) arise from the three-dimensional graphite needles.

High resolution transmission electron microscopy (HRTEM) provides elegant confirmation of the X-ray diffraction results. Figure 4.5 shows the HRTEM of a portion of a GCB particle.

The individual colloidal units are seen as rough hexagons 100–300 Å across, composed of "booklets" of graphitic sheets. The sheets are 3.44 Å apart, exactly as expected for a two-dimensional graphite. The outside faces of the colloidal particles are mostly straight and the different booklets are joined at angles of 120°. In places some bending of the sheets is noted, although the overriding impression is of booklets of flat sheets. This is a typical turbostratic or two-dimensional graphite as described by Biscoe and Warren (5). The physical fragility of GCB is due to the individual colloidal units, clearly seen in the HRTEM, being weakly interconnected, and probably held together only by van der Waals forces.

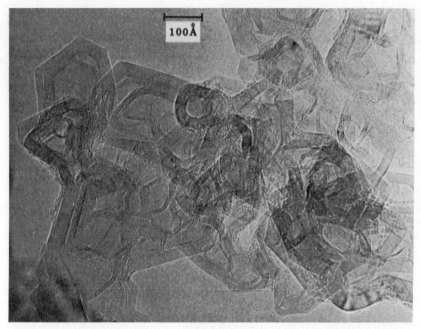

Figure 4.5. High-resolution electron micrograph of GCB, showing individual hexagonal colloidal particles. [From Ref. (9). Reproduced by permission of the *Journal of Chromatography*.]

Figure 4.6 shows the corresponding HRTEM of a PGC sample. The structure now consists of intertwined two-dimensional graphitic ribbons rather than separate hexagonal units. The layer spacing is still 3.44 Å and the angle between ribbons 120°. Some curvature is noted, but the ribbons are predominantly straight. The ribbons and booklets in Figs. 4.5 and 4.6 are 30–60 Å thick, in close agreement with the crystallite dimensions determined from the widths of the 002 X-ray diffraction peaks using the equation of Scherrer (32). The distinction between the ribbonlike structure of PGC and the particulate nature of GCB is striking, and undoubtedly explains the high mechanical strength of PGC.

Typical values for the surface area, pore volume, particle porosity, and mean pore diameters of graphitic carbons are compared in Table 4.1 with those of silica gels.

The main difference between PGC and a typical HPLC silica gel, such as Hypersil, is that PGC is much more porous. This arises from the procedure used to produce PGC: when phenol-formaldehyde resin is pyrolyzed to give carbon, about 50% of the original weight is lost, while the

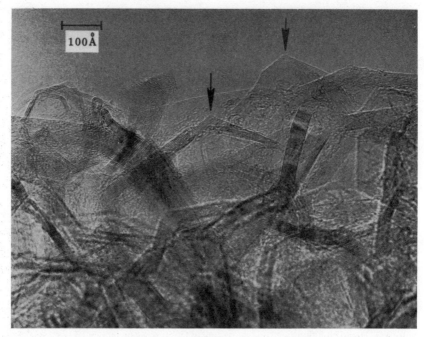

Figure 4.6. High-resolution electron micrograph of PGC, showing intertwined graphitic ribbons. [From Ref. (9). Reproduced by permission of the *Journal of Chromatography*.]

density of the carbonaceous material increases from 1.2 g/mL for the polymer to about 2.2 g/mL for the two-dimensional graphite. The final graphite therefore occupies only about 30% of the volume of the original polymer. Since the silica gel template preserves the original shape and form of the particle, the final porosity is about 75%. PGC is therefore

Table 4.1. Physical Properties of Silica Gels and Graphitic Carbons

Material	Specific Surface Area (m²/g)	Specific Pore Volume (mL/g)	Porosity (%)	Mean Pore Diameter (Å)
Hypersil	170	0.68	60	100
Wide-pore silica gel and template for making PGC	50	1.4	75	300
PGC 1000°C	350	1.4	75	300 + micropores
PGC 2500°C	120	1.4	75	300
Carbopack B	80	0.6	60	200

significantly weaker than a typical silica gel, whose porosity is about 60%, but of about the same strength as a silica gel of 75% porosity. The surface area of PGC per unit particle volume is slightly more than that of the original silica gel template, suggesting that there is some separation of ribbons during grahitization.

The surface of graphite is unique in that it consists essentially of gigantic aromatic molecules which possess no functional groups except possibly at their edges. The main part of the surface will thus be uniform and totally unreactive. It will interact with adsorbate molecules by dispersion forces only. It will show little selectivity according to functionality, but marked stereoselectivity for isomers. The unique adsorptive properties of graphite are exemplified by the data obtained by Isirikyan and Kiselev (16) on the heats of adsorption of benzene and hexane on several different Sterling FT carbon blacks, as illustrated in Fig. 4.7.

We first note the remarkable consistency of the data which were obtained on four different samples of GCB from different sources and having significantly different surface areas. Except at the lowest coverage, where there are noticeable differences in the heats of adsorption, the ΔH versus coverage plots for the four samples are identical. The differences in ΔH at low coverage arise from surface defects, presumably at the edges of the graphitic sheets, which differ from one sample to another. If these defect sites can be deactivated by prior adsorption of, say, a high molecular weight additive, a totally homogeneous and reproducible surface remains. For this reason it is predicted that graphitic adsorbents should be highly reproducible with adsorptive properties more or less independent of their source.

A second important feature shown by Fig. 4.7 is the qualitative difference between the ΔH versus coverage plots for benzene and hexane. The curve for benzene is flat up to near monolayer coverage, whereas that for hexane rises to a maximum just before this point. Kiselev concluded that adsorbed benzene molecules were localized and immobile (as indeed one might expect on a hexagonally structured surface), while adsorbed hexane molecules were mobile and could mutually interact. Their interaction augmented the energy of adsorption as the surface coverage increased.

We also note that the heats of adsorption of both benzene and hexane are 20–50% larger than their heats of condensation, making graphite an extremely strong nonpolar adsorbent. The great strength of adsorption by graphite is a key factor in its exploitation as a coated adsorbent in gas chromatography [see (6), for example], where submonolayer coatings show extraordinary thermal stability because of their very high heats of adsorption. At the same time adsorption of monolayers produces sub-

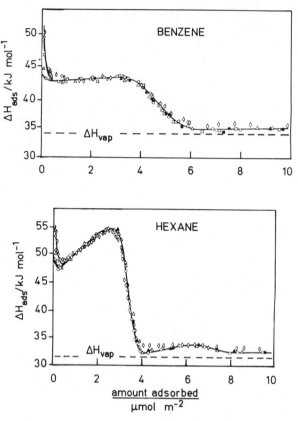

Figure 4.7. Dependence on surface coverage of the heats of adsorption of benzene (upper) and hexane (lower) on four different Sterling FT GCBs. Data from Isirikyan and Kiselev (Ref. 16). *Note:* Monolayer coverage occurs at around 4 mol/m², the effects of surface heterogeneity are observable at coverages below about 0.5 mol/m².

stantial changes in the retentive behavior of graphite. Undoubtedly similar modification of the surface of graphite by adsorptive coverage can be exploited in liquid chromatography, and will lead to new and unique separation capabilities.

As pointed out by Kiselev and Yashin (33), the flat graphite surface can provide unique stereochemical selectivity and there is a good probability that this can be predicted reliably. As an example, they cite the reversal of the common elution order of xylenes when separated on GCB. When a xylene molecule is adsorbed onto graphite, the two large methyl groups must always be in close contact with the surface, and they prevent many of the ring CH groups from touching the surface. Thus with 1,3-xylene, only the single CH group at position 5 can lie on the surface, whereas with the 1,2- and 1,4-xylenes two CH groups can lie on the surface. Accordingly *m*-xylene with three C atoms touching the surface is eluted before *o*- and *p*-xylenes with four C atoms touching the surface. This simple model was extended to explain the elution order of some much more complex terpene isomers by Belyakova et al. (18).

4.6. CHROMATOGRAPHIC PROPERTIES OF UNCOATED GRAPHITIC CARBON

4.6.1. General Chromatographic Properties

Since a perfect graphitic surface will act as an ideal van der Waals adsorbent, the chromatographic behavior of graphitic carbon is expected to be broadly similar to that of reversed-phase bonded silica gels. But there will be significant differences arising from the following factors. (1) Bonded silica gels possess residual functionality arising from the underivatized silanol groups, whereas graphites possess minimal adventitious functional groups, almost exclusively associated with the edges of the graphitic sheets. (2) The heats of adsorption of nonpolar molecules from the gas phase onto graphite are substantially greater than their heats of condensation, whereas their heats of partition into, say, the hydrocarbon monolayer of an ODS bonded silica are close to their heats of condensation. Graphite is therefore a much stronger reversed-phase adsorbent than an ODS silica gel. (3) Because the surface of graphite is flat, it shows specific stereoselectivity, especially for isomers, unlike an ODS silica gel. The differences in the chromatographic behavior of graphite and ODS silica gel arising from these factors are illustrated by a series of comparative chromatograms, published by Knox, Kaur, and Millward (9). Three of these pairs are shown in Figs. 4.8–4.10.

While the reversed-phase nature of elution from PGC is confirmed by Fig. 4.8, it is notable that the ethers, which on ODS silica gel elute after the three phenols, come significantly earlier with graphitic carbon. Figure 4.9 shows broadly similar elution orders for the methyl benzenes but significant reversals (as discussed above) for the xylenes and trimethyl benzenes. Figure 4.10 shows the substantial differences in the elution pattern and peak symmetry for amines. With PGC excellent symmetry is obtained, while with ODS silica the peak shapes are poor due to the presence of underivatized silanol groups.

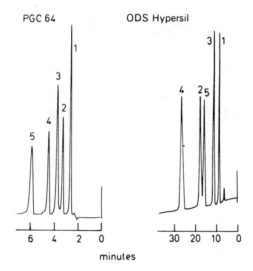

Figure 4.8. Comparative chromatograms of phenols and aromatic ethers on PGC and ODS Hypersil (ODS-HS). Eluents: methanol:water (95:5 for PGC, 60:40 for ODS-HS). Solutes: 1) phenol; 2) anisole; 3) 4-cresol; 4) phenetole; 5) 3,5-xylenol. [From Ref. (9). Reproduced by permission of the *Journal of Chromatography*.]

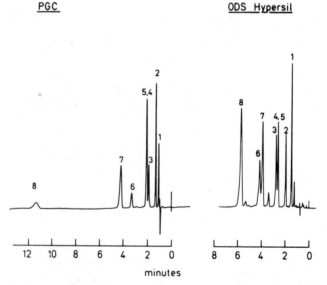

Figure 4.9. Comparative chromatograms of methyl benzenes on PGC and ODS-HS. Eluents: methanol:water (95:5 for PGC, 70:30 for ODS-HS). Solutes: 1) benzene; 2) toluene, 3) 3-xylene; 4) 2-xylene; 5) 4-xylene; 6) 1,3,5-TMB; 7) 1,2,4-TMB; 8) 1,2,4,5-TeMB. [From Ref. (9). Reproduced by permission of the *Journal of Chromatography*.]

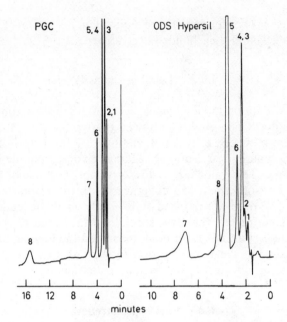

Figure 4.10. Comparative chromatograms of amines on PGC and ODS-HS. Eluents—as for Fig. 4.9. Note the better peak symmetry on PGC than on ODS-HS. Solutes: 1) aniline; 2) pyridine; 3) 2-methylpyridine; 4) methylaniline; 5) dimethylaniline; 6) ethylaniline; 7) diethylaniline; 8) benzylamine. [From Ref. (9). Reproduced by permission of the *Journal of Chromatography*.]

4.6.2.   Column Efficiency

The column efficiencies, in terms of reduced plate height, which Colin et al. (7, 22) determined for their carbon blacks coated with pyrocarbon, were disappointing. For ungraphitized material they ranged from 20 to 100, and after graphitization from 30 to 40. Graphitization was most effective in improving the HETP of late eluting solutes. These high HETPs can be attributed mainly to thermodynamic peak broadening and consequent asymmetry arising from nonlinear adsorption isotherms.

The calcined active carbons of Unger and coworkers (24) gave much better reduced plate heights of 4 for solutes having k' values of 0.22–0.66.

Ciccioli et al., with fractionated Carbopack B (21), obtained symmetrical peaks up to high k' values with negligible dependence of h upon k'. At optimum eluent velocities, their reduced plate heights at high k' were 3.5 for 75–88-μm particles, 4.0 for 33–45-μm particles, and 3.5 for 25–33-μm particles. Kaur et al. (see Figs. 4.8–4.10) obtained similar values with 6-μm PGC particles. Two-dimensional graphites therefore show mass transfer properties which are equivalent to those of high-grade HPLC silica gels and their bonded derivatives.

4.6.3. Loading Capacity

There are few data on the loading capacity of graphitic carbons in liquid chromatography. Colin et al. (7) showed that with a graphitized sample of pyrocarbon on GCB, k' was reduced by 10% for a load of only 5 μg on a column containing 2.2 mL of packing, giving a loading capacity of around 1.5 μg/g. This value is very low, probably because of severe nonlinearity of the isotherm at low coverage. With the graphitized material of Unger et al. (24), k' was reduced by 10% only when the loading reached 50 μg/g, in spite of the very low specific surface area of the material. Neither Ciccioli et al. nor Kaur et al. measured the loading capacities of their materials, but there is no reason to suppose that for graphite it is any less than for silica gel based packing materials.

4.6.4. Eluotropic Strength

The concept of an eluotropic series has been widely used to organize elution data for liquid chromatography on silica gel and alumina, and it is widely assumed that a similar series exists for carbon [see, for example, (22, 34)], only that the order of elution strength is reversed. Recent work suggests that this view is incorrect.

The usefulness of the concept of an eluotropic series depends critically on whether or not a single eluotropic order applies to all solutes. According to the widely accepted Snyder–Soczewinski (S-S) theory, adsorbed solvent molecules are displaced to make way for solute molecule at the adsorbent surface. The strength of retention of a solute then depends on the standard free energy change $\Delta G°$ for this displacement, and differences in the eluotropic strengths of different solvents are considered to arise exclusively from the different standard free energies of adsorption per unit area of these solvents. This highly simplified model (as freely acknowledged by Snyder) assumes that the effects of specific solute–solvent interactions in the liquid phase and on the adsorbent surface are unimportant. The great merit of the S-S theory is that it provides a simple expression for the column capacity ratio and explains its dependence on the nature of the eluent and to some extent on the composition of mixed eluents. The Snyder equation is

$$\log_{10}k' = \log_{10}(V_s/V_m) + \beta(S_X^0 - A_X\epsilon_M^0)$$

where V_s and V_m are the volumes of a monolayer of adsorbed eluent and of bulk eluent, β is a surface activity factor between 0 and 1, S_X^0 is the reduced free energy of adsorption of the solute X from the standard eluent, A_X is the molecular area of X in units of 8.5 Å^2, and ϵ_M^0 is the reduced free energy of adsorption of enough eluent (again from the standard eluent) to cover an area of 8.5 Å^2. The standard area of 8.5 Å^2 is one-sixth of the area of a molecule of benzene when adsorbed on an oxide surface. The reduced free energies of adsorption are the relevant values of $\Delta G°/2.303RT$. The Snyder equation when applied to silica gel and alumina conveniently provides values of ϵ in the range of zero to unity, taking pentane as the standard eluent for which ϵ is zero.

For oxide adsorbents, the key assumption is not too seriously in error, and the equation gives a semiquantitative explanation of differences in eluting power of typical liquid chromatography solvents. However, when the differences in the free energies of adsorption per unit area of different eluents become small, the effects of the assumptions made by Snyder are amplified and the order of eluting power of different solvents becomes dependent on the nature of the solute. There can then be no such thing as a unique eluotropic series applicable to a majority of solutes, and the concept of an eluotropic series loses much of its usefulness.

The eluotropic strength parameter ϵ is measured directly by determining k' values for a range of solutes with known A values and then applying the Snyder equation, taking ϵ as zero for some agreed standard solvent. Alternatively, ϵ may be obtained by measuring the k' values for the members of a homologous series. The logarithmic dependence of k' on the number of C atoms normally provides a straight line. The gradient of this line is inversely proportional to ϵ, whose evaluation finally requires the A value for the methylene group. More generally, ϵ can be found from the selectivity ratio α for any two compounds, provided that the difference in their A values is known.

Colin et al. (22, 35) used a combination of the first two, and their results are presented in Table 4.2 along with some theoretical values obtained from surface tension measurements using the theory of Eon (36). Unger et al. (24) also determined eluotropic strengths from retention data using a selection of aromatic solutes. Since they used a different standard area from that of Snyder, their values have been multiplied by 5.1 to bring them to the same basis. Table 4.2 shows that they found considerably larger variations in ϵ values between solutes than did Colin et al.

Dias, Kaur, and Knox (37) determined ϵ values on PGC by a number of different methods, and the values given in Table 4.2 are those which they obtained for methylene from plots of log k' against n for the n-alkyl benzenes. The values for alumina are taken from Snyder (34). It is clear

Table 4.2. Eluotropic Strengths on Carbons with Methanol as Standard Solvent

Solvent	ε Values on Carbon				ε Values on Alumina
	Colin et al. (35) Exp.	Eon (36) Theor.	Unger et al. (24) Exp.	Dias et al. (37) Exp.	Snyder (34) Lit.
Water	−0.35	−0.3	—	—	High
Methanol (standard)	0.00	0.00	0.00	0.00	0.95
Acetonitrile	0.04	—	0.00–0.10	−0.01	0.65
Ethanol	0.05	0.07	—	—	0.88
Propanol	0.07	0.095	—	—	0.82
Methyl acetate	0.08	—	—	—	0.60
Ethyl acetate	0.09	—	—	0.04	0.58
Butanol	0.09	0.12	—	—	—
Pentanol	0.11	0.13	—	—	0.61
Propyl acetate	0.115	—	—	—	—
Butyl acetate	0.12	—	—	—	—
Tetrachloromethane	—	—	0.12	—	0.18
Hexane	0.125	0.09	0.11	0.07	0.00
Pentyl acetate	0.13	—	—	—	—
Butyl chloride	0.135	—	—	0.09	0.30
Tetrahydrofuran	0.135	—	0.13	0.09	0.45
Dichloromethane	0.135	—	0.10–0.19	0.07	0.42
Heptane	0.14	0.11	—	—	0.00
Octane	0.15	0.135	—	—	0.01
Nonane	0.16	0.14	—	—	0.03
Methyl sulfoxide	—	—	0.16	—	0.62
Hexyl acetate	0.17	—	—	—	—
Dioxan	—	—	0.17	0.14	0.56
Chloroform	0.20	—	0.00–0.20	0.12	0.40
Benzene	0.20	0.215	—	—	0.32
Xylene	0.24	0.315	—	—	0.26

that the range of eluotropic strength for organic solvents on carbon is small, only about 0.2 unit compared with nearly 1 unit for alumina. (Silica gel has values similar to those for alumina.) When compared with the values for alumina, there is little correlation, and the order for carbon is certainly not the reverse of that for alumina. The order of ε values in the first column is that found by Colin et al. (35) and is slightly different from that obtained using the theory of Eon (36), although the values fall within the same range. Unger's values differ significantly from those of Colin et al., although they again fall within the same range. The values of Dias et al. fall in the same order as those of Colin et al., but are significantly smaller. The combined data of Colin et al., Unger et al., and Dias et al. give some support to the idea of a unique eluotropic series for carbon, but the small range and the inconsistencies of the values cast doubt on the general validity of this idea. These doubts are amplified by the further data of Dias, Kaur, and Knox (37). When they measured eluotropic strengths on PGC directly for a variety of simple solutes, contrary to what might be deduced from the data of Table 4.2, they found little correlation between the ϵ values obtained for different solute systems. The lack of correlation is shown in Fig. 4.11, where the ϵ values for different solutes are plotted against the mean ϵ values obtained from the homologous series. Their data imply that there is no universal order of eluotropic strength for carbon, although there is general agreement that methanol and acetonitrile are weak solvents in nearly all instances.

In the same work Dias et al. used gas chromatography to determine absolute heats of adsorption ΔH for a number of typical eluents from the gas phase. According to the S-S theory, these should correlate directly with the ϵ values found by liquid chromatography. Figure 4.12 shows that the enthalpy of adsorption of various typical eluents from the gas phase onto PGC is directly proportional to their molecular areas A as computed

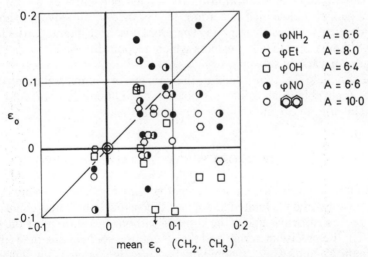

Figure 4.11. Plot of eluotropic strength parameter ϵ on PGC for various simple aromatic compounds against the mean eluotropic strength on PGC for methyl and methylene groups obtained from homologous series, showing the lack of any correlation (methanol taken as standard solvent). [From Ref. (37)].

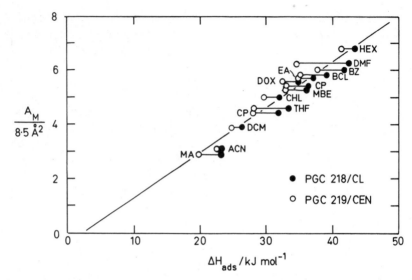

Figure 4.12. Dependence on molecular areas of heats of adsorption onto PGC from the gas phase of common liquid chromatography eluents. Molecular areas as computed from Snyder (Ref. 34). Open and closed points refer to different samples of PGC (Ref. 37).

from the tables of Snyder (34). Since the ϵ value in S-S theory is defined as the free energy of adsorption of the solvent per unit area of surface, it now becomes clear that according to this theory, all solvents should have identical ϵ values on graphite. Therefore it is those interactions which the S-S theory specifically neglects which account for the differences in ϵ values for various solvents.

We conclude that, for graphite, the concept of a universal eluotropic series, applicable to all solutes, is not tenable. Other concepts will have to be called in to organize elution data on this material.

4.7. APPLICATIONS OF UNCOATED GRAPHITIC CARBONS

The number of published liquid chromatography applications using modern forms of carbon is limited and it is still possible to be comprehensive. They are listed below under the type of carbon; the authors of the application; the precise material; and the eluent, the solutes separated, their maximum k', and, where a homologous series is involved, the value of α. α is given by the following equation, where n is the number of repeated groups (normally methylene or methyl) in the homologue:

$$\alpha = \log_{10}(k'_{n+1}/k'_n)$$

Special features are also be noted. "Chr" means that a chromatogram is shown in the original publication; "Plot" means that the data are included in a plot, such as a plot of log k' against n for a homologous series.

Pyrocarbon-Coated Carbon Black

These materials are ungraphitized and tend to give rather tailed peaks. Nevertheless their general chromatographic characteristics are similar to those of graphitized materials, such as GCB and PGC.

Colin et al. (7)   Black Pearls L (carbon black) + 25% pyrocarbon, specific surface area before deposition of pyrocarbon 110 m^2/g, particle size 15–20 μm

Acetonitrile

Benzene + phenol to trimethylmethyl phenols, $k' = 2.0$, Chr

Alkanols from C_5 to C_{12}, $k' = 1.8$, Chr

Naphthalene derivatives from tetralin to 1,3,7-trimethyl naphthalene, $k' = 7.5$, Chr

Acetonitrile:water (46:54 v/v)

Benzene to *n*-butyl benzene, $k' = 22$, last peak very wide, Chr

Methanol

Alkyl benzenes, $n = 1–8$, $k' = 9$, $\alpha = 0.25$, Plot

Alkyl bromides, $n = 3–7$, $k' = 2.2$, $\alpha = 0.25$, Plot

Alkyl chlorides, $n = 3–7$, $k' = 0.5$, $\alpha = 0.25$, Plot

Alkanols, $n = 3–11$, $k' = 2.5$, $\alpha = 0.25$, Plot

Colin et al. (22)   Black Pearls L (carbon black) + 40% pyrocarbon, specific surface area before deposition 110 m^2/g, particle size 25–32 μm

Acetonitrile

Homologous series of alkyl benzenes, alkyl bromides, alkyl chlorides, and alkanols, k' and α values very similar to those immediately above, obtained with methanol, Plots

Methyl phenols, $n = 1–4$, $k' = 5$, $\alpha = 0.43$, Plot

Methyl benzenes, $n = 2–5$, $k' = 6$, $\alpha = 0.43$, Plot, Chr. These solutes gave the best chromatogram of any mixture on pyrocarbon-coated carbon black (see Fig. 4.1)

Benzene

Polynuclear aromatics up to pyrene, $k' = 3.6$, Chr

Ethyl acetate

Diphenol derivatives from 1,3-dihydroxy-5-propyl benzene to 1,3-dihydroxy-2,5-dimethyl-2-carboxyethyl-benzene, $k' = 11$, Chr

Colin et al. (22)   Sterling FTFF (carbon black) + 14% pyrocarbon, specific surface area before deposition ≈ 10 m²/g, particle size 15–20 μm
  Acetonitrile
    Methyl naphthalenes, n = 0–3 (n = number of methyl groups), k' = 8, α = 0.43, Plot, Chr
  Acetonitrile:water (41:59 v/v)
    Benzene, toluene, nitrobenzene, nitrotoluene, k' = 4, Chr

Purified Graphitized Active Carbon

Unger and Roumeliotis (24)   Active carbon heated to 1800°C (carbon #6), specific surface area 1–5 m²/g, particle diameter 9 μm
  Methanol:water (80:20 v/v) buffered to pH 11.2
    Alkaloids: ephedrine, codeine + anaestesin, theobromine, caffeine, strychnine, k' = 4.5, Chr

Unger and Roumeliotis (24)   Purified coke heated to 2800°C (carbon #2), specific surface area 1–5 m²/g, particle diameter 6 μm
  Methanol
    Naphthalene to pyrene, k' = 20, Plot
  Methanol:water (55:45 v/v)
    Toluene to 1,3,5-trimethylbenzene, k' = 11, Plot
    Dimethyl- to dicyclohexyl-phthalate, k' = 33, Plot

Knox, Unger, and Mueller (38)   Active carbon heated to 1800°C, specific surface area 5 m²/g, particle diameter 9 μm
  Methanol:water (30:70 v/v) pH 6.0 with 5 mM tetrabutyl ammonium ion
    Benzene pentacarboxylic acid to 2,3-dihydroxy-benzoic acid, k' = 10, Chr, low plate efficiency

Graphitized Carbon Black

Ciccioli, Tappa, di Corcia, and Liberti (21)   Carbopack B, specific surface area 80 m²/g, particle diameter 25–33 μm
  Pentane
    Methyl naphthalene, k' = 3.1, 3,6-dimethylnaphthalene, k' = 7.0, HETP plots, Chr, excellent peak symmetry
    PCB mixture (Fenchlor 64): chlordane to Metoxichlor, k' = 2, Chr
  Methanol
    PCB mixture (Fenchlor 64): as immediately above, k' = 3, Chr, excellent peak shape and resolution

Dimethyl- to dioctyl phthalates, k' = 11, Chr

Triazine isomers, k' = 5, Chr, excellent peak shape

Phenylacetamide, phenacetin, caffeine, k' = 25, Chr, excellent peak shape (see Fig. 4.2)

Phenylalanine and tryptophan, k' = 3.5, Chr

Porous Graphitic Carbon

Gilbert, Knox, and Kaur (8) Spherical particles, specific surface area 380 m²/g, particle diameter 6 μm

Methanol

Toluene to 1,2,4,5-tetramethylbenzene, k' = 15, α = 0.4, Plot, Chr

Phenol to trimethyl phenol, k' = 10, α = 0.4, Plot

Alkyl benzenes n = 7, k' = 18, α = 0.23, Plot

Dichloromethane + 0.1% 1,3-terphenyl

Naphthalene, acenaphthene, fluorene, k' = 10, Chr

Knox, Kaur, and Millward (9) Spherical particles, specific surface area 150 m²/g, particle diameter 6 μm, chromatograms on PGC compared with those on ODS Hypersil

Methanol:water (95:5 w/w)

Phenol, anisole, *p*-cresol, phenetole, 3.5-xylenol, k' = 2, Chr, excellent peak shape (see Fig. 4.8)

Benzene to 1,2,4,5-tetramethylbenzene, k' = 12, Chr, excellent peak shape

Monofunctional benzene derivatives: benzene to methylbenzoate, k' = 3, Chr, excellent peak shape (see Fig. 4.9)

Anilines to diethylaniline, benzylamine, k' = 6.5, Chr, excellent peak shape (see Fig. 4.10)

Phenacetin, paracetamol, caffeine, k' = 10, Chr, excellent peak shape (see Fig. 4.3).

Methanol:water (90:10 w/w)

Benzoic, toluic, and salicylic acids, k' = 2.2, Chr

Methanol:water:acetic acid (94:5:1 w/w/w)

Phenacetin, aspirin, paracetamol, salicylic acid, k' = 3, Chr

Van Zoonen et al. (39) Material as for previous group

Acetonitrile:0.05 M Tris buffer, pH 8, 0.001 M hydrogen peroxide in water (80:20)

4-Isopropyl aniline, *N,N*-dimethylaniline, *N*-ethyl-3-toluidine, *N,N*-dipropylaniline, k' = 1.5, Chr

De Biasi et al. (40)   Material as for previous group
   Methanol:0.5 M aqueous HCl (90:10)
    Diastereoisomers of dipeptide BRL36378

4.8. CHROMATOGRAPHIC PROPERTIES AND APPLICATIONS ON COATED GRAPHITIC CARBON

Whereas the silica surface can be derivatized to form chemically bonded silica gels such as ODS silica gel, the graphite surface is chemically unreactive and cannot be derivatized directly, apart possibly from the nongraphitic functional groups which may be present at the edges of the graphitic sheets. However, graphite is an extremely strong adsorbent and high molecular weight substances can be almost irreversibly adsorbed to form monolayers on the surface. The adsorptive properties of such monolayers are significantly different from those of graphite itself, and retention by them is generally weaker. Monolayers can in principle be stabilized by cross-linking to render them insoluble in eluents. The process can be taken further by polymerizing and cross-linking substantially thicker layers. In this way the initial adsorptive properties of graphite may be radically modified, and even masked completely. There is thus a continuous spectrum of adsorptive modification which can be applied to graphite from the reversible adsorption of additives such as ion-pairing agents to the formation of permanent polymeric coatings. To date only a few of these possibilities have been explored.

4.8.1. Modification by Adsorption of Polar Additives

In general, the adsorption of an additive of relatively high molecular weight from an eluent will markedly reduce the k' values of solutes. When an eluent containing modifier is first supplied to the column, a front first forms, which gradually moves down the column as the modifier is extracted from the eluent by adsorption. The movement of the front may be observed by determining the change in the k' of a test solute during this adsorption period. The k' value should change linearly with the volume of the eluent passed until, when the front reaches the end of the column, k' becomes constant. The modifier adsorbed on the surface is then in equilibrium with the modifier dissolved in the eluent. Provided that the modifier concentration is not too high, this procedure will deposit a monolayer which will often be so strongly adsorbed that it is not readily washed off even by eluent containing no modifier.

Figure 4.13 shows the reduction in the k' of test solutes eluted from

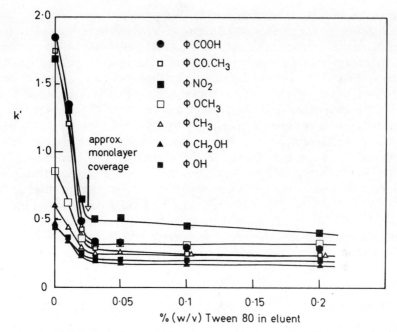

Figure 4.13. Dependence of k' on concentration of Tween 80 (polyoxyethylene sorbitol stearate). Packing: PGC. Eluent: methanol:water (95:5). Note the constancy of k' when Tween concentration exceeds about 0.025%, corresponding to monolayer coverage of PGC surface (Ref. 41).

PGC when Tween 80 (polyoxyethylene sorbitol stearate) at various concentrations is present in a typical eluent, methanol:water (95:5) (41). The k' values are progressively reduced as the concentration of Tween 80 in the eluent increases from zero to 0.025%, and then become more or less constant, indicating that a monolayer of Tween 80 has probably been formed at this eluent concentration. The surface concentration of Tween may be obtained from the dependence of k' on the volume of eluent passed during the initial loading period. In this case it is about 2.5% by weight of PGC. The selectivity of the Tween-coated PGC is significantly different from that of bare PGC. Evidently surface coating is a powerful method for modifying the adsorptive properties of graphite in liquid chromatography, confirming what is well known from studies in gas chromatography (6).

4.8.2. Modification by Adsorption of Ion-Pairing Agents

The retentive behavior of graphite like that of reversed-phase silica gels can be dramatically altered by the addition of ion-pairing agents to the eluent, especially for ionized solutes. With graphite there are two competing effects. The masking of the surface by the ion-pairing agent tends to reduce the retention of all solutes, while the ion-pairing effect enhances the retention of solute ions having opposite charge to the pairing ion. The overall effect of an ion-pairing agent may therefore be to either increase or decrease the retention of an ionized solute. This is illustrated in Fig. 4.14. These data, obtained by Ghaemi and Stewart (42), show the effect of added tetrabutyl ammonium ion (TBAI) on the elution of phenols from PGC by 80:20 methanol:water. In order to ionize the phenols an eluent with pH 12 was used, emphasizing the potential of carbon adsorbents for use under strongly alkaline conditions, unlike silica gel based packings which would be unusable. The k' values rise to maxima for three of the four phenols studied, in accord with standard ion-pairing theory, but for 2,3-xylenol k' is more or less independent of the TBAI concentration. For unionized solutes k' values decreased with increasing addition of TBAI.

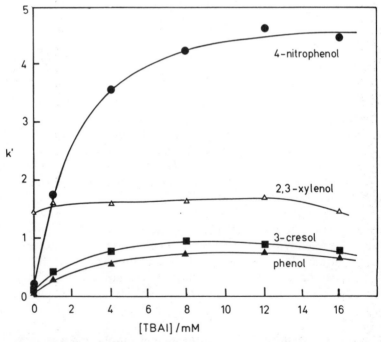

Figure 4.14. Dependence of k' of phenols on concentration of ion-pairing agent tetrabutyl ammonium iodide (TBAI). Packing: PGC. Eluent: methanol:water (80:20); pH 12, adjusted by addition of sodium hydroxide (Ref. 42).

4.8.3. Polymer Coating

The very strong adsorption of organic species by graphite makes it possible to mask the original characteristics of graphite completely by multilayer polymer coating. This is illustrated by the size exclusion chromatography calibration curve for polystyrenes shown in Fig. 4.15.

The data were obtained on a sample of PGC, which had been coated with approximately 100% of its weight of a hydrophilic polymer (43). The exclusion of polystyrenes of high molecular weight by coated PGC shows that this material has now no adsorptive affinity for polystyrene when tetrahydrofuran is used as eluent. The mean pore diameter obtained from the size of the polymer which shows 50% exclusion is about 300 Å. The

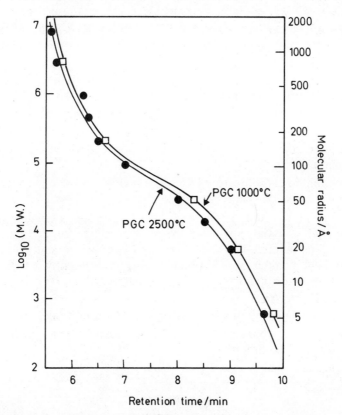

Figure 4.15. Size exclusion chromatography calibration curve for polystyrene standards. Packing: PGC heated to temperatures of 1000°C and 2500°C, respectively, and subsequently coated with ≈100% w/w of a hydrophilic polymer. Eluent: tetrahydrofuran. [From Ref. (43). Reproduced by permission of Polymer Laboratories Ltd., Church Stretton, UK.]

same material has been used to separate biopolymers by aqueous size-exclusion chromatography.

These experiments show that it is possible to mask the adsorptive properties of graphite totally by polymer coating. A route has therefore been opened for modifying graphite to provide a range of novel functionalities. In particular, the possibility now exists for using polymer coated graphites as a base for making strong robust packings for affinity chromatography, and for separations of sensitive biological compounds.

4.9. CONCLUSIONS

1. Graphitic carbons have now been developed with suitable specific surface area, porosity, and mechanical strength for use in HPLC.
2. Graphitic carbon acts as a very strong reversed-phase adsorbent, which shows unusual stereoselectivity owing to the ordered crystalline nature of its internal surface.
3. Graphitic carbons provide roughly the same plate efficiencies as do bonded silica gels.
4. The loading capacity of graphitic carbon is similar to that of other adsorbents with energetically homogeneous surfaces.
5. Graphitic carbon can be used in eluents of all pH and is chemically unreactive.
6. Graphitic carbons are likely to be highly reproducible since they all possess the same crystalline surface.
7. Graphitic carbons can be modified by adsorption of high molecular weight materials such as nonionic detergents, ion-pairing agents, and polymers. Adsorbed polymers can subsequently be cross-linked to render them insoluble in eluents. Adsorptive modification can provide a wide spectrum of separation characteristics.
8. Suitably coated graphitic carbon can be used for size exclusion chromatography of synthetic polymers and of biopolymers.
9. Graphitic carbons are likely to become major HPLC packing materials in the future.

REFERENCES

1. A. Standen, Ed., *Kirk-Othmer Encyclopedia of Chemical Technology*, 2nd ed., Wiley, New York, 1964.
2. M. J. Telepchak, *Chromatographis, 6,* 234 (1973).

3. G. M. Jenkins and K. Kawamura, *Polymeric Carbons*, Cambridge University Press, Cambridge, UK, 1976.

4. A. Tiselius, *Adv. in Colloid Sci.*, *1*, 81 (1941).

5. J. Biscoe and B. E. Warren, *J. Appl. Phys.*, *13*, 364 (1942).

6. A. Di Corcia and A. Liberti, *Advances in Chromatography*, vol. 14, Dekker, New York, 1976, p. 305.

7. H. Colin, C. Eon, and G. Guiochon, *J. Chromatogr.*, *119*, 41–54, (1976).

8. M. T. Gilbert, J. H. Knox, and B. Kaur, *Chromatographia*, *16*, 138–146 (1982).

9. J. H. Knox, B. Kaur, and G. R. Millward, *J. Chromatogr.*, *352*, 3–25 (1986).

10. E. Lederer and M. Lederer, *Chromatography*, Elsevier, London, 1953.

11. A. Tiselius, *Kolloid-Z.*, *105*, 101 (1943).

12. A. Tiselius, *Adv. Protein Chem.*, *3*, 67–93 (1947).

13. R. L. M. Synge and A. Tiselius, *Acta Chem. Scand.*, *3*, 231 (1949).

14. A. Tiselius and S. Claesson, *Arch. Kemi. Mineral. Geol.*, *15B* (18), (1942).

15. S. Claesson, *Arch. Kemi. Mineral. Geol.*, *23A* (1) (1946).

16. A. A. Isirikyan and A. V. Kiselev, *Russ. J. Phys. Chem.*, *36*, 618–622 (1962).

17. A. V. Kiselev and D. P. Poshkus, *J. Chem. Soc., Faraday Symp.*, *15*, 13–24 (1980).

18. L. D. Belyakova, A. V. Kiselev, and N. V. Kovaleva, *Russ. J. Phys. Chem.*, *42*, 1204–1208 (1968).

19. A. V. Kiselev, G. M. Petrov, and K. D. Shcherbakova, *Russ. J. Phys. Chem.*, *41*, 751–756 (1967).

20. Supelco Catalog, Supelco, Inc., Bellefonte, PA 16823, 1986.

21. P. Ciccioli, R. Tappa, A. Di Corcia, and A. Liberti, *J. Chromatogr.*, *206*, 35 (1981).

22. H. Colin, C. Eon, and G. Guiochon, *J. Chromatogr.*, *122*, 223–242 (1976).

23. H. Colin and G. Guiochon, *J. Chromatogr.*, *137*, 19–33 (1977).

24. K. Unger, P. Roumeliotis, H. Mueller, and H. Goetz, *J. Chromatogr.*, *202*, 3–14 (1980).

25. Z. Plzak, F. P. Dousek, and J. Jansta, *J. Chromatogr.*, *147*, 137–142 (1978).

26. E. Smolkova, J. Zima, F. P. Dousek, J. Jansta, and Z. Plzak, *J. Chromatogr.*, *191*, 61–69 (1980).

27. J. Zima and E. Smolkova, *J. Chromatogr.*, *207*, 79–84 (1981).

28. J. H. Knox and M. T. Gilbert, U.K. patent 7939449, 1979; U.S. patent 4263268, 1979; F.R.G. patent P2946688-4, 1979.

29. I. Novak and D. Berek, Czech. patent 221197, 1982.

30. E. Skutchanova, L. Feltl, E. Smolkova-Keulemansova, and J. Skutchan, *J. Chromatogr.*, *292*, 233–239 (1984).

31. J. D. Bernal, *Proc. R. Soc.* (*London*), *A100*, 749 (1924).

32. P. Scherrer, *Nachr. Ges. Wiss. Göttingen*, *2*, 98 (1918).

33. A. V. Kiselev and Y. I. Yashin, *Gas Adsorption Chromatography*, J. E. S. Bradley, Transl., Plenum, London, 1969.

34. L. R. Snyder, *Principles of Adsorption Chromatography*, Dekker, New York, 1968.

35. H. Colin, G. Guiochon, and P. Jandera, *Chromatographia, 15,* 133–139 (1982).

36. C. H. Eon, *Anal. Chem., 47,* 1871–1873 (1975).

37. H. Dias, B. Kaur, and J. H. Knox, University of Edinburgh, Edinburgh, unpublished data.

38. J. H. Knox, K. K. Unger, and H. Mueller, *J. Liq. Chromatogr., 6,* 1–36 (1983).

39. P. Van Zoonen, D. A. Kamminga, C. Gooijer, N. H. Velthorst, and R. W. Frei, *Anal. Chem., 58,* 1245–1248 (1986).

40. V. De Biasi, M. B. Evans, and W. J. Lough, Abstract for 11th ISCLC, Amsterdam, July 1–5, 1987.

41. H. Dias and J. H. Knox, University of Edinburgh, Edinburgh, unpublished data.

42. Y. Ghaemi and M. Stewart, University of Edinburgh, Edinburgh, unpublished data.

43. M. Wareing, Polymer Laboratories Ltd., Church Stretton, UK, unpublished data.

CHAPTER

5

ORGANIC POLYMERIC STATIONARY PHASES

DONALD J. PIETRZYK

Department of Chemistry
University of Iowa
Iowa City, Iowa

5.1. INTRODUCTION

5.1.1. Historical

Liquid chromatography (LC) in a column is a strategy that was first introduced by M. Tswett at the turn of the century. Subsequent development of this strategy, which is more familiarly known as *liquid–solid*

223

column chromatography (or *adsorption column liquid chromatography*) was accomplished primarily with silica and alumina as stationary phases. While organic materials such as activated charcoal, sucrose, starch, and more recently polyamides were evaluated and shown to be useful, the major stationary phases in adsorption liquid chromatography (LC) in a column were silica and alumina. Interest in solid, insoluble organic materials as stationary phase particles primarily grew with the emergence of liquid–liquid column chromatography, (or partition column LC). In partition LC in a column, an "inert" particle is coated with a liquid layer, and this coated material serves as the stationary phase. While alumina, silica, charcoal, and many other inorganic-type materials were used successfully as inert supports to hold the liquid layer, organic polymeric materials tended to be more versatile since nonpolar-type organic polymers could be used to hold nonpolar liquid stationary phases and polar-type organic polymers could be used to hold polar liquid stationary phases.

During this period of growth in column LC, ion exchangers were also being developed as stationary phases for column LC. A major breakthrough in ion exchange LC in columns occurred in the 1930s when the first organic polymeric-type ion exchangers were synthesized. Following this and up to the 1960s, essentially all organic polymeric ion exchangers were cross-linked polyelectrolytic gel-type polymers which possessed no clearly defined internal structure. The porelike structure in these types of exchangers, which is small, dependent on the degree of bead swelling, and generally less than 30 Å, is determined by the distances between the polymer chains and the cross-links. These distances in turn are determined by swelling factors associated with the ionogenic site and its environment. For example, increasing the solvent polarity, the number of ionogenic groups, the ionic strength, and the temperature and decreasing the cross-linking increases the porouslike character of the gel exchanger.

In the 1960s a major development in ion exchange technology occurred when macroporous (macroreticular) ion exchangers were synthesized and subsequently became commercially available in large quantities (1–4). This was a significant advance in ion exchange technology because it represented the first major development in the synthesis of ion exchangers that focused on physical structure.

The macroporous ion exchanger is a highly cross-linked copolymer containing chemically attached ionogenic groups that are composed into a network of microsphere gel beads joined to form a single larger bead. In these structures the pores and the polymer matrix are present as continuous phases. The more permanent porous character (large pores) arises from the interconnected free space in the network of the microsphere gel beads. The size of the large pores is determined by the size of the microspheres and by the polymerization strategy, including the kind and amounts of monomers and the kind and amount of solvents used in the polymerization. It is not unusual to prepare macroporous ion exchangers that have pore sizes well beyond 50 Å. Furthermore, the pore sizes of the macroporous ion exchanger are more permanent and less susceptible to the effects of swelling factors than in the case of the microporous ion exchanger. The macroporous ion exchanger, which often is considered to be a rigid structure, is, however, not completely free from the swelling factors since it still is composed of microsphere gel beads, and their presence and contribution must always be considered.

5.1.2. Macroporous Adsorbents

A significant spin-off from the synthesis of the macroporous ion exchangers was the observation that the copolymeric matrix of the macroporous ion exchanger in the absence of ionogenic groups retained its high porosity and surface area and exhibited excellent characteristics as an adsorbent (5, 6). Subsequently polymerization strategies were improved and modified and were used to produce macroporous, copolymeric, nonionic adsorbents. These macroporous adsorbents are like the macroporous ion exchangers in that they are also composed of a network of microsphere gel beads joined together to provide a continuous pore and polymer phase, and differ in that they contain no chemically attached ionogenic groups. Because of the macroporous structure, these nonionic adsorbents like the macroporous ion exchanger possess several key improvements over gel-type polymers.

1. The contribution of the permanent pore structure permits their use in an anhydrous, nonpolar environment since solvent swelling is no longer a major requirement.
2. Resistance to osmotic and physical stress is significantly improved.
3. The large pore structure widens the scope of a filtering and/or exclusion property.
4. The copolymer structure can be further chemically modified with little or no effect on the porous character.
5. Macroporous structures of a wide range of pore sizes and surface areas can be prepared.

The initial development of the macroporous nonionic adsorbents was to complement the classical adsorbents, such as activated carbon, silica, and alumina. Since commercial large-scale applications were the goal, spherical, large particles (20–50 mesh or ≈1000-μm average) were prepared in order to facilitate flow characteristics. While these particle sizes are favorable for large-scale operations, they are not suitable as stationary phases for modern column LC at an analytical loading level. The absence of readily available microparticles in a size range that would yield a favorable column efficiency clearly handicapped the acceptance of macroporous polymers and their development as modern stationary phases for high performance liquid chromatography (HPLC). However, this has changed and recent commercial interest by chromatographic suppliers has resulted in the availability of nonionic, macroporous organic polymeric particles that are spherical, of uniform size, and are in a particle size that yields favorable column efficiencies in HPLC applications.

Organic polymeric stationary phase particles have also been developed for applications in other types of chromatography. For example, they can be used as adsorbents and inert supports for liquid layers in thin-layer and gas chromatography, as stationary phases for size-exclusion chromatography, and as matrix material in affinity chromatography. No attempt is made to consider these important applications. Several of these are discussed in subsequent chapters. The focus of this chapter is on the use of organic polymers, particularly those derived from the copolymerization of styrene and divinylbenzene, as stationary phase adsorbents, as an inert support matrix for partition chromatography, and as stationary phases for the HPLC separation of organic nonionic and ionic analytes. While the scope of the applications of organic polymeric stationary phases in adsorption and partition is wide and varied, and they even have commercial use, particularly as adsorbents, the major emphasis in this chapter is on the latter application, namely, as stationary phases in HPLC.

5.2. PHYSICAL PROPERTIES OF MACROPOROUS ORGANIC POLYMERS

The physical properties of the macroporous polymeric adsorbents are best described by considering their pore structure and surface area. The average size of the pores and the surface area of the polymer are determined directly by the internal gel structures due to the microspherical gel beads and indirectly by the types and amounts of monomers and solvents used in the polymerization. Porosities of macroporous polymeric adsorbents can be very high. For example, for the XAD copolymers listed in Table 5.1, porosities are in the range of 0.37–0.55 mL of pore volume per milliliter of dry polymer, or 0.60–1.08 mL of pore volume per gram of dry polymer. The interconnecting polymeric and pore phase characteristics

Table 5.1. Commercially Available Macroporous Copolymers

Trivial Name*	Polymer Type	Particle Size (μm)	Average Surface Area (m²/g)	Average Pore Diameter (Å)
XAD-1	Styrene	300–1200	100	205
XAD-2	Styrene	300–1200	300	90
XAD-4	Styrene	300–1200	725	50
XAD-7	Acrylic ester	300–1200	450	90
XAD-8	Acrylic ester	300–1200	160	235
AD-71	Acrylic ester	50–100, 80–160	500	250
AD-161	Styrene	50–100, 80–160	875	20–300
PRP-1	Styrene	5, 10	415	75
PRP-3	Styrene	10	100	300
PLRP-S 100	Styrene	5, 10	550	100
PLRP-S 300	Styrene	8, 10	384	300

* Amberlite XAD and Amberchrom AD—Rohm and Haas Co., XAD copolymers are also available commercially from chromatographic suppliers under other trade names; PRP— Hamilton Co.; PLRP-S—Polymer Laboratories, Inc.

of the macroporous polymers are shown in Fig. 5.1, which is a scanning electron photomicrograph of an XAD-2 particle. The spherical-like microbeads are gellular, while the spacings between the microgellular beads make up the continuous permanent porous characteristic.

The first commercially available macroporous, nonionic, organic polymers were synthesized by Rohm and Haas Company (1, 3, 5, 6) and were designated under the trivial name Amberlite XAD copolymers. Several XAD adsorbents were and still are available and are listed in Table 5.1. Since they were initially synthesized primarily for commercial large-scale applications, the spherical particles were large and covered a wide range of sizes. Table 5.1 indicates typical average sizes for the XAD absorbents. A styrene based copolymer Amberchrom AD-161, and a methacrylate based copolymer, Amberchrom AD-71, have recently been made available in a smaller range of spherical particles. About 15 years later Hamilton Company introduced a styrene-based macroporous copolymer, listed as PRP-1 in Table 5.1. It was a spherical microparticle of uniform size distribution (8, 9). Shortly thereafter Polymer Laboratories also introduced several styrene-based macroporous polymers of uniform spherical microparticles that differ in average pore sizes. These are designated by the trivial name PLRP-S in Table 5.1. The PRP-1 and PLRP-S adsorbents, unlike the early XAD adsorbents, are capable of delivering favorable column efficiencies in HPLC applications.

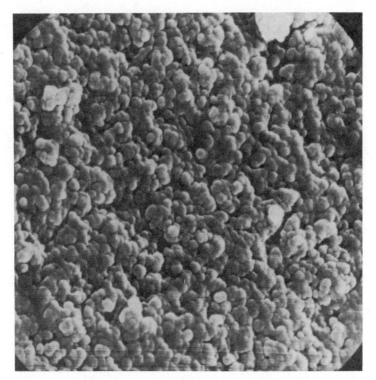

Figure 5.1. Scanning electron photomicrograph of XAD-2. [Reprinted from Ref. (7) by permission of Institution of Mining and Metallurgy.]

A major physical characteristic of all of these organic macroporous nonionic adsorbents is that they can be prepared with high average surface areas and porosities. A summary of these important characteristics, which were provided by the manufacturers, is also given in Table 5.1. Furthermore, these properties can be systemically altered synthetically. The

Table 5.2. Typical Surface Areas of Adsorbents

Porous		Nonporous	
Adsorbent	Typical Surface Area (m^2/g)	Adsorbent	Typical Surface Area (m^2/g)
Carbon, granular	500–2000	Carbon black	100
Silica gel	600	TiO_2 pigment	70
Soils	10–100	ZnO_2 pigment	1–10
Asbestos	20		
Polymeric adsorbents	<10 to >1000		

polymeric adsorbents are essentially a new class of adsorbents and in terms of surface area compare favorably with other older well-known adsorbents. This is illustrated in Table 5.2, where typical surface areas of other adsorbents are listed.

5.3. CHEMICAL STRUCTURE OF MACROPOROUS ORGANIC POLYMERS

Many experimental polymerization techniques are available which can be used to synthesize organic polymers that range in properties from a complete gellular structure to a macroporous structure. In general, the macroporous types that have proven to be the most useful at present as adsorbents and as stationary phases in column LC are the styrene-divinylbenzene- and acrylic-based polymers (see Table 5.1). Synthetic procedures for their preparation are reviewed elsewhere (1–6, 10).

Styrene-divinylbenzene-based polymers can be classified as one of three general types (10).

1. Gel-type polymers made by block copolymerization of monomers
2. Gel-modified (macroporous) polymers made by polymerization in an inert solvent (If the solvent dissolves the monomer and solvates the polymer, a telogenated polymer is formed. If it also precipitates out, it does so with macroporous structure.)
3. Specialty cross-linked polymers (In these types of polymers the cross-linking agent is chosen based on its ability to act as a spacer as well as a cross-linking agent.)

The organic polymeric adsorbents are nonionic, and their adsorptive properties are totally dependent on their surface characteristics. The pore structure that results in the macroporous polymers tends to be heterogeneous. Consequently, physical properties (see Table 5.1), such as surface area, pore diameter, pore volume, pore size distribution, and swelling characteristics, are treated in their determination and application as average properties. This is in contrast to the organized structure possessed by most other adsorbents, particularly inorganic ones.

The common organic polymeric adsorbents listed in Table 5.1 are nonpolar or of modest polarity. The nonpolar or styrene-based group are copolymers of styrene and are heavily cross-linked with divinylbenzene (see Fig. 5.2a). The monomer dipole moment for this group is on the order of about 0.3 debye. Thus this cross-linked copolymer provides an extremely hydrophobic surface, which is able to bind organic analytes by

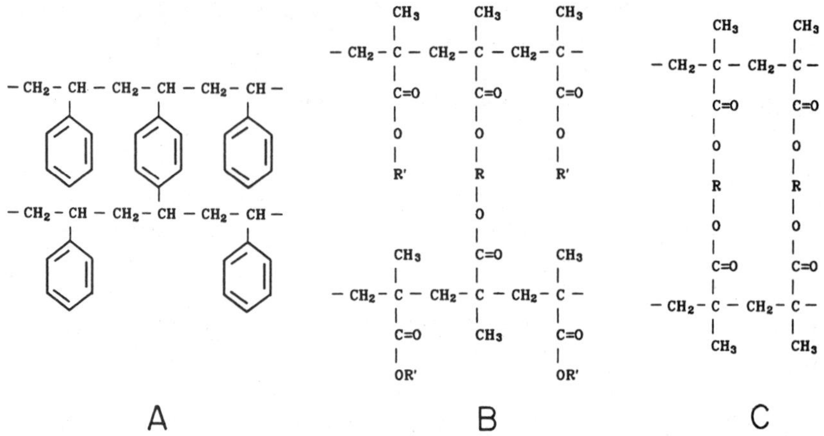

Figure 5.2. Structures of commercially available (a) styrene- and (b), (c) acrylic-based macroporous copolymers.

van der Waals attractive forces. The acrylic-based polymers are cross-linked polymethacrylate structures (see Fig. 5.2b and c). The monomer dipole moments are approximately 1.8 debyes, and while the surface is not truly polar, it is much more polar than are the styrene-based adsorbents. Other macroporous polymers can be prepared; however, these are not commercially available (1–6, 10, 11).

All the polymeric adsorbents listed in Table 5.1 are insoluble in water and common polar and nonpolar organic solvents. Because of the hydrophobic surface property, the styrene- and acrylic-based adsorbents are not readily water wetted and do not settle rapidly when immersed in water. If the beads are first treated with a few milliliters of water-miscible organic solvent, such as alcohol, water wetting is facilitated. Similarly, immersion of the beads in a water–organic solvent mixture leads to a wetted surface. The macroporous beads in Table 5.1 are chemically inert, in particular the styrene-based adsorbents, physically strong, and rigid. Although swelling and contracting will take place due to the gel microbead phase, this property is much less than that observed with the corresponding gel-type polymer. Consequently, bead fracture is less of a problem with the macroporous adsorbent. The XAD adsorbents, depending on the source, may require a cleanup procedure to remove monomer and low molecular weight polymer. Extraction by a Soxhlet extractor using alternating alcohol and benzene is often appropriate (12, 13). In contrast the PRP-1 and PLRP-S HPLC stationary phase adsorbents are free of impurities and can be used as received.

5.4. PHYSICAL AND CHEMICAL PROPERTIES OF MACROPOROUS POLYMERS

5.4.1. Thermodynamic Considerations

The XAD copolymeric adsorbents bind an adsorbate to the polymeric surface by van der Waals or physical-type forces. Other factors, such as $\pi-\pi$, dipole–dipole, and hydrogen bonding, can also contribute to binding. Adsorption isotherm data have been determined for the retention of water-soluble organic acids (phenols, carboxylic acids, and sulfonic acids) on the macroparticle size XAD copolymers under various concentration and temperature conditions (6, 14). The isotherm data are consistent with retention by hydrophobic interaction. For example, for a homologous series of test adsorbates, the degree of adsorption increases as the molecular weight of the test adsorbates and also the surface area of the absorbent increase.

Removal of the adsorbate from the macroparticle sized XAD adsorbents is rapid and complete, indicating that diffusion into and out of the macroporous XAD structural network is favorable. This also suggests that the microgellular phase interior is not appreciably involved in the retention since diffusion into and out of this would be significantly slower. Modeling experiments, where both spherical and theoretical cylindrical filament forms of macroporous polymers were considered, are in general agreement with this view (15).

Several workers have determined heats of adsorption for the binding that occurs between XAD adsorbates and organic adsorbates. For XAD-4 and dichlorophenols the heat of adsorption was found to be -4 to -6 kcal/mol (6), for XAD-8 and phenol it was -11.9 kcal/mol (14), and for XAD-2 and butyric acid, it was -2 to -4 kcal/mol (5, 16). In general, heats of adsorption of this magnitude are consistent with a physical adsorption type of binding.

5.4.2. Solvent Uptake

The permanent porous character of the macroporous polymers as well as their solvent affinity become apparent when solvent uptake data are examined (17). In Table 5.3 solvent uptake data for XAD-1 and XAD-2 are compared to a gel-type 8% cross-linked polystyrene-divinylbenzene copolymer designated as P-8D. Neither the macroporous nor the gel polymeric adsorbents show an appreciable uptake or affinity for the hydrophilic water. This is consistent with the useful adsorption generality of "like likes like." Since the polymeric adsorbents have nonpolar surfaces,

Table 5.3. Solvent Uptake by Organic Polymeric
Adsorbents

Solvent	Solvent Uptake (g/g) of Dry Adsorbent		
	XAD-1	XAD-2	P-8D
Water	0.061	0.072	0.054
Methanol	0.488	0.699	0.050
Ethanol	0.507	0.719	0.062
n-Propanol	0.535	0.757	0.068
Isopropanol	0.509	0.721	0.131
Acetonitrile	0.688	0.743	0.171
Dioxane	1.21	0.960	0.881
Chloroform	1.59	1.22	1.40
Benzene	0.993	0.779	0.724
Hexane	0.515	0.540	0.015

SOURCE: Taken in part from Ref (17).

their preference should be toward organic solvents, particularly nonpolar ones.

The XAD polymers with their permanent pore phases, unlike the P-8D gel polymer which relies on solvent swelling to produce a porous like character, however, show different characteristics toward the organic solvents. For example, all organic solvents are taken up by the XAD polymer, consistent with its permanent pore characteristic. For the P-8D only solvents that tend to swell or show affinity for the P-8D are taken up; thus hydrophilic solvents such as the alcohols are not appreciably taken up. Both the macroporous XAD and the gel P-8D polymers show a high uptake for the hydrophobic solvents, except for saturated hydrocarbons, where the P-8D shows little tendency to swell in saturated hydrocarbons.

The differences in the XAD-1 and XAD-2 are not uniform. In general, XAD-2 with its larger surface area and smaller average pore diameter has a higher uptake for the more polar solvents while XAD-1 with its lower surface area and larger average pore diameter has a higher uptake for the nonpolar solvents. Since the XAD-1 average pore diameter is larger, it will be less dependent on swelling, which still can occur within the interconnecting microsphere gel phase and this could account for nonpolar solvent preference. Swelling of the microporous structure of PRP-1 (18) and the PLRP-S copolymers (19) and how this can influence the peak shape from mixed solvents has been reported. The best analyte chromatographic peak shape in a reversed-phase column elution can be obtained by using a ternary mixture of water and a strong and poor solvent (in terms of eluting power) at a composition that maintains the desired relative solvent strength (19).

5.4.3. Solvent Strength

A semiquantitative 39-solvent eluotropic series for XAD-2 was established using benz[a]anthracene and benzo[a]pyrene as the test analytes (13) and pure solvent as the mobile phase. Table 5.4 provides a comparison of the eluotropic scales on XAD-2 to other nonpolar stationary phase materials. [Only the solvents for the XAD-2 that can be directly compared to the other stationary phases are listed. The experimentally determined eluent strength values from Ref. (13) are listed in parentheses; other solvent ϵ^0 values for XAD-2 are available in Ref. (13).]

Qualitatively, the eluotropic trend on all the nonpolar or reversed stationary phases is similar and follows the order water < alcohols < acetonitrile < aliphatic hydrocarbons < ethers < aromatic solvents. Recent studies (19) with the PLRP-S adsorbents and a much smaller group of solvents—a 1:4 (v/v) water–organic solvent mobile phase mixture was used—suggested differences between the eluotropic series for XAD-2 and the PLRP-S. For example, DMSO was suggested to be a much stronger solvent on XAD-2 (13) than what was found with the PLRP-S (19). However, for the more common organic modifiers used in HPLC, such as simple alcohols and acetonitrile, differences were not significant, particularly when taking into account that mixed-solvent and pure-solvent mobile phases were being compared.

5.4.4. Factors that Influence Adsorption

The solvent uptake and eluotropic data are indications of the wide range of organic molecules that are readily retained by the macroporous polymeric adsorbents. Table 5.5 lists distribution coefficient data K_D for the retention of a series of selected organic compounds on XAD-2 as a function of the EtOH-H_2O ratio (12). In these experiments a batch equilibration technique was used in which 100–200 mesh XAD-2 particles were equilibrated with a solvent mixture containing the test compound.

Many general conclusions can be drawn from the retention data shown in Table 5.5. All the organic molecules are retained, and retention decreases as the EtOH concentration increases in the solvent mixture; this is consistent with eluotropic solvent behavior. The variation in retention for the widely different organic molecules suggests that organic structure has a large influence on the adsorption process. In general, as the analyte

Table 5.4. Comparison of Eluotropic Scales for Different Nonpolar Solvents*

Styrene-Divinylbenzene	Reversed Phase C-18 or C-8			Carbon	
XAD-2†	Permaphase ODS	C-8 and C-18 on LiChrosorb	LiChrosorb RP-8	Carbon Black	Charcoal
Water	Ethylene glycol	Water	Methylene chloride	Methanol	Water
Methanol (0)	Methanol	Methanol	Dimethyl sulfoxide	Acetonitrile	Methanol
Ethanol (0.014)	Dimethyl sulfoxide	Acetic acid	Methanol	Ethanol	Ethanol
n-Propanol (0.015)	Ethanol	Ethanol	Acetonitrile	n-Hexane	Acetone
n-Butanol (0.030)	Acetonitrile	Acetonitrile	Diethyl ether	Ethyl acetate	n-Propanol
Cyclohexane (0.063)	Dioxane	2-Propanol	N,N-Dimethylformamide	n-Heptane	Diethyl ether
n-Nonane (0.067)	2-Propanol	N,N-Dimethylformamide	Tetrahydrofuran	n-Butyl chloride	Butanol
Acetonitrile (0.072)		Acetone		n-Octane	Ethyl acetate
n-Heptane (0.078)		n-Propanol		Tetrahydrofuran	Hexane
Hexanes (0.079)		Dioxane		Methylene chloride	Benzene
Dimethyl sulfoxide (0.097)				n-Nonane	
Acetone (0.105)				Chloroform	
Diethyl ether (0.157)				Benzene	
N,N-Dimethylformamide (0.167)				m-Xylene	
Chloroform (large)					
Methylene chloride (large)					
Ethyl acetate (large)					
Tetrahydrofuran (large)					
Benzene (large)					
o-Xylene (large)					

Increasing ϵ^0 →

* See Ref. (13) for references and experimental conditions.

† Solvent strength ϵ^0 is given in parentheses (13).

Table 5.5. Batch Distribution Coefficients for Selected Compounds as a Function of Ethanol–H$_2$O Concentration on XAD-2

Compound	Ethanol (%)				
	10	25	50	75	100
Benzene	329	134	0.2	0	0
Benzoic acid	120	44	34	0	0
Methyl benzoate	1240	259	27	5.4	1.0
Chlorobenzene	824		21		0
p-Nitroaniline	118		3.7		1.4
Acetophenone	403	114	14	2.6	1.2
Phenol	42		0		0
p-Nitrophenol	65	36	4.6	0.1	0
p-Chlorophenol	179	68	6.8	0	0
Picric acid	13	10	0	0	0
p-Chlorophenoxyacetic acid	175	69	7.8	0.6	0
o-Chlorophenoxyacetic acid			5.8		
Pyridine	19	6.7	0	0	0
p-Toluenesulfonic acid	1.6	0.1	0	0	0

SOURCE: Taken from Ref. (11).

polarity increases, retention decreases. If the polar groups present are acidic or basic, retention is further reduced as the groups' ability to dissociate increases. Offsetting this, however, is the presence of side groups or the remaining portions of the molecule. If the side group increases in hydrophobicity, its effect is to increase analyte retention. For example, in Table 5.5 phenol, the weaker acid, is less retained than p-chlorophenol, which is a stronger acid.

It becomes quite evident that small changes in hydrophobicity within the organic molecules can lead to a significant difference in the level of retention of nonionic, acidic, or basic organic molecules. Figure 5.3, where batch distribution coefficient data for the retention of p-chloro-phenol on XAD-2 are plotted as a function of organic solvent–water mixtures, confirms that subtle changes in retention are possible by using other water-miscible organic solvents as mobile phase modifiers. Furthermore, the level of analyte retention from the mixed solvent can be conveniently altered by switching from one macroporous polymer to another one. This is shown in Fig. 5.4, where batch distribution coefficient data for the retention of p-chlorophenol on the Amberlite XAD copolymers are shown. If solvent composition is changed, as was done in Fig. 5.3, p-chlorophenol retention on the different XAD copolymers is altered accordingly. In Fig. 5.4, as the surface area of the styrene (XAD-1, -2, -4)

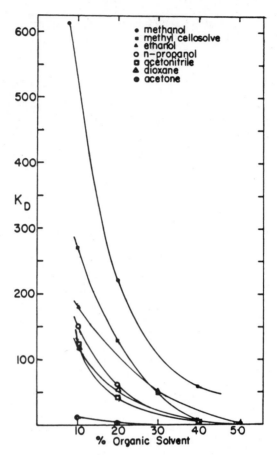

Figure 5.3. Retention of *p*-chlorophenol on XAD-2 as a function of organic solvent–water ratio. [Reprinted from Ref. (20) by permission of the American Chemical Society.]

and acrylic (XAD-7, -8) based macroporous adsorbents increases, *p*-chlorophrenol retention increases. Also, the higher retention of *p*-chlorophenol is on the acrylic polymers, and this is consistent with the high polymer surface area and the polar character of these polymers and the analyte.

Retention of organic molecules that contain acidic or basic groups on the macroporous adsorbents will be altered sharply by adjusting the pH to a condition where the acidic or basic group is dissociated. Thus retention of organic acids is high in acid solution where association is favored and low in base solution where dissociation is favored. For organic bases the reverse would occur. This is illustrated in Fig. 5.5, where batch distribution coefficient data for the retention of several organic acids and bases on XAD-2 are shown. Furthermore, it is possible (12) to quantitively correlate retention expressed as K_D to H^+ concentration and the K_a or K_b value for the organic acid or base. This expression is simplified if the organic solvent concentration is high enough so that retention of the dissociated form of the analyte is or approaches zero. Subsequent studies demonstrated that correlation is possible in terms of column experiments where the capacity factor k' is determined, and for conditions where both the neutral and the charged forms of the analyte are retained (22). In Fig. 5.5 the solid lines are determined by the equations and favorably fit the experimental results indicated by the data points. If organic solvent concentration is increased, the entire curve of retention as a function of pH

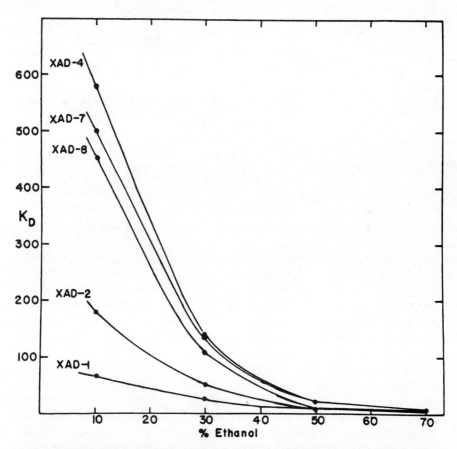

Figure 5.4. Comparison of p-chlorophenol retention on XAD styrene- and acrylic-based macroporous copolymers as a function of EtOH:H₂O ratio. [Reprinted from Ref. (20) by permission of the American Chemical Society.]

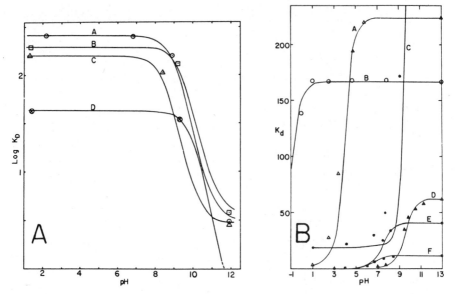

Figure 5.5. Effects of mobile phase pH on retention of (*a*) organic acids and (*b*) organic bases on XAD-2. Fig. (*a*): Line A, 3-chlorophenol. B, 4-chlorophenol. C, 2-chlorophenol. D, phenol. Figure (*b*): Line A, 4-chloroaniline. B, 4-chloro-2-nitroaniline. C, 1,3-diphenylguanidine. D, benzylamine. E, 2-amino-4-methylpyridine. F, 2-aminopyridine. [Reprinted from Ref. (12) (Fig. 5*a*) and (21) (Fig. 5*b*) by permission of the American Chemical Society.]

is shifted downward. Similarly, switching from one organic solvent to another in a water–organic solvent mixture will shift the entire retention–pH curve up or down, depending on the eluotropic strength of the organic solvent. When different Amberlite XAD adsorbents are used, the entire retention–pH curve is also shifted according to the adsorptive power exhibited by the XAD adsorbent. For example, for the XAD-1 to -4 series, retention of the analyte, whether dissociated or undissociated, is highest for XAD-4 and lowest for XAD-1; this trend is consistent with the surface area provided by these adsorbents.

If electrolyte is present in modest concentrations, the retention of non-electrolyte on the macroporous adsorbents is not affected to an appreciable extent when the electrolyte concentration is changed. Similarly, the retention of organic acids or bases in their undissociated forms is not appreciably affected by a modest electrolyte concentration change. In contrast, the retention of the dissociated forms of organic acids and bases is affected, and it increases as electrolyte concentration increases (20, 23, 24). When studying organic analyte ion elution behavior in the presence of electrolyte, peak widths were reduced and peak shapes were improved for these kinds of analytes, and the improvement was found to increase with increasing electrolyte concentration.

5.5. MACROPOROUS ORGANIC POLYMERS AS ADSORBENTS

Activated carbon is the oldest and still probably the most widely used adsorbent in large-scale commercial, manufacturing, and cleanup processes, and perhaps even in small-scale laboratory applications. However, the macroporous organic polymers, since their introduction in the 1960s, have become an attractive alternative to carbon. Consequently, their applications as industrial adsorbents have increased sharply. The polymers, like the carbons, show a wide affinity for all kinds of organic molecules (5, 6, 12). There are thus major advantages offered by the polymers. (1) They are often more easily chemically regenerated for reuse, and (2) through control of the parameters they can be made to be more selective in the kinds of organic molecules that are retained. No attempt will be made to survey manufacturing, waste water cleanup, environmental cleanup, and other important large-scale applications of macroporous organic polymers as adsorbents; these are conveniently reviewed elsewhere (25–27). However, the following briefly focuses on the scope of using the macroporous organic adsorbents in small-scale organic analyte isolation or adsorption applications where organic analyte isolation, concentration, or removal from interferences is the major goal. Often these are key steps in the sample preparation for subsequent chromatographic separation and analysis by gas chromatography (GC) or HPLC, or for characterization by gas chromatography-mass spectrometry (GC-MS).

Most applications are based on the isolation of organic compounds, particularly nonpolar ones, from an aqueous or a gas environment. The former are usually physiological, pharmaceutical, environmental water, and aqueous synthetic samples, while the latter are environment air and head space samples. Either a batch or a column procedure can be used. In a batch strategy the sample (liquid solution) is equilibrated with a quantity of the organic adsorbent, the adsorbent is separated from the solution, and the sought after analytes either remain in the solution or are collected on the adsorbent. In a column procedure the sample (a liquid solution or a gas mixture) is passed through a column of the polymeric adsorbent either continuously or in an aliquot form, using either a liquid eluent or a carrier gas as a mobile phase. If a sample aliquot is used, the sample may occupy a relatively large volume since concentration may also be part of the procedure. Following this collection and isolation step, the concentrated adsorbed organic components are rapidly removed from the column, usually with a very strong eluent mixture in order to maintain the concentrating effect. The solvent in this concentrated, eluted sample may then be removed, reduced in volume for further concentration, or replaced, and often this sample is then subjected to GC, HPLC, or GC-MS separation and analysis or to spectroscopic examination and analysis.

Conditions for a favorable adsorption and elution are indicated by the studies described in the previous section. For nonionic analytes favorable adsorption is achieved from an aqueous solution or one that is low in organic solvent concentration. If the organic analytes are acidic or basic, the favorable adsorption condition is when they are undissociated or at an acidic pH for acids and a basic pH for bases. Strong eluting conditions are favored by high organic solvent concentration in order to keep the adsorbed sample concentrated. In some cases pure volatile organic solvent is preferred as the strong eluent. Since the solvent is volatile, it can be easily removed or reduced in volume for a concentrating effect. The eluotropic series shown in Table 5.4 and Fig. 5.3 is very useful in choosing the appropriate strong elution solvent or solvent mixture.

It is not unsual to be able to take complex, ultra trace, organic mixtures in complex sample matrices and concentrate them by many orders of magnitude into sample sizes that are handled by modern chromatographic and spectroscopic techniques. Table 5.6, which is not intended to be complete, surveys the major kinds of solution samples that have benefited from the isolating and concentrating effects that can be achieved by using organic polymeric absorbents.

The macroporous XAD copolymers have also been used extensvely to isolate and concentrate trace organic compounds from gas samples. This type of application has been particularly useful in atmospheric, environmental, and head space analyses. For example, XAD-2 coated with benzaldehyde has been used to isolate and concentrate sub-ppm levels of hydrazine (89). Similarly, coating XAD-2 with 2,4-dinitrophenylhydrazine has been used to isolate trace quantities of formaldehyde, acrolein, and glutaraldehyde, while an amine coating can be used to isolate gaseous diisocyanates. The hydrophobic adsorptive power of the XAD copolymers is sufficiently strong that many trace organics, particularly aromatic compounds, can be stripped and concentrated from a gaseous sample by hydrophobic interactions, and this approach is widely used in environmental [(90–94) and references within] and head space analyses [(95, 96) and references within].

In many studies, comparisons between the polymeric adsorbents and other kinds of adsorbents in terms of percent isolation and recovery, efficiency of the isolation and recovery, selectivity of the isolation, and

Table 5.6. Selected Applications of Porous Organic Polymers as Adsorbents in the Isolation, Stripping, or Concentration of Organic Analytes

Polymer	Adsorbed Sample Type	Reference
	A. Drugs	
XAD-2	Drugs of abuse isolated from urine	(28–33)
XAD-2	Drug isolation from blood	(34)
XAD-2	Drug isolation from autopsy samples	(35)
XAD-2	Drug isolation from pharmaceutical syrups	(36)
XAD-2	Anthracene aglycone and glycoside drug derivatives	(37)
XAD-2	Methaqualone in blood plasma	(38)
XAD-2	Drugs from physiological samples	(39)
XAD-2	Thiazinamium sulfoxide derivatives in urine	(40)
XAD-4	Phentermine in blood	(41)
XAD-2	Hemoperfusion for acute drug intoxication	(42–44)
	B. Biological Compounds	
XAD-2	Urinary steroids	(45–48)
XAD-2	Polar steroid conjugates	(49)
XAD-2	Anabolic steroids	(50)
XAD-2	Steroids in plasma and milk	(51)
XAD-2, -4	p-Nitrophenyl glucuronide derivatives	(52)
XAD-2	Quaternary ammonium compounds in urine	(53)
XAD-2	Mutagenic activity of smokers' urine	(54)
XAD-2	Prostaglandin isolation	(55)
XAD-2	Vitamin B_{12} and other corrinoids	(56, 57)
XAD-2	Plant nucleotide extracts	(58)
XAD-2	Flavonoids	(59, 60)
XAD-4	Phenolics, terpenes, quinones from plant enzymes	(61)
XAD-4	Purines, pyrimidine, nucleosides (desalting)	(62)
XAD-4	3:5′ and 2:3′ cyclic ribonucleotides	(63)
XAD-2, -4, -7, -8	Sympathomimetic central nervous stimulants	(64)
XAD-7	Mitochondria from tissue	(65)
XAD-7	Endogenous plant growth hormones	(66)
	C. Surfactant Removal from Protein Solutions	
XAD-2	Triton X-100 removal	(67, 68)
XAD-2	NP-40 removal	(69)

Table 5.6. (*Continued*)

Polymer	Adsorbed Sample Type	Reference
XAD-2	Emulgen 9-11	(70)
XAD-2, -4	Deoxycholate	(71, 72)

D. Organics in Water Samples

XAD-2	Trace organics in well, waste, and sea-water	(73–75)
XAD-2, -4	Trace organics in water	(76)
XAD-2, -4	Pesticides in water	(77, 78)
XAD-4	Chlorinated insecticides	(79, 80)
XAD-4	Metals in water	(81)
XAD-2, -4, -7, -8	Chlorinated phenoxyacetic acids in water	(82)
XAD-2, -4, -7	Fenitrothion derivatives in water	(83)
XAD-8	Humic substances in water	(84)

E. Trace Organics

XAD-2	Ionic and nonionic organic compounds	(5, 12)
XAD-2	Aromatic organic compounds	(85)
XAD-2	Aromatic hydrocarbons in mineral oil	(86)
Polystyrene, polyacrylic ester, polyhydroxystyrene	Organic compounds	(87)
PRP-1	Chlorinated phenols and phenoxyacetic acids in water	(88)

other pertinent experimental parameters have been made. Often the organic polymers tend to be the more versatile in both the isolation and the removal steps, and consequently they are widely and frequently recommended as the preferred adsorbents.

5.6 LIQUID–LIQUID COLUMN PARTITION CHROMATOGRAPHY

5.6.1. Partition Strategies

In liquid–liquid column partition chromatography, procedures are based on one of two general strategies. One involves the use of a stationary aqueous phase held up on an inert support and a nonaqueous immiscible liquid phase. This is referred to as liquid–liquid normal-phase column partition chromatography. The other is based on a nonaqueous immiscible stationary phase held up by the inert support and an aqueous mobile phase. This approach is referred to as liquid–liquid reversed-phase column partition chromatography. In both cases extractant additives cause the analytes to distribute between the two phases as the analytes pass through the column. Differences in extractant equilibrium behavior toward the analytes are generally the basis for the separation. Also, it is still possible to carry out the partition process, if the two phases are partially miscible, provided the inert support has a reasonable affinity for one of the phases.

Liquid–liquid column partition chromatography can be applied to the separation of both organic and inorganic analytes. Even though it is possible to achieve favorable column efficiencies relative to modern HPLC, partition chromatography of this type is not widely practiced, particularly in the case of organic analyte separations. In general, the major limitation is that it is difficult to maintain a uniform, constant stationary phase liquid layer during the course of elution. Even though solvents may be classified as immiscible, low-level equilibrium distribution of one into the other occurs, which even under the best circumstances produces a low-level bleeding of the stationary liquid phase.

5.6.2. Inorganic Analyte Separation and Concentration

Column partition chromatography, particularly of the reversed-phase type, when applied to inorganic metal ion separations, however, is still a useful separation strategy. While column efficiencies are not high relative to modern HPLC, separations are favorable because of selectivity. This is achieved through the use of coordination and pH, which often strongly influences the coordination. Thus selectivity is a reflection of the ligand that is used as the extractant and its coordination properties toward the metal ion, pH, which can often alter the coordination, and the differences in the distribution of the metal–ligand complexes between the two phases.

The styrene-based organic macroporous polymers, as indicated in Table 5.3, are able to hold up large quantities of nonpolar organic solvents, including those that are favorable in partition procedures. Consequently, these polymers have been used successfully in reversed-phase liquid–liquid column partition chromatography of metal ions. Other kinds of organic polymers are also useful. In general, a successful organic polymer should possess reasonable porosity, and should have a high surface area and a nonpolar surface. Table 5.7, which is a partial survey, illustrates the scope of the reversed-phase partition procedures developed for metal

Table 5.7. Selected Applications of Porous Organic Copolymers as Support Materials for the Partition Chromatographic Separation of Metal Ions

Metals	Porous Polymer	Extraction System	Ref.
27 metal ions	XAD-2	Isopropyl ether/HCl	(97)
		Isobutylmethyl ketone/HCl	
		Trioctylphosphine oxide/HCl	
Ga, In, Th	XAD-2	Isobutylmethyl ketone/HBr	(98)
U(IV)	XAD-2	Dioctylsulfoxide/1,2-dichloroethane	(99)
Mo(VI), W(VI), V(VI)	XAD-2	Aliquat 336 liquid anion exchanger/toluene/ H_2SO_4	(100)
Cu	XAD-2	Aliphatic α-hydroxyoximes/toluene	(101)
Mo	XAD-2	5,8-Diethyl-7-hydroxydecan-6-one oxime/toleune	(102)
Au	XAD-7	HCl/H_2O	(103)
Fe, Cu	XAD-7	Kelex 100/H_2O	(104)
Zn, Cd, Hg	Macroporous polystyrene-divinylbenzene	Dithiozone-dibutyl phthalate	(105)
Ni, Fe, Co	Ethylstyrene-divinylbenzene	Monothiodibenzoyl-methane/heptane	(106)

244

ion separations where macroporous styrene- and acrylic-based copolymers are used as the inert support.

Many other kinds of organic polymers can and have been used successfully as inert supports for holding up the stationary liquid phase. For example, esterified ion exchangers, rubbers, polyvinylchloride–vinyl acetate copolymers, polyethylenes, halogenated polyethylene, and fluorinated polymers in addition to the styrene- and acrylic-based copolymers have been used the most in metal ion separations (107). In general, these have been used as semimicro, most often irregularly shaped particles in column experiments, where column efficiencies usually are on the order of plate heights of about 0.2 mm or larger. Another approach is to physically impregnate the polymer with an extractant-type material. This can be done before or after the polymerization process (108). In general, these kinds of modifications have been used in hydrometallurgical and water treatment processes as well as in analytical applications.

It is important to note that a partition strategy is valuable in the separation of complex mixtures of metal ions and for the collection and concentrating of dilute samples of metal ion mixtures prior to their determination. The following example focuses on a procedure that illustrates the first type of application. Table 5.8 gives a flowchart that outlines a separation procedure that uses partition chromatography in addition to ion exchange to achieve metal ion group separations (97). Metal ions within these groups are subsequently separated by manipulation of the mobile phase conditions. Usually these are mobile phase changes that alter coordination, which results in a selectivity change and subsequent

Table 5.8. Flowchart for Separation Scheme*

Sample in $8M$ HCl

$\downarrow$

1. IPE on XAD-2 column (T 5 mL; M 6 mL + 1 mL/additional 0.1 mmol)

   Sb(V) ⎫
   Ga(III) ⎬   Strip with $0.1M$ HCl in MeOH
   Fe(III) ⎭

   $\downarrow$

2. IBMK on XAD-2 column (T 5–10 mL; M 15 mL + 2 mL/additional 0.1 mmol)

   Mo(VI)   Elute with $1M$ HCl–$3M$ H_2SO_4
   Sn(IV)   Strip with $0.1M$ HCl in MeOH

   $\downarrow$

Table 5.8. (*Continued*)

3. Amberlyst A-26 column (T 4–5 mL; M 6 mL + 1 mL/additional 0.1 mmol)

Co(II)	Elute with 0.5M HCl in 65% EtOH
Cu(II)	Elute with 2.5M HCl
U(VI)	Elute with 1.0M HCl
Zn(II)	Elute with 0.05M HBr
Cd(II)	Elute with 1.0M HNO$_3$–0.01M HBr
Bi(II)	Elute with 2.0M HClO$_4$

↓

4. TOPO–cyclohexane on XAD-2 column (T 5–10 mL; M 15 mL + 2 mL/ additional 0.1 mmol)

Ti(IV) ⎱ Sc(III) ⎰	Elute with 5.0M HNO$_3$
Th(IV)	Elute with 12M HCl
Zr(IV) ⎱ Hf(IV) ⎰	Elute with 1.0M HCl

↓

Evaporate to near dryness; add HNO$_3$ and reevaporate; dissolve in 20 mL of H$_2$O

↓

If Cr present add citrate, pH4, boil, cool, adjust to pH 2

↓

5. Dowex 50W-X8 column (T 4 mL; M 6 mL; + 1 mL/additional 0.1 mmol)

V(IV)	Elute with 1% H$_2$O$_2$–0.01M HClO$_4$
Pb(II)	Elute with 0.6M HBr
Mn(II)	Elute with 1.0M HCl in 92% acetone
Al(III)	Elute with 0.3M HF
Ni(II) ⎱ Mg(II) ⎰	Elute with 3.0M HCl in 60% EtOH
Ca(II) ⎫ Ba(II) ⎪ Sr(II) ⎬ RE(III) ⎭	Strip with 4.0M HNO$_3$

SOURCE: Taken from Ref. (97).

* IPE—isopropyl ether; IBMK—isobutyl methyl ketone; TOPO—trioctylphosphine oxide. Data in parentheses are recommended column bed volumes for trace (T) and macro (M) amounts of metal ions.

separation and elution. The XAD-2 particles used in Table 5.8 were crushed and sieved and used as irregularly shaped 100–200-mesh particles. No attempt was made to separate all 26 metal ions since this is an unlikely mixture. However, several synthetic and standard metal mixtures, which often contained one or more metal ions per group, were separated successfully and the analytes determined. In general, analyte loading was about 0.1 mmol of each cation.

A partition type of separation strategy for the concentration or determination of inorganic cations appears to be reliable, convenient, and reasonably time effective. Improvements in these factors and in column efficiency should also be realized by using modern, uniform microparticle styrene-based macroporous copolymers and modern detection strategies which would permit lower detection limits and sample loading.

5.7. MACROPOROUS POLYMERS IN HPLC

5.7.1. Semimicro XAD Particles

The initial studies with the macroporous polymers as adsorbents [see (5, 6, 12) and Table 5.6] and as stationary phase adsorbents for the low-efficiency separation of organic analytes (12, 20–24, 109–119) suggested that these polymers would be useful stationary phases for more efficient organic analyte separations, provided columns packed with smaller, narrower range particles were to be used. In other studies Poropak Q, also a polystyrene-divinylbenzene copolymer, was used to separate amino acids and oligopeptides (118), while Poragel PN, a gel polystyrene-divinylbenzene copolymer, and Poragel PS, a gel polystyrene-pyridine-type copolymer, were used to separate peptides (119).

Figure 5.6 illustrates a separation of an eight-component mixture of amines on a semimicroparticle XAD-2 column using a decreasing pH gradient for their elution. As the pH is decreased, the amines are converted into cations, which sharply reduces their retention, according to their K_b values. The XAD-2 particles used in Fig. 5.6 were irregular, crushed, and sieved 45–65 μm in size. As indicated in Fig. 5.7, plate heights for separation with columns of XAD-2 particles of this size were about 10 mm at 1 mL/min; column inlet pressures, which were sometimes excessive, were due to the difficulty in removing XAD-2 dust particles during the sieving process. Similar separations, typical of reversed phase, of organic acids and nonelectrolytes were also reported during this period (12, 20–24, 109–117).

As a chromatographic stationary phase, the XAD macroporous poly-

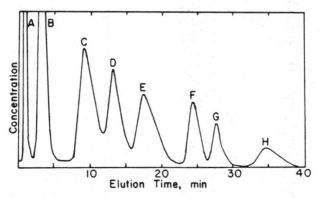

Figure 5.6. Separation of an amine mixture on semimicro XAD particles. A 300-mm × 1.9-mm, 45 to 65 μm XAD-2 column with a gradient made from a 30.0 mL tartrate buffer of pH 5.06 in 1:9 EtOH:H$_2$O in a fixed mixing chamber and added 0.20M HCl, 1:9 EtOH:H$_2$O at 1.0 mL/min. *A*—2-aminopyridine; *B*—pyridine; *C*—aniline; *D*—4-methylaniline; *E*—2-methylaniline; *F*—4-chloroaniline; *G*—3-chloroaniline; *H*—2-chloroaniline. [Reprinted from Ref. (21) by permission of the American Chemical Society.]

mers, particularly the XAD-1, -2, and -4, possess many distinguishing characteristics. They are low in cost, are useful throughout the entire pH range of 1–14 unlike the silica-based stationary phases, are compatible with virtually all solvents, do not swell in water, swell modestly in the chromatographically useful solvents (see Table 5.3), and exhibit relatively strong adsorbent properties. Since the XAD-1, -2, and -4 possess a hy-

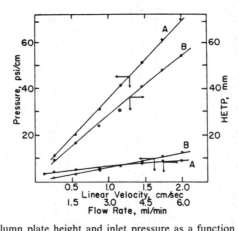

Figure 5.7. XAD-2 column plate height and inlet pressure as a function of mobile phase linear velocity and XAD-2 particle size. Column *A*—120 mm × 2.5 mm, 45–65 μm; Column *B*—500 mm × 2.5 mm, 75–150 μm. [Reprinted from Ref. (21) by permission of the American Chemical Society.]

drophobic surface, they should provide reversed-phase properties and be applicable in many of the same applications where reverse-bonded phase packings, such as alkyl-modified silica, are used. Like these latter reversed phases, the macroporous polymers should be useful stationary phases for the separation of ionic organic analytes. These analytes must also contain hydrophobic centers. Organic acids and bases can be analyzed through suppressed dissociation, while organic and inorganic analyte ions can be separated through use of ion interaction reagents (hydrophobic or pairing ions) as mobile phase additives. Furthermore, because of the pH stability, very basic mobile phases (up to pH 14), which are not compatible with the silica reversed phases, can be used advantageously in ion interaction (ion pairing) chromatographic strategies.

The macroporous polymers (see Table 5.1) are well characterized adsorbents in that they possess a true adsorbent surface. They are uniform and do not suffer from changes in adsorbent properties due to physical imperfections in their surface or in localized structural or chemical changes throughout the polymer matrix.

5.7.2. Microparticle Macroporous Polymers

A major drawback to the development and application of the XAD macroporous polymers up to 1980 was the unavailability of microparticles that were capable of delivering high-efficiency columns with plate numbers comparable to those that could be routinely obtained with commercially available silica-based reversed-phase bonded phase packings. Crushed XAD particles [see Fig. 5.6 and, for example, (12, 20–24, 109–117)] in the 45–65-μm range or larger, while providing favorable selectivity in organic analyte separations, were not satisfactory in terms of column efficiency.

In 1974 Hitachi Gel 3010, which is also a polystyrene-divinylbenzene copolymer that has a high surface area and porosity and was available in a spherical particle size range of 20–30 μm, was used as a reversed-phase adsorbent (120). In general, the chromatographic performance of the Hitachi gel is much like that of the XAD-2 and functions as a reversed-phase adsorbent. A major advantage achieved with the Hitachi copolymer was that it was spherical and small in particle size, unlike the previously used XAD particles, and this resulted in improved column efficiency.

In 1980 it was reported that XAD-2 columns could be prepared that would yield column efficiencies greater than 20,000 plates per meter (121). In these experiments macrosized XAD-2 particles were ground, sieved, and treated by solvent elutriation to obtain 3.6–8.4-μm irregularly shaped XAD-2 microparticles. These particles were subsequently slurry packed into 15-cm columns. Figure 5.8 shows that favorable plate heights of less than 0.05 mm at k' values of 0.9 and 2.1 were obtained over typical mobile phase linear velocities. Figure 5.8 also illustrates the magnitude in improved column efficiency since experiments were also carried out with a large XAD-2 particle size for comparative purposes. Peak tailing, based on asymmetry factor calculations, compared favorably with slurry packed columns of bonded phase microparticles. Column permeability calculations indicated low compressibility of the macroporous XAD-2 copolymer beads over the column inlet pressures studied. However, the permeability was smaller than were typical values for commercially available reversed-phase bonded columns. This was partly due to the particle size range of the XAD-2. While this was small, it was still larger than that of the commercially available reverse bonded phase. A second contributing factor was the low-level swelling of the XAD-2 in the mobile phase.

Many applications of the semimicro- ($\sim$50-μm) and microparticle ($\sim$6-μm) XAD-2 and XAD-4 particles were reported. Mixtures of organic bases (21, 22), organic acids and phenols (12, 21, 111), drugs in pharmaceutical syrups (23), benzoate esters (113, 121), amino acids and small chain peptides (114), sulfas (117), aromatic hydrocarbons (13), and hydrolysis products of nucleic acids (115) were separated successfully. These studies and others, where XAD copolymers were examined as adsorbents for organic molecules, clearly established the key mobile phase parameters that influence organic analyte retention and elution. These are summarized in the following.

The effect of organic modifier is consistent with reversed-phase chromatography. As the organic modifier concentration in a water–organic solvent mobile phase mixture increases, organic analyte retention decreases; the effect of the organic modifier follows the eluotropic order listed in Table 5.4. The more hydrophobic the stationary phase surface is or the larger its surface area, the greater the retention. For organic acidic or basic analytes mobile phase pH and ionic strength become important. In the undissociated form organic acid or base analyte retention is high, while at pH conditions where the acid or base is ionized, retention is low. The presence of electrolyte in the mobile phase has little effect on the retention of nonelectrolytes or undissociated acids or bases but causes the dissociated forms of the acids or bases to increase in retention as electrolyte concentration in the mobile phase is increased. If the organic modifier in the mobile phase is increased, retention of both the neutral and the dissociated forms of acidic and base analytes will decrease.

5.7.3. Effect of Mobile Phase pH on Analyte Retention

Understanding the effect of protonic ionization on the retention of organic acids and bases in reversed-phase chromatography is important for two

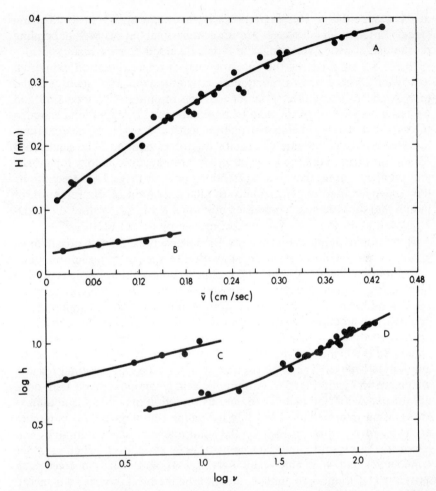

Figure 5.8. Effect of XAD-2 particle size on plate height as a function of mobile phase linear velocity (curves *A* and *B*) and log reduced plate height as a function of mobile phase log reduced linear velocity (curves *C* and *D*). Curves *A*, *D*—150 mm × 4.1 mm, 15.6–44 μm; curves *B* and *C*—150 mm × 4.1 mm, 3.6–8.4 μm. [Reprinted from Ref. (8) by permission of the American Chemical Society.]

reasons. (1) It contributes to the understanding of the nature of the interactions in reversed-phase chromatography and (2) manipulation of the ionization by pH adjustment is a powerful tool in affecting selectivity and subsequently resolution. The porous polymers are ideal stationary phases to study the effect of mobile phase pH because they are stable throughout the entire pH range. Thus retention of even weak organic acids and bases can be studied since the porous polymers will withstand strongly acidic and basic conditions, which are required to dissociate weak bases and acids, respectively. This is particularly favorable in the basic direction since bonded reversed phases are stable only to about pH 8–9, depending on the use of precolumns and the type of bonded-phase silica.

Table 5.9 summarizes the equations which relate retention expressed as capacity factor k' as a function of ionization constants, acidity of the mobile phase, and capacity factors for the retention of dissociated and undissociated forms of the acid or base analyte (22). Equations can also be derived in terms of batch distribution coefficients (12). In the equations in Table 5.9 k_0 is the capacity factor for the retention of the uncharged or neutral form and k_{-1}, k_{-2} and k_1, k_2 are capacity factors for mono- and divalent anion and cation retention, respectively. These equations have been verified experimentally by determination of the retention of mono- and diprotic acids and bases of known K_a and K_b values on XAD-2, as well as PRP-1, as a function of mobile phase pH (22).

Retention of ampholytes on a reversed phase presents two possibilities since certain ampholytes will form zwitterion species at an intermediate pH while others will not. This will produce a sharp difference in retention because the former, although electrostatically neutral, is a highly charged species and sharply reduces retention. A nonzwitterionic ampholyte in contrast has no charge at the intermediate pH and is charged in acidic (a cationic analyte) and basic (an anionic analyte) solution. Thus its retention should be the highest at the intermediate pH. The equation for a diprotic ampholyte listed in Table 5.9 is consistent with the experimental results, which are shown in Fig. 5.9. Anthranilic acid (o-aminobenzoic acid) does not form a zwitterion like valine, and its retention on XAD-2 is at a maximum at an intermediate pH while for valine retention is at a minimum at the zwitterion pH. In Fig. 5.9 the solid lines are calculated and compared to the experimental data points. Other zwitterions, such as other amino acids and short-chain peptides (22, 114), and nonzwitterions, such as sulfas (117), undergo similar retention behavior. Tyrosine, which is a triprotic ampholyte containing a carboxyl, a phenolic, and an amine group, and which exhibits zwitteronic properties, undergoes a series of changes in retention on XAD-2 due to the different charged forms that are present as a function of pH. An equation describing this retention as a function of pH and its verification have been described (114).

Understanding the effect of mobile phase pH quantitatively facilitates the optimization of mobile phase conditions for the separation of organic acids, bases, and ampholytes, such as amino acids and peptides. Furthermore, determining chromatographic retention as a function of pH permits the determination of ionization constants for weak organic acids and bases (122). If a high column efficiency is maintained, this HPLC method of pK_a determination offers the major advantage of requiring only small amounts of the acid or base to be available.

Table 5.9. Equations Correlating Chromatographic Retention to Ionization and Mobile Phase pH

	Acid	Base	Ampholyte
Monoprotic	$k' = \dfrac{k_0}{1 + K_a/[H^+]} + \dfrac{k_{-1}}{1 + [H^+]/K_a}$	$k' = \dfrac{k_1}{1 + [OH^-]/K_b} + \dfrac{k_0}{1 + K_b/[OH^-]}$	$k' = \dfrac{k_0 + k_1\,[H^+]/K_{a1} + k_{-1}\,K_{a2}/[H^+]}{1 + [H^+]/K_{a1} + K_{a2}/[H^+]}$
Diprotic	$k' = \dfrac{k_{-1} + k_0\,[H^+]/K_{a1} + k_{-2}\,K_{a2}/[H^+]}{1 + [H^+]/K_{a2} + K_{a2}/[H^+]}$	$k' = \dfrac{k_1 + k_0\,[OH^-]/K_{b1} + k_2\,K_{b2}/[OH^-]}{1 + [OH^-]/K_{b1} + K_{b2}/[OH^-]}$	

SOURCE: See Ref. (22).

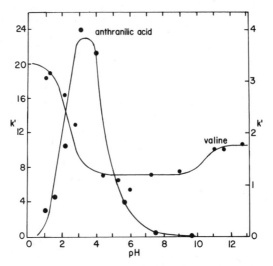

Figure 5.9. Effect of mobile phase pH on retention of zwitterion and nonzwitterion ampholytes. Valine—300-mm × 2.36-mm, 45–65-μm XAD-4 column and 0.020M phosphate buffer at 0.20M ionic strength (NaCl) mobile phases at 0.50 mL/min. Anthranilic acid—250-mm × 2.36-mm, 45–65-μm XAD-2 column and 0.010M phosphate buffer at 0.10M ionic strength (NaCl) mobile phases at 0.94 mL/min. [Reprinted from Ref. (22) by permission of the American Chemical Society.]

5.7.4. Charged Analyte Retention

Adsorption of nonionic analytes on the Amberlite XAD macroporous co-polymers and similar porous polystyrene-divinylbenzene copolymers is through hydrophobic interactions between the analyte and the nonionic surface of the macroporous copolymer. Charged analytes are also retained and their retention increases with increasing mobile phase ionic strength. The effect of ionic strength and the retention of charged analytes can be explained quantitatively by the Stern–Goüy–Chapman electrical double-layer theory. This was demonstrated (24, 123, 124) by using adsorption isotherm data and microelectrophoresis measurements to show that an organic analyte ion is adsorbed into the stationary phase surface as a primary layer. A diffuse layer of counterions of opposite charge constitutes the secondary layer of the electrical double layer. Retention studies of organic cations, such as tetraalkylammonium salts and amine hydro-chlorides (123), and anions, such as aromatic sulfonates (24), on XAD-2 were consistent with this interaction model. Not only does this model explain satisfactorily mobile phase ionic strength effects on the retention of organic ions, but it also accounts for the enhanced retention of simple organic and inorganic analyte ions on the copolymeric reversed phases from a mobile phase containing a hydrophobic ion of opposite charge as a mobile phase additive or interaction ion (see Section 5.7.5.6).

5.7.5. Advantages of Microparticle Copolymers

With the commercial availability of microparticles of the macroporous styrene-based copolymers (see Table 5.1), application of these reversed stationary phases in the separation of organic nonionic and ionic analytes at high column efficiency increased sharply. It is not unusual to be able to obtain with the microparticle copolymers column efficiencies of 40 000 plates per meter or better, depending on the analytes and mobile phase conditions.

In addition to the improvement in column efficiency due to the commercial availability of small uniform particles, other useful characteristics are retained. For example, the modern particles (1) are mechanically stable to column inlet pressures exceeding 4000 psi; (2) are compatible with all common HPLC useful organic modifiers; (3) are chemically stable for pH 1–14; (4) provide favorable reversed-phase selectivity that is sensitive to small changes in analyte structure; (5) permit a high analyte loading; (6) are stable to 80°C temperatures; (7) provide a homogeneous hydrophobic surface; (8) can have a ratio of small pore size–high surface area to large pore size–low surface area which permits separation of macromolecules; and (9) provide reproducible, selective chromatography with long column life at a modest cost. The following sections focus on the impact of several of these advantages in reversed-phase HPLC.

5.7.5.1. Mobile Phase Solvent

Figure 5.10 illustrates the effect of solvent mixture on the separation of a series of alkyl phenyl ketones. As alkyl hydrophobicity increases, retention increases. Because the nonionic analytes have a high hydrophobicity, a weak eluent mixture, such as 5:1 MeOH–H₂O, leads to a high analyte retention. Switching to 5:1 CH₃CN–H₂O, which is consistent with eluotropic power (see Table 5.4), sharply reduces retention while still maintaining a favorable selectivity and resolution of the mixture. In general, like other reversed phases, the more nonpolar the nonionic analyte is, the greater the analyte retention and the more nonpolar the eluent must be in order to lower the retention.

5.7.5.2. Mobile Phase pH

Mobile phase pH can be used to suppress ionization and increase the retention of organic acids and bases. Adjustment of the mobile phase

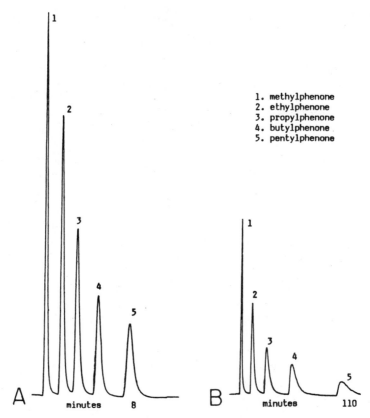

1. methylphenone
2. ethylphenone
3. propylphenone
4. butylphenone
5. pentylphenone

Figure 5.10. HPLC separation of alkylphenols on PLRP-S 100 Å. A 150-mm × 4.6-mm, 5-μm PLRP-S 100-Å column and a 5:1 $CH_3CH:H_2O$ (Figure 5.10*a*) and 5:1 $MeOH:H_2O$ (Figure 5.10*b*) mobile phase. [Reprinted by permission of Polymer Laboratories, Ltd.]

solvent mixture to affect retention is made according to the polarity of the undissociated acid or base. Alternatively, a mobile phase pH can be used to dissociate organic acids and bases. The charged form of these analytes reduces their retention, and separations are then carried out with a weak eluent solvent mixture.

The availability of the basic pH range allows even weak acids to be dissociated and separated as anions. Figure 5.11 illustrates the separation of chlorophenols on a PRP-1 column using a basic eluent that converts all the phenols to phenolate anions (88, 125). Since the retention is reduced significantly, a predominantly aqueous mobile phase can be used for their elution. If separation is to be attempted from an acidic mobile phase where the phenols are undissociated, a predominantly nonaqueous mobile phase

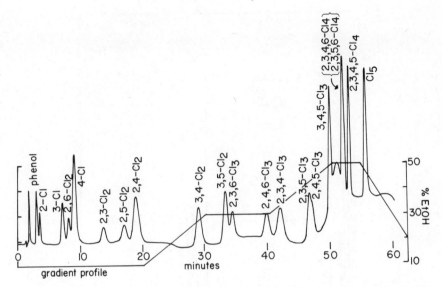

Figure 5.11. HPLC separation of complex mixture of chlorinated phenols on PRP-1. 150-mm × 4.1-mm, 10-μm PRP-1 column and EtOH:H₂O gradient at constant 0.010M NaOH, 0.050M NaCl at 1.0 mL/min. [Reprinted from Ref. (88) by permission of Preston Publications, Inc.]

would be required for elution. As the number of Cl groups increases in the phenol, retention increases. The successful elution of a complex mixture is therefore achieved by using a gradient at constant pH, where EtOH concentration is increased. It is also possible to load an aqueous, dilute, large-volume acidified sample onto a short PRP-1 precolumn. At the acidic condition the chlorophenols are highly retained and the sample is isolated and concentrated, thus permitting the separation and determination of ultratrace quantities of phenols (88). Switching to the basic mobile phase will sweep the phenols off the short column into the analytical PRP-1 column. The gradient is started and the chlorinated phenol mixture is subsequently separated. Chlorophenoxyacetic acids can also be handled in a similar approach (88).

The option of carrying out a separation at an acidic or a basic condition is particularly important when separating ampholytes. In many cases elution orders are quite different if the ampholytes are separated as cations (acidic mobile phase) compared to separation as anions (basic mobile phase). In Fig. 5.12, where sulfas are separated on PRP-1, retention between the two conditions is different. Because sulfas are not zwitterions, their retention is at a maximum at an intermediate pH where these ampholytes are uncharged. Thus depending on the sulfa K_a values, the elu-

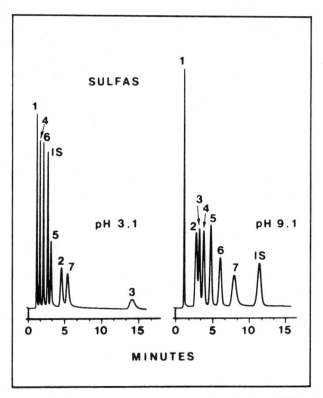

Figure 5.12. Effect of mobile phase pH on HPLC separation of sulfas on PRP-1. 1—sulfanilic acid. 2—sulfamerazine. 3—sulfisoxazole. 4—sulfaguanidine. 5—sulfathiazole. 6—sulfanilamide. 7—sulfamethazine. IS—internal standard benzamide. A 150-mm × 4.1-mm, 10-μm PRP-1 column, pH = 3.1, 3.5:1 $H_2O:CH_3CN$ $0.10M$ NaH_2PO_4 buffer, and pH = 9.1, 12:1 $H_2O:CH_3CN$ $0.10M$ Na_3BO_3 buffer at 1.0 mL/min. [Reprinted by permission of Hamilton Co.; see also Ref. (8).]

tion order will change even at intermediate pH. For mixtures of acids and bases the elution order will also change with mobile phase pH. This is shown in Fig. 5.13, where nucleoside and base separations on PLRP-S in acidic and basic mobile phase are compared.

For ampholytes that are zwitterions intermediate pH, in general, is not the most favorable elution condition because at this condition, although neutral, zwitterions are highly charged and retention is low. Often the better conditions are from an acidic or basic mobile phase where the ampholytes are cations or anions. Furthermore, the elution order often differs between the two conditions. For example, from an acidic mobile phase (pH 1.6) the dipeptide Phe-Gly elutes before Gly-Phe, while from a basic mobile phase (pH 11.0) the reverse elution is obtained (126). The dipeptide side chain structure (see the next section) is also a major contributing factor to the reversal in elution as pH is changed. In general, when other factors are equal, the closer the hydrophobic center is to the charge center the lower its retention becomes.

5.7.5.3. *Analyte Structure*

Small changes in analyte structure, if hydrophobicity is affected, will alter retention and resolution on the macroporous polymers, which is typical of reversed-phase HPLC. In Fig. 5.10 the alkyl side chain in the phenol derivatives increases in hydrophobicity and retention increases. Similar increases in retention are found for other homologous series where hy-

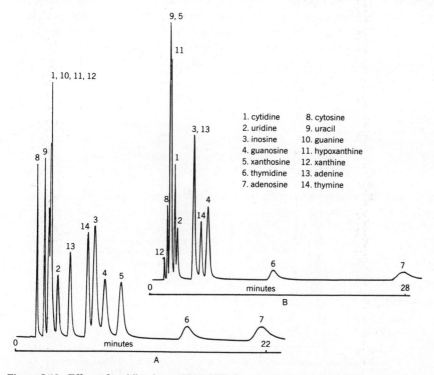

1. cytidine   8. cytosine
2. uridine   9. uracil
3. inosine   10. guanine
4. guanosine   11. hypoxanthine
5. xanthosine   12. xanthine
6. thymidine   13. adenine
7. adenosine   14. thymine

Figure 5.13. Effect of mobile phase pH on HPLC separation of nucleosides and bases on PLRP-S 100 Å. A 150-mm × 4.6-mm, 5-μm PLRP-S 100-Å column and (Figure 5.13*a*) a pH = 4.25 100:1 H_2O:CH_3CN ammonium formate buffer or (Figure 5.13*b*) a pH = 8.5 100:1 H_2O:CH_3CN ammonium formate buffer at 1.0 mL/min. [Reprinted by permission of Polymer Laboratories, Ltd.]

drophobicity increases in the series. For example, retention of substituted benzoate esters, alkyl-substituted phenols, benzaldehydes, nitrobenzenes, benzenes, polyaromatics, benzoic acids, and anilines and amino acids are consistent with this trend. Correlation of retention and/or retention indices to substituents, side chains, and/or chain length is possible (114, 126–130).

Small differences in hydrophobicity that occur in small-chain peptide diasteromers are often sufficient enough to permit a separation. Figure 5.14 shows a separation of a complex mixture of dipeptide diasteromers on PRP-1 (126). In all cases the L,L and D,D dipeptides coelute before the coelution of the L,D and D,L dipeptides. In longer chain peptides the location of a single side chain can be instrumental in determining the level of retention. This is illustrated in Fig. 5.15, where the effect of the hydrophobic side chain of an L-Phe subunit in a series of L-Phe-(Gly)$_4$ pentapeptides on retention on PRP-1 is shown (129). In an acidic mobile phase where the terminal amine group is a cation, retention is the highest when the side chain is at position 5 or furthest from the charge center. In a basic mobile phase the highest retention is when the side chain is at position 1 or furthest from the terminal carboxylate anion site. At zwitterion

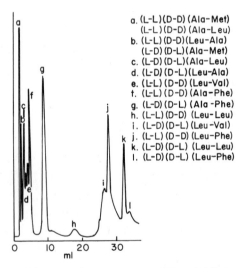

a. (L-L)(D-D)(Ala-Met)
   (L-L)(D-D)(Ala-Leu)
b. (L-L)(D-D)(Leu-Ala)
   (L-D)(D-L)(Ala-Met)
c. (L-D)(D-L)(Ala-Leu)
d. (L-D)(D-L)(Leu-Ala)
e. (L-L)(D-D)(Leu-Val)
f. (L-L)(D-D)(Ala-Phe)
g. (L-D)(D-L)(Ala-Phe)
h. (L-L)(D-D)(Leu-Leu)
i. (L-D)(D-L)(Leu-Val)
j. (L-L)(D-D)(Leu-Phe)
k. (L-D)(D-L)(Leu-Leu)
l. (L-D)(D-L)(Leu-Phe)

Figure 5.14. Separation of dipeptide diastereomers on PRP-1. A 150-mm × 4.1-mm, 10 μm PRP-1 column and gradient elution of 20 min 3.3:96.7 CH$_3$CH:H$_2$O, pH = 5.25, phosphate buffer, 0.10M ionic strength, followed by linear change over 3 min to 15:85 CH$_3$CN:H$_2$O, pH = 5.25, and then a constant elution at this condition with 1.0 mL/min flow rate. [Reprinted from Ref. (126) by permission of the American Chemical Society.]

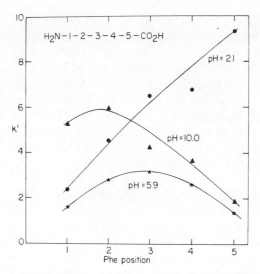

Figure 5.15. Effect of location of Phe subunit on retention of a series of L-Phe-(Gly)$_4$ peptides on PRP-1 as a function of mobile phase pH. 150 mm × 4.1 mm, 10 μm PRP-1 column and a 5:95 CH$_3$CN:H$_2$O, phosphate buffer, 1.0M ionic strength (NaCl) mobile phase at 1.0 mL/min. [Reprinted from Ref. (129) by permission of M. Dekker, Inc.]

conditions where the —NH$_3^+$ and —CO$_2^-$ groups are present retention is at a maximum when the side chain is at unit 3 or midway between the two terminal charge centers. The combined effect of pH, which can be used to control the location of charge at the terminal groups or at side chain acidic and basic groups, and the presence and location of polar and nonpolar side chains are major factors in determining small-chain peptide retention and retention order.

5.7.5.4. Preparative Chromatography

The macroporous copolymers, because of the high surface area, will exhibit a high analyte loading capacity and are favorable stationary phases for preparative HPLC. Furthermore, their stability in a basic mobile phase provides preparative conditions that are not readily available with other reverse stationary phases that are also suitable to preparative HPLC. For example, for an 8.0-mm id × 25-cm 37–44-μm and a 20.5-mm id × 32-cm 75–105-μm XAD-4 column a column mass overload, which is defined as a 10% decrease in column efficiency, was obtained for $k' \approx 2$ at about 15 mg of anlayte and 100 mg of analyte injected, respectively (131). Higher overloads are still possible, depending on the mixture, since a reasonable number of plates are still available. Also, improved performance should be expected if spherical particles, such as the Amberchrom copolymers, or microparticles of uniform size, such as the PRP or PLRP-S copolymers (see Table 5.1), are used in the prep column. In general, analytes that are readily separated quantitatively on the macroporous copolymers at an analytical level, depending on the mixture, the amount, and its complexity, can often be handled in a preparative manner (131, 132). XAD columns as large as 58-mm id have even been used in preparative HPLC, and depending on the mixture, purification of multigram quantities is possible with these columns (132).

5.7.5.5. Stationary Phase Pore Size Effects

Macroporous polymers at microparticle size are not commercially available in assorted pore sizes and surface areas. Two styrene-based copolymers are commercially available that differ in pore size and surface area (see Table 5.1) and these have been examined. In Fig. 5.16 the ability of the copolymers to separate polystyrene standard MW 480 oligomers as a function of pore size and surface area is shown. The solid copolymer is a low porous, low surface area copolymer and is incapable of retaining or resolving the mixture. The others in contrast show favorable resolving power depending on pore size and surface area. The PLRP-S 300 Å is commercially available and is capable of separating macromolecules. This is illustrated in Fig. 5.17, where a reversed-phase protein separation on PLRP-S 300 Å using an acidic acetonitrile–H_2O gradient is shown. PRP-3, which is also a wide pore styrene based copolymer (see Table 5.1), has also been shown to be useful in protein separations (133). Furthermore, it was suggested that protein recoveries from PRP-3 were favorable and that the column could be freed of protein contaminants that are introduced after repeated applications by hydrolytic treatment. Such a procedure cannot be used with a silica based bonded reversed phase because of the reactivity of silica in a highly acidic environment.

5.7.5.6. Interaction Reagents as Mobile Phase Additives

The macroporous copolymers are suitable reversed stationary phases for the chromatographic separation of analyte anions and cations through the use of interaction reagents (pairing ions) of opposite charge as mobile phase additives to augment the retention and selectivity of the analytes. The question of how the analyte ion and the interaction reagent of opposite charge, such as an R_4N^+ salt or an RSO_3^- salt, increases analyte retention has been a subject of wide controversy. The various mechanistic inter-

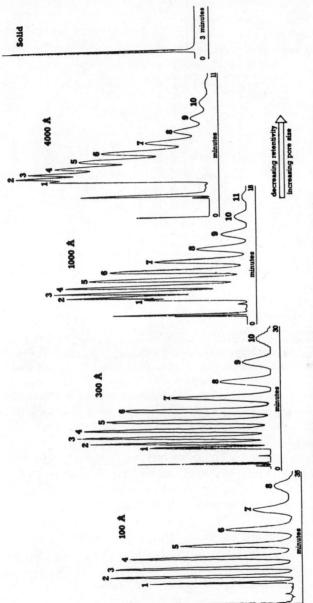

Figure 5.16. HPLC separation of polystyrene standard MW-480 oligomers as a function of stationary phase pore size and surface area. 150 mm × 4.6 mm PLRP-S columns and 3:2 THF:H₂O mobile phase at 1.0 mL/min. [Reprinted by permission of Polymer Laboratories, Ltd.]

PLRP-S (solid)	5 μm	3 m²/g
PLRP-S 100 Å	6 μm	414 m²/g
PLRP-S 300 Å	8 μm	384 m²/g
PLRP-S 1000 Å	8 μm	267 m²/g
PLRP-S 4000 Å	10 μm	139 m²/g

263

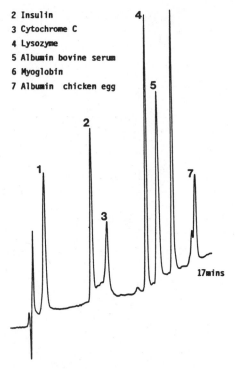

2 Insulin
3 Cytochrome C
4 Lysozyme
5 Albumin bovine serum
6 Myoglobin
7 Albumin chicken egg

17mins

Figure 5.17. HPLC separation of proteins on PLRP-S. 150-mm × 4.6-mm, 6-μm PLRP-S 300-Å column and gradient elution with eluent *A* of 0.1% TFA in 95:5 $CH_3CN:H_2O$ and eluent *B* of 0.1% TFA in H_2O using linear gradient of 20–60% A over 22 min at 1.0 mL/min. [Reprinted by permission of Polymer Laboratories, Ltd.]

pretations are reviewed in detail elsewhere (134) and are not considered here.

The macroporous polystyrene-divinylbenzene copolymers developed for HPLC applications provide a homogeneous hydrophobic surface in terms of interaction sites. Under defined experimental conditions, where an interaction reagent is used as a mobile phase additive, experimental results are consistent with a Stern–Goüy–Chapman electrical double-layer-type interaction (24, 123, 124), that is, the interaction reagent is retained on the stationary phase surface forming a primary charged layer and is accompanied by a secondary diffuse layer of counterions of opposite charge. The amount of interaction ion and its counterion retained is an equilibrium amount consistent with the concentration of the interaction reagent in the mobile phase. Other key mobile phase factors, which influence the magnitude of this equilibrium amount, are the type of counterion, the presence of organic modifier, and the amount of buffer or ionic strength salt in the mobile phase. Analyte ions thus compete with the counterions in the diffuse layer; the more favorable the competition is, the higher the analyte ion retention. The parameters that determine these equilibria are subsequently manipulated and optimized to enhance analyte retention and bring about resolution of the analytes.

If R_4N^+ salts are used as the mobile phase additive, organic and inorganic analyte anions can be separated (135), while with RSO_3^- salts analyte cations (136) can be separated. Similar separations can be carried out using other reversed stationary phases with these interaction reagents. However, the macroporous polymers have the advantage of basic pH stability, which allows the use of basic mobile phases to bring about weak acid analyte ionization, retention, and resolution. Strong acidic mobile phases, which permit the ionization of weak bases, their separation, and resolution, can also be used. Furthermore, a rapid, reversible conditioning and removal of the equilibrated interaction ion from the polymer surface takes place. This latter property contributes to good reproducibility and long column life.

Figure 5.18 illustrates the separation of inorganic anions (137) and organic acid anions (138) on PRP-1 using the iron (II) complex of 1,10-phenanthroline, $Fe(1,10\text{-phen})_3^{2+}$, as the interaction ion. Not only are column efficiency, selectivity, and resolution very favorable, but the $Fe(1,10\text{-phen})_3^{2+}$ presence in the mobile phase is used for the indirect detection of the analyte ions. Under favorable chromatographic conditions detection limits are 5–20 ng of analyte anion or better, depending on the absorbance detector employed in the separation.

Table 5.10 focuses primarily on analytical LC on porous polymers. Porous polymers are also useful stationary phases in preparative LC. Applications of these kind should continue to grow as the commercial availability of particles suitable to preparative LC in bulk form and prepacked columns is realized.

5.7.6. Survey of HPLC Applications

The nonpolar macroporous polymeric adsorbents have been applied to a wide variety of separation problems. Table 5.10 surveys many of these applications according to area and analyte class. In general, the column chromatographic studies, where the XAD copolymers are used, were done with the semimicroparticles, and consequently they yield only modest column efficiencies. On the other hand column studies with the microparticle PRP-1 and PLRP-S copolymers are at high column efficiencies, and these are more comparable to applications with other microparticle reversed stationary phase.

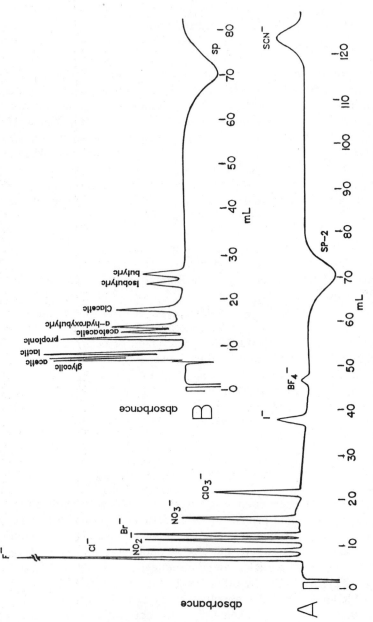

Figure 5.18. HPLC separation of inorganic and organic analyte anions on PRP-1 using ion interaction reagent as mobile phase additive. 150 mm × 4.6 mm, 10 μm PRP-1 column and $1.0 \times 10^{-4} M$ Fe(1,10-phen)$_3$(ClO$_4$)$_2$, $1.0 \times 10^{-4} M$ succinate, pH = 6.1 mobile phase at 1.0 mL/min. [Reprinted from Ref. (137) (*a*) and (138) (*b*) by permission of the American Chemical Society.]

Table 5.10. Applications of Macroporous Polymers as Stationary Phases in HPLC

Sample	Polymer	Reference
A. Biological Compounds		
Amino acid derivatives	XAD-2, -4	(114, 128)
	PRP-1	(125)
Deoxyribonucleotides	PRP-1	(139)
Fatty acids	Hitachi Gel 3011	(127)
Glucuronides	XAD-2	(140)
Nucleosides	PRP-1	(9, 63)
Purines, pyrimidines	XAD-4	(63)
Peptides	XAD-2, -4	(114, 118)
	PRP-1	(125, 141)
	Hitachi Gel 3013	(142, 143)
Proteins	PLRP-S 300A	(132)
Pyrrolizidine alkaloids	PRP-1	(144)
Thiochrome derivatives of thiamine and phosphate esters	PRP-1	(145)
B. Organic Compounds		
Aromatic derivatives	XAD-7	(146)
Benzene and naphthalene derivatives	Styrene-based gel	(128, 130)
Alkyl benzenes	Hitachi Gel 3010	(147)
Nitrobenzene derivatives	Styrene-based gel	(148)
Anilines	Macroporous styrene-based copolymer	(120)
	XAD-2	(47)
	Hitachi Gel 3010	(149)
Carboxylic acids	XAD-2	(20, 47)
	Macroporous styrene-based copolymer	(120)
	XAD-7,-8	(150)
	PRP-1	(151)
Chlorophenols	XAD-2	(20)
	PRP-1	(88, 125)
Chlorophenoxyacetic acids	XAD-2	(20
	PRP-1	(88)
Esters	XAD-2	(47, 110, 113)
	Macroporous styrene-based copolymer	(120)

267

Table 5.10. (*Continued*)

Sample	Polymer	Reference
	Hitachi Gel 3010	(146)
Flavonols	PRP-1	(152)
Phenols	XAD-2	(47, 58)
	XAD-7, -8	(111, 150)
	Hitachi Gel 3010	(149)
	PRP-1	(50, 88)
Quaternary ammonium salts	Hitachi Gel 3010	(153)
Sulfonic acids	XAD-2	(20, 109)
C. Pharmaceuticals		
Alkaloids	Hitachi Gel 3010	(154)
Amphetamines	Hitachi Gel 3010	(152)
Analgesics	XAD-7	(116)
Antibiotics	XAD-2	(155)
Antiepileptics	PRP-1	(8)
Erythromycins	PRP-1	(156)
Hydroxybenzoates	XAD-2	(113)
Hydroxychloroquine derivatives	PRP-1	(157)
Penicillians and cephalosporins	XAD-4	(158)
Steroids	XAD-2	(20, 48, 50)
Sulfas	XAD-2	(20, 117)
	PRP-1	(8)

5.7.7. Other Polymers and Their Future in HPLC

This chapter, for the most part, has focused on macroporous styrene-based polymers as reverse stationary phase adsorbents. Other kinds of organic polymers have been examined and shown to provide a favorable chromatographic performance. Often these polymers have to be synthesized, or commercial products are crushed into small irregularly shaped particles. Thus a major limitation in the development of organic polymers as LC stationary phases, in general, is that they are not conveniently available in a small uniform particle size and shape suitable for modern HPLC. Table 5.11 is a brief listing of other polymers that have been used in column chromatography. Several new and commercially available polymeric materials are the Toya Soda TSK/PW stationary phases in Table 5.11. These are hydroxylated polyether materials that show promise

Table 5.11. Several Other Organic Polymers Used as Stationary Phases in Column Liquid Chromatography

Polymer	Application	Reference
1. Ethylene glycol bis methacrylate and hydroxyethyl methacrylate copolymers (Spheron)	Proteins	(159)
2. Polyacrylamide C_{18} modified	Reversed phase	(160)
3. Polyamide on glass	Amine and hydroxy compounds and others that form H bonds	(161, 162)
4. Polyvinylpyrrolidone	Benzaldehyde derivatives	(163)
	Chlorinated phenols	(164)
	Nitrated triazines and tetrazines; preparative and analytical	(165)
	Polycyclic aromatic hydrocarbons	(166)
5. Styrene-vinyl copolymers (Porogel PN)	Peptides	(119)
6. Styrene-vinylpyridine copolymers (Porogel PS)	Peptides	(119)
Styrene-vinlypyridine copolymers (ACT-2)	Bases, cation exchange, reversed phase	(167)
7. Vinyl-hydroxylated polyether copolymer (TSK-Gel PW series)	Proteins (hydrophobic interaction chromatography)	(168)
8. Phenyl modified TSK-Gel PW	Proteins (hydrophobic interaction chromatography)	(168)
9. Polybutadiene on alumina	Organics, reversed phase	(170)

as stationary phases for hydrophobic interaction chromatography and reversed-phase chromatography, in particular for protein separations.

Much success has been achieved with organic polymers as stationary phases for LC. However, much synthesis work still needs to be done, with a major objective being the preparation of a physical form that is suitable for modern HPLC. A second major objective should be to take advantage of the uniqueness that organic polymers can provide as stationary phases. On the one hand, polymer chemical structure can be altered widely and systematically. This can be achieved by introducing different levels of polarity into the polymer matrix by the careful selection of suitable monomers that yield a given kind of polar or π-electron center within the polymer linkage. Alternatively, chromatographically active layers can be deposited synthetically onto other polymer beads or beads of inert material. One such example, which yields a favorable mass transfer and sample capacity, has already been described. In this case long poly (oxyethylene) chains were deposited onto macroporous and microporous cross-linked polystyrene (169). The terminal groups were then derivatized with fatty acid residues. Improved sample loading capacity and mass transfer properties were obtained with this column compared to silica-based materials. A second example is the polymerization of butadiene to form a polymer coating onto a micro alumina particle (170). This stationary phase acts as a reversed phase and is useful throughout the entire pH range because of the pH stability of alumina.

Another feature of organic polymers that has great potential in HPLC is that many of the polymers are still chemically active, and consequently their surfaces can be altered to provide a variety of surface functionalities. Classical examples are the sulfonation and quaternization of polystyrene-divinylbenzene to produce cation and anion exchangers, respectively, and the introduction of chelating groups to produce metal ion selective stationary phases. Other kinds of functionality can be introduced into this copolymer. One such example (ACT-I) is already commercially available (167), where a C_{18} group is covalently bonded to the aromatic units within the polystyrene-divinylbenzene copolymeric matrix. The presence of the C_{18} group apparently reduces the hydrophobicity of the polymeric surface, which reduces retention and therefore allows elution with weaker eluents. A systematic control of the C_{18} amount on the surface should produce stationary phases of controlled hydrophobicity. Introduction of other functionalities at low levels would likely achieve the same result, but they could also provide an additional kind of interaction center. Thus a stationary phase yielding mixed modes of interaction could be obtained.

REFERENCES

1. R. Kunin, E. F. Meitzner, and N. Bostnick, *J. Am. Chem. Soc., 84*, 305 (1962).
2. J. R. Millar, D. G. Smith, W. E. Marr, and T. R. E. Kressman, *J. Chem. Soc.*, 218 (1963); British patent 900,496 (1962).
3. R. Kunin, E. F. Meitzner, J. A. Oline, S. A. Fisher, and N. W. Frisch, *Ind. Eng. Chem. Prod. Res. Develop., 1*, 140 (1962).

4. R. M. Wheaton and M. J. Hatch, in J. A. Marinsky, Ed., *Ion Exchange*, vol. 2, Dekker, New York, 1969, p. 191.

5. R. L. Gustafson, R. L. Albright, J. Heisler, J. A. Lirio, and O. T. Reid, Jr., *Ind. Eng. Chem. Prod. Res. Develop., 7,* 107 (1968).

6. J. Paleos, *J. Colloid. Interface Sci., 31,* 7 (1969).

7. A. Warshawsky and H. Bercovitz, *Trans. Inst. Min. Metal, 88c,* 31 (1979).

8. D. P. Lee, *J. Chromatogr. Sci., 20,* 203 (1982).

9. D. P. Lee and J. H. Kindsvater, *Anal. Chem., 52,* 2425 (1980).

10. V. A. Davankov, S. V. Rogozhin, and M. P. Tsyurupa, in J. Marinsky and Y. Marcus, Eds., *Ion Exchange and Solvent Extraction*, vol. 7, Dekker, New York, 1977, p. 29.

11. N. Ohsawa, *Saikinno Kagasku Kogaku*, 81 (1976).

12. M. D. Grieser and D. J. Pietrzyk, *Anal. Chem., 45,* 1348 (1973).

13. J. L. Robinson, W. J. Robinson, M. A. Marshall, A. D. Barnes, K. J. Johnson, and D. S. Solas, *J. Chromatogr., 189,* 145 (1980).

14. D. S. Farrier, A. L. Hines, and S. W. Wang, *J. Colloid. Interface Sci., 69,* 233 (1979).

15. J. M. Brown and D. J. Wilson, *Sep. Sci. Technol., 15,* 1533 (1980).

16. M. A. Hasanian and A. L. Hines, *Ind. Eng. Chem. Prod. Res. Develop. 20,* 621 (1981).

17. D. J. Pietrzyk, *Talanta, 16,* 169 (1969).

18. F. Nevejans and M. Verzele, *Chromatographia, 20,* 173 (1985).

19. L. D. Bowers and S. Pedigo, *J. Chromatogr., 371,* 243 (1986).

20. D. J. Pietrzyk and C. H. Chu, *Anal. Chem., 49,* 757, 860 (1977).

21. C. H. Chu and D. J. Pietrzyk, *Anal. Chem., 46,* 330 (1974).

22. D. J. Pietrzyk, E. P. Kroeff, and T. D. Rotsch, *Anal. Chem., 50,* 497 (1978).

23. H. Y. Mohammed and F. F. Cantwell, *Anal. Chem., 50,* 491 (1978).

24. T. D. Rotsch, W. R. Cahill, Jr., D. J. Pietrzyk, and F. F. Cantwell, *J. Can. Chem., 59,* 2179 (1981).

25. C. R. Fox, in F. L. Slejko, Ed., *Adsorption Technology*, vol. 19, Chemical Industries Ser., Dekker, New York, 1985, p. 167.

26. S. D. Faust and O. M. Aly, *Adsorption Processes for Water Treatment*, Dekker, New York, 1986.

27. R. L. Gustafson and J. Paleos, in S. J. Faust and J. V. Hunter, Eds., *Organic Compounds in Aquatic Environment*, Dekker, New York, 1971, p. 213.

28. N. Weissman, M. L. Lowe, J. M. Beattie, and J. A. Demetriou, *Clin. Chem., 17,* 875 (1971).

29. D. N. Osborne and B. H. Gore, *J. Chromatogr., 77,* 233 (1973).

30. M. L. Bastos, D. Jukofsky, and S. J. Mule, *J. Chromatogr., 81,* 93 (1973).

31. M. P. Kullberg and C. W. Gorodetzky, *Clin. Chem., 20,* 184 (1974).

32. G. J. Digregorio and C. O'Brien, *J. Chromatogr., 101,* 424 (1974).

33. D. L. Roerig, D. Lewand, M. Mueller, and R. I. H. Wang, *J. Chromatogr.*, *110*, 349 (1975).
34. H. P. Gelbke, T. H. Grell, and G. Schmidt, *Arch. Toxicol.*, *39*, 211 (1978).
35. M. Bogusz, J. Gierz, and J. Bialka, *Arch. Toxicol.*, *41*, 153 (1978).
36. H. Y. Mohammed and F. F. Cantwell, *Anal. Chem.*, *50*, 491 (1978).
37. R. Denee and H. J. Huizing, *J. Nat. Prod.*, *44*, 257 (1981).
38. R. A. Hux, H. Y. Mohammed, and F. F. Cantwell, *Anal. Chem.*, *54*, 113 (1982).
39. G. Ibrahim, S. Andryauskas, and M. L. Bastos, *J. Chromatogr.*, *108*, 107 (1975).
40. J. H. G. Jonkman, J. Wijsbeek, J. E. Greving, R. E. Van Gorp, and R. A. De Zeeuw, *J. Chromatogr.*, *128*, 208 (1976).
41. W. Vycudilik, *J. Chromatogr.*, *111*, 439 (1975).
42. J. L. Rosenbaum, M. S. Kramer, R. Raja, and C. Boreyko, *New Engl. J. Med.*, *284*, 874 (1971).
43. J. L. Rosenbaum, *Ind. Eng. Chem. Prod. Res. Develop.*, *14*, 99 (1975).
44. J. L. Rosenbaum, M. S. Kramer, and R. Raja, *Arch. Int. Med.*, *136*, 263 (1976).
45. H. L. Bradlow, *Steroids*, *11*, 265 (1968).
46. C. H. L. Shackleton, J. Sjoevall, and O. Wilson, *Clin. Chim. Acta*, *27*, 354 (1970).
47. S. Levy and T. Schwartz, *Clin. Chem.*, *19*, 167 (1973).
48. M. Matsui, M. Hakozaki, and Y. Kinuyama, *J. Chromatogr.*, *115*, 625 (1975).
49. H. L. Bradow, *Steroids*, *30*, 581 (1977).
50. R. Verbeke, *J. Chromatogr.*, *177*, 69 (1979).
51. M. Axelson and B. L. Sahlberg, *Anal. Lett.*, *14*, 771 (1981).
52. J. D. White and D. P. Schwartz, *J. Chromatogr.*, *196*, 303 (1980).
53. J. P. Franke, J. Wijsbeek, J. E. Greving, and R. A. De Zeeuw, *Arch. Toxicol.*, *42*, 115 (1979).
54. G. Caderni and P. Dolara, *Pharmacol. Res. Commun.*, *15*, 775 (1983).
55. F. F. Sun and J. E. Stafford, *Biochim. Biophys. Acta*, *369*, 95 (1974).
56. H. Vogelman and F. Wagner, *Biotechnol. Bioeng. Symp.*, vol. 4, p. 2, 1974, p. 959.
57. T. Kamikubo and H. Narahara, *Bitamin*, *37*, 225 (1968).
58. R. H. Nieman, D. Pap, and R. A. Clark, *J. Chromatogr.*, *161*, 137 (1978).
59. K. H. Rosler and R. S. Goodwin, *J. Nat. Prod.*, *47*, 188 (1984).
60. M. Hori, *Bull. Chem. Soc. Jpn.*, *42*, 2333 (1969).
61. W. D. Loomis, J. D. Lile, R. P. Sandstrom, and A. J. Burbott, *Phytochemistry*, *18*, 1049 (1979).
62. G. C. Mills, *J. Chromatogr.*, *355*, 193 (1986).

63. T. Uematsu and T. Sasaki, *J. Chromatogr., 179,* 229 (1979).

64. F. T. Delbeke and M. Debackere, *J. Chromatogr., 136,* 385 (1977).

65. K. Kohmoto, K. Akimitsu, T. Kohguchi, H. Otani, and J. M. Gardner, *Plant Cell Rep., 5,* 54 (1986).

66. B. Andersson and K. Andersson, *J. Chromatogr., 242,* 353 (1982).

67. P. W. Holloway, *Anal. Biochem., 53,* 304 (1973).

68. F. Bonomi and D. M. Kurtz, *Anal. Biochem., 142,* 226 (1984).

69. T. Momoi, *Biochem. Biophys. Res. Commun., 87,* 541 (1979).

70. G. G. Gibson and J. B. Schenkman, *J. Biol. Chem., 253,* 5957 (1978).

71. R. C. Garland and C. F. Cori, *Biochemistry, 11,* 4712 (1972).

72. D. J. Lorusso and F. A. Green, *Science, 188,* 66 (1975).

73. A. K. Burnham, G. V. Calder, J. S. Fritz, G. A. Junk, H. J. Svec, and R. Willis, *Anal. Chem., 44,* 139 (1972).

74. W. H. Glaze, J. E. Henderson, IV, J. E. Bell, and V. A. Wheeler, *J. Chromatogr. Sci., 11,* 580 (1973).

75. G. Osterroht, *J. Chromatogr., 101,* 289 (1974).

76. G. A. Junk, J. J. Richard, M. D. Grieser, D. Witiak, J. L. Witiak, M. D. Arguello, R. Vick, H. J. Svec, J. S. Fritz, and G. V. Calder, *J. Chromatogr., 99,* 745 (1974).

77. E. E. McNeil and R. Osten, *J. Chromatogr., 132,* 277 (1977).

78. K. M. S. Sundaram, S. Y. Szeto, and R. Hindle, *J. Chromatogr., 177,* 29 (1979).

79. P. R. Musty and G. Nickless, *J. Chromatogr., 89,* 185 (1974).

80. D. Levesque and V. N. Mallet, *Int. J. Envirn. Anal. Chem., 16,* 139 (1983).

81. D. J. Mackey, *J. Chromatogr., 237,* 79 (1982).

82. V. Niederschulte and K. Ballschmiter, *Fresenius Z. Anal. Chem., 269,* 360 (1974).

83. G. Volpe and V. N. Mallet, *Chromatographia, 14,* 333 (1981).

84. P. MacCarthy, M. J. Peterson, R. L. Malcolm, and E. M. Thurman, *Anal. Chem., 51,* 2041 (1979).

85. L. M. Jahangir and O. Samuelson, *J. Chromatogr., 193,* 197 (1980).

86. J. L. Robinson, P. E. Griffith, D. S. Salas, and G. E. Sam, *Anal. Chim. Acta, 149,* 11 (1983).

87. K. Urano, H. Furuune, and K. Hayashi, *Bull. Chem. Soc. Jpn., 55,* 2248 (1982).

88. R. L. Smith and D. J. Pietrzyk, *J. Chromatogr. Sci., 21,* 282 (1983).

89. K. Andersson, C. Hallgren, J. O. Levin, and C. A. Nilsson, *Anal. Chem., 56,* 1730 (1984).

90. J. C. Harris, E. V. Miseo, and J. F. Piecewicz, EPA-60017-82-052, ORD Tech. Publ., CERI, USEPA, 26 W. St. Clair Ave., Cincinnati, OH 45268, 1982.

91. J. Namieshik and E. Kozlowski, *Fresenius Z. Anal. Chem., 311,* 581 (1982).

92. *Quality Assurance Handbook for Air Pollution Measurement Systems*, vols. 1–3, EPA-600/9-76-005, EPA-600/4-77-027a, EPA-600/4-77-027b, ORD Tech. Publ., CERI, USEPA, 26 W. St. Clair Ave., Cincinnati, OH 45268, 1976–1977.

93. R. Syder and D. J. Pietrzyk, *Anal. Chem., 50,* 1842 (1978).

94. J. L. Robinson, M. A. Marshall, M. E. Draganjac, and L. C. Noggle, *Anal. Chim. Acta, 115,* 229 (1980).

95. A. A. Nicholson, O. Meresz, and B. Lemyk, *Anal. Chem., 49,* 814 (1977).

96. D. A. Withycombe, B. D. Mookherjee, and A. Hruza, in G. Charalambous, Ed., *Analysis of Foods and Beverages, Headspace Techniques*, Academic, New York, 1978, p. 81.

97. J. S. Fritz and G. L. Latwesen, *Talanta, 17,* 81 (1970).

98. J. S. Fritz, R. T. Frazee, and G. L. Latwesen, *Talanta, 17,* 857 (1970).

99. J. S. Fritz and D. C. Kennedy, *Talanta, 18,* 837 (1971).

100. J. S. Fritz and J. J. Topping, *Talanta, 18,* 865 (1971).

101. J. S. Fritz, D. R. Beuerman, and J. J. Richard, *Talanta, 18,* 1095 (1971).

102. J. S. Fritz and D. R. Beuerman, *Anal. Chem., 44,* 692 (1972).

103. J. S. Fritz and W. G. Miller, *Talanta, 18,* 323 (1971).

104. J. R. Parrish, *Anal. Chem., 49,* 1189 (1977).

105. A. Sugii, N. Ogawa, T. Hida, and H. Imamura, *Chem. Pharm. Bull., 25,* 1899 (1977).

106. A. Sugii, N. Ogawa, and H. Yamamura, *Talanta, 29,* 697 (1982).

107. G. S. Katykhin, *Zhur. Anal. Khim., 27,* 758 (1972) (Engl. transl.); *Zhur. Anal. Khim., 27,* 849 (1972).

108. A. Warshawsky, in J. A. Marinsky and Y. Marcus, Eds., *Ion Exchange and Solvent Extraction*, vol. 8, Dekker, New York, 1981, p. 229.

109. M. W. Scoggins and J. W. Miller, *Anal. Chem., 40,* 1155 (1968).

110. M. W. Scoggins, *Anal. Chem., 44,* 1285 (1972).

111. J. S. Fritz and R. B. Willis, *J. Chromatogr., 79,* 107 (1973).

112. H. Joshua and C. Deber, in J. Meienhofer, Ed., *Chemistry and Biology of Peptides, Proc. 3rd Am. Peptide Symp.*, Ann Arbor Sci. Publ., Ann Arbor, MI, 1972, p. 67.

113. F. F. Cantwell, *Anal. Chem., 48,* 1854 (1976).

114. P. Kroeff and D. J. Pietrzyk, *Anal. Chem., 50,* 502 (1978).

115. T. Uematsu and R. T. Suhadolnik, *J. Chromatogr., 123,* 347 (1976).

116. R. G. Baum and F. F. Cantwell, *Anal. Chem., 50,* 280 (1978).

117. T. D. Rotsch, R. J. Sydor, and D. J. Pietrzyk, *J. Chromatogr. Sci., 17,* 339 (1979).

118. A. Niederweiser, *J. Chromatogr., 61,* 81 (1971).

119. J. J. Hansen, T. Greibrokk, B. L. Currie, K. Nils-Gunnar Johansson, and K. Folkers, *J. Chromatogr., 135,* 155 (1977).

120. H. Takahagi and S. Seno, *J. Chromatogr. Sci., 12,* 507 (1974).

121. R. G. Baum, R. Saetre, and F. C. Cantwell, *Anal. Chem., 52,* 15 (1980).

122. D. Palalikit and J. H. Block, *Anal. Chem., 52,* 624 (1980).

123. F. F. Cantwell and S. Puon, *Anal. Chem., 51,* 623 (1979).

124. F. F. Cantwell, in J. A. Marinsky and Y. Marcus, Eds., *Ion Exchange and Solvent Extractions,* vol. 9, Dekker, New York, 1985, p. 339.

125. C. E. Werkhoven, W. M. Boon, A. J. J. Praat, R. W. Frei, U. A. T. Brinkman, and C. J. Little, *Chromatographia, 16,* 53 (1982).

126. Z. Iskandarani and D. J. Pietrzyk, *Anal. Chem., 53,* 489 (1981).

127. M. Uchida and T. Tanimura, *J. Chromatogr., 138,* 17 (1977).

128. M. Popl, V. Dolansky, and J. Fahnrich, *J. Chromatogr., 148,* 195 (1978).

129. D. J. Pietrzyk, R. L. Smith, and W. R. Cahill, Jr., *J. Liq. Chromatogr., 6,* 1645 (1983).

130. R. M. Smith, *J. Chromatogr., 291,* 372 (1984).

131. D. J. Pietrzyk and J. D. Stodola, *Anal. Chem., 53,* 1822 (1981).

132. D. J. Pietrzyk, W. J. Cahill, Jr., and J. D. Stodola, *J. Liq. Chromatogr., 5,* 443 (1982); *ibid., 5,* 781 (1982).

133. D. P. Lee, *J. Chromatogr., 443,* 143 (1988).

134. W. R. Melander and C. Horvath, in M. T. W. Hearn, Ed., *Ion Pair Chromatography,* Chromatographic Science ser., vol. 31, Dekker, New York, 1985, p. 27.

135. Z. Iskandarani and D. J. Pietrzyk, *Anal. Chem., 54,* 2427 (1982).

136. R. L. Smith and D. J. Pietrzyk, *Anal. Chem., 56,* 1572 (1984).

137. P. G. Rigas and D. J. Pietrzyk, *Anal. Chem., 58,* 2226 (1986).

138. P. G. Rigas and D. J. Pietrzyk, *Anal. Chem., 59,* 1388 (1987).

139. S. Ikuta, R. Chattopadhyaya, and R. E. Dickerson, *Anal. Chem., 56,* 2253 (1984).

140. W. Dieterle, J. W. Faigle, and H. Mory, *J. Chromatogr., 168,* 27 (1979).

141. T. Sasagawa, L. Ericsson, D. C. Teller, K. Titani, and K. A. Walsch, *J. Chromatogr., 307,* 29 (1984).

142. T. Isobe, Y. Kurosu, Y. Fang, N. Ishioka, H. Kawasaki, and N. Takai, *J. Liq. Chromatogr., 7,* 1101 (1984).

143. Y. Kurosu, H. Kawasaki, X. C. Chen, Y. Amano, Y. Fang, T. Isobe, and T. Okuyama, *Bunseki Kagaku, 33,* E301 (1984).

144. H. S. Ramsdell and D. R. Buhler, *J. Chromatogr., 210,* 154 (1981).

145. J. Bontemps, L. Bettendorff, J. Lombet, C. Grandfils, G. Dandrifosse, E. Schoffeniels, F. Nevejans, and J. Crommen, *J. Chromatogr., 295,* 486 (1984).

146. T. G. McRae, R. P. Gregson, and R. J. Quinn, *J. Chromatogr. Sci., 20*, 475 (1982).
147. A. Nakae and G. Muto, *J. Chromatogr., 120*, 47 (1976).
148. F. Smejkal, M. Popl, A. Cihova, and M. Zazvorkova, *J. Chromatogr., 197*, 147 (1980).
149. H. Takahagi and S. Seno, *J. Chromatogr., 108*, 354 (1975).
150. M. A. Curtis and L. B. Rogers, *Anal. Chem., 53*, 2347 (1981).
151. T. Hanai and J. Hubert, *Chromatographia, 17*, 633 (1983).
152. J. G. Buta, *J. Chromatogr., 295*, 506 (1984).
153. G. Muto and A. Nakae, *Chem. Lett.*, 549 (1974).
154. K. Aramaki, T. Hanai, and H. F. Walton, *Anal. Chem., 52*, 1963 (1980).
155. Y. Kimura, H. Kitamura, A. Hisami, N. Toyoshige, K. Noguchi, M. Baba, and M. Hori, *J. Chromatogr., 206*, 563 (1981).
156. I. O. Kibwage, E. Roets, J. Hoogmartens, and H. Vanderhaeghe, *J. Chromatogr., 330*, 275 (1985).
157. S. E. Tett, D. J. Cutler, and K. F. Brown, *J. Chromatogr., 344*, 241 (1985).
158. F. Salto and J. Prieto, *J. Pharm. Sci., 70*, 994 (1981).
159. O. Mikes, P. Strop, J. Zbrozek, and J. Coupek, *J. Chromatogr., 119*, 339 (1976).
160. J. V. Dawkins, L. Lloyd, and F. P. Warner, *J. Chromatogr., 352*, 157 (1986).
161. F. M. Rabel, *Anal. Chem., 45*, 957 (1973).
162. H. Szumillo, E. Soczewinski, and W. Golkiewicz, *Chem. Anal. (Warsaw), 26*, 131 (1981).
163. G. Alibert and J. L. Puech, *J. Chromatogr., 124*, 369 (1977).
164. L. Olsson, N. Renne, and O. Samuelson, *J. Chromatogr., 123*, 355 (1976).
165. D. H. Freeman, R. M. Angeles, and I. G. Poinesen, *J. Chromatogr., 118*, 157 (1976).
166. G. Goldstein, *J. Chromatogr., 129*, 61 (1976).
167. Interaction Chemicals, Inc., Mountain View, CA.
168. Toyo Soda Manufacturing Co., Tokyo, Japan.
169. E. Bayer, presented at the 10th Int. Symp. on Column Liquid Chromatography (San Francisco, May 1986), paper 504.
170. Alco Laboratories, New Kensington, PA.

CHAPTER

6

SIZE-EXCLUSION LIQUID CHROMATOGRAPHY

W. W. YAU, J. J. KIRKLAND, and D. D. BLY

E. I. du Pont de Nemours & Co.
Central Research and Development Department
Wilmington, Delaware

6.1. INTRODUCTION

Size-exclusion chromatography (SEC) is a highly predictable, convenient method for rapidly separating simple mixtures where the components are sufficiently different in molecular size. SEC can quickly indicate the complexity of a sample mixture, and at the same time it can conveniently supply approximate molecular weight values MW of sample components. Initial exploratory separations of unknown samples are easily performed with SEC to gain an overall picture of sample composition with minimum method development.

    SEC is relatively easy to comprehend and use. Molecules in solution are separated according to size by means of a column packed with porous particles. Retention and separation are determined by the molecular hydrodynamic radius of the dissolved solute, which is related to the logarithm of the molecular weight. Figure 6.1 illustrates the basis for SEC separation. The smaller solute molecule A in Fig. 6.1 is able to sample a larger effective volume within the particle pore, relative to the larger solute molecules B, as suggested by the corresponding dashed lines. Therefore, in the same time period, larger particles B move further downstream than the smaller particles A, which spend a longer time within the pore. Sample molecules that are too large to enter the pores at all are totally excluded from the particle and appear first in the chromatogram. Because solvent molecules usually are smallest, they emerge last, and the entire sample normally elutes prior to this solvent dead-time t_0 peak. This chromatographic behavior is fundamentally different from the other liquid chromatography (LC) methods, where sample components elute after the t_0 peak. Theory and experiment show that a single pore size packing is capable of separating molecules over about a two-decade range of molecular weight.

The distinct property of SEC is that separation occurs solely as a result of a difference in the molecular size of sample components. By utilizing proper combinations of packing material and mobile phases, undesired retention of solutes by other mechanisms (such as adsorption) can usually

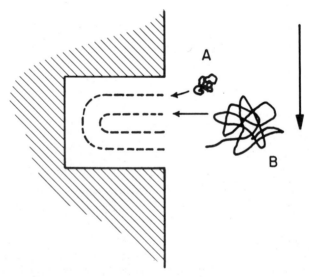

Figure 6.1. Size exclusion effects in single pore. Larger solute sees smaller effective pore volume.

be avoided. There are several advantages to SEC. First, each solute is retained as a relatively narrow band, which facilitates solute detection with detectors of only moderate sensitivity. Second, unknown samples elute within a known separation time, since the beginning and the end of every chromatogram are predictable and constant for a particular set of operating conditions. Third, since retention in SEC is determined by molecular size, the retention of any compound is predictable, and this is an advantage for sample identification. Fourth, SEC is a very gentle separation technique. Rarely does sample loss or reaction occur during the separation when proper operating conditions are employed. Finally, column deactivation problems normally do not occur in SEC, so that the special handling techniques that are often required to protect columns with other LC methods generally are unnecessary.

On the other hand, there are certain disadvantages as a result of the unique characteristics of SEC. For example, only a few separated bands can be accommodated within the SEC range of separation, that is, columns exhibit limited peak capacity. Thus complete resolution of complex multicomponent samples normally is not achieved. (However, molecular weight distribution analyses of polymers are readily accomplished by SEC, since individual components of these samples need not be resolved completely.) Another disadvantage of SEC is that the method is inapplicable to certain samples, particularly those which adsorb onto column packings, or those of similar size. For small molecules in SEC, a *size difference* of more than about 10% is required for acceptable resolution of individual components, even when highly efficient columns are used. For macromolecules a two-fold difference in molecular weight is necessary for significant separation.

Separation by SEC is illustrated in Fig. 6.2. In this example, the large solute molecules travel faster than the small ones, because they spend less time in the pore of the packing. Based on this effect, the two compounds elute from the column at different times. A more detailed look at size exclusion in a porous packing can be obtained by examining Fig. 6.3. It shows an electron micrograph of an actual SEC packing particle containing representations of two coiled polymer molecules of different sizes. From the relative size of the molecules and the pores, it is apparent that small molecules ($<10^5$ MW) will fit into most of the pores. However, the number of available pores for the large molecules (10^6 MW) is much more limited, and any molecule larger than 10^6 MW would have difficulty permeating the packing pores.

SEC is widely used to separate and identify macromolecules of biological origin using aqueous mobile phases and hydrophylic packings. This form of SEC traditionally has been known as gel filtration chromatography

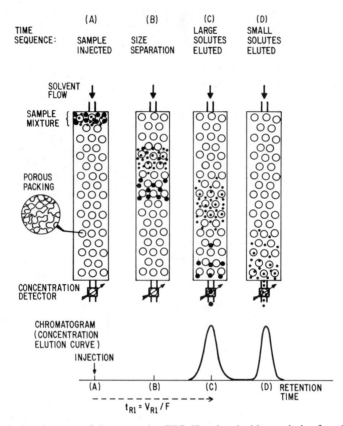

Figure 6.2. Development of size separation SEC. [Reprinted with permission from Ref. (1).]

(GFC). An example of a high performance gel filtration chromatography separation is shown in Fig. 6.4, in which a complex mixture of protein standards has been completely resolved within a few minutes. This approach represents a considerable saving in analysis time over traditional gravity-fed gel filtration chromatography.

The other SEC technique, termed gel permeation chromatography (GPC), normally employs organic mobile phases and lipophilic (hydrophobic) packings. It is used for separating small organic molecules, or for obtaining molecular weight distribution information on polymers. For example, Fig. 6.5 shows a 6-min chromatogram of an epoxy resin. The peaks of even-numbered oligomers elute in the order of decreasing molecular weight. The information in this chromatogram provides a useful tool for studying the curing of this thermosetting resin. Typically, gel permeation

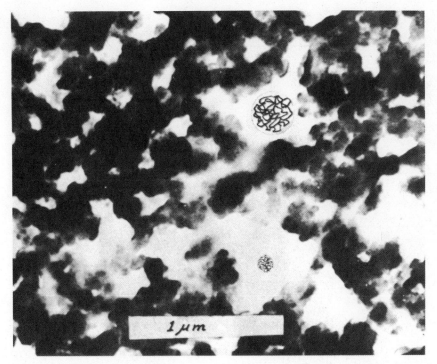

Figure 6.3. Comparison of effective sizes of polystyrene molecules in pores. MS $= 10^5$ and 10^6 on electron micrograph of Styragel packing. [Reprinted with permission from Ref. (2).]

chromatography is also used to determine the molecular weight distribution difference between polymer samples, as illustrated in Fig. 6.6, where several polyethylene samples of different distributions (and end-use properties) exhibit noticeably different elution profiles.

This chapter is devoted primarily to an overview of high performance SEC using columns of small particles and high inlet pressures for fast separations. Low-pressure SEC involving soft packing particles and relatively slow separations is not stressed. Development of small, rigid packing particles of about 5–10-μm diameter has made possible the preparation of efficient columns to provide fast, high-resolution SEC separations like those illustrated in Figs. 6.4–6.6. The effect of packing particle size on analysis time is clearly demonstrated in Fig. 6.7, where the reduction of particle diameter d_p from 120 to 6 μm provides time savings of several orders of magnitude without loss in resolution. The SEC analysis in Fig. 6.7a (120 μm) required several hours, versus less than a minute in the high performance separation (6 μm) at the bottom of the figure.

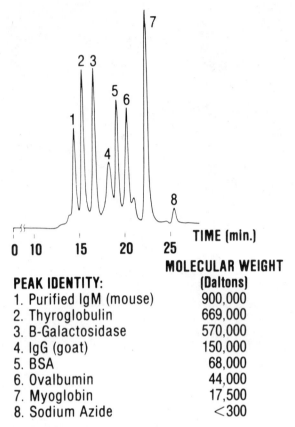

PEAK IDENTITY:	MOLECULAR WEIGHT (Daltons)
1. Purified IgM (mouse)	900,000
2. Thyroglobulin	669,000
3. B-Galactosidase	570,000
4. IgG (goat)	150,000
5. BSA	68,000
6. Ovalbumin	44,000
7. Myoglobin	17,500
8. Sodium Azide	<300

Figure 6.4. Separation of proteins on small-particle SEC columns. Column—Zorbax Bio Series GF-450, GF-250, 25 × 9.4 mm each; mobile phase—0.2M Na$_2$HPO$_4$, pH 7.5; flow rate—1.0 mL/min; temperature—ambient; detector—UV, 280 nm. [Reprinted with permission from Ref. (3).]

6.1.1.   Theory

In the other LC methods, sample components are retained by the column packing and elute after the unretained solvent. However, SEC is unique in that solutes are excluded from the packing according to size and elute ahead of the smaller solvent peak. In SEC this "unretained solvent" peak is referred to as the total permeation peak. Band broadening characteristics in SEC also are unique. Normally in the other LC methods, instrumental band broadening increases with solute retention time, but in SEC, the early eluting peaks of larger solutes often are the broadest because reduced solute diffusion rates lead to poor mass transfer.

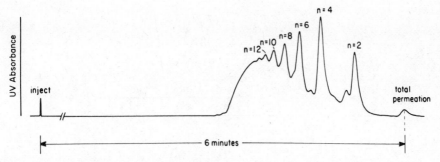

Figure 6.5. Separation of epoxy resin oligomers with porous silica. Column—25 × 0.62-cm Du Pont-SE 60; mobile phase—tetrahydrofuran; flow rate—0.9 mL/min; temperature—ambient; detector—UV, 254 nm. [Redrawn from Ref. (4).]

6.1.2. Retention

For a given SEC column packing, the retention of sample components according to molecular size can be represented by a graph of molecular size versus retention volume V_R, as illustrated in Fig. 6.8. The exclusion limit A in Fig. 6.8 for this packing is 10^5 MW; larger molecules cannot enter the pores of the packing. With this particular packing, all totally excluded compounds ($>10^5$ MW) will coelute as a single band A at V_0.

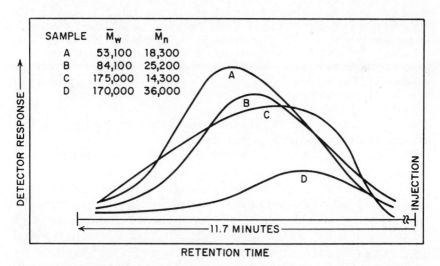

Figure 6.6. High performance–high-temperature chromatograms of linear polyethylenes. Three Du Pont SE columns in series—1000 Å, 500 Å, 100 Å; solvent—*o*-dichlorobenzene, 135°C; flow rate—1.3 mL/min; IR detection at 3.4 μm. [Redrawn from Ref. (5).]

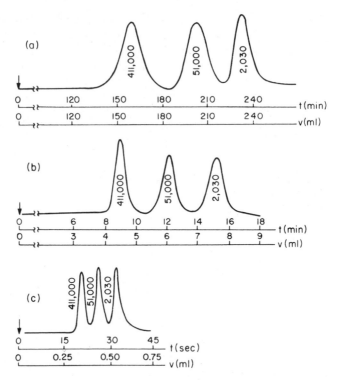

Figure 6.7. Effect of particle size on separation time. (a) d_p = 120 μm; 1 column each—4 ft × 0.305 in of CPG 10–2000, -1250, -700, -370, -240; mobile phase—toluene; flow rate— 1.0 mL/min. (b) d_p = 44 μm; 1 column each—50 × 0.26 cm of CPG 10–700, -350, -125; mobile phase—tetrahydrofuran; flow rate—0.5 mL/min. (c) d_p = 6 μm; 1 column—50 × 0.2 cm porous silica microspheres; mobile phase—tetrahydrofuran; flow rate—1.0 mL/min. [Redrawn from Ref. (6).]

This interparticle volume V_0 characterizes the solvent volume that fills the space between the packing particles.

The permeation limit D in Fig. 6.8 is 10^3 MW; smaller molecules all have equal access to the particle pores. Compounds with MW ≤ 10^3 elute as a single band at D, with a retention volume V_t (corresponding to the column dead volume or the LC unretained peak) that is equivalent to the *total* volume of solvent within the column. The intraparticle volume of the column V_i equals $V_t - V_0$; this is the volume of solvent held within the packing particles.

Thus in Fig. 6.8, compounds with molecular weights that decrease with retention volumes from A to D spend more time within the column packing pores as a result of their increasingly smaller size, and elute later. The intermediate molecular weight range (10^5–10^3) is the fractionation range and represents the useful range of SEC separation for a given packing.

Due to the unique features of peak retention and band broadening, the descriptive terminology used for other LC methods is inappropriate in SEC, that is,

$$k' = \frac{t_R - t_0}{t_0} = \frac{V_R - V_0}{V_0} \tag{1}$$

Since in SEC solutes elute before the solvent peak (i.e. $t_R < t_0$), Eq. 1 provides negative values of the capacity factor k' for all separations. SEC

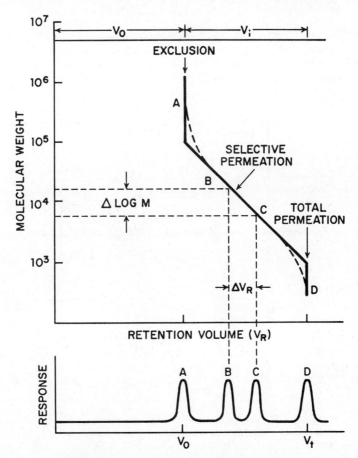

Figure 6.8. SEC calibration and separation range.

retention is therefore usually described by the solute distribution coefficient K_{SEC} (see below). It likewise follows that the separation factor α, defined as a ratio of k' values, is also not useful in SEC.

Size separation in SEC is the direct result of differential solute distribution between the moving mobile phase outside the particle pores and the stagnant mobile phase inside the pores, as illustrated in Fig. 6.9a. A large (wormlike) macromolecule permeates less space in the pores than smaller solutes. This mechanism is in sharp contrast to that for liquid-partition LC (Fig. 6.9b), where solute distribution occurs between the mobile phase and a second immiscible stationary liquid phase. In Fig. 6.9b, retention in liquid partition is shown to be the dominant factor where retention by the size-exclusion effect is relatively insignificant. A generalized account of retention in LC is given by

$$V_R = V_0 + K_{SEC}V_i + K_{LC}V_s \qquad (2)$$

where K_{SEC} is defined as the SEC solute distribution coefficient, V_i is the pore volume, K_{LC} the LC distribution coefficient, and V_s the actual, or

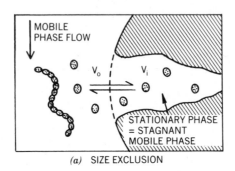

(a)   SIZE EXCLUSION

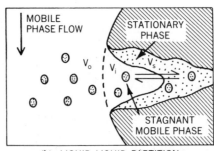

(b)   LIQUID–LIQUID PARTITION

Figure 6.9. Solute distribution between chromatographic phases.

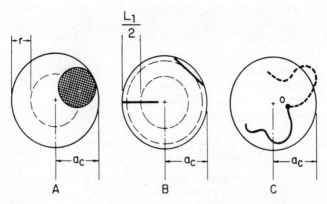

Figure 6.10. Exclusion effect in cylindrical void of radius a_c. A—hard sphere of radius r; B—thin rod of length L_1 in two orientations in the plane of the cross section; C—random flight chain with one end at point 0, showing allowed conformation (solid curve) and forbidden conformation (dashed curve). [Redrawn from Ref. (7).]

equivalent, volume of LC stationary liquid phase. In separations where only the SEC effect occurs (no adsorption, etc.),

$$V_R = V_0 + K_{SEC}V_i \qquad (3)$$

and

$$K_{SEC} = \frac{V_R - V_0}{V_t - V_0} \qquad (4)$$

K_{SEC} is a function of solute molecular size and of the column packing pore size. K_{SEC} values vary between zero and unity, corresponding to the total exclusion and total permeation limits of the SEC column, respectively.

Molecular retention in an SEC column is a complex function of the shape and size of both the solute molecules and the particle pores. However, useful insights about SEC retention have been developed from simple solute exclusion models, such as shown in Fig. 6.10. Shown here are a cylindrically shaped pore and representations of three different-shaped solutes: hard sphere, rigid rod, and random coil. With this pore model, a quantitative theory for predicting values of K_{SEC} has been developed (7, 8) using statistics to evaluate the number of allowed solute configurations within the pore cavity. Note that some solute configurations are excluded from the pores. The constraints presented by the cavity walls correspond to a reduction in the configurational entropy of the solutes. A

Table 6.1. Thermodynamics of LC Retention

Size Exclusion (GPC, GFC)

$K_{SEC} = e^{-\Delta G/RT} \simeq e^{\Delta S/R}$
Entropy S controlled process
ΔS is negative for all solutes; $S_{stationary} < S_{mobile}$
$K_{SEC} < 1$; solute elutes before solvent; k' negative
Temperature independent

Other LC Methods (Partition, Adsorption, Ion Exchange)

$K_{LC} = e^{-\Delta G/RT} \simeq e^{-\Delta H/RT}$
Enthalpy H controlled process
ΔH is negative for most solutes; $H_{stationary} < H_{mobile}$
$K_{LC} > 1$; solute elutes after solvent; k' positive
Temperature dependent; $\Delta G = \Delta H - T\Delta S$

---

larger solute undergoes a larger entropy decrease in a pore, which results in a smaller K_{SEC} value and earlier elution.

The fundamental thermodynamic differences between SEC retention and other LC methods are summarized in Table 6.1. Basically, SEC retention is an entropy-controlled equilibrium process. Under normal experimental conditions, K_{SEC} is independent of flow rate and temperature.

6.1.3. Peak Broadening

The extent of peak broadening in SEC columns is highly dependent on the value of K_{SEC} and the rate of lateral diffusion of the solutes in and out of the packing pores (see, for example, Fig. 6.9a). This effect exists to a lesser degree in the other LC methods [and in gas chromatography (GC)], where it is known as stagnant mobile phase mass transfer [see Ref. (1), chap. 3]. The slow diffusion of large solutes in SEC is a major cause of instrumental band broadening.

It is more difficult to develop a quantitative expression for peak broadening in SEC than for the other LC methods and for GC. Because of the large differences in solute sizes and diffusion coefficients commonly encountered in SEC, band broadening due to slow mass transfer varies significantly for different bands within a chromatogram. In LC (and GC) the broadening of eluting peaks usually increases in proportion to the retention volume, as illustrated in Fig. 6.11, that is, relatively constant plate number and plate height values are found in LC and GC. Because of this, a single value of the plate number N or plate height H can be used to express the column efficiency for any peak. In contrast, plate number and plate height in SEC often vary as a function of the solute. Greater band broadening (smaller N) is generally found for larger size molecules which comprise the earlier eluting SEC peaks (except those very near the exclusion limit) (Fig. 6.11). For reasons described later in this chapter, band broadening in SEC is often reported in terms of peak volume standard deviation values σ_v. These values are one-fourth the baseline peak width.

For permeating polymers ($K_{SEC} \neq 0$) the effect of band broadening as a function of mobile phase flow rate is given by a straight-line relationship between reduced plate height h and reduced velocity v (see curve A in Fig. 6.12). Band broadening in SEC increases at higher flow rates and generally increases with solute molecular weight. However, very large solutes elute at the total exclusion volume as relatively sharp bands, since band broadening from lateral diffusion within the pores is essentially eliminated for these compounds. As demonstrated by plots B and C in Fig. 6.12, similar plate height plots are observed for large nonpermeating solutes with both porous and nonporous packings. Plate height curves B and C exhibit the unique shape typical of the coupling between extraparticle eddy diffusion and mobile phase mass transfer [Ref. (1), chap. 3].

Longitudinal diffusion as a contribution to SEC band broadening is very small for large molecules and can be completely ignored. Actually, a solute band trapped in an SEC column will maintain its original band width for hours or days without noticeable broadening. Therefore, one can always expect less band broadening in SEC separations when flow rates are decreased, as suggested by curve A in Fig. 6.12.

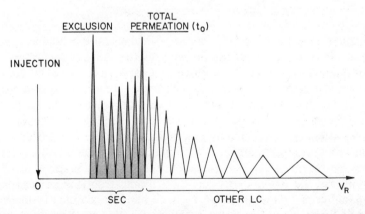

Figure 6.11. Characteristics of column dispersion and peak capacity in SEC and other LC methods. [Reprinted with permission from Ref. (1).]

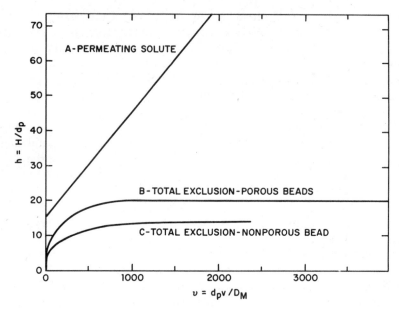

Figure 6.12. Effect of permeation on reduced plate height versus reduced velocity. [Redrawn from Ref. (8).]

6.1.4. Resolution

The resolution factor R_s can be expressed in SEC as (5)

$$R_s = \frac{\Delta V_R}{4\sigma_v} \tag{5}$$

The quantity ΔV_R is the difference in V_R values for two adjacent bands. The value of R_s is a measure of how well solute peaks are separated. SEC resolution depends equally on both peak separation ΔV_R and column dispersion σ_v; therefore any operating parameter that affects σ_v or ΔV_R also influences resolution. The expected effects of various SEC separation parameters on separation are summarized in Table 6.2. In general, parameters affecting σ_v are mainly kinetic factors (such as diffusion rates). On the other hand, parameters affecting ΔV_R are mainly factors that influence the extent of solute permeation into the packing–pore structure. For example, the strong dependence of column resolving power on particle diameter is evident in Fig. 6.7. Since band broadening is much less for small SEC packing particles because of better mass transfer, a proportionately smaller SEC peak-to-peak separation ΔV_R is needed to main-

Table 6.2. Influence of Operating Parameters on SEC Performance*

Parameters	Peak Separation D_2	Band Broadening σ	Resolution and Accuracy R_{sp}
Column volume	+ +	+ +	+ +
Particle size	−	+ +	+ +
Particle porosity	+	−	+
Particle shape	−	−	−
Pore size distribution	+	−	+
Pore shape	−	−	−
Solute conformation	+ +	+	+ +
Flow rate	−	+ +	+ +
Solvent viscosity	−	+	+
Temperature	−	+	+

* *Key:*
−   Negligible (or unsubstantiated) effect.
+   Moderate effect.
+ + Large effect.

tain the same resolution. Thus relatively short columns of small volumes can be used with small particle packings to achieve fast SEC separations without sacrificing resolution.

6.2.   EXPERIMENTAL

6.2.1.   Equipment

Except in special cases, the usual LC instrumentation is applicable for SEC. Equipment for gradient elution is not needed. SEC separations are strictly carried out in an isocratic (or constant solvent composition) mode. However, as in LC, minimal extracolumn band broadening is mandatory in high-performance SEC. Care should be taken to minimize injector and detector cell volumes and volumes associated with connecting tubing and fittings. High-efficiency SEC columns produce very sharp peaks, and extracolumn band broadening effects can seriously degrade the quality of the separation.

Accurate mobile-phase flow rates are very important in SEC. This is especially true of molecular weight assays, where the flow rate is assumed constant in converting retention time data to retention volume units for comparison with standard calibration data. Variations in pump output which occur during an SEC separation or between different runs can cause large errors in calculated sample molecular-weight values (9). Pulsations in pump output should be minimal to ensure good detector baseline stability.

Heated column compartments and detector cells are needed for the analysis of polymers that are soluble only at high temperatures.

Ultraviolet (UV) photometers and spectrophotometers are widely used as detectors in SEC for UV-absorbing compounds, particularly biological materials. However, the general differential refractometer (RI) detector is commonly used in the analysis of synthetic polymers. The RI detector measures the refractive index difference between the mobile phase and the column effluent containing the eluting solutes. A major limitation of the RI detector is its modest sensitivity. Infrared (IR) absorption detectors have also been used successfully for certain polymer analyses (such as polyolefins), but they are not suitable for use with aqueous mobile phases. The UV, RI, and IR detectors are all concentration dependent. Their response is proportional to the weight concentration of solutes in the detector cell.

Molecular-weight-specific detectors based on low-angle laser light scattering (LALLS) or continuous viscometry [Ref. (1), chap. 5] are useful for certain polymer molecular weight studies. These devices are more sensitive toward high molecular weight solutes, in contrast to concentration detectors (such as RI, UV, and IR), which respond only to weight concentration regardless of differences in component molecular weights. An example with the LALLS detector involving a polymer branching study is shown in Fig. 6.13a; dual traces of IR and LALLS detector responses for low-density polyethylenes are compared. The fact that the LALLS trace shows a large peak at the void volume is an indication of the presence of very large branched molecules. This effect is in contrast to the similar response of IR and LALLS detectors to a linear (high-density nonbranched) polymer, shown in Fig. 6.13b. When a sensitive viscosity detector is coupled with the usual SEC concentration detector, two SEC elution plots are obtained which provide information for determining sample molecular weight distribution via the universal calibration method discussed later in this chapter. Practical designs of the SEC viscosity detector are now available (11, 12).

More selective detectors (such as fluorescence and reaction detectors) are often used to measure single molecular weight species. Detection with fluorometers and spectrofluorometers is based on the fluorescent energy emitted from certain UV-light-activated solutes. Picogram amounts of solute can sometimes be detected, and this very high sensitivity can be particularly useful for biological compounds occurring at very low concentrations.

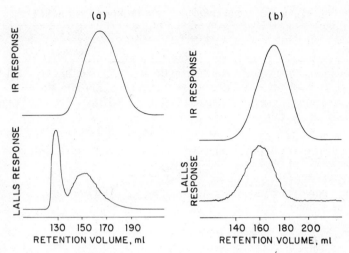

Figure 6.13. Comparison of IR and LALLS detector responses. (*a*) Branched (low-density) polyethylene SRM 1476. (*b*) Linear (high-density) polyethylene SRM 1475. [Redrawn from Ref. (10).]

Reaction detectors use color-forming reactions between an externally added reagent and the eluting solute. These devices can be arranged to function either as general or as selective detectors, and final measurement is usually by absorption photometry. With macromolecules, in contrast to small molecules, reaction detectors have thus far been utilized only for the measurement of end groups.

6.2.2. Columns

Column and column-packing technology are the same for SEC as for the other LC techniques. Typically, 0.6 to 0.8-cm id columns are used with 15- to 50-cm lengths, as this combination appears to be a good compromise between overall column performance and the quantity of mobile phase and column packing required. Column performance often is better with larger internal diameters than with smaller ones. An advantage of larger internal diameter, larger volume SEC columns is that extracolumn band broadening effects associated with the equipment are minimized.

A variety of column packings with semirigid organic gels and rigid inorganic particles are commercially available. Table 6.3 includes examples of such packings, plus compatible polymer types and effective solvent systems. The manufacturer normally supplies approximate calibration data for a given SEC packing. However, when accurate molecular weight information is required, exact calibration curves should be deter-

Table 6.3. Typical Solvents and Packings Used in SEC

Polymer Types	Typical Solvent System	Typical Column Packings†	Supplier
1. Proteins, polypeptides	Aqueous buffers	Zorbax Bio Series GF	Du Pont
2 Biopolymers, viruses, DNA, RNA	Aqueous buffers	TSK-G-PW	Toya Soda
		Superose	Pharmacia
3. Cellulose derivatives, polyvinyl alcohol, polysaccharides	Aqueous buffer, salts	TSK-G-SW	Toya Soda
4. Many polar noncrystalline synthetic polymers, some crystalline polymers, small molecules	Tetrahydrofuran	Ultra-Styragel	Waters
5. Nonpolar, noncrystalline synthetic polymers, hydrocarbon polymers, low MW polymers	Toluene (benzene* or chloroform*)	LiChrospher Si	Merck
6. Polar crystalline polymers (e.g., polyamides and polyesters)	m-Cresol (hot) or hexafluoroisopropanol (cold)	PL gel	Polymer Labs
7. Nonpolar crystalline polymers (e.g., polyethylene and stereoregular polyhydrocarbons)	1,2,4-Trichlorobenzene (hot) or 1,2-dichlorobenzene (hot)	Zorbax PSM (bimodal)	Du Pont

† These are merely examples to illustrate commercially available packings. Each packing often can be used for many polymer types. See Ref. (1), chap. 6 and Ref. (13).
* Suspected carcinogen

mined for each column, using appropriate macromolecular standards. Figure 6.14 shows typical calibration plots for some microparticular hydrophilic packings (TSK gels) for gel filtration, using proteins as solutes with an aqueous mobile phase.

Optimum performance by an SEC column is characterized by high resolution, low surface adsorption, and low back pressure, plus good mechanical, chemical, and thermal stabilities. A well-made column will withstand a wide variety of experimental conditions, including solvent changes, high flow rates, high pressure, and varying temperatures.

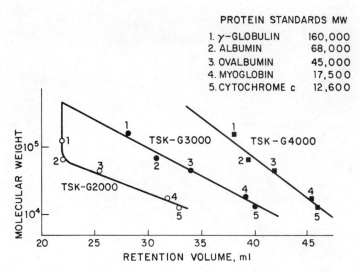

PROTEIN STANDARDS MW

1. γ-GLOBULIN   160,000
2. ALBUMIN   68,000
3. OVALBUMIN   45,000
4. MYOGLOBIN   17,500
5. CYTOCHROME c   12,600

Figure 6.14. Calibration plots for aqueous SEC packings. 2 columns—60 cm Spherogel TSK SW. [Redrawn from Ref. (12).]

Columns of rigid particles have several advantages over soft gels, including mechanical stability, which make them better suited for high-pressure applications. A potential disadvantage for rigid inorganic particles is the surface adsorptive characteristics toward some solutes. However, particle surfaces can be deactivated by chemical modification. While soft organic gel packings have been used traditionally in gel filtration separations, rigid inorganic and organic particles are now available for modern high-pressure applications. For example, the SW-type TSK-Gel (Toyo Soda) is a rigid silica-based hydrophilic packing developed primarily for use with aqueous mobile phases.

6.2.3. Operating Conditions

The effects of separation variables in SEC are similar to those for the other LC methods, although there are some differences, which are stressed in this section.

6.2.3.1. Flow Rate Effects

The efficiency of SEC columns can be very much affected by the mobile phase velocity or flow rate. As illustrated in Fig. 6.15, the increase in plate height (lowering of column efficiency) with increased mobile phase

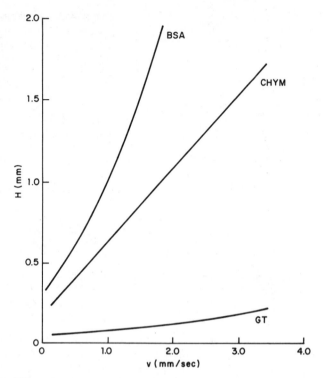

Figure 6.15. Effect of molecular weight on plate height versus velocity. Column—25 × 1.0 cm, SynChropak GPC 100; mobile phase—0.1M phosphate buffer, pH 7. BSA—bovine serum albumin; CHYM—chymotrypsinogen; GT—glycyltyrosine. [Redrawn from Ref. (14).]

velocity (or flow rate) is substantially greater for higher molecular weight materials than for lower molecular weight components. For lower molecular weight solutes such as glycyltyrosine (GT, MW = 132), this column of small particles exhibits a relatively small increase in plate height with increased mobile phase velocity. However, as solute molecular weight increases [for example, chymotrypsinogen (CHYM, MW = 23,000) and bovine serum albumin (BSA, MW = 65,000)], diffusion becomes poorer and the plate height increases rapidly with increasing mobile phase velocity. Because of this effect, SEC columns normally are operated at low mobile phase velocities, typically 0.2 cm/s or less (measured with a totally excluded solute). This translates to flow rates of about 2 mL/min for a column of 0.8-cm id.

The use of low flow rates is especially important for very high molecular weight solutes, because of the likelihood of mechanical shear degradation in a packed column at higher flow rates. Shear degradation has been observed for some polymers as small as 10^6 MW (15). To reduce the possibility of shear, some workers are using columns of particles in the 15–20-μm diameter range, rather than columns of 5-μm particles that are often preferred because of superior resolution characteristics.

6.2.3.2. Temperature Effects

In contrast to the other LC methods, varying the temperature is not an effective means of creating differences in SEC separation selectivity. Temperature is increased in SEC to enhance solubility or to improve column efficiency by decreasing solvent viscosity. However, most SEC separations are carried out at room temperature. When characterizing certain synthetic polymers, such as higher molecular weight polyolefins, temperatures higher than 100°C are required; these materials cannot be dissolved at the lower temperatures. If a sample of biochemical origin is sufficiently stable, SEC separation can also be carried out at higher temperatures (for example, 50–70°C) to improve column efficiency and resolution. Even with marginally stable materials, separations can be carried out fast enough so that sample degradation is usually insignificant (16).

6.2.3.3. Sample Size

Sample volume effects in SEC are similar to those in the other LC methods (1). For maximum resolution, sample volumes should not be more than one-third of the baseline volume of a monomer peak injected at a very small volume. Since one-half of the injected sample volume contributes to the measured retention volume, it is also important to use a constant sample injection volume for calibration and analysis. Sample volumes of 25–50 μL are typically used for single column lengths of 0.8-cm id, and larger volumes can be tolerated for connected column sets.

Higher sample concentrations in SEC generally cause increased retention volumes, rather than a decrease, as found for other LC methods. Therefore, the retention volumes of eluting peaks may not accurately reflect true sample molecular weight information when the column is overloaded by large samples. Typically, sample concentrations of 0.05–0.3 wt% are used.

Increasing the sample load can also result in decreased column efficiency, as indicated in Fig. 6.16. Here we see the dependence of plate height on sample loading for bovine serum albumin, using a column of silica deactivated by a chemically bonded stationary phase. Additional details on sample size effects are given in Ref. (1).

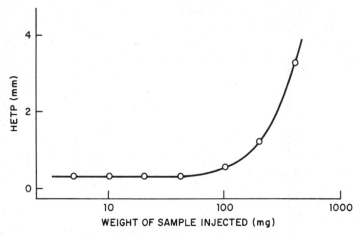

Figure 6.16. Dependence of plate height on sample loading. 2 columns—60 cm, Toya Soda G3000 SWG; mobile phase—0.1M phosphate buffer with 0.3M NaCl, pH 7; flow rate—8 mL/min; sample—bovine serum albumin. [Redrawn from Ref. (17).]

6.2.4. Mobile Phases

In SEC the mobile phase is selected for its ability to dissolve the sample, for compatibility with the detector and the packing, and for low viscosity (high solute diffusivity). In contrast to the other LC methods, the mobile phase composition in SEC generally does not influence solute retention or the shape of the elution curve. Solvent choice in SEC is sometimes limited by the solubility of certain large molecular weight polymers. Several common SEC mobile phases are listed in Table 6.3. When soft or semirigid organic gel packings are used in SEC, it is required that the solvent first swell the gel to open the packing pore structure for effective separations.

The mobile phase in SEC should also be chosen to prevent interaction of the sample components with the surface of the packing by adsorption or other effects. Silica-based packings present a particular problem, and a good rule of thumb with unmodified silicas is to add other components to the mobile phase that will be more strongly adsorbed than the sample. Sample adsorption is indicated if sample components elute after the total permeation volume of the column. Peak tailing in SEC may not mean sample adsorption, since polymeric samples can have asymmetric molecular weight profiles.

To eliminate adsorption on silica-based packings in aqueous mobile phases, ionic strengths in the 0.01–0.5M range often are satisfactory (18).

Addition of highly polar molecules (for example, 0.01–0.1% of polyethylene glycol or ethylene glycol) is also effective in reducing the adsorption of solutes onto silica packings. Here the modifier is strongly adsorbed to the silicious support, and solutes are unable to compete effectively for active sites on the packing.

Because of their acidic surface, some unmodified silica-based packings also can behave as weakly acidic ion exchangers. Sodium, ammonium, potassium, and tetraalkylammonium ions are increasingly effective in eliminating undesired adsorptive ion exchange effects (19). Tetraalkylammonium cations may cause the silica surface to become hydrophobic, and reversed-phase retention may then occur. In this case, the addition of organic solvents such as methanol to the mobile phase prevents this undesirable effect (19). Mobile phases of pH ≪ 3 are also effective in minimizing the ion exchange effect of silica surfaces.

In the separation of polyelectrolytes, a change in the ionic strength of the mobile phase usually affects the separation drastically. If additional salt is added to the mobile phase, electrostatic repulsive forces along the polyelectrolyte chain are reduced by ionic shielding. Salt addition also decreases osmotic forces. Overall, these effects cause a reduction of the hydrodynamic volume of the polymer molecule, which results in its greater permeation into the packing pores. Figure 6.17 is an example of the effect of mobile phase ionic strength on the elution of carboxymethylcellulose. Higher ionic strength increases the population of late-elut-

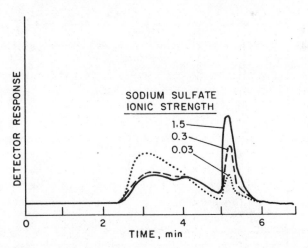

Figure 6.17. Effect of mobile phase ionic strength on elution of carboxymethylcellulose. Column—25 × 0.41 cm, SynChropak 500; mobile phase—sodium sulfate solution; flow rate—0.5 mL/min. [Redrawn from Ref. (20).]

ing (smaller) components, and decreases the amount of early-eluting (larger) materials.

Silica packings that are surface modified with organic functional groups are substantially less likely to exhibit solute adsorption. For example, to minimize adsorption and ion exchange, silica-based column packings used in aqueous mobile phases are often modified with glyceryl groups:

$$\overset{\displaystyle O}{\underset{\displaystyle O}{\big|}}$$

$$Si{-}O{-}\underset{\displaystyle |}{\overset{\displaystyle |}{Si}}{-}(CH_2)_3{-}O{-}CH_2{-}\overset{\displaystyle OH}{\underset{\displaystyle |}{CH}}{-}CH_2OH$$

With such chemically modified hydrophylic packing materials, separation of a wide range of water-soluble synthetic polymers, proteins, enzymes, nucleic acids, polynucleotides, and similar materials is feasible (15). For example, Fig. 6.18 shows the purification of crude β-galactosidase on a hydrophylic-modified silica-based packing. In this case, the distribution of the total proteins was monitored by UV absorbance and the elution of β-galactosidase was identified by enzymatic activity. Recoveries of this protein were in the 69–93% range, which is typical for

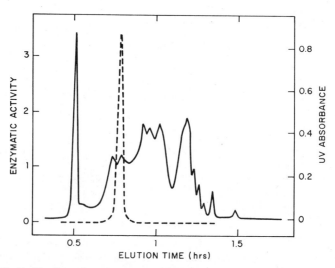

Figure 6.18. Purification of crude β-galactosidase on chemically modified porous silica. Separation conditions same as for Fig. 6.16. [Redrawn from Ref. (17).]

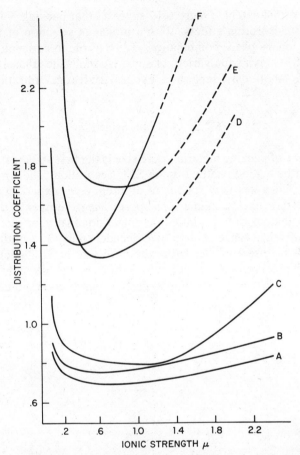

Figure 6.19. Dependence of distribution coefficient of lysozyme on mobile phase ionic strength. Mobile phase—phosphate buffer, pH 7.05 at ionic strength indicated. Packings: A—TSK-SW-2000; B—TSK-SW-3000; C—LiChrosorb Diol; D—Shodex OHpak B-804; E—Waters I-125; F—SynChropak GPC 100. [Redrawn from Ref. (21).]

samples of this type separated on properly prepared hydrophylic-modified SEC packings. Corresponding SEC separations on unmodified silica packings give much lower recoveries.

Silica-based SEC packings modified with hydrophylic groups may still exhibit residual cation exchange activity. For example, Fig. 6.19 illustrates the dependence of the distribution coefficient for the strongly basic enzyme lysozyme on the ionic strength of the mobile phase for a number of packing types. Lysozyme is a sensitive indicator of the anionic and hydrophobic characteristics of column packings, and not all proteins are adsorbed to this extent. Figure 6.19 suggests that packing *A* is the least affected by adsorption effects. With a judicious selection of the mobile phase–stationary phase combination, it is feasible to carry out high performance SEC on a wide variety of sensitive molecules that traditionally have been laboriously separated by gravity-fed gel filtration chromatography.

6.2.5. Column Selection

In the choice of packing material, pore size is the first and most important decision to be made in setting up an SEC separation. The simple model of Fig. 6.1 illustrates how solutes of different sizes experience different degrees of steric exclusion in a single pore. The experimental data in Fig. 6.20 verify that an SEC packing of a single nominal pore size separates a molecular weight range of about two decades, regardless of the absolute pore size. The effect of increasing the pore size in SEC packings is to move the log MW versus V_R calibration curves up the MW axis as shown; the slope and the length of the calibration curve corresponding to the

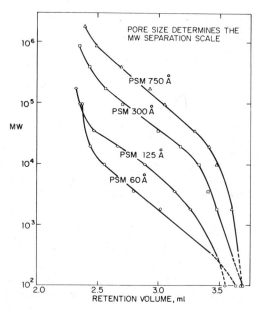

Figure 6.20. Molecular weight calibration range as a function of pore size. Columns—10 × 0.78 cm of each porous silica microsphere; mobile phase—tetrahydrofuran, 22°C; flow rate—2.5 mL/min; detector—UV, 254 nm; sample—25-μL solutions of polystyrene standards. [Redrawn from Ref. (22).]

linear molecular weight separation range are not affected. It is especially important to note that the characteristic slope of the calibration curve arises naturally from a single pore size exclusion effect and is *not* the result of a range of pore sizes in the SEC packing.

Both the molecular weight level and range of the samples to be analyzed are important considerations in selecting the column packing. Pore size differences among single pore size columns will change the accessible molecular weight level, but not the length of the molecular weight range for a given SEC separation. The data in Fig. 6.21, which show the conventional gel filtration chromatography separation of a biological sample, illustrate how pore size can affect the quality of the SEC chromatogram. Important concepts regarding column selection strategy are demonstrated in this figure. For example, solute elution moves from the total exclusion limit to near total permeation when columns of increasingly larger pore sizes are used [from top to bottom in Fig. 6.21 (23)], and poor separations were obtained with excessively large or small pore size columns. In this case, with single pore sizes the molecular weight separation range of the packing does not totally overlap that of the sample. Optimum resolution is always obtained for components which elute within the linear molecular weight portion of the calibration curve (Bio-Gel P-200 in Fig. 6.21).

Once a given pore size is selected for a separation, resolution can be improved by connecting two columns of the same pore size, as illustrated in Fig. 6.22. Note that while the range of molecular weight of the separation remains unchanged, the separation volume ΔV_R is increased, which provides higher resolution. This principle also holds for connected identical column sets containing more than one pore size.

Often the molecular weight separating range of a single pore size column is too narrow to analyze complex sample mixtures. This is especially true in polymer molecular weight distribution analyses. Most commercial polymers have components covering more than two decades of molecular weight, and columns of different pore sizes usually are placed in series to provide a broader separation range for such samples. When different pore size columns are connected, the shape of the resultant log calibration curve will vary. To obtain a wide molecular weight separating range, the preferred approach is to combine two different pore size columns to produce a *bimodal set*. Columns for a bimodal set should be selected so that the two pore sizes differ by about one decade in size and have comparable pore volumes. In this way, the molecular weight ranges of the individual calibration curves are closely adjacent but not overlapping.

A properly designed bimodal pore size column set provides a highly linear log MW versus V_R calibration plot with a range of molecular weight of about four decades, as illustrated in Fig. 6.23. This optimum bimodal set was obtained by coupling the individual 750-Å and the 60-Å PSM columns shown earlier in Fig. 6.20. The wide linear calibration plot of bimodal columns is especially desirable, since it corresponds to a more even resolution of components throughout the separation range of the

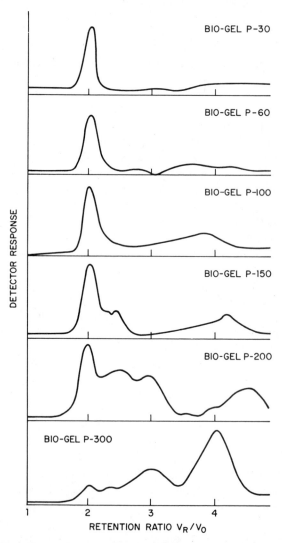

Figure 6.21. Effect of pore size on erythrocyte elution profile. Bio-Gel columns; mobile phase—0.3% sodium dodecyl sulfate (SDS), pH 9; sample—human erythrocyte membrane and 0.5% SDS. [Redrawn from Ref. (23).]

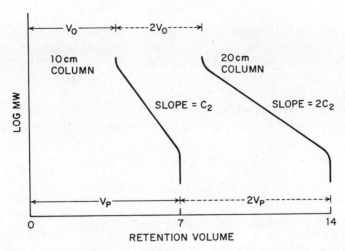

Figure 6.22. Effect of column pore volume on separation capacity (slope of molecular weight calibration curve). [Redrawn from Ref. (1).]

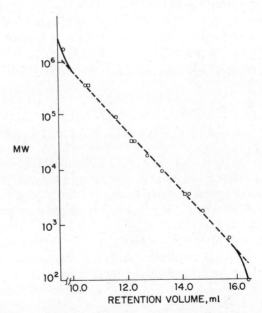

Figure 6.23. Broad range linear calibration for polystyrene standards using bimodal pore size principle. Four 10 × 0.78-cm porous silica microsphere packings, 60 Å, 60 Å, 750 Å, 750 Å; mobile phase—tetrahydrofuran, 22°C; flow rate—2.5 mL/min, detector—UV, 254 nm; sample—250 μL polystyrene standard solution. [Reprinted with permission from Ref. (22).]

305

column set. This arrangement permits a better visual comparison of elution curve profiles for samples of different molecular weights and molecular weight distributions. More importantly, the linear calibration curve provides a convenient approach to treating data quantitatively for accurate polymer molecular weight distribution analyses (24).

Standard reference materials are needed both to evaluate SEC system performance and to calibrate column retention in terms of molecular weight.

6.3. APPLICATIONS

6.3.1. Separation of Small Molecules

Small molecules (up to about 5000 MW) can be separated rapidly and conveniently by SEC, without extensive method development, using well-packed columns of small particles (5 or 10 μm). A variety of problems can be solved when there is sufficient difference in solute size to permit separation. Widespread use of SEC is now being made for separating small molecules, not only for the initial separation of unknown samples, but also for quality control and routine quantitation of certain materials.

A basic limitation to the application of SEC for small-molecule separations is the low peak capacity n, which can be estimated by

$$n = 1 + 0.2N^{1/2} \tag{6}$$

where N is the column plate count (25). For example, a set of columns having a total plate number of about 30,000 would have a theoretical peak capacity of only 35 bands for very small molecules; that is, in theory, 35 peaks of equal width could be placed between the total permeation and the total exclusion volumes at unity peak resolution (see Fig. 6.11). This is relatively limiting when compared to the larger peak capacity of the other LC methods.

Typically, a total of 10,000–30,000 theoretical plates are used for the SEC separation of small molecules, and the columns used must have the smallest available pore size, generally 60 Å or less. The main purpose in separating small molecules by SEC is to determine the number of components of different molecular weights in a sample, or to identify and measure one or more specific compounds. A molecular weight calibration for small molecules is sometimes useful for predicting the identity of unknown compounds in a particular series. By way of example, Fig. 6.24 illustrates the separation of a series of plasticizers on an organic gel col-

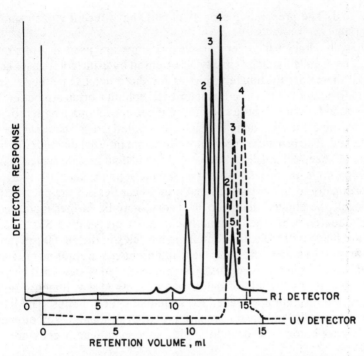

Figure 6.24. SEC separation of small molecules. Column—61 × 0.8 cm TSK G2000 H8; mobile phase—tetrahydrofuran; flow rate—0.5 mL/min; detectors—RI and UV at 254 nm. Samples: 1—epoxidized soya bean oil; 2—didodecyl phthalate; 3—didecyl phthalate; 4—di-2-ethylhexylphthalate; 5—dibutyladipate. [Redrawn from Ref. (28).]

umn, which in turn facilitated the identification of these compounds, because of the approximate molecular weight values obtained by this approach. For a tabulation of more than 50 papers dealing with the characterizations of fossil fuels, by-products, plastics additives, fats and oils, refined products, and various other kinds of samples, see Refs. (26) and (27).

6.3.2. Preparative SEC

While previous sections have emphasized analytical aspects, SEC can also be used as a preparative technique to conveniently isolate relatively large amounts of purified components (such as for molecular weight standards, for extensive characterization, or for biological testing). Conditions and techniques for preparative separations are somewhat different than those for analysis since with preparative work sample loading capacity is the goal. The preparative aspects of SEC have recently been summarized (29).

Typically, low-cost larger particle packings are used in preparative SEC. The sample loading capacity is increased by using columns of larger internal diameter. Fortunately, just as for the other LC methods, separation efficiencies with large-diameter SEC columns often are better than those found for narrow-bore columns of the same column length provided that the same ratio of sample weight to column cross-sectional area is maintained. The practical upper limit of column internal diameter for SEC has not yet been established, but 0.8–2.3-cm id columns are typically used for preparative purposes. Since the sample loading capacity increases proportionally to the cross-sectional area (regardless of the packing particle size), the column diameter, in theory, may be further increased to accommodate even larger sample loads. Most preparative SEC separations are made with columns containing particles in the 30–60-μm range.

In preparative SEC the equipment requirements are not as critical as for analysis. In general lower cost, less sophisticated systems can be used. Pumping systems capable of delivering solvents at a volumetric output of up to 100 mL/min are needed for wide-diameter columns. The RI detector is generally suitable for preparative SEC, although UV photometric and spectrophotometric detectors with short-path-length cells are also useful.

Large sample volumes are conveniently introduced into large-diameter preparative columns with a sampling valve. Sample volumes of 10–50 mL are typical, but very large volumes (>200 mL) are needed for some applications. These very large samples often are metered into the column with an external pump. As in analytical separations, the volumetric flow rate in preparative SEC columns must be increased proportionally to the column cross-sectional area to maintain the same linear flow velocity. At small sample loads, the plate number N decreases with an increase in flow rate. However, this effect is less important at high sample loadings.

Manual collection of fractions is usually adequate in preparative SEC when only a few components are to be isolated. However, an automatic fraction collector is advantageous when long, repetitive runs are to be made.

An example of the use of preparative SEC for isolating significant amounts of small molecules in the high-resolution mode is shown in Fig. 6.25. Here 150 mg each of three components was collected from a mixture separated on an organic gel column of small pores. Even larger amounts of these materials could have been isolated by injecting a larger size sample and taking a "heart cut" of each peak to ensure high component purity. Preparative SEC also has been used to prepare narrow molecular weight

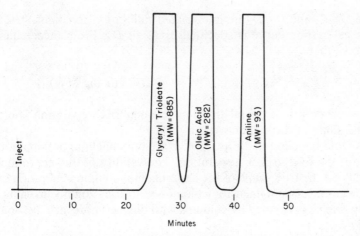

Figure 6.25. Preparative SEC of small molecules. Column—150 × 0.8 cm μ-Styragel 100 Å; mobile phase—tetrahydrofuran; flow rate—1.0 mL/min; detector—RI; sample—150 μg each compound. [Reprinted with permission from Ref. (30).]

distribution standards of polymers for use in molecular weight calibration studies.

6.3.3.   Molecular Weight Calculations

6.3.3.1.   Calibration Curve

In SEC the profile of the chromatogram contains information about the sample composition or molecular weight distribution. Often precise and accurate sample molecular weight information is sought, and a calibration plot relating retention volume to molecular weight often is the first step in interpreting SEC data.

The basic features of an SEC calibration curve (commonly plotted as the logarithm of molecular weight MW versus V_R) are shown in Fig. 6.8. Between total exclusion A and total permeation D (Fig. 6.8) lies the useful molecular weight separation region. For solutes retained in this region (such as solutes B and C in Fig. 6.8) a simple proportional relationship between V_R and log MW exists. This linear portion of the calibration curve can be described by

$$V_R = C_1 - C_2 \log MW \tag{7}$$

or

$$MW = D_1 e^{-D_2 V_R} \tag{8}$$

where C_1, C_2 and D_1, D_2 are interrelated calibration constants. The slope of the calibration curve is specified by C_2 or D_2. From these equations it follows that

$$\Delta V_R = -C_2 \, \Delta \log \text{MW} = -2.303 \, (\Delta \log \text{MW})/D_2 \qquad (9)$$

Equation 9 describes the relationship between SEC separation selectivity and the slope of the calibration curve.

It should be noted that Eqs. 7 and 8 are only linear approximations for a part of the overall SEC separation. The validity of this approximation depends on the true shape of the experimental column calibration curve. Best accuracy is obtained for columns with long log-linear calibration ranges, such as provided by columns with bimodal pore size distributions (see Fig. 6.23).

The range of the log-linear separation needed to measure the molecular weight distribution of a sample accurately depends on the nature of the sample and the expected molecular weight ranges of its components. Prediction of the required separation range can often be based on the origin of the sample or its polymerization kinetics. The theoretical molecular weight distribution curve shapes for three different polymerization processes are shown in Fig. 6.26. Typical condensation polymers (such as polyesters and polyamides) have a theoretical polydispersity value of d = 2, where d is equal to the ratio of the weight-average molecular weight $\overline{M}_w$ to the number-average molecular weight $\overline{M}_n$,

$$\overline{M}_n = \frac{\sum_i N_i M_i}{\sum_i N_i} \qquad (10)$$

$$\overline{M}_w = \frac{\sum_i W_i M_i}{\sum W_i} \qquad (11)$$

and

$$d = \frac{\overline{M}_w}{\overline{M}_n} \qquad (12)$$

For free radical polymers (such as polystyrene and polyethylene) a value of d = 1.5 is typical. In ionic polymerization, where all polymer molecules are initiated simultaneously and terminated only after all reactive monomer is consumed, a very sharp, nearly monodisperse molecular weight distribution ($d \sim 1$) is predicted. Single macromolecular species of biological origin also exhibit a polydispersity of unity.

If narrow molecular weight distribution standards of known molecular

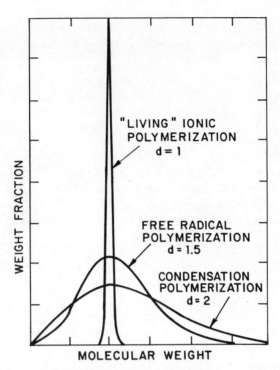

Figure 6.26. Molecular weight distributions for typical condensation, free-radical, and "living" ionic polymerizations. [Redrawn from Ref. (31).]

weight are available, as is the case for polystyrene or proteins, the SEC calibration curve can be obtained by plotting the known log MW values against the corresponding peak position retention volumes V_R. When several standards are used, this peak position point-by-point plotting method provides the complete detail of the true calibration curve shape, and when molecular weights are computed, the most accurate molecular weight is obtained for samples. To determine the molecular weight of unknown single molecular species, elution volumes of sample peaks are compared directly to the calibration curve. For example, the molecular weights of unknown proteins can be estimated with the peak position calibrations in Fig. 6.14, provided that the unknown components are of the same molecular conformation.

The peak position method is also useful for determining the molecular weight or the size of small molecules. The logarithm of molecular weight, molecular length, and molar volume all appear to correlate well with the retention volume V_R. Figure 6.27 shows the calibration plot of molecular

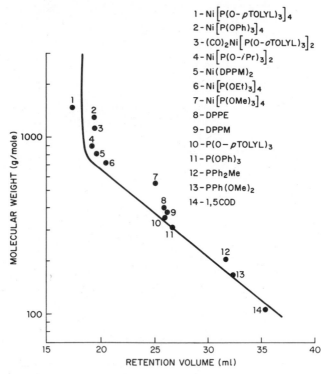

1 - Ni[P(O- pTOLYL)$_3$]$_4$
2 - Ni[P(OPh)$_3$]$_4$
3 - (CO)$_2$Ni[P(O-oTOLYL)$_3$]$_2$
4 - Ni[P(O-iPr)$_3$]$_2$
5 - Ni(DPPM)$_2$
6 - Ni[P(OEt)$_3$]$_4$
7 - Ni[P(OMe)$_3$]$_4$
8 - DPPE
9 - DPPM
10 - P(O- pTOLYL)$_3$
11 - P(OPh)$_3$
12 - PPh$_2$Me
13 - PPh(OMe)$_2$
14 - 1,5COD

Figure 6.27. Peak position versus molecular weight calibration for nickel complexes. Column—120 × 0.8 cm, μ-Styragel 100 Å; mobile phase—dry toluene; flow rate—3.4 mL/min; detector—RI; sample—25 μL, 2% solution in THF. [Redrawn from Ref. (32).]

weight versus retention volume for certain nickel complexes and other solutes containing no nickel.

Narrow synthetic polymer standards are only available for a few polymers. Calibration for other polymer types must be obtained by other means (1). These alternatives include (*a*) the peak-position universal calibration method, (*b*) the broad standard linear calibration method, and (*c*) the general broad standard calibration method (34). The first method utilizes universal calibration, which is based on the hydrodynamic volume concept of polymers. The size of the molecules can be expressed as the product of solute molecular weight M and intrinsic viscosity $[\eta]$ and can be uniquely defined as the radius of gyration of the molecule. This $[\eta]M$ value serves as a universal molecular size parameter for all polymer types. As illustrated in Fig. 6.28, SEC calibration curves for different polymer types merge into a single calibration plot when log $[\eta]M$ is plotted versus retention volume. With a sensitive on-line viscometer detector, the intrinsic viscosity values at the different retention volumes can be monitored for the unknown polymer sample. By dividing out these viscosity calibration data from the universal calibration, the desired molecular weight calibration curve can be obtained for the particular polymer of interest.

The second method, broad standard linear calibration, uses no polystyrene data but assumes the linear calibration approximation of Eq. 7 for the SEC separating system of interest. With a computer this is a convenient and accurate method, especially when combined with bimodal column sets that provide long log-linear calibration curves [Ref. (1), chap. 9]. The third method, a general broad standard calibration, uses a search for the nonlinear SEC calibration curve. The shape of the nonlinear cal-

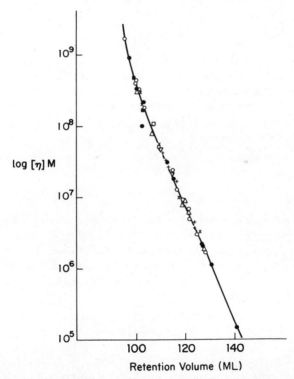

Figure 6.28. SEC universal calibration plot. Solvent—THF; ○—PS; ●—PS (comb.); +— PS; △—heterograft polymer; x—poly(methacrylate);⊖—poly (vinylchloride); ▽—graft copolymer PS/PMMA; ■—poly(phenylsiloxane; □—polybutadiene. [Redrawn from Ref. (33).]

ibration curve is first established by using available narrow standards of a commercially available type. The final calibration is accomplished by utilizing the average molecular weight values of the broad standard of the desired polymer type (34).

6.3.3.2. Molecular Weight Distribution Calculations

Once the calibration curve for a particular polymer system is established, the molecular weight distribution plots for samples of the same polymer type can be obtained directly from the elution curves using a point-by-point V_R-to-MW conversion. Sample molecular weight average values are calculated from the sample elution curves utilizing Eqs. 13 and 14,

$$\overline{M}_w = \frac{\sum h_i M_i}{\sum h_i} \tag{13}$$

$$\overline{M}_n = \frac{\sum h_i}{\sum h_i/M_i} \tag{14}$$

where h_i represents the SEC chromatographic height at retention volume $V_{R,i}$ and the summations are carried out in equal retention volume increments. This is a rather tedious procedure if performed manually, and more convenient computer systems have been described for this operation [see Refs. (24) and (1), chap. 9].

6.3.4.   SEC Column Performance

General performance specifications used in other chromatographic methods do not accurately describe SEC. Therefore, special SEC resolution and molecular weight accuracy criteria have been developed [see Ref. (1), chap. 4]. The concepts assume a linear calibration (Eqs. 7 and 8) and that the effect of sample molecular weight must be a part of the resolution criteria. SEC resolution can be defined by combining Eqs. 5 and 9,

$$R_s = \frac{\Delta V_R}{4\sigma} = \frac{0.58}{\sigma D_2} \cdot \Delta \log \text{MW} \tag{15}$$

It is more useful to define a special SEC column performance parameter, specific resolution R_{sp}, as the resolution for a pair of solutes which differ by a factor of 10 in molecular weight,

$$R_{sp} = \frac{0.58}{\sigma D_2} \tag{16}$$

The R_{sp} values of SEC column sets are much more meaningful than simple plate number calculations. Values for R_{sp} take into account not only the sharpness of the peak (as measured by the σ value), but also the slope of the calibration curve D_2, which describes the ability of the column to discriminate between components of different molecular sizes.

REFERENCES

1. W. W. Yau, J. J. Kirkland, and D. D. Bly, *Modern Size-Exclusion Liquid Chromatography*, Wiley, New York, 1979.

2. F. W. Billmeyer, Jr., and K. H. Altgelt, in K. H. Altgelt and L. Segal, Eds., *Gel Permeation Chromatography*, Dekker, New York, 1971.

3. Du Pont Biotechnology Systems Div. *Forum Features, 1*(1), 6 (1985).

4. Du Pont Instruments Products Div., Bull. E-14063, 1977.

5. *Liquid Chromatography Review: Size-Exclusion Chromatography*, Du Pont Instruments Products Div., 1977.

6. E. P. Otocka, *Acc. Chem. Res., 6,* 348 (1973).

7. E. F. Casassa, *J. Phys. Chem., 75,* 3929 (1971).

8. J. C. Giddings, L. M. Bowman, Jr., and M. N. Meyers, *Macromolecules, 10,* 443 (1977).

9. D. D. Bly, H. J. Stoklosa, J. J. Kirkland, and W. W. Yau, *Anal. Chem., 47,* 1810 (1975).

10. M. L. McConnell, Chromatix, Sunnydale, CA, private correspondence, 1980.

11. M. A. Haney, *Am. Lab.* (Apr. 1985).

12. S. D. Abbott and W. W. Yau, U.S. patent 4,578,990, Apr. 1, 1986.

13. H. G. Barth, W. E. Barber, C. H. Lochmuller, R. E. Majors, and F. E. Regnier, *Anal. Chem., Fundam. Rev., 58,* 211A (1986).

14. F. E. Regnier and K. M. Gooding, *Anal. Biochem., 103,* 1 (1980).

15. H. G. Barth and F. J. Carlin, Jr., *J. Liq. Chromatogr., 7,* 1717 (1984).

16. *Altex Chromatogram, 4*(1), 1 (1980).

17. Y. Kato, Y. Sawada, H. Sasaki, and T. Hashimoto, *J. Chromatogr., 190,* 305 (1980).

18. L. R. Snyder and J. J. Kirkland, *Introduction to Modern Liquid Chromatography*, Wiley, New York, 1979, chap. 12.

19. F. A. Buytenhuys and F. P. B. Van der Maeden, *J. Chromatogr., 149,* 489 (1978).

20. H. G. Barth, *J. Chromatogr. Sci., 18,* 409 (1980).

21. E. Pfannkoch, K. C. Lu, F. E. Regnier, and H. G. Barth, *J. Chromatogr. Sci., 18,* 430 (1980).

22. W. W. Yau, C. R. Ginnard, and J. J. Kirkland, *J. Chromatogr., 149,* 465 (1978).

23. S. Bakerman and G. Wasemillers, *Biochemistry, 6,* 1100 (1967).
24. W. W. Yau, H. J. Stoklosa, and D. D. Bly, *J. Appl. Polym. Sci., 21,* 1911 (1977).
25. J. C. Giddings, *Anal. Chem., 39,* 1027 (1967).
26. V. F. Gaylor and H. L. James, *Anal. Chem., 48,* 44A (1976).
27. A. Krishen, *ACS Symp. Ser., 245,* 241 (1984).
28. A. Krishen and R. G. Tucker, *Anal. Chem., 49,* 498 (1977).
29. J. Lesec, *J. Liq. Chromatogr., 8,* 875 (1985).
30. A. P. Graffeo, presented at the Association of Official Analytical Chemists' Meeting, Washington, DC, Oct. 19, 1977.
31. C. Tanford, *Physical Chemistry of Macromolecules*, Wiley, New York, 1964.
32. C. A. Tolman and P. E. Antle, *J. Organomet. Chem., 159,* C5 (1978).
33. Z. Grubisic, R. Rempp, and H. Benoit, *J. Polym. Sci. B, 5,* 753 (1967).
34. W. W. Yau, *preprints, 44th Ann. Techn. Conf., Society of Plastic Engineers* (1986), p. 461.

CHAPTER

7

HIGH PERFORMANCE AFFINITY CHROMATOGRAPHY

IRWIN M. CHAIKEN,* GIORGIO FASSINA, and PAOLO
CALICETI

National Institute of Diabetes and Digestive and Kidney Diseases
National Institutes of Health
Bethesda, Maryland

7.1. INTRODUCTION

The development of affinity chromatography as a separation method to purify biological macromolecules and small molecules has been stimulated by the specificity of surface recognition of biomolecules. The basic approach has been to attach a known molecular interactor to an insoluble matrix and then use the matrix to adsorb selectively a sought-after molecular species, multimolecular assembly, or even cell from mixtures eluted through the affinity matrix. The affinity chromatographic approach has been particularly important for the separation of biological macromolecules, which coexist in biological sources with a large number of other molecules with chemically complex surfaces. While conventional chromatographic methods are very often inadequate for separating a unique species from such complex mixtures, the biospecific recognition

* Present address: Department of Macromolecular Sciences, Smith Kline and French Research and Development, Box 1539, King of Prussia PA 19406-0939.

317

ites on the surfaces of folded macromolecules make such selective separations possible by using an appropriate immobilized molecule as affinity adsorbant. By the late 1960s, with the development of matrix chemistry (1, 2), the use of immobilized ligands to purify enzymes (3), and an elevated understanding of macromolecular structure in general, the explosive growth of affinity chromatography as a preparative method started in earnest [see (4, 5) and references within]. Furthermore, beginning in the early 1970s, the growing force of its reliability as a biospecific separation method led to an awareness that affinity chromatography also had potential use as an analytical method to study mechanisms of the interactions between mobile and immobilized biomolecules (6–15).

While biochemical separations have been carried out increasingly in the high performance liquid chromatography (HPLC) mode, affinity chro-

A: **Stepwise Elution Affinity Chromatography**

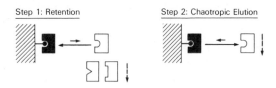

B: **Isocratic Elution Affinity Chromatography**

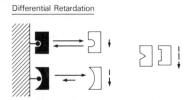

Figure 7.1. Schematic diagram comparing (*a*) the stepwise elution approach of preparative affinity chromatography with (*b*) the isocratic elution approach of analytical affinity chromatography. (*a*) A mobile macromolecule binds tightly enough to the affinity matrix to be retained (step 1). It is rapidly eluted from the matrix (long dashed arrow) in purified form by abrupt change to a chaotropic elution condition (step 2). (*b*) The mobile macromolecule is eluted (different-length dashed arrows denote differential rates of elution) on the affinity matrix at particular column and buffer conditions, including nature of immobilized interactor, designed to effect retardation but not retention. Isocratic elution allows resolution of differentially interacting molecules from one another as well as quantitation of matrix interaction molecules with each mobile interactor resolved by differential affinity. Competitors or other interaction modulators can be included in elution buffers, either at constant concentration or in gradients, as variations to this overall continuous elution approach.

matography has entered the HPLC field more slowly (16–19). This has been due partly to the fact that the most common affinity support, agarose, has been eminently successful for many preparative needs, has required only simple instrumentation, and has been employed in a compressible porous gel form not amenable to the mechanical forces of HPLC. Moreover, the intrinsic specificity of a particular immobilized ligand for only a limited number of mobile molecules makes it difficult to produce a limited set of high performance affinity matrices for all chromatographic needs. Nonetheless, noncompressible affinity supports suitable for high performance affinity chromatography (HPAC) offer an important opportunity to scale up separations, a possibility more difficult with compressible gel supports. Furthermore, HPAC offers several advantages for microscale uses, including the potential for improved resolution and data acquisition in analytical separations by isocratic or gradient elutions versus the stepwise elutions used more commonly in early porous gel affinity chromatography (Fig. 7.1). These forces have helped stimulate HPAC development in general and the early development of analytical HPAC.

7.2. DEVELOPMENT OF HPAC MATRICES

Based on the success of classical (soft gel) affinity matrices for macromolecular separation, the standard approach in designing idealized high performance affinity supports has been to combine hydrophilicity and absence of charged groups on the matrix surface, to avoid nonbiospecific (non-"affinity") interactions, with a chemically and mechanically stable core particle which can withstand prolonged use in aqueous buffers at moderate or high pressure. In order to provide maximum flexible use for a wide range of biomolecular separations, matrices most often are designed to contain reactive or activatable groups to which interactive ligands can be covalently attached. Silica, glass beads, highly cross-linked agarose, and methacrylate gel all have become available for HPAC use (Table 7.1). Matrix and linker methodology is likely to continue expanding as the overall usage of HPAC is adopted.

The first support used for high performance affinity chromatography was silica (23). It is currently available in a wide range of particle sizes and has a high porosity, which allows unhindered access of large macromolecules to its surface. Unmodified silica is not suitable for chromatography of biological macromolecules due to the presence of charged silanol groups on its surface. To convert silica to a usable affinity support, the silica particles are coated with organic modifiers which suppress these charged groups, provide a hydrophilic character to the matrix surface,

Table 7.1. Some Representative Chromatographic Supports in High Performance Affinity Chromatography

Affinity Support	Reactive Group on Matrix	Immobilized Ligand	Ref.
Silica	Epoxy	Concanavalin A	(20)
	N-Hydroxysuccinimide	Bovine neurophysin II	(21)
	Tresyl*	Alcohol dehydrogenase	(22)
	Aldehyde	Adenosine monophosphate	(23)
Glass Beads	Thionyl[†]	Bovine neurophysin II	(24)
	EDC[‡]	Bovine neurophysin II	(24)
Methacrylate gel	Epoxy	e-Aminocaproyl-L-Phe-D-Phe-OCH$_3$	(25)
Highly cross-linked agarose	Tresyl	Bovine neurophysin II	(21)

* 2,2,2-Trifluoroethanesulfonyl.
† Activation of succinamidopropyl glass by thionyl chloride.
‡ Activation of carboxyl groups of succinamidopropyl glass by EDC (1-ethyl-3-[3-(dimethylamide) propyl] carbodiimide.

and contain functional groups which can be derivatized for the subsequent coupling of biospecific ligands. Coating procedures also can impart a greater stability of the support in alkaline media than that possessed by the silica core (26). Early work (27) demonstrated that, among the different silylation agents available, glycidoxypropyl silane was a particularly suitable modifier of silica, due to the stability and ease of derivatization of the epoxy silica material produced. The oxirane moiety introduced on the matrix surface can react with molecules bearing amine, thiol, and hydroxyl groups. Unreacted epoxy groups can be neutralized by conversion to diols with acid or base hydrolysis or by treating with mercaptoethanol. After ligand coupling, unreacted epoxy groups can be blocked by endcapping reagents such as glycinamide and ethanolamine. Improved coupling yields to silica have been obtained by activating diol silica with tresyl chloride (22). Diol silica also can be activated by converting the hydroxyl groups to the aldehyde form by periodate oxidation (23). Silica also can be preactivated for the direct coupling with nucleophiles by coating the particle surface with a thin skin of cross-linked polymer carrying functional groups, such as N-hydroxysuccinimide ester (28).

Use of glass beads for the immobilization of enzymes was first reported in 1969 (29). Glass as a biomolecular support benefits from its structural stability, inertness to microbial attack, low cost, and availability in a wide range of particle sizes and porosities (including nonporous) (30). Many of the coating and ligand immobilization chemistries used for silica also are applicable for glass beads. Although a number of organic functional groups can be grafted onto the surface of glass, all typically involve initial reaction with 3-aminopropyl silane, thus introducing primary amino groups that can be used for subsequent derivatization procedures (24, 31, 32).

Hydroxyalkyl methacrylate gels have found applicability in high performance affinity chromatography (33). This support is mechanically and chemically stable and resistant to the action of microbial enzymes. Ligands bearing amino groups can be coupled easily following activation of the support by cyanogen bromide (34), as with agarose, or by benzoquinone (35). Alternatively, the hydroxyl groups of the copolymer can be reacted with epichlorohydrin in alkaline media, thus introducing epoxide groups which can react with amino, carboxyl, hydroxyl, and sulfhydryl groups and also with some aromatic groups, such as indole and imidazole.

Stimulated by the vast usage of agarose in pre-HPLC affinity chromatography, highly cross-linked forms of the polymer have become available for HPAC. These more rigid forms provide increased mechanical stability while still allowing immobilization chemistries, choice of ligands, and elution conditions similar to those used with soft agarose. Thus affinity chromatography systems previously made with soft agarose can be adapted directly to HPAC using more rigid agarose as the starting material.

7.3.   BIOMOLECULAR SEPARATIONS USING HPAC

7.3.1.   Group-Specific Affinity Matrices—General Ligand Chromatography

Most early uses of HPAC for biomolecular separation have utilized group-specific adsorbents such as lectins, boronic acid, triazine dyes, and protein A. A major reason for this trend, which has been occurring during the period of high performance affinity matrix development, may well be the compulsion of investigators to maximize the preparative usefulness of the matrices made to separate as broad a range as possible of macromolecules and small molecules. Group-specific affinity matrices can recognize a large set of molecular species with common structural elements, but often with different affinities, so that adsorption followed by elution with gradients of a chaotropic agent or soluble competitor will effect useful separation. The approach found early and wide application with soft gel affinity chromatography (4, 36, 37), and its implementation with HPAC methodology has had little inertia barrier.

Results with immobilized plant lectins are prototypic. Lectins can interact with mono- or oligosaccharides, both free and conjugated to other molecular structures (38). While the potential is real to bind many glycosylated molecular species to a lectin affinity matrix, different lectins vary in their specificities; judicious choice of the particular lectin to be immobilized thus can allow the optimization of preparative efficiency for a particular glycosylated species (39). Studies with immobilized concanavalin A-Sepharose have shown that the specific glycosyl chain structure mannosyl α1-6(mannosyl α1-3)mannose, in which at least one hydroxyl group at the C-3 position of C-6 linked mannose is free, binds strongly to the affinity matrix with a dissociation constant below $3.4 \times 10^{-7}M$ (40). Concanavalin A immobilized on silica has been used for the successful purification of sugars (20), serum glycoproteins (41), and membrane glycoproteins (42).

With some overlapping specificity with the lectins, boronic acid also has found wide application as a general ligand for high performance affinity chromatography due to its ability to recognize the vicinal diol structure present in nucleosides, nucleotides, carbohydrates, and catecholamines (43). Vicinal diols reversibly form cyclic boronate esters with the boronate anion at high pH; and formation of these complexes is dependent not only on pH but also on ionic strength, on temperature, and, in the case of nucleic acids, on the structure of the base (44). Compared to boronic acid cellulose, separation times with boronic acid silica are significantly reduced without deterioration of resolution. Boronic acid silica has been used diagnostically for serum and urine analysis, allowing deviations from the normal content of nucleosides to be detected rapidly, thus minimizing delay between diagnosis and clinical treatment (43).

Reactive triazine dyes and protein A also have received wide use as general ligands. While triazine dyes, including the often used Cibacron Blue F3G-A, may be considered generally to be mimics of naturally occurring heterocycles (45), the actual mechanisms of interaction of dyes with macromolecules continue to be topics of lively discussion and investigation (46). Whatever the interaction mechanisms, the separatory power of reactive dyes is impressive, as observed first with soft gels and now increasingly with rigid matrices [see Refs. (47, 48) for reviews]. Protein A, a coat protein from Cowan strain *Staphylococcus aureus*, has been immobilized to microparticulate silica, and used to fractionate immunoglobins (Fig. 7.2). Furthermore, since protein A binds to the tail, or Fc, portion of the immunoglobin G (IgG) molecule, IgG can be bound and cross-linked to immobilized protein A, thus creating matrices for an extensive series of antigen separations (50). The quantitative diagnostic use of this approach is discussed below.

7.3.2. Molecule-Specific Affinity Supports

In contrast to group-specific affinity matrices, which have grown in use with soft gel affinity chromatography and have been adapted readily to high performance applications, single-purpose molecule-specific affinity matrices have been rampant with soft gels since their inception, but slower in being used in HPAC. Among the early applications, Nilsson and Larsson (51) have prepared horse liver alcohol dehydrogenase immobilized on tresyl chloride activated glyceropropyl silica and used this matrix to fractionate adenine nucleosides, adenine nucleotides, and triazine dyes. Silica-immobilized proteins, as highly chiral species, have been employed in HPAC for the resolution of enantiomeric molecules from racemic mixtures (51–54). And, very recently, Goldman et al. (55) isolated antibenzodiazepine antibodies using a thyroglobulin-benzodiazepine derivative immobilized on a prepacked, epoxy-activated silica matrix. This latter type of molecule-specific isolation, so rampant with preparative affinity chromatography on compressible gels, is likely to increase in frequency for HPAC as rigid matrices become increasingly available.

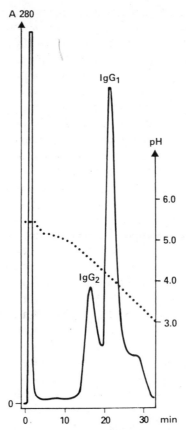

Figure 7.2. Separation of immunoglobulin G subclasses on silica–protein A. Human serum (160 μL) was fractionated on SelectiSpher-10™ protein A (10 × 0.5-cm id) in 0.1M citrate buffer, pH gradient 5.5 to 3.0, at 22°C. Flow rate was 3 mL/min for 0.5 min, then 1 mL/min for 5–35 min. [Figure was kindly provided as a personal communication by Dr. Sten Ohlson, Pierce AB (Lund, Sweden), and reprinted with permission from Ref. (49).]

7.4. USING HPAC AS AN ANALYTICAL TOOL

Previous experience has shown that simple theoretical relationships can be used to describe affinity chromatographic elution behavior of macromolecules quantitatively (10, 11, 13, 49, 56). When a zone of mobile macromolecule is eluted on an affinity matrix under binding conditions but adjusted to allow moderate retardation (Fig. 7.1) instead of retention (very strong retardation), the extent of retardation during isocratic elution can be used as a measure of the binding affinity of mobile and immobilized interactors. [For a development of the theory and full derivations of equations used, see Ref. (56).] The effect on the elution volume of soluble molecules competing with immobilized ligands can be used to evaluate binding properties of the mobile macromolecule fully in solution. Beyond allowing the evaluation of macromolecular interaction properties chromatographically, a useful research goal in itself, the quantitative analysis allows the specificity of affinity matrices to be measured and compared to solution interactions, thus providing an evaluation of the degree of biospecificity of an affinity support and therein of its reliability as a separatory tool. The theory developed for analytical affinity chromatography on classical compressible gels has been adapted straightforwardly for analytical high performance affinity chromatography [for a review, see Ref. (49)].

Several interacting systems, with well-defined solution characteristics and in some cases previously defined elution behavior on compressible affinity matrices, have been used to develop and evaluate matrices for analytical HPAC. As in preparative use, the group-specific case of concanavalin A has been emphasized in several laboratories. Muller and Carr (57) have measured equilibrium binding constants of a series of closely related p-nitrophenyl saccharides on silica-immobilized Con A, using competitive zonal elution techniques. As shown by the representative data in Table 7.2, they generally found good agreement between chromatographically derived equilibrium constants and those determined for the solution interaction. Similar agreement has been found by Anderson and Walters (58). The elution of saccharides on Con A silica also has been used to analyze interaction kinetics (57, 58).

Studies with the multimolecular interacting complexes of neurophysin and the neurohypophysial hormone oxytocin and vasopressin also have been used extensively to develop analytical HPAC methods, which then have been employed to characterize macromolecular interactions (see Table 7.2). The first high performance affinity supports used for characterizing this system were succinamidopropyl derivatives of glass. Bovine neurophysin II (BNP II) was covalently immobilized on both controlled-pore glass (CPG) and nonporous glass (NPG), and the interaction behavior of [^{3}H]Arg^8-vasopressin (AVP) was measured by zonal analysis (24, 59). In general, the extent of retardation was dependent on the amount of mobile component in the zone, especially for nonporous matrix, as shown by the data of Fig. 7.3. This dependence as a rule has not been found with agarose-based supports and becomes evident with high performance matrices of relatively low capacity. Given the dependence of

Table 7.2. Some Interacting Macromolecular Systems Evaluated by Analytical HPAC

Affinity Matrix M	Mobile Molecule P	K_{MIP} (M)	Related K_d in Solution (M)	Ref.
[ConA]silica	p-Nitrophenylmannoside	4.5×10^{-5}	11.5×10^{-5}	(58)
	p-Nitrophenylmannoside	6.2×10^{-5}	11.5×10^{-5}	(57)
[ADH]silica	AMP	13.4×10^{-5}	7.0×10^{-5}	(22)
[BNPII]CPG	AVP	1.1×10^{-5}	$1.6-2.0 \times 10^{-5}$	(59)
[BNPII]NPG	BNPII	1.7×10^{-4}	$1.2-1.7 \times 10^{-4}$	(59)
[BNPII]silica	AVP	1.0×10^{-5}	$1.6-2.0 \times 10^{-5}$	(21)
[BNPII]HXL-agarose	AVP	1.1×10^{-5}	$1.6-2.0 \times 10^{-5}$	(21)

Key: Con A = concanavalin A; ADH = alcohol dehydrogenase; HXL = highly cross-linked.

V on the micrograms of AVP in the injected zone, $K_{M/P}$ values have been determined from $1/(V - V_0)$ extrapolated to 0 μg based on the equation

$$\frac{1}{V - V_0} = \frac{K_{M/P} + [P]}{M_T} \tag{1}$$

where $K_{M/P}$ is the dissociation constant of the complex of matrix-immobilized interactant M and mobile molecule P, V is the elution volume of retarded molecule P, V_0 is the elution volume of an unretarded molecule, $[P]$ is the concentration of the mobile interactant, and M_T is the total amount of immobilized interactant. ($M_T = [M]_T$ [accessible volume], with $[M]_T$ being the concentration of immobilized interactant.) In the case of the NPG affinity matrix, equilibrium binding constants so derived and those determined fully in solution were quite similar (Table 7.2).

Elution characteristics of AVP also have been observed by frontal anal-

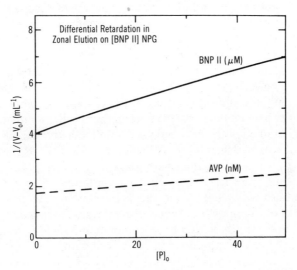

Figure 7.3. Comparison of retardation of BNPII and AVP on [BNPII]NPG. AVP profile [redrawn from Ref. (59)]: Zonal elution of tritium-labeled Arg[8]-vasopressin (^{3}H-AVP) on [BNPII]NPG showing dependence of the extent of retardation $[1/(V - V_0)]$ on initial concentration of AVP ($[AVP]_0$) in injected zone of 200 μL. The column (25 × 0.46 cm id) was equilibrated and eluted at room temperature and a flow rate of 0.2 mL/min in 0.4M ammonium acetate, pH 5.7, using a Varian 5000 HPLC system. The straight line was obtained by fit to Eq. 1. BNPII profile [redrawn from Ref. (59)]: Data obtained by zonal elution of ^{125}I-BNPII with the same [BNPII]NPG column and elution conditions as for ^{3}H-AVP above except using nM instead of μM concentrations; the slightly downward curved line through the data was obtained by fit to a model which accounts for both soluble BNPII–matrix BNPII interaction and soluble BNPII self-association.

ysis of broad chromatographic elutions (24). In the concentration range studied, the dependence of V on the concentration of peptide in the zone applied conforms to Eq. 2,

$$\frac{1}{\overline{V} - V_0} = \frac{K_{M/P}}{M_T} + \frac{[P]_T}{M_T} \tag{2}$$

where $\overline{V}$ is the first moment of the mobile solute front in broad-zone elution and corresponds to the elution volume V in small-zone elution (Eq. 1), $M_T = [M]_T$ [accessible volume], and $K_{M/P}$ and $[P]_T$ are as before. These broad-zone data have been useful to determine matrix capacity, expressed as either M_T or $[M]_T$.

The dimerization properties of neurophysin also were studied by zonal elution of [^{125}I]BNPII and [BNPII]NPG (Fig. 7.3). In this case, again, there was a concentration dependence of retardation; and extrapolation to zero mobile protein concentration in the injected zone was used to calculate a dimerization binding constant that was in excellent agreement with the value calculated from equilibrium sedimentation studies (Table 7.2).

The neurophysin–hormone system has been used to evaluate the utility of a number of matrices other than glass beads for HPAC by immobilizing BNPII and then measuring the biospecificity of interaction with mobile interactors. As judged by the values of the chromatographic equilibrium dissociation constants determined, two matrices with adequate biospecificity were obtained with N-hydroxysuccinimide-activated silica and tresyl chloride-activated highly cross-linked agarose (21). Higher coupling yields with these matrices than with glass beads have produced affinity supports of high capacity, allowing the retardation of large amounts of applied peptides and proteins and their consequent detection by on-line monitoring of ultraviolet absorbance. Equilibrium binding constants were determined upon extrapolation to zero mobile interactant concentration, with the values calculated agreeing quite well with those determined by zonal affinity chromatography with [BNPII]NPG. The ability to increase capacity by the amount of neurophysin immobilized also has allowed analytical HPAC to be used to detect and evaluate low-affinity binding interactions between BNPII and a series of synthetic hormone analogues and chemically modified hormone derivatives (60).

7.5. MICROSCALE MOLECULAR DIAGNOSIS

The development of quantitative affinity chromatography has broadened our expectations for biomolecular affinity separations from two-step batch procedures, which separate binding from nonbinding molecules, to more complete chromatographic separations of many molecular species of different functional binding properties by isocratic, competitive or more elaborate gradient elutions. Such expectations suggest that, as with other HPLC tools (such as reversed phase, ion exchange, or size exclusion), high performance affinity chromatography can be used diagnostically to determine the presence of specific components of biomolecular families in mixtures, including biological fluids and cell and tissue extracts. Such analytical applications as those for clinical diagnosis seem likely not only to further stimulate the use of HPLC technology for analytical affinity chromatography, but also to stimulate instrumentational innovation tailored to the needs of this chromatographic mode.

Compared with standard techniques for the clinical determination and quantitation of biomolecules, analytical HPAC has a specificity and speed of operation that can permit rapid detection and simultaneous binding activity characterization of a whole set of molecules which are related functionally to one another. Batch separation procedures already have been used, for example, to determine immunoglobulin levels in blood sera using a protein A–silica affinity column (61). Good agreement was observed between levels found by HPAC and those obtained using radioimmunoassay. Immobilized protein A matrices also have been used for analyzing molecules other than antibodies, by binding and cross-linking specific antibodies to the matrix via their F_c tails (62) and then using the resultant immunoaffinity matrices to analyze specific antigens (50, 63). Batchwise affinity separation also has been used as a preliminary separation step in the molecular diagnosis of cortisol in urine samples without interference from other corticosteroids and polar metabolites of cortisol (64). The prototypic method devised included specific absorption on an anticortisol silica matrix, which was placed in the loop of an HPLC instrument upstream from a reversed-phase column. After a washing step, the fraction bound to the affinity matrix "loop" was eluted off and fractionated directly by the reversed-phase HPLC mode. The promising results of batchwise affinity chromatography analysis of sera and urine underscore the potential diagnostic usefulness of analytical HPAC.

Yet, as with analytical affinity chromatography in general, it is probably the isocratic zonal elution mode (Fig. 7.1) that will be most productive for clinical diagnosis, since such an elution profile provides a direct assessment not only of the amounts of separated binding molecules but also, by the elution position and peak shapes, of their functional (binding) properties (at least equilibrium, perhaps kinetic also). For example, the differential retardation behavior of neurophysin and AVP on immobilized BNPII (Fig. 7.3) shows the potential usefulness of the latter matrix to determine both the number and the amounts of neurophysin-binding molecules in mixtures obtained from tissue extracts and their equilibrium affinity constants for neurophysin interaction. A prototypic example of such a diagnostic separation is shown in Fig. 7.4.

Given the potential usefulness of isocratic separations as diagnostic tools, we recently initiated an effort to test immobilized antibodies for the molecular profiling of antigens by isocratic elution analytical immuno-HPAC. Anti-BNPII was attached to protein A–silica by carbodiimide cross-linking after noncovalent binding. Radiolabeled N^{ϵ}-acetimidated antigen, [^{14}C-diAcet]BNPII, was eluted on the affinity matrix to establish that retardation during isocratic elution would occur under binding (non-chaotropic) conditions. Both zonal and frontal elutions of [^{14}C-di-Acet]BNPII can be accomplished, with the zonal data shown in Fig. 7.5A.

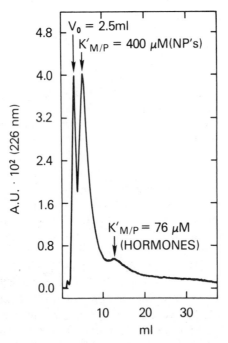

Figure 7.4. HPAC analysis of neurophysin-binding molecules in crude extract of bovine posterior pituitary. A 50-μL sample containing about 300 μg of polypeptide material was injected on a [BNPII]HXL-agarose column (75 × 6.6 mm id) equilibrated with 0.4M ammonium acetate, pH 5.7, at a flow rate of 0.5 mL/min. The elution was monitored by UV absorbance at 226 mm, 0.08 absorbance units (A.U.) full scale. $V_0 = 2.5$ mL. Two retarded peaks were detected and identified by $K'_{M/P}$ values (apparent $K_{M/P}$, determined at a finite value of [P] and not by extrapolation to [P] = 0) calculated from the observed elution volumes. [Reprinted from Ref. (49) with permission.]

Isocratic Elution on [Anti-BNPII] Protein A-Silica

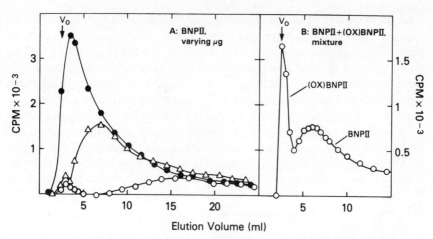

Figure 7.5. Isocratic immuno-HPAC separation of neurophysins on [anti-BNPII]protein A-silica. The column was 66 mm × 9 cm; bed volume was 3.35 mL. The amount M_T of functional polyclonal antibody, as judged by BNPII capacity analysis using frontal elutions, was 0.78×10^{-6} mmol. (*a*) Zonal elution of different amounts of [^{14}C-diAcet]BNPII, in $0.1M$ phosphate–$0.15M$ NaCl, pH 7.2. Amounts in eluted peaks: ●—5.2 μg; △—3.5 μg; ○—1.3 μg. (*b*) Separation of [^{14}C-diAcet]BNPII from the more weakly immunoreactive performic acid oxidized derivative. Buffer conditions as in (*a*). In addition to the oxidized BNPII derivative, BNPI also was found to be essentially unretarded on the lightly loaded polyclonal antibody matrix used here, indicative of the substantially lower affinity of BNPI than that of BNPII for anti-BNPII (65). [From Ref. (66).]

The $K_{M/P}$ of the [antigen]–[immobilized antibody] interaction estimated from the chromatographic data is in the range of $10^{-7}M$. The extent of retardation depends on the concentration of the protein antigen being eluted, a phenomenon predicted by Eq. 1 and similar to that in Fig. 7.3. The results suggest that microscale HPAC with immobilized antibodies, perhaps especially when the latter are attached by directed immobilization (such as via protein A) to maximize antibody functionality, could be used analytically to provide a diagnostic map of the number and functional nature of antigens in mixtures. As a test, a mixture of [^{14}C-diAcet]BNPII and its performic acid oxidized derivative [expected to exhibit greatly reduced antibody binding (65)] were separated on the [anti-BNPII]–protein A–silica matrix as shown in Fig. 7.5*B*. Beyond the neurophysin test case, the potential to use the chromatographic approach with an expanding array of immobilized antibodies, monoclonal as well as polyclonal, could make analytical HPAC a valuable immunodiagnostic technique.

7.6. CONCLUSIONS

The development of HPAC in general is still in its early stages. Nevertheless, currently available matrices and instrumentation already have been adapted straightforwardly to HPAC as both a preparative and an analytical tool. Preparative isolation on both very small (micro) and very large (macro) scales should benefit from HPAC development. In addition, it is expected that analytical HPAC can provide significant microscale precision and experimental convenience for the characterization of macromolecular interactions in biochemical and biomedical research and could lead to diagnostic methods of clinical and commercial use. Given biomolecular recognition as a common theme in both biology and biotechnology, developing HPAC methods for biomolecular separations promises to be an active area of future research.

REFERENCES

1. R. Axen, J. Porath, and S. Ernbach, *Nature, 214,* 1302–1304 (1967).
2. J. Porath, R. Axen, and S. Ernbach, *Nature, 215,* 1491–1492 (1967).
3. P. Cuatrecasas, M. Wilchek, and C. B. Anfinsen, *Proc. Nat. Acad. Sci. (U.S.), 61,* 636–643 (1968).
4. C. R. Lowe and P. D. G. Dean, *Affinity Chromatography*, Wiley, London, 1974.
5. W. B. Jakoby and M. Wilchek, Eds., *Methods in Enzymology*, vol. 34, Academic, New York, 1974.
6. B. M. Dunn and I. M. Chaiken, *Proc. Nat. Acad. Sci. (U.S.), 71,* 2382–2385 (1974).
7. L. W. Nichol, A. G. Ogston, D. J. Winzor, and W. H. Sawyer, *Biochem. J., 143,* 435–443 (1974).
8. B. M. Dunn and I. M. Chaiken, *Biochemistry, 14,* 2343–2349 (1975).
9. K. Kasai and S. Ishii, *J. Biochem., 77,* 261–264 (1975).
10. I. M. Chaiken, *Anal. Biochem., 97,* 1 (1979).
11. D. J. Winzor, in P. D. G. Dean, W. S. Johnson, and F. A. Middle, Eds., *Affinity Chromatography—A Practical Approach*, IRL Press, Oxford, 1984, pp. 149–168.
12. D. M. Abercrombie and I. M. Chaiken, in P. D. G. Dean, W. S. Johnson, and F. A. Middle, Eds., *Affinity Chromatography—A Practical Approach*, IRL Press, Oxford, 1984, pp. 169–189.
13. I. M. Chaiken, *J. Chromatogr., 376,* 11–32 (1986).
14. K. I. Kasai, Y. Oda, M. Nashikata, and S. I. Ishii, *J. Chromatogr., 376,* 33–47 (1986).

15. I. M. Chaiken, Ed., *Analytical Affinity Chromatography*, CRC Press, Boca Raton, FL, 1987.

16. P. O. Larsson, M. Glad, L. Hansson, M. O. Mansson, S. Ohlson, and K. Mosbach, *Adv. Chromatogr., 21,* 41–85 (1983).

17. I. M. Chaiken, M. Wilchek, and I. Parikh, Eds., *Affinity Chromatography and Biological Recognition*, Academic, New York, 1983.

18. P. O. Larrson, in *Methods in Enzymology*, vol. 104, Academic, New York, 1984, pp. 212–223.

19. J. Turkova, I. M. Chaiken, and M. T. W. Hearn, Vol. Eds., *J. Chromatogr., 376* (1986).

20. A. Borchert, P. O. Larsson, and K. Mosbach, *J. Chromatogr., 244,* 49–56 (1982).

21. G. Fassina, H. E. Swaisgood, and I. M. Chaiken, *J. Chromatogr., 376,* 87–93 (1986).

22. K. Nilsson and P. O. Larsson, *Anal. Biochem., 134,* 60–72 (1983).

23. S. Ohlson, L. Hansson, P. Larsson, and K. Mosbach, *FEBS Lett., 93,* 5–9 (1978).

24. H. E. Swaisgood and I. M. Chaiken, *J. Chromatogr., 327,* 193–204 (1985).

25. J. Turkova, K. Blaha, J. Horacek, J. Vajcner, A. Frydrycxhova, and J. Coupek, *J. Chromatogr., 215,* 165–279 (1981).

26. A. J. Alpert and F. E. Regnier, *J. Chromatogr., 185,* 375–392 (1979).

27. S. H. Chang, K. M. Gooding, and F. E. Regnier, *J. Chromatogr., 120,* 321–333 (1976).

28. J. Schutyser, J. Buser, D. Van Olden, H. Tomar, F. Van Houdenhoven, and A. Demen, in T. C. J. Gribnau, J. Visser, and R. J. F. Nivard, Eds., *Affinity Chromatography and Related Techniques*, Elsevier, Amsterdam, 1982, pp. 143–153.

29. H. H. Weetall, *Science, 166,* 615–616 (1969).

30. W. Haller, in W. H. Scouten, Ed., *Solid Phase Biochemistry*, John Wiley and Sons, New York, 1983, p. 535.

31. V. G. Janolino and H. E. Swaisgood, *Biotechnol. Bioeng., 24,* 1069–1080 (1982).

32. G. E. DuVal, H. E. Swaisgood, and H. R. Horten, *J. Appl. Biochem., 6,* 240–250 (1984).

33. J. Coupek, M. Krivakova, and S. Pokorny, *J. Polym. Sci., Polym. Symp., 42,* 185–190 (1973).

34. J. Turkova, *Meth. Enzymol., 44,* 66 (1976).

35. N. Stambolieva and J. Turkova, *Collect. Czech. Chem. Commun., 45,* 1137–1143 (1980).

36. P. O'Carra, in R. Epton, Ed., *Chromatography of Synthetic and Biological Polymers*, vol. 2, *Hydrophobic, Ion Exchange and Affinity Methods*, Ellis Horwood, Chichester, England, 1978, p. 131.

37. O. Hoffman-Ostenhof, M. Breitenbach, F. Koller, D. Kraft, and O. Scheiner, Eds., *Affinity Chromatography*, Pergamon, Oxford, 1978.
38. R. Lotan and G. L. Nicolson, *Biochem. Biophys. Acta, 559,* 329–376 (1979).
39. H. Lis and N. Sharon, *Ann. Rev. Biochem., 42,* 541–574 (1973).
40. Y. Ohyama, K. I. Kasai, H. Nomoto, and Y. Inoue, *J. Biol. Chem., 260,* 6882–6887 (1985).
41. C. A. K. Borrebaeck, J. Soares, and B. Mattiasson, *J. Chromatogr., 284,* 187–192 (1984).
42. D. Renauer, F. Oesch, J. Kinkel, K. K. Unger, and R. J. Wieser, *Anal. Biochem., 151,* 424–427 (1985).
43. M. Glad, S. Ohlson, L. Hansson, M. Mansson, and K. Mosbach, *J. Chromatogr., 200,* 254–260 (1980).
44. H. L. Weith, L. S. Wiebers, and P. T. Gilham, *Biochemistry, 9,* 4396–4401 (1970).
45. C. R. Lowe, D. A. P. Small, and T. Atkinson, *Int. J. Biochem., 13,* 33–40 (1981).
46. J. F. Biellmann, J. P. Samana, C. I. Branden, and M. Eklund, *Eur. J. Biochem., 102,* 107–110 (1979).
47. C. R. Lowe, M. Glad, P. O. Larsson, S. Ohlson, D. A. Small, T. Atkinson, and K. Mosbach, *J. Chromatogr., 215,* 303–316 (1981).
48. C. R. Lowe, S. J. Burton, J. C. Pearson, and Y. D. Clonis, *J. Chromatogr., 376,* 121–130 (1986).
49. G. Fassina and I. M. Chaiken, *Adv. Chromatogr., 27,* 247–297 (1987).
50. T. M. Phillips, N. S. More, W. D. Queen, and A. M. Thompson, *J. Chromatogr., 327,* 205–211 (1985).
51. K. Nilsson and P. O. Larsson, *Anal. Biochem., 134,* 60–72 (1983).
52. S. Allenmark, *Chem. Scripta, 20,* 5–10 (1982).
53. S. Allenmark, B. Bomgren, and M. Horen, *J. Chromatogr., 264,* 63–68 (1983).
54. S. Allenmark, B. Bomgren, M. Horen, and P. O. Lagerstrom, *Anal. Biochem., 136,* 293–297 (1984).
55. M. E. Goldman, R. J. Weber, A. H. Newman, K. C. Rice, P. Skolnick, and S. M. Paul, *J. Chromatogr., 382,* 264–269 (1986).
56. H. E. Swaisgood and I. M. Chaiken, in I. M. Chaiken, Ed., *Analytical Affinity Chromatography*, CRC Press, Boca Raton, FL, 1987.
57. A. J. Muller and P. W. Carr, *J. Chromatogr., 284,* 33–51 (1984).
58. D. J. Anderson and R. R. Walters, *J. Chromatogr., 376,* 69–85 (1986).
59. H. E. Swaisgood and I. M. Chaiken, *Biochemistry, 25,* 4148–4155 (1986).
60. G. Fassina, Y. Shai, and I. M. Chaiken, *Fed. Proc., 45,* 1944 (1986).
61. T. M. Phillips, N. S. More, W. D. Queen, T. V. Holshan, N. C. Kramer, and A. M. Thompson, *J. Chromatogr., 317,* 173–179 (1984).
62. A. Forsgren and J. Sjoquist, *J. Immunol., 97,* 822–827 (1966).

63. T. M. Phillips, W. D. Queen, N. S. More, and A. M. Thompson, *J. Chromatogr., 327*, 213–219 (1985).

64. B. Nilsson, *J. Chromatogr., 276*, 413–417 (1983).

65. D. M. Abercrombie, S. Angal, R. P. Sequeira, and I. M. Chaiken, *Biochemistry, 21*, 6458–6465 (1982).

66. P. Caliceti, G. Fassina, and I. M. Chaiken, *Appl. Biochem. Biotechnol.*, in press.

CHAPTER

8

SEPARATION OF CHIRAL COMPOUNDS WITH α₁-ACID GLYCOPROTEIN AS SELECTOR

JÖRGEN HERMANSSON

Apoteksbolaget AB, Central Laboratory
Department of Biomedicine
Stockholm, Sweden

GÖRAN SCHILL

Department of Analytical Pharmaceutical Chemistry
University of Uppsala
Uppsala, Sweden

8.1. INTRODUCTION

Optical isomers (enantiomers) often have widely different biological activities, and methods for the isolation of enantiomers in pharmaceutical formulations and biological material are of great importance. The possibilities of such separations have increased drastically during the last

Table 8.1. Separation of Enantiomeric Compounds

Principle	Chiral Selector	Binding of Substrate	Solid Phase
Diastereomeric derivative	Precolumn	Covalent	Nonchiral
Diastereomeric complex	In liquid phase	Reversible	Nonchiral
Diastereomeric complex	Bound to solid phase	Reversible	Chiral

decade, mainly due to the development of new liquid chromatographic separation principles. The separation of amino acids as chiral metal chelates by Gil-Av et al. (1) and the development of chiral solid phases by Pirkle and Finn (2) have been of particular importance for the rapid growth of this new field.

Enantiomeric compounds cannot be separated directly in normal chromatographic systems since the groups around the chiral atom are equally accessible to binding. However, if the enantiomers are combined with an enantiomeric reagent (a chiral selector), two diastereomers are formed which might have such differences in binding properties that they can be separated in chromatographic systems. The basic principles for such separations are summarized in Table 8.1.

8.1.1. Diastereomeric Derivatives

The formation of diastereomeric derivatives with covalent bonds between the solute and the selector was for a long time the dominating technique. This apprroach can, however, create problems. The selector should be a pure enantiomer and the reaction must be free from racemization. If the chiral selector $B(+)$ is contaminated by its antipode $B(-)$, its reaction with the racemic solute $A(+)/A(-)$ will give four compounds, as demonstrated in Fig. 8.1. In a nonchiral chromatographic system, they are separated into two peaks, each containing two enantiomeric compounds.

Such a process can only be used for quantification if the content of the contaminant $B(-)$ is known and the reaction rates of all the processes are the same. The latter is not always the case since the reaction takes place in an asymmetric environment (3).

The consequences of using an impure reagent can be particularly serious if one of the enantiomers is in large excess, which might be the case in studies on biological material or with purity testing of drug substances (4). An illustration is given in the following example.

If the reagent $B(+)$ contains 5% of $B(-)$ as an impurity and all the reactions with analytes $A(+)$ and $A(-)$ occur with the same rate, 95%

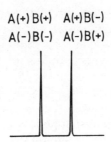

$$A(+) \begin{array}{l} + B(+) \longrightarrow A(+)B(+) \\ + B(-) \longrightarrow A(+)B(-) \end{array}$$

$$A(-) \begin{array}{l} + B(+) \longrightarrow A(-)B(+) \\ + B(-) \longrightarrow A(-)B(-) \end{array}$$

$$\begin{array}{ll} A(+)B(+) & A(+)B(-) \\ A(-)B(-) & A(-)B(+) \end{array}$$

Figure 8.1. Separation of diastereomeric compounds with two asymmetric centers on non-chiral column.

of each of the analytes will react with $B(+)$ and 5% with $B(-)$. The concentration ratios between the products are

$$\frac{[A(+)B(+)]}{[A(+)B(-)]} = \frac{[A(-)B(+)]}{[A(-)B(-)]} = \frac{95}{5}$$

The separation of the total amounts of the analytes $m_{A(+)}$ and $m_{A(-)}$ into two peaks (Fig. 8.1) is then

peak 1: $A(+)B(+) + A(-)B(-) = 0.95m_{A(+)} + 0.05m_{A(-)}$

peak 2: $A(-)B(+) + A(+)B(-) = 0.95m_{A(-)} + 0.05m_{A(+)}$

In a racemate $m_{A(+)}$ is equal to $m_{A(-)}$, and the degrees of impurity in the two peaks are the same. However, if $m_{A(-)} = 0.1m_{A(+)}$, the enantiomers are present in the peaks in the following ratios:

peak 1: $\dfrac{C_{A(+)}}{C_{A(-)}} = \dfrac{0.95m_{A(+)}}{0.05 \times 0.1m_{A(+)}} = 190$

peak 2: $\dfrac{C_{A(-)}}{C_{A(+)}} = \dfrac{0.95 \times 0.1m_{A(+)}}{0.05m_{A(+)}} = 1.9$

In the absence of the contaminant $B(-)$, peak 1 contains pure $A(+)$ and peak 2 pure $A(-)$. However, the presence of 5% of $B(-)$ as an impurity has a disastrous effect on the isolation of $A(-)$, which now is contaminated by 35% $A(+)$ in peak 2.

8.1.2. Diastereomeric Complexes

The rapid development of the chromatographic separation of enantiomers came with the introduction of methods based on the isolation as diastereomeric complexes. The retention of an enantiomeric solute $A(+)$ will depend on the concentration of the selector $B(+)$ and the distribution constant of the complex $A(+)B(+)$. If the selector is contaminated by its antipode $B(-)$, a further complex $A(+)B(-)$ is formed, and the retention will then depend on the concentrations and constants of both selector antipodes. This means that the impurity of the selector will affect the retention of the enantiomeric solute, but each of the separated peaks will only contain one of the enantiomeric solutes. The process has been clearly demonstrated in a study by Pettersson (5).

The advantages of methods based on the formation of chiral complexes are obvious, but the separation efficiencies might not be as good as in procedures based on the synthesis of diastereomeric derivatives and nonchiral phases. Methods with the chiral selector bound to the stationary phase have practical advantages over techniques with the selector in the mobile phase, such as a more free choice of detection technique and a lower consumption of expensive or unique reagents.

A number of chiral stationary phases are now commercially available. Some examples are given in Table 8.2. Many of the phases are only applicable to separations of uncharged enantiomeric compounds with nonpolar mobile phases, and hydrophilic and protolytic compounds can as a rule only be separated after masking of the protolytic group. Charged and hydrophilic compounds can be separated in underivatized form on the systems that work in the reversed-phase mode, such as those with proteins or a ligand exchange function, as selectors.

The chiral phases have been used for a rather short period of time, and advanced mechanism studies of the chiral recognition have been performed to a limited extent (12). Detailed recommendations of applications and choice of separation conditions are as a rule not available.

The present review only discusses the properties and application of α_1-acid glycoprotein (AGP) as chiral selector.

Table 8.2. Chiral Selector Bound to Solid Phase

Principle	Chiral Selector (Example)	Ref.	Mobile Phase	Substrate (Examples)
Charge transfer interaction	(R)-N-(3,5-dinitroebenzoyl) phenylglycine	(2)	Organic	Amides, cyclic imides, carbamates
Inclusion complexation	Triacetylcellulose	(6)	Organic	Aromatic compounds
	β-Cyclodextrine	(7)	Polar	Amides, imides
Helicity	Polymetacrylate	(8)	Organic	Aromatic compounds
Protein binding	Albumin	(9)	Aqueous	Acidic compounds
	α_1-Acid glycoprotein	(10)	Aqueous	Amines, carboxylic acids, nonprotolytes
Ligand exchange	L-proline amide + Cu(II)	(11)	Aqueous	Amino acids

8.2. α_1-ACID GLYCOPROTEIN AS SILICA-BONDED SELECTOR

α_1-Acid glycoprotein is isolated from human plasma. It has a molecular weight of about 41,000 daltons (13) and its isoelectric point is 2.7 in phosphate buffer. The molecule contains a peptide chain with 181 amino acid units, and five carbohydrate units are linked to the chain via asparagine residues. The protein has a fairly hydrophilic character owing to the presence of 14 sialic acid units in the carbohydrate moieties, which also contain hexosamines and neutral hexoses. There are numerous asymmetric centers in the carbohydrate units and in the peptide chain, but the chiral binding principles are so far not elucidated.

Hermansson developed in 1983 a chiral stationary phase based on the use of α_1-acid glycoprotein as the selector (14). The characterization of the phase and its application have been presented in a series of publications (15–20). The chiral binding of amines with AGP as complexing agent in the mobile phase has also been investigated (21).

The first chiral separations were made with a phase having AGP covalently bound to silica. The phase had good stereoselectivity but low binding capacity and fairly low stability at pH ≥ 7.0. A phase with considerably better properties in both respects was obtained by binding the protein by charge forces to diethylaminoethyl silica and immobilizing it by cross-linking (10). The cross-linking procedure involves an oxidation of alcohol groups to aldehydes, coupling to primary amines (Schiff base formation), followed by reduction to secondary amines. It can be assumed that the bonds are given mainly by the ϵ-amino groups in the 16–17 lysine residues.

The protein is coated on the diethylaminoethyl silica in a monomolecular layer. The mean pore diameter is about 250 Å and the total loading of AGP is about 180 mg/g. The diethylaminoethyl silica has a very low tendency to bind cationic analytes, and they are significantly retained only after the binding of AGP to the silica matrix (10).

8.3. STABILITY OF THE AGP COLUMN

Long-term stability studies have been performed with mobile phases containing lower alcohols, ketones, and cyclic ethers as modifiers (10, 19). A column with a mobile phase of 6% 2-propanol in phosphate buffer pH 7.0 that had been stored for 12 months at ambient temperature (20–25°C) showed changes in retention of less than 10% for three enantiomeric test solutes. A run with mobile phases containing lower alcohols and ketones in phosphate buffers of pH 6.0–7.5 for 11 weeks at a rate of about 2 L per week resulted in almost insignificant changes in retention and stereoselectivity for the enantiomers of disopyramide. The long-term stability is very good in 2-propanol + water (1 + 1), and the columns are normally stored in this solvent.

Temperatures of up to 80°C have been used for shorter periods of time without noticeable deterioration of the columns. Eluents with pH outside the range of 3.0–7.5 should be avoided, and the stability decreases above pH 7.0 owing to the dissolution of the silica matrix. A certain decrease of the stability has also been observed when the mobile phases contain charged modifiers and amines in particular, but there are no indications that the AGP columns are more sensitive than other silica-based solid phases. The rather wide pores and the large pore volume of the silica matrix make it somewhat pressure sensitive. The first sign of column degradation is usually a widening of the peaks, whereas the chiral selectivity is almost unchanged. It indicates that it is the silica matrix and not the protein that is the stability limiting component in the packing.

8.4. REGULATION OF RETENTION AND STEREOSELECTIVITY BY MODIFIERS

In nonchiral reversed-phase systems the retention can be regulated in many different ways. Alkyl or aryl-bonded absorbents of different hydrophobicities can be used, and the binding of the solute to the mobile and the solid phases can be varied by the addition of modifiers or, for protolytes, by a change of pH. On the chiral AGP column it is also possible to obtain large changes in the retention by modifiers, but their effects might in part be of another kind than in the nonchiral systems. The retaining part of the solid phase is a protein, and its charge changes with pH, which will affect its binding ability. The modifier can of course act as a competitor to the solute, but its effect might be selective since the protein has different kinds of binding groups. It is furthermore possible that the modifier can give rise to conformational changes of the protein, which might have a significant influence on its binding properties. There are so many modes of action that it is understandable that certain kinds of modifiers can give rise to quite remarkable improvements of the stereoselectivity.

8.4.1. Uncharged Modifiers

Monovalent alcohols are as a rule efficient means for a reduction of the retention. Some examples with 2-propanol as modifier and tertiary amines as solutes are given in Table 8.3 (10). The alcoholic modifiers decrease the separation factors for most enantiomeric pairs, but the influence is highly dependent on the structure of the solute: for chlorpheniramine the decrease of α is rather drastic on the addition of 2-propanol, whereas α of the homologues mepivacaine and bupivacaine is completely unchanged. Improvements of the chiral selectivity have also been observed after addition of uncharged modifiers. Differences in the effects of a change of the chromatographic conditions depending on solute structures are rather common on the AGP phase. The background is so far not elucidated, but it might be due to sites with different binding properties.

Diols have also been tested as moderators since they give low energy interactions with proteins and might affect the retention in a different way than the monoalcohols. Table 8.4 (20) presents some results obtained with ethylene glycol, 1,2-butanediol, and ethanol as additives to the mobile phase. The effects depend, as expected, on the structure of the solute. Ethylene glycol decreases the retention for Labetalol A and B less than ethanol, but it is preferable as moderator since the influence on the separation factor is limited. Ethanol is on the other hand more suitable for cyclopentolate, whereas 1,2-butanediol gives the highest chiral selectivity for methylhomatropine. Uncharged amino acids have also been tried as moderators. In concentrations up to $0.1M$ they have given a rather limited decrease of the retention without significant change of the separation factor (20).

8.4.2. Charged Modifiers

Ionic additives to the mobile phase can sometimes have more favorable effects on the retention of the solutes than uncharged modifiers. They

Table 8.3. Influence of 2-Propanol (%) on Retention and Stereoselectivity

	1%		2%		4%		6%		8%	
Solute	k'_1	α	k'_1	α	k'_1	α	k'_1	α	k'_1	α
Disopyramide	—	—	—	—	8.51	3.70	3.62	3.37	1.77	3.20
Chlorpheniramine	11.2	2.34	7.35	1.71	4.59	1.38	—	—	—	—
Mepensolate	—	—	6.35	1.54	2.65	1.40	1.42	1.38	1.00	1.21
Mepivacaine	26.0	1.36	10.7	1.31	4.42	1.33	2.48	1.36	1.58	1.35
Bupivacaine	—	—	—	—	18.6	1.70	8.84	1.72	5.01	1.74
3-PPP	4.85	1.76	—	—	1.43	1.59	1.03	1.39	0.90	1.17

Solid phase: AGP 180mg/g silica. Mobile phase: 2-propanol in phosphate buffer, pH 7.2. k'_1, k'_2—capacity ratios of first and second eluted enantiomers; α—separation factor, = k'_2/k'_1.

Table 8.4. Influence of Alcohols as Modifiers

Modifier	Cyclopentolate		Labetalol A		Labetalol B		Methylhomatropine	
	k'_1	α	k'_1	α	k'_1	α	k'_1	α
Ethylene glycol 0.32M	23	1.73	27	1.69	25	1.37	1.8	2.65
Ethylene glycol 1.29M	12	1.46	12	1.64	14	1.30	1.3	2.75
1,2-Butanediol 0.25M	7.5	1.86	8.4	1.27	9.6	1.17	1.5	3.81
Ethanol 0.44M	21	1.95	15	1.45	19	1.28	1.6	3.10
Ethanol 1.74M	3.2	1.81	5.2	1.12	—	—	0.7	2.00

Solid phase: AGP 180mg/g silica. Mobile phase: modifier + NaCl 0.1M in 0.02M phosphate buffer, pH 7.0.

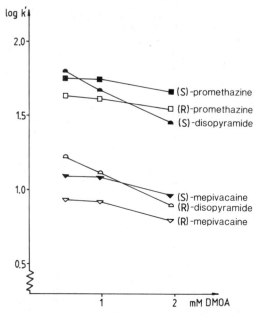

Figure 8.2. Influence of dimethyloctylammonium (DMOA) on cationic enantiomers. Solid phase: AGP 180mg/g silica. Mobile phase: DMOA in 2-propanol + phosphate buffer, pH 7.2 (2 + 98).

cannot only give a decrease of the retention, but the change in retention can sometimes be combined with an improvement of the chiral selectivity.

A series of interesting results have been achieved with organic ammonium ions as modifiers. The effect of N,N-dimethyloctyl ammonium (DMOA) in a mobile phase of pH 7.0 containing 2% of 2-propanol is shown in Fig. 8.2 (10). The retention of the cationic solutes decreases significantly, but the separation factors are almost unchanged. A further increase of the modifier concentration might, however, give rise to changes in the chiral selectivity, as observed when tetrabutylammonium is used as modifier (20). The separation factor is unaffected at concentrations below 3mM, but decreases at higher modifier concentrations.

The hydrophobicity of the ammonium ion modifier is highly important for its influence on the retention as well as on the chiral selectivity. Some results obtained with cationic solutes are given in Table 8.5. It is obvious that the effect on the retention increases with the hydrophobicity of the modifier. The enantiomers of the solutes are, however, affected quite differently, and significant improvements of the separation factors can be obtained by a proper choice of modifier, as demonstrated for methyl-

Table 8.5. Influence of the Hydrophobicity of Cationic Modifiers

Modifier	Tocainide		Metoprolol		Methylhomatropine		Phenmetrazine	
	k'_1	α	k'_1	α	k'_1	α	k'_1	α
NaCl 0.1M	4.4	1.31	6.1	1.64	4.7	2.65	8.3	1.0
Tetraethylammonium bromide 0.01M	—	—	5.5	1.41	4.1	2.58	4.1	1.28
Tetrapropylammonium bromide 0.003M	2.7	1.30	2.4	1.41	2.1	4.17	2.5	1.25
Tetrabutylammonium bromide 0.001M	1.7	1.29	1.7	1.0	1.8	3.19	2.1	1.0
Dimethylethylamine 0.1M	1.7	1.27	1.8	1.64	2.0	2.57	1.6	1.19

Solid phase: AGP 180mg/g silica. Mobile phase: modifier in 0.02M phosphate buffer, pH 7.0

347

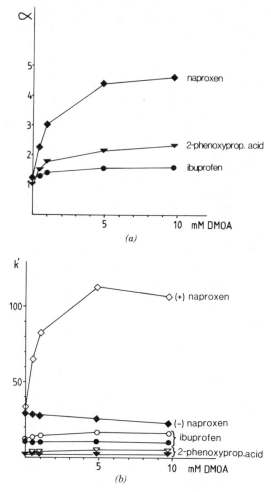

Figure 8.3. Influence of dimethyloctylammonium (DMOA) on anionic enantiomers: (*a*) Solid phase: AGP 180mg/g silica. (*b*) Mobile phase: DMOA in 2-propanol + phosphate buffer, pH 7.0 (1 + 199).

homatropine and phenmetrazine. It is, however, not possible to give clear guidelines for this choice, since it has not been possible to find any relationship between solute structure and retention behavior. It is assumed, so far, that the behavior differences are due to different modes of binding to the chiral phase.

Anionic solutes are also highly affected by cationic modifiers, as demonstrated in Fig. 8.3 (19). The solutes are 2-arylpropionic acids, and the presence of DMOA in the mobile phase has a particularly strong influence on naproxen: the separation factor increases from 1.1 to 4.5 on addition of 10mM DMOA. A closer study of the retention pattern of the two enantiomers shows that DMOA gives rise to an increase of the retention of the (+) form, in accordance with an ion-pair retention model, whereas the retention of the (−) form decreases slightly, indicating dominance of a competitive effect. The influence of DMOA on the chiral selectivity of 2-phenoxypropionic acid and ibuprofen is less dramatic, but it should be observed that no chiral selectivity is obtained in the absence of DMOA, whereas the presence of 10mM DMOA in the mobile phase gives separation factors of 1.6 and 2.3, respectively, and possibilities to a complete resolution of the enantiomers.

Hydrophobic carboxylic acids are also highly interesting as modifiers for cationic solutes. The retention of the solutes usually decreases with increasing hydrophobicity of the acids, but this is in some cases combined with a strong improvement of the chiral selectivity, as shown in Table 8.6. The optimum effect is for atropine and methylphenidate obtained with octanoic acid, but there are no general rules for the effects of these modifiers as well as for the cationic additives.

A further illustration of the specific effects of the hydrophobic carboxylic acids is given in Table 8.7 (20), which shows its effects on some solutes related to terbutalin. Two modifiers have been tested, decanoic acid 0.005M and 6-aminohexanoic acid 0.10M, the latter usually having a rather small influence on the retention. Decanoic acid gives a significant improvement of the separation factors for the monohydroxy derivatives, but it has a negative effect on the chiral resolution of the dihydroxy derivatives and bambuterol, which have higher separation factors in mobile phases containing 6-aminohexanoic acid. It should also be noted that decanoic acid gives a particularly strong reduction of the retention of the 2-hydroxyphenyl and 3-hydroxy-5-dimethylcarbamyl derivatives.

Table 8.6. Influence of the Hydrophobicity of Anionic Modifiers

	Ephedrine		Atropine		Methylphenidate		Phenmetrazine	
Modifier	k_1'	α	k_1'	α	k_1'	α	k_1'	α
NaCl 0.1M	1.4	1.24	2.8	1.10	—	—	8.3	1.0
Butyric acid 0.05M	0.90	1.83	5.1	1.23	12.5	1.29	3.0	1.42
Octanoic acid 0.01M	0.85	1.71	6.3	1.64	7.5	1.89	2.9	1.57
Decanoic acid 0.005M	0.64	1.76	5.3	1.25	3.3	1.13	1.5	1.55

Solid phase: AGP 180mg/g silica. Mobile phase: modifier in 0.02M phosphate buffer, pH 7.0

Table 8.7. Separation of Enantiomeric Compounds Related to Terbutaline

$$OH$$
$$|$$
$$R_1—C \cdot CH_2 \cdot NH \cdot C(CH_3)_3$$
$$|$$
$$H$$

Solute R$_1$	6-Aminohexanoic Acid 0.10M		Decanoic Acid 0.005M	
	k_1'	α	k_1'	α
3,5-Dihydroxyphenyl*	0.9	1.33	1.1	1.21
2,5-Dihydroxyphenyl	1.2	1.47	0.8	1.0
4-Hydroxyphenyl	0.8	1.0	0.9	2.00
2-Hydroxyphenyl	6.7	1.0	1.6	1.78
Phenyl	—	—	1.7	1.13
3,5-Bis(dimethylcarbamyl) phenyl†	2.2	1.50	1.2	1.0
3-Hydroxy-5- dimethylcarbamylphenyl	37	1.16	2.3	1.32

Solid phase: AGP 180mg/g silica. Mobile phase: modifier in 0.02M phosphate buffer, pH 7.0

* Terbutaline.

† Bambuterol.

The carboxylic acids can sometimes give rise to rather strong effects, as shown (Fig. 8.4) by some studies of chiral separations of the diastereomeric amino alcohols ephedrine ($RS;SR$) and pseudoephedrine ($RR;SS$) (20). The separation factors of the two compounds are only slightly affected by the uncharged modifier 6-aminohexanoic acid, the cationic tetrapropylammonium, and the anionic aspartic acid. A drastic change is, however, obtained with octanoic acid. The chiral separation of the enantiomers of ephedrine is significantly increased, whereas the separation of the pseudoephedrine enantiomers is decreased. In addition there is a reversal of the retention order between the enantiomers of pseudoephedrine, the L form now having the shorter retention. Studies of the influence of modifiers on the retention order have been made with a limited number of compounds, but no further reversals have so far been observed, although enantiomeric compounds with so different structures as tocainide, atropine, nadolol, and methorphan have been included in the investigation.

A further drastic illustration of the highly complex relationship between solute structure, modifier properties, and chiral selectivity is given in Fig.

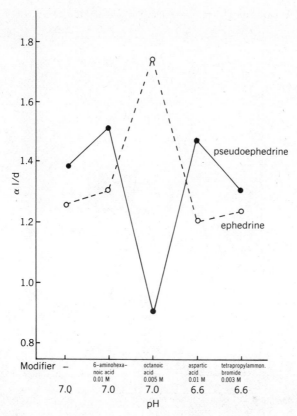

Figure 8.4. Influence of modifier on diastereomers ephedrine and pseudoephedrine. Solid phase: AGP 180mg/g silica. Mobile phase: modifier in 0.02M phosphate buffer, pH 7.0.

8.5. It shows the effect of various mobile phase additives on the chiral separations of three closely related compounds: the tertiary amine atropine, its N-methyl derivative (methylatropine), and N-methylhomatropine, which differs from methylatropine only by a methylene group in a side chain attached to the chiral carbon.

N-Methylhomatropine can easily be separated into the enantiomeric forms with and without a modifier in the mobile phase. The maximum chiral separation has been obtained with tetrapropylammonium bromide as the modifier. It is interesting to observe that the separation factor decreases on the addition of octanoic acid. Atropine shows a quite different response to the mobile phase additives. Very low chiral separation is obtained with uncharged or cationic modifiers, but the addition of octanoic acid to the mobile phase increases the separation factor to more

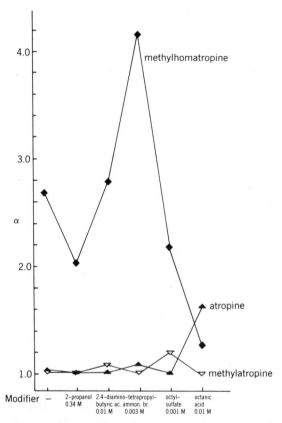

Figure 8.5. Chiral separation of atropine, methylatropine, and methylhomatropine. Solid phase: AGP 180mg/g silica. Mobile phase: modifier in 0.02M phosphate buffer, pH 7.0.

Table 8.8. Influence of pH on Chiral Selectivity and Retention of Uncharged Enantiomers

Solute	pK_a	pH 3.92		pH 4.95		pH 6.02		pH 7.02	
		k'_1	α	k'_1	α	k'_1	α	k'_1	α
Bendroflumethiazide	8.5	16.4	1.46	18.5	1.67	20.4	1.89	22.9	2.16
Hexobarbital	8.5	2.95	1.34	3.62	1.53	4.43	1.77	5.24	2.13
Ethotoin	—	1.70	2.13	1.69	2.18	1.82	2.11	1.74	2.11
Mandelic acid methyl ester	—	1.36	1.21	1.71	1.32	1.84	1.34	2.05	1.33
Mandelic acid ethyl ester	—	2.76	1.51	3.70	1.64	3.99	1.68	4.53	1.72

Solid phase: AGP 180mg/g silica. Mobile phase: 1% 2-propanol in phosphate buffer. (The final concentration of phosphate was 6.6–10mM.)

than 1.6, and a complete resolution of the enantiomers can then be obtained easily.

The corresponding quaternary ammonium ion methylatropine is very difficult to separate into enantiomeric forms. It is notable that neither octanoic acid nor tetrapropylammonium improve the separation factor, although they are highly suitable for the closely related atropine and N-methylhomatropine, respectively. A maximum α value of 1.2–1.3 is obtained in the presence of octylsulfate or cyclohexylsulfamate.

8.5. pH EFFECTS

The isoelectric point of α_1-acid glycoprotein, which constitutes the stationary phase, is about 2.7 and its negative charge increases with increasing pH. A change of pH can, however, also give rise to conformational changes which can affect its binding properties. Conformational changes are known to occur with proteins in solution, and the behavior might be different when it is present in cross-linked form as in the AGP column. Anyhow it is quite clear that a pH change has a considerable influence on the retaining properties and the chiral selectivity of the AGP column.

The changes of the retention and the chiral separation of some uncharged solutes in the pH range of 3.9–7.5 are demonstrated in Table 8.8 (14). The retention and the stereoselectivity increase with pH for all the solutes except ethotoin. The effect of pH on the retention is much larger for cationic solutes, but the magnitude of the change is highly dependent on the structure of the solute (14). Some examples in Table 8.9 (18) show results obtained in the pH range of 6.5–7.5 with 2-propanol as modifier. Increasing pH improves the chiral selectivity in some of the cases, but the reverse effect was also observed, namely, for disopyramide.

The effect of a pH change depends not only on the structure of the solute but also on the properties of the additives in the mobile phase. Some results with cationic solutes are given in Table 8.10 (20). The influence of a pH increase from 6.0 to 7.0 with tetrabutylammonium as the modifier is in some cases drastic: chlorpheniramine and methylphenidate show no chiral separation at pH 6.0, but the separation factors are higher than 2 at pH 7.0. Another group of compounds, for example, cyclopentolate, show a significant decrease of the chiral separation on an increase of pH, whereas no change is observed for others like methylhomatropine and methorphan. The pH effect can, however, be quite different with other modifiers. The chiral separation of methorphan increases strongly with pH when octanoic acid is used as modifier, and cyclopentolate shows the same kind of pH dependence when 2-propanol is the modifier (Table 8.9).

Table 8.9. Influence of pH on Cationic Solutes: Uncharged Modifiers

pH	Methylhomatropine			Cyclopentolate			Disopyramide			Methadone		
	Pr(%)	k'_i	α	Pr(%)	k'_i	α	Pr(%)	k'_i	α	Pr (%)	k'_i	α
6.5	0	2.4	2.38	2	4.8	1.79	8	1.9	2.77	8	5.0	1.51
7.0	0	4.5	2.47	2	8.9	1.89	8	2.7	2.70	8	6.5	1.59
7.5	0	5.0	2.64	2	14.0	1.96	8	3.3	2.67	8	7.4	1.57

Solid phase: AGP 180mg/g silica. Mobile phase: 2-propanol (Pr) in 0.02 M phosphate buffer.

Table 8.10. Influence of pH on Cationic Solutes: Charged Modifiers

		α	
Modifier	Solute	pH 6.0	pH 7.0
Tetrabutylammonium bromide 0.003M	Chlorpheniramine	1.0	2.26
	Labetalol A	1.46	2.10
	Methylphenidate	1.0	2.17
	Cyclopentolate	2.09	1.70
	Doxylamine	1.37	1.23
	Tocainide	1.54	1.25
	Methorphan	2.69	2.72
	Methylhomatropine	3.10	3.02
Octanoic acid 0.005M (10°C)	Atropine	1.32	1.66
	Ephedrine	1.10	1.51
	Methorphan	1.23	2.26

Solid phase: AGP 180mg/g silica. Mobile phase: modifier in 0.02M phosphate buffer.

Anionic solutes get increased retention on a decrease of pH and behave thus opposite to the cations. Only small changes of the separation factors with pH have so far been observed (Table 8.11) (19, 20).

8.6.   INFLUENCE OF TEMPERATURE

The retention and the chiral selectivity on the AGP column are usually strongly influenced by temperature. The retention changes of two enantiomers in the temperature range of 25–80°C are illustrated in Fig. 8.6. The chiral selectivity increases with decreasing temperature; the separation factor increases for disopyramide from 1.8 to 3.2, that is, by about 80%, and for RAC 109 the degree of increase is even higher, from 1.3 to

Table 8.11. Influence of pH on Anionic Solutes

	2-Phenylbutyric Acid		2-Phenylpropionic Acid		2-Phenoxypropionic Acid	
pH	k'_1	α	k'_1	α	k'_1	α
6.1	10.8	1.65	4.5	1.16	4.3	1.22
6.6	5.3	1.70	2.6	1.15	2.4	1.26

Solid phase: AGP 180mg/g silica. Mobile phase: 0.003M tetrapropylammonium bromide in 0.02M phosphate buffer.

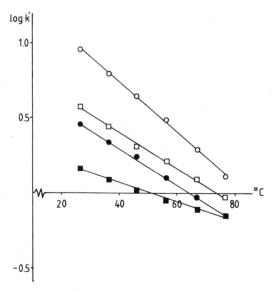

Figure 8.6. Influence of temperature on retention and stereoselectivity. Solid phase: AGP 180mg/g silica. Mobile phase: DMOA in 2-propanol 7% in 0.02M phosphate buffer, pH 7.1. ●—disopyramide (R), ○—disopyramide (S), ■—RAC 109 I; □—RAC 109 II.

2.5 (19). The change of the separation factor with temperature can also be dependent on pH, as shown in Table 8.12 (18).

8.7. CHROMATOGRAPHIC PROPERTIES OF THE AGP PHASE

The AGP phase is prepared by coating a monomolecular layer of α_1-acid glycoprotein on a silica matrix with a mean particle diameter of 5–10 μm. The protein coating corresponds to 4.5×10^{-6} mol per gram of solid phase, and the binding capacity of the phase is thus fairly low. Overloading effects such as a slight peak asymmetry and loading dependent retention can often be observed when more than 5 nmol of solute is injected on a normal commercial column (100-mm length × 4-mm id). An illustration is given in Table 8.13 (18). The limited loading is, however, somewhat dependent on the composition of the mobile phase. A minor peak asymmetry is usually only disturbing when the enantiomers are present in highly unequal amounts, that is, for studies of enantiomeric purity.

The separating efficiency is highly dependent on the mobile phase speed. An example is given in Fig. 8.7, which shows the increase of the reduced plate height, H/d_p, with increasing mobile phase speed for the

Table 8.12. Influence of pH and Temperature

Solute	pH	Temperature (°C)	k_1'	α
Cyclopentolate	6.5	10	3.9	1.94
		20	4.8	1.79
	7.0	10	10.0	1.96
		20	8.9	1.89
	7.5	10	15.0	2.02
		20	14.0	1.96
Methylhomatropine	6.5	10	3.3	2.82
		20	2.4	2.38
	7.0	10	5.5	3.00
		20	4.5	2.47
	7.5	10	5.2	3.10
		20	5.0	2.64

Solid phase; AGP 180mg/g silica. Mobile phase: NaCl $0.1M$ in phosphate buffer.

enantiomers of two acids, phenoxypropionic acid and ibuprofen (19). The strong increase of h with the mobile phase speed indicates a slow phase transfer.

The reduced plate height is also highly dependent on the structure of the solute, as demonstrated in Table 8.14 (20). The solutes are cationic and anionic solutes of widely different structures. The retention of the solutes varies over a wide range, with k' values of 0.5–9, but there is no correlation between k' and the magnitude of h. The results in the table

Table 8.13. Influence of Sample Loading

Loading (nmol)	t_{RI} (s)	α	asf_1	asf_2
1.5	1022	2.12	1.2	1.2
3	1007	2.11	1.2	1.5
6	964	2.11	1.7	2.5
9	950	2.07	2.0	2.3
12	935	2.06	2.0	2.4

Solid phase: AGP 180mg/g silica.
Mobile phase: NaCl $0.1M$ in phosphate buffer, pH 7.5
Flow: 0.30 mL/min.
Substrate: methylhomatropine.
asf_1 and asf_2 = asymmetry factors of first and second eluted enantiomer, respectively.

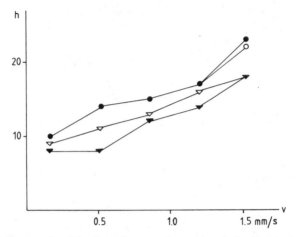

Figure 8.7. Influence of mobile phase flow speed on reduced plate height *h*. Solid phase: AGP 180mg/g silica. Mobile phase: 2-propanol 0.5% in 0.02*M* phosphate buffer, pH 7.0. ●—ibuprofen I; ○—ibuprofen II; ▼—2-phenoxypropionic acid I; ▽—2-phenoxypropionic acid II.

do not indicate that the properties of the mobile phases have an influence on the separating efficiencies and a pH change in the range of 6.0–7.0 did not result in any significant changes of *h*. A decrease of the temperature from 20 to 10°C increased *h* by about 20%.

The reason for the much larger *H* values for ephedrine, for example, than for atropine is so far not elucidated but it might be due to differences in the binding mechanisms. The observed differences in *h* for the first and second peaks are small and do not indicate that the enantiomers of any of the compounds in the table have different retention mechanisms. There are, however, observations of large deviations in the shapes for the first and second eluted enantiomers. Some examples are given in Table 8.15 (18). Methadone gives a first peak with good symmetry and a second peak that displays significant tailing. It is also important to note that the asymmetry can change with pH. Propoxyphene gives a second peak with extreme tailing at pH 7.0 and 7.5. However, at pH 6.5 this peak is almost symmetrical.

8.8. SOLUTE STRUCTURE AND STEREOSELECTIVE SEPARATION

The stereoselective separation is based on the fact that the enantiomeric forms of the solute form complexes with different stability with the chiral selector in the solid phase. Interaction between the solutes and the se-

Table 8.14. Separating Efficiency h for Enantiomeric Solutes

Solute	Tetrapropylammonium Bromide 0.003M, pH 6.6		Octanoic Acid 0.005M, pH 7.0		Without Modifier, pH 6.6	
	h_1	h_2	h_1	h_2	h_1	h_2
2-Phenylbutyric acid	8	8	—	—	8	7
2-Phenoxybutyric acid	10	9	—	—	9	7
Atropine						
Methylhomatropine	—	—	11	11	11	—
Tocainide	13	12	10	9	13	12
Nadolol A	11	9	—	13	—	10
Ephedrine	15	12	16	—	12	—
Metoprolol	20	17	19	20	25	—
	26	20	—	—	28	21

Solid phase: AGP 180mg/g silica.

Mobile phase: 0.02M phosphate buffer with and without modifier.

h_1 and h_2 are reduced plate heights of first and second peak, respectively.

lector in at least three points is generally assumed to be needed, and the interacting groups must furthermore have certain defined properties to obtain stereoselective retention. Our studies with AGP as the selector have indicated that the solute should have two binding groups close to the chiral center and bulky or rigid structures in the vicinity of these

Table 8.15. Peak Symmetry Deviations

Solute	pH	k_1'	α	asf_1	asf_2
Propoxyphene	6.5	3.2	1.52	1.2	0.8
	7.0	4.4	2.3	0.9	8
	7.5	5.1	2.4	1.0	*
Methadone	6.5	5.0	1.51	0.9	2.0
	7.0	6.5	1.59	1.0	1.7
	7.5	7.4	1.57	1.5.	2.0

Solid phase: AGP 180mg/g silica.

Mobile phase: 0.1M NaCl in 0.02M phosphate buffer + 8% 2-propanol.

Loading: 2 nmol.

asf_1 and asf_2 are ratios of back half-peak area to front half-peak area for first and second peak, respectively.

* Unmeasurable.

Table 8.16. Separation of Enantiomeric Compounds Related to Metroprolol

$$CH_3OCH_2CH_2 - \langle O \rangle - OCH_2\ \underset{\underset{OH}{|}}{CH}\ (CH_2)_n\ NHR_1$$

R_1	n	α
i-Propyl*	1	1.49
i-Propyl	2	1.14
i-Propyl	3	1.0
t-Butyl	1	1.73
n-Propyl	1	1.23
Ethyl	1	1.12
Methyl	1	1.0
H	1	1.17

Solid phase: AGP 180mg/g silica.

Mobile phase: 0.001M tetrapropylammonium bromide in 0.02M phosphate buffer, pH 7.0.

* Metoprolol.

groups. These assumptions have in part been developed by studies on groups of structurally related compounds.

A physiologically important group of compounds with a chiral carbon between two binding moieties are the amino alcohols with betareceptor blocking effect. Metoprolol belongs to this group, and some results obtained with substances related to that compound are given in Table 8.16 (18). The hydroxyl is coupled directly to the chiral center, and chiral selectivity is very good when the secondary amine group is separated from the chiral carbon by one alkyl carbon. The separation factor decreases, however, drastically when the chain is lengthened by one or two carbons. The bulkiness of the substituent at the amine group is also highly important with a maximum α for a t-butyl and a minimum for a methyl substituent.

A series of compounds related to the primary amine tocainide give another approach to the relationship between structure and chiral selectivity. The hydrogen bonding moiety is an amide group and it is, like the amine, directly coupled to the chiral center (Table 8.17) (18). An increase of the length of the alkyl chain, which is the third group around the chiral center, gives a strong increase of the separation factor and so does the introduction of a methylene group between the amine and the asymmetric carbon. If the methylene group is placed on the other side, that is, the

Table 8.17. Separation of Enantiomeric Compounds Related to Tocainide

$$R_1 \cdot N \cdot CO \cdot R_3 \cdot NH_2$$
$$R_2$$

R_1	R_2	R_3	α
2,6-Xylyl*	H	$CH(CH_3)$	1.45
2,6-Xylyl	H	$CH(C_2H_5)$	3.31
2,6-Xylyl	H	$CH(CH_3) \cdot CH_2$	2.30
2,6-Xylyl	H	$CH_2 \cdot CH(CH_3)$	1.0
2,6-Xylyl	CH_3	$CH(CH_3)$	1.0
2,4,6-Mesityl	H	$CH(CH_3)$	1.24
2-Tolyl	H	$CH(CH_3)$	1.0
Phenyl	H	$CH(CH_3)$	1.0
Benzyl	H	$CH(CH_3)$	1.0

Solid phase: AGP 180mg/g silica.
Mobile phase: $0.002M$ tetrabutylammonium bromide in $0.02M$ phosphate buffer, pH 6.0.
* Tocainide.

chiral center is separated from the amide by one methylene group, the chiral selectivity disappears. A change of the hydrogen-bonding character of the amide by methylation has the same effect. The substitution in the aromatic ring is also important, and the removal of one or both of the methyl groups in the 2 and 6 positions seems to destroy the possibilities to chiral separation.

The absence of detailed knowledge of the character of the chiral binding groups in the AGP phase makes it difficult to reach general conclusions about the structural features that are needed for a chiral separation of enantiomeric compounds. Studies have been performed on substances with widely different structures, in many cases with an extensive variation of the separation conditions. Some of the results that have been achieved with 50 cationic solutes are summarized in Table 8.18. It gives the highest separation factors that have so far been achieved and lists the separation conditions used.

The empirical conclusions about structural features needed for chiral separations that can be reached on the basis of these results are summarized in Table 8.19. The basic features are the two binding moieties and a bulky or rigid structural element, and it seems that these groups should be located very close to the chiral center. The deviations from the basic rules in the table (see comments) are very few. Methadone has the hydrogen bonding moiety as well as the bulky group not directly coupled to the chiral center. Cyclopentolate and dimethindene, with complex ring configurations close to the chiral center, have more than three atoms

Table 8.18. Separation Factors for Cationic Racemic Drugs

Solute	α	Conditions*	Reference
Atropine	1.64	16	(18)
Bromdiphenhydramine	1.17	13	(18)
Brompheniramine	1.50	22	(18)
Bupivacaine	1.74	8	(10)
Butorphanol†	1.99	6	(18)
Carbinoxamine	1.33	24	(18)
Chlorpheniramine	2.34	2	(10)
Clidinum	1.21	4	(18)
Cocaine	1.46	23	(18)
Cyclopentolate	3.86	19	(18)
Dimethindene	1.53	14	(18)
Diperodone	1.47	25	(18)
Disopyramide	3.70	5	(10)
Dobutamine	1.56	18	(20)
Doxylamine	1.37	21	(18)
Ephedrine	1.83	15	(18)
Ephedrine, pseudo-	1.34	15	(18)
Homatropine	1.63	16	(18)
Ketamine	2.48	9	(10)
Ketoprofen	1.53	10	(19)
Labetalol A	2.10	22	(18)
Labetalol B	1.36	15	(18)
Mepensolate	1.54	3	(10)
Mepivacaine	1.36	7	(10)
Methadone	1.59	14	(18)
Methorphan	2.54	17	(18)
Methylatropine	1.27	17	(18)
Methylhomatropine	4.2	19	(18)
Methylphenidate†	1.70	22	(18)
Metoprolol	1.64	12	(18)
Nadolol A	3.98	20	(18)
Nadolol B	3.03	20	(18)
Oxyphencyclimin	1.42	14	(18)
Oxprenolol	1.25	13	(18)
Pentazocine	1.51	9	(10)
Phenmetrazine†	1.57	16	(18)
Phenoxybenzamine	1.37	24	(18)
3-PPP	1.76	2	(10)
Promethazine	1.73	11	(15)
Propiomazine	1.60	1	(14)
Pronethalol	1.26	24	(18)
Propoxyphene	2.30	14	(18)
Propranolol	1.13	25	(18)
Terbutalin	1.22	19	(18)
Tetrahydrozoline	1.66	26	(20)

Table 8.18. (*Continued*)

Solute	α	Conditions*	Reference
Tocainide	1.44	18	(18)
Tridihexethyl	1.64	25	(18)
Verapamil	1.75	27	(20)

* Conditions: H_3PO_4 or NaOH added to give the indicated pH of the mobile phases containing basic or acidic modifiers, respectively. The results presented in this chapter are obtained on the first generation $α_1$-AGP column, EnantioPac. The second generation $α_1$-AGP column, CHIRAL-AGP, may give α-values that differ from the above reported, since CHIRAL-AGP, is based on another silica matrix with another surface chemistry and a modified immobilization technique.

1—0.065M 2-propanol, pH 6.5, 0.01M phosphate buffer
2—0.013M 2-propanol, pH 7.0, 0.01M phosphate buffer
3—0.26M 2-propanol, pH 7.2, 0.01M phosphate buffer
4—0.33M 2-propanol, pH 7.0, 0.02M phosphate buffer
5—0.05M 2-propanol, pH 7.2, 0.01M phosphate buffer
6—0.67M 2-propanol, pH 7.0, 0.02M phosphate buffer
7—0.78M 2-propanol, pH 7.2, 0.01M phosphate buffer
8—1.04M 2-propanol, pH 7.2, 0.01M phosphate buffer
9—0.00195M dimethyloctylamine (DMOA), pH 7.0, 0.01M phosphate buffer
10—0.065M 2-propanol, 0.0049M DMOA, pH 7.0, 0.01M, phosphate buffer
11—0.33M ethanol, 0.00195M DMOA, pH 7.0, 0.01M phosphate buffer
12—0.1M NaCl, pH 7.0, 0.02M phosphate buffer
13—1.74M ethanol, 0.1M NaCl, pH 7.0, 0.02M phosphate buffer
14—1.33M 2-propanol, 0.1M NaCl, pH 7.0, 0.02M phosphate buffer
15—0.05M butyric acid, pH 7.0, 0.02M phosphate buffer
16—0.01M octanoic acid, pH 7.0, 0.02M phosphate buffer
17—0.25M cyclohexylsulfamic acid, pH 7.0, 0.02M phosphate buffer
18—0.001M tetrapropylammonium bromide (TPrABr), pH 6.0, 0.02M phosphate buffer
19—0.003M TPrABr, pH 7.0, 0.02M phosphate buffer
20—0.001M tetrabutylammonium bromide (TBuABr), pH 6.0, 0.02M phosphate buffer
21—0.003M TBuABr, pH 6.0, 0.02M phosphate buffer
22—0.003M TBuABr, pH 7.0, 0.02M phosphate buffer
23—0.001M DMOA, pH 7.0, 0.02M phosphate buffer
24—0.17M 2-propanol, 0.001M DMOA, pH 7.0, 0.02M phosphate buffer
25—0.33M 2-propanol, 0.002M DMOA, pH 7.0, 0.02M phosphate buffer
26—0.1M dimethylethylammonium, pH 7.0, 0.02M phosphate buffer
27—0.005M decanoic acid + 0.05 tetraethylammonium bromide, pH 7.0, 0.02M phosphate buffer

† One of the diastereomers (for which there are no generally accepted names).

Table 8.19. Structural Features for Chiral Separations: Cationic Solutes

Character	Interacting Group		Comments
	Distance from Chiral Center (Ch)	Properties	
H bonding (HB)	Coupled directly	Single groups or aromatic rings	1–2 carbons between HR and Ch in some cases
Ammonium (N^+)	Usually ≤ 3 atoms	α increases with size of alkyl substituent	>3 atoms between N^+ and Ch in some cases (e.g., >1 HB or N^+ in ring)
Bulky or rigid structures	Comprising or coupled directly to Ch or HB	Minimum benzene ring	Some bulky rings have H bonding properties

between this center and the amine group. It seems that the distance between the hydrogen bonding moiety and the chiral center can be very critical. Chlorpheniramine and dimethindene, which are 2-pyridine derivatives with one carbon between the hydrogen bonding group and the chiral carbon, show good chiral separation, whereas nicotine, which is a 3-pyridine derivate with two carbons between the H bonding and the chiral atoms, cannot be separated into enantiomers. The simple empirical rules cannot explain why some simple structural changes give rise to a loss of chiral separation. Terbutalin with a tertiary butyl substitutent on the amine group and hydroxyls in the 3 and 5 positions in the aromatic ring shows good chiral separation, whereas isoproterenol with an isopropyl substituent on the amine and hydroxyls in the 3 and 4 positions have so far not been separated into enantiomeric forms. Diethylpropion shows very low chiral separation in spite of having a keto group directly coupled to the chiral center and a benzene ring.

The character of the hydrogen bonding moiety seems not to be critical. Even weakly hydrogen accepting units such as nitrile, ether, and methoxyphenyl are sufficient to give good chiral separations under proper chromatographic conditions if the properties and positions of other important structures are favorable.

The influence of the structures on the chiral separations of anionic solutes are demonstrated in Table 8.20 (19, 20). 2-Phenylbutyric acid shows a remarkably high chiral separation with regard to the fact that a weakly hydrogen accepting phenyl is the only bonding moiety beside the carboxyl group. The only additional substituent at the chiral center is an ethyl group, and obviously this group is sufficient to improve the chiral selectivity since 2-phenylpropionic acid and 3-phenylbutyric acid with methyl groups in the same position have much lower separation factors. When the phenyl group is substituted for a phenoxy group, a considerable improvement of the chiral selectivity is obtained and an increase of the bulkiness of the aromatic group has the same effect. The strong influence of the naphthyl group in naproxen is particularly noteworthy.

The chiral separation of uncharged compounds has only been studied to a limited extent, but it seems that the structural requirements are analogous to those valid for charged solutes: the compounds should have two hydrogen bonding groups and a rigid or bulky structure close to the chiral center (see Tables 8.8 and 8.21) (17, 19).

Derivatization with nonchiral reagents can sometimes give rise to considerable improvements of the stereoselectivity (17). Some interesting results obtained with propanolamine derivatives used as betablocking agents are given in Table 8.21. The propanolamines have been transformed to oxazolidones by reaction with phosgene, and drastic improve-

Table 8.20. Separation of Enantiomeric Carboxylic Acids

Modifier	Solute	k_1'	α
Dimethyleethylamine 0.1M	2-Phenylbutyric acid	3.4	1.97
	3-Phenylbutyric acid	2.5	1.20
	2-Phenylpropionic acid (0-PA)	1.7	1.15
	2-Phenoxypropionic acid	1.3	1.99
	Ibuprofen (4-i-bu-0-PA)	8.3	1.90
Dimethyloctylamine 0.001M	2-Phenoxypropionic acid	1.9	1.73
	Ketoprofen (3-benzoyl-0-PA)	25.5	1.32
	Ibuprofen (4-i-bu-0-PA)	9.9	1.40
	Naproxen (6-methoxy-2-naphthyl-PA)	27.4	3.00

Solid phase: AGP 180mg/g silica.
Mobile phase: modifier in phosphate buffer, pH 7.0.

Table 8.21. Chiral Separation of Beta-Receptor-Blocking Propanolamine Derivatives

Solute	A as Oxazolidone		B Underivatized	
	k'_1	α	k'_1	α
Alprenolol	9.0	1.57	—	—
Metoprolol	3.0	1.95*	6.1	1.64*
Oxprenolol	28	1.64	11.2	1.25
Pindolol	4.6	1.65	—	—
Propranolol	27	4.7	27	1.12

Solid phase—AGP on silica (A: 150 mg/g; B: 180 mg/g); mobile phase—phosphate buffer $0.02M$, pH 7.0, containing modifier (A: 2-propanol 10%; B: NaCl $0.1M$ + ethanol 8%).
* No organic modifier

ments of the separation factors have been obtained in several cases: propranolol can after derivatization be separated into enantiomeric forms by a column with only about 50 theoretical plates. The high separation factors have been obtained in the presence of 10% 2-propanol, which is remarkable since such a high content of alcohol usually reduces the stereoselectivity.

The propanolamines lose their base properties by the transformation to oxazolidones, but a ring structure is obtained that comprises the chiral center and a hydrogen bonding moiety. Another example of a favorable effect of a transformation into uncharged form is given by mandelic acid. This hydroxy acid shows no stereoselectivity in underivatized form, possibly due to internal hydrogen bonding. However, after transformation to methyl or ethyl ester it can easily be separated into enantiomeric forms, as shown in Table 8.8 (17). Some examples of stereoselective separations are given in Figs. 8.8–8.12.

8.9. BIOANALYTICAL APPLICATIONS

The enantiomeric forms of a drug that interacts with an asymmetric receptor in the human organism can, from a pharmacological standpoint, be considered as different substances. Methods for determining the isolated enantiomers are therefore highly important for the development of an adequate therapy. Indirect techniques based on the preparation of diastereomeric derivatives have previously dominated, but it is often difficult to decide on the accuracy of such methods, as has been discussed.

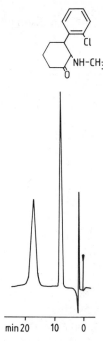

Figure 8.8. Resolution of racemic ketamine. Column: AGP 136mg/g silica (100 × 3.0mm i.d.). Mobile phase: 0.01M phosphate buffer, pH 7.0 with addition of 1.95mM N,N,-dimethyloctylamine. Flow rate: 0.5 mL/min.

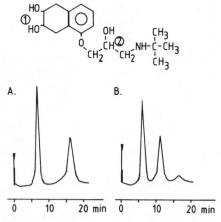

Figure 8.9. Separation of enantiomeric cations. Column: AGP 180mg/g silica (100 × 4.0 mm i.d.). Mobile phase: tetrabutylammonium bromide 0.001M in 0.02M phosphate buffer, pH 6.0. Solutes: A, Nadolol ($R,S;S,R$); B, Nadolol ($R,R;S,S$). Reprinted from Ref. 18 p. 651 by courtesy of Marcel Dekker Inc.

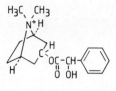

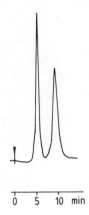

0   5   10   min

Figure 8.10. Separation of the enantiomers of methylhomatropine. Column: AGP 180mg/g silica (100 × 4.0 mm i.d.). Mobile phase: 0.003M tetrabutylammonium bromide in phosphate buffer, pH 6.0.

Direct methods based on the use of chiral selectors in the stationary or the mobile phase are usually preferable.

The AGP column has been applied in several studies of this kind (16, 22), and the determination of (R)- and (S)-disopyramide in human plasma (16) can be given as an example. The enantiomers of the unchanged drug and of the main metabolite (the monodesisopropyl homologue) are included in the determination. Disopyramide and the metabolite are first separated on a short precolumn containing LiChrosorb RP-2. The enantiomers of the two compounds are then separated on the AGP column and quantified by UV detector at 261 nm. Linear response was obtained in the range of 0.15–2.0 µg/mL. Chromatograms of authentic and spiked plasma samples are given in Fig. 8.13.

8.10. RETENTION MODELS

α_1-Acid glycoprotein has a negative charge at a pH higher than 2.7. The protolytic groups are of different kinds. Carboxylic groups are present in

Figure 8.11. Resolution of ethotoin enantiomers. Column: AGP 180mg/g silica (100 × 4.0mm i.d.). Mobile phase: 1% 2-propanol in 0.01M phosphate buffer, pH 7.15. Flow rate: 0.9 mL/min.

Figure 8.12. Resolution of racemic 2-phenoxypropionic acid. Mobile phase: 0.5% 2-propanol and 4.9mM N,N-dimethyloctylamine in 0.01M phosphate buffer, pH 7.0; Flow rate: 0.9 mL/min.

370

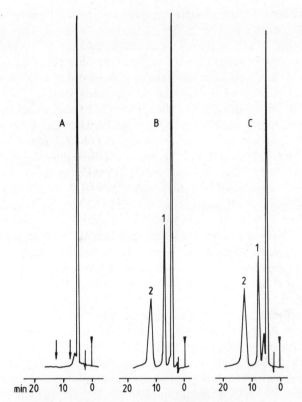

Figure 8.13. Isolation of enantiomers of disopyramide from plasma. Column: AGP 180mg/ g silica (100 × 3.0 mm id). Mobile phase: N,N-dimethyloctylamine 1.95mM and 2-propanol 4.3% in 0.02M phosphate buffer; ph 6.2. A, blank plasma; B, blank plasma spiked with racemic disopyramide 1.5 μg/mL; C, plasma sample obtained after administration of racemic disopyramide; peaks: 1—(R)-disopyramide, 2—(S)-disopyramide.

the 14 sialic acid moieties in the carbohydrate units, and the peptide chain contains 10–12 aspartic acid residues and a terminal serine group. Amino groups are present in the 13–15 arginine and histidine residues and in the hexosamine moieties. The cross-linking procedure that immobilizes the protein on the silica matrix has transformed a large part of the primary amino groups in the 16–17 lysine residues to secondary amines. The protein has a significant hydrophobic character owing to its content of hydrophobic amino acids, such as tryptophane, phenylalanine, leucine, and isoleucine, 45 residues in all.

The protein contains numerous asymmetric centers and there are groups with chiral binding abilities in the peptide chain as well as in the carbohydrate units. The stereoselective binding to the native form of AGP has been determined by chromatographic and nonchromatographic techniques (21, 23, 24).

An increase of pH increases the retention of cationic compounds, but the effect on the stereoselectivity depends on the structure of the solute, and a decrease as well as an increase have been observed (Tables 8.9 and 8.10). The retention of the anionic solutes presented in Table 8.11 decreases with increasing pH, but the effect on the stereoselectivity has so far not been studied in detail. The retention changes can be related to the protolysis of carboxylic and amine groups in the AGP phase, but pH can affect the binding properties of the protein also in other ways. It has thus been observed that the retention and the stereoselectivity of uncharged enantiomers increase with increasing pH (Table 8.8). It is well known that a change of pH can affect the conformation of proteins in solution, with changes in their binding properties as a result. It is probable that the observed pH effects on the AGP column, which without exception are reversible, have the same background. The protein is attached to the silica matrix by charge forces and cross-linking, and it is likely that the protein molecules still have a certain degree of mobility and can change their conformation.

Retention mechanism studies have been performed on cationic compounds such as metoprolol with a low concentration of tetrabutylammonium bromide in the mobile phase. The results show that the retention follows a model for ion pair retention to a homogeneous uncharged surface (18). Studies on uncharged compounds have indicated similar retention mechanisms: ion pairs from the mobile phase compete for the binding surface, which in this case seems to have two kinds of binding sites (19).

The highly complex structure of the binding protein can, however, give rise to quite different binding mechanisms. A study on atropine and homatropine with $0.01–0.05M$ octanoic acid in the mobile phase has shown that they are retained according to an ion exchange model (18). This might be due to a saturation of the binding surface with octanoic acid, but it is also possible that the acid has given rise to such conformational changes of the protein that its carboxylic groups now can interact with the solute molecules.

The assumption of a conformational change is supported by studies on ephedrine and pseudoephedrine presented in Fig. 8.4. They show that the addition of a low concentration of octanoic acid gives rise to a reversal of the retention order between the enantiomers of pseudoephedrine. A change of this kind can be due to a selective blocking of one of several chiral binding groups with different structures, but it is also possible that the added octanoic acid can induce such conformational changes that new chiral binding groups become available for interaction with the solutes.

The applicability of the AGP column to stereoselective separations of molecules of widely different charges indicates that the chiral binding can be due to different mechanisms. The remarkable effects of pH, temperature, and charged modifiers on the stereoselective properties of the AGP phase indicate that a conformational change of the protein can be one of the causes of the wide applicability.

8.11. CHOICE OF SEPARATION CONDITIONS

The AGP phase has a moderately hydrophobic character, which means that it cannot be applied to very hydrophilic or strongly hydrophobic compounds. Retention values that are suitable for analytical purposes can be obtained if the number of aryl or alkyl carbons are within the limits of 8–18 for amines and 8–15 for carboxylic acids. The exact limits depend of course on the number and character of other hydrophilic groups in the molecule.

The retention and the chiral selectivity can be regulated by the chromatographic conditions, and the possibilities of variations are so many that a systematic approach is highly needed. The following suggestions might serve as a guideline.

1. Start at ambient temperature using an aqueous mobile phase of 0.02M phosphate buffer, pH 7.0. Acids with a weakly hydrophobic character might need pH 6.0 to get sufficient retention.
2. If the retention is too high, add 1–8% of 2-propanol, depending on the hydrophobicity of the solute.
3. If no chiral separation is achieved, try instead a charged modifier such as
   a. Tetrapropylammonium bromide 0.003M or
   b. Octanoic acid 0.005M
4. If an incomplete separation is obtained, try
   a. Lower temperature (down to 5°C)
   b. Lower or higher concentration of the modifier
   c. pH lower than 7

Consult the literature for separation conditions for structurally similar compounds. Separation conditions are given for cations in Table 8.18, for anions in Table 8.20, and for uncharged compounds in Tables 8.8 and 8.21.

REFERENCES

1. E. Gil-Av, A. Tishbee, and P. E. Hare, *J. Am. Chem. Soc.*, *102*, 5115 (1980).

2. W. H. Pirkle and J. Finn, in J. D. Morrison, Ed., *Asymmetric Synthesis*, Academic, New York, 1983, pp. 87–124.

3. J. D. Adams, Jr., T. F. Woolf, A. J. Trevor, L. R. Williams, and N. Castagnoli, Jr., *J. Pharm. Sci.*, *71*, 658 (1982).

4. J. Hermansson and C. von Bahr, *J. Chromatogr.*, *221*, 109 (1980).

5. C. Pettersson, presented at the 10th Int. Symp. on Column Liquid Chromatography, San Francisco, CA, May 18–23, 1986.

6. H. Hähli and A. Mannschreck, *Angew. Chem.*, *89*, 419 (1977).

7. D. W. Armstrong and W. DeMond, *J. Chromatogr. Sci.*, *22*, 411 (1984).

8. G. Blaschke, *Angew. Chem.*, *92*, 14 (1980).

9. S. Allenmark, B. Blomgren, and H. Borén, *J. Chromatogr.*, *237*, 473 (1982).

10. J. Hermansson, *J. Chromatogr.*, *298*, 67 (1984).

11. V. A. Davankov, A. A. Kurganov, and A. S. Bochkov, in *Adv. Chromatogr.*, *21*, 71 (1984).

12. W. Pirkle et al., *J. Pharm. Biomed. Anal.*, *2*, 173 (1984).

13. K. Schmid in F. W. Putnam, Ed., *The Plasma Proteins*, Academic, New York, 1975, pp. 184–222.

14. J. Hermansson, *J. Chromatogr.*, *269*, 71 (1983).

15. J. L. G. Nilsson, J. Hermansson, U. Hacksell, and S. Sundell, *Acta Pharm. Suecia*, *21*, 309 (1984).

16. J. Hermansson, M. Eriksson, and O. Nyquist, *J. Chromatogr.*, *336*, 321 (1984).

17. J. Hermansson, *J. Chromatogr.*, *325*, 379 (1985).

18. G. Schill, I. Wainer, and S. Barkan, *J. Liq. Chromatogr.*, *9*, 641 (1986).

19. J. Hermansson and M. Eriksson, *J. Liq. Chromatogr.*, *9*, 621 (1986).

20. G. Schill, I. Wainer, and S. Barkan, *J. Chromatogr.*, *365*, 73 (1986).

21. J. Hermansson, *J. Chromatogr.*, *316*, 537 (1984).

22. D. Ofori-Adjei, Ö. Ericsson, B. Lindström, J. Hermansson, K. Adjepon-Yamoah, and F. Sjöqvist, *Therapeutic Drug Monitoring*, *8* 457 (1986).

23. U. K. Walle, T. Walle, S. A. Bal, and L. S. Olanoff, *Clin. Pharmacol. Ther.*, *34*, 718 (1983).

24. F. Albani, R. Riva, M. Cintin, and A. Baruzzi, *Brit. J. Clin. Pharmacol.*, *18*, 244 (1984).

CHAPTER

9

HIGH SPEED LIQUID CHROMATOGRAPHY

RICHARD C. SIMPSON*

Merck Sharp and Dohme Research Laboratories
Rahway, NJ

9.1. INTRODUCTION

The recent trend toward using packing materials of decreased diameter has resulted in a dramatic increase in the number of theoretical plates

---

* Current Address: *Department of Chemistry, University of Maryland, Baltimore County Campus, Baltimore, Maryland 21228.*

obtained per unit length of column, with values of well over 100,000 plates per meter routinely achieved. The use of standard-length (25- or 30-cm) columns which are packed with small-diameter (3- or 5-μm) materials provides the chromatographer with very high efficiencies, capable of resolving extremely complex sample mixtures into their individual components. However, many sample mixtures are not extremely complex and may require only 5000–10,000 theoretical plates to achieve adequate resolution. In such a situation, use of a column with an extremely high number of theoretical plates will result in excessive resolution of the sample components. Since resolution is achieved at the expense of time, the resulting separation will be unnecessarily long due to the column efficiency–sample matrix complexity mismatch.

However, the use of short (3- or 5-cm) columns of the same diameter and packed with the same small-diameter (3- or 5-μm) material will provide a reduced, but adequate, column efficiency and provide a closer match of the column efficiency requirements to the sample complexity. The resulting separation will still possess sufficient resolution and will be completed in a much shorter period of time.

Such an approach is the general basis for high speed liquid chromatography (HSLC) or "fast" LC. The use of short columns packed with 3- or 5-μm materials can provide efficiences comparable to 25-cm-length columns packed with 10-μm material. However, the separations obtained when using short columns are much more rapid due to the decreased column length. In this chapter some of the basic concepts and requirements for HSLC are discussed from both theoretical and practical viewpoints. A review of general advantages and disadvantages concludes the chapter. Abbreviations used throughout the chapter are defined in Table 9.1. They are listed in the order of introduction in the chapter.

9.2. THEORETICAL CONSIDERATIONS

9.2.1. System Variance

The effective number of theoretical plates in a chromatographic system can be described as

$$N = \frac{V_r^2}{\sigma_{sys}^2} \tag{1}$$

To maximize the effective number of theoretical plates, the system variance must be minimized. Assuming a Gaussian peak shape, the variance

Table 9.1. List of Symbols

HSLC	High speed liquid chromatography
N	Effective number of theoretical plates
t_r	Retention time
σ_{sys}^2	System variance
σ_{col}^2	Column variance
σ_{ex}^2	Extracolumn variance
d_c	Column diameter
ϵ	Column porosity (typical value, 0.7)
L	Column length
h	Reduced plate height
d_p	Diameter of packing material
σ_{inj}^2	Injection volume variance
σ_{det}^2	Detector variance
σ_t^2	Connecting tubing variance
V_s	Injection volume
K	Constant
V_{sm}	Maximum allowable injection volume
H	Height equivalent to a theoretical plate
$\sigma_{(inj)m}^2$	Maximum allowable injection variance
σ_{dv}^2	Detector cell volume variance
σ_{dt}^2	Detector time variance
V_d	Detector cell volume
F	Volumetric flow rate
V_{dm}	Maximum allowable detector cell volume
σ	Standard deviation of Gaussian peak; 4σ—peak width at base of triangulated Gaussian peak
t_d	Detector time constant
σ_t^0	Time standard deviation of nonretained peak
t_m	Elution time of nonretained peak
t_{dm}	Maximum allowable detector time constant
$\sigma_{(det)m}^2$	Maximum allowable detector variance
r	tubing radius
l	Length of connecting tubing
D_m	Solvent diffusion coefficient
θ^2	Maximum allowable fractional loss in efficiency
k'	Capacity factor
V_r	Retention volume
ΔP	Pressure drop across column
ϕ	Flow resistance factor
η	Mobile phase viscosity
u	Mobile phase linear velocity
C_m	Solute concentration at peak maximum
m	Mass of sample injected

Table 9.1. (*Continued*)

C_0	Original solute concentration
t_a	Data acquisition system time constant
IBW	Instrumental bandwidth
W	Peak width at half-height
CS	Chart speed

is an additive property and the system variance may be broken down into contributions from the column variance and the extracolumn variance,

$$\sigma_{sys}^2 = \sigma_{col}^2 + \sigma_{ex}^2 \tag{2}$$

Thus to minimize σ_{sys}^2, both σ_{col}^2 and σ_{ex}^2 must be minimized.

9.2.1.1. Column Variance

Equation 3 indicates that the column variance may be reduced by use of a short column which is well packed with small-diameter material,

$$\sigma_{col}^2 = \frac{(\pi d_c \epsilon)^2}{16} Lhd_p \tag{3}$$

When using commercially packed high speed columns, the chromatographer has little control over the variables in Eq. 3. However, commercial columns are generally of very high quality, with σ_{col}^2 typically on the order of 20 μL^2 or less, based on the volume of a nonretained peak.

9.2.1.2. Extracolumn Variance

Since there is little flexibility in decreasing σ_{col}^2, the chromatographer's efforts should be focused on minimizing σ_{ex}^2. As stated previously, σ_{ex}^2 is an additive property and may be broken down into its major components,

$$\sigma_{ex}^2 = \sigma_{inj}^2 + \sigma_{det}^2 + \sigma_t^2 \tag{4}$$

a. Injection Volume Variance. The value of σ_{inj}^2 has previously been described (1, 2) as

$$\sigma_{inj}^2 = \frac{V_s^2}{K^2} \tag{5}$$

where K is dependent on the injection technique. For a normal injection, K has a value of approximately 2. Guiochon and Colin (3) have subsequently shown that the maximum allowable injection volume is

$$V_{sm} = 0.25\sqrt{LH}\, d_c^2 \qquad (6)$$

or for a well-packed column with h approximately equal to 2,

$$V_{sm} \cong 0.36\sqrt{Ld_p}\, d_c^2 \qquad (7)$$

Thus as the column length and the packing material diameter are decreased, the maximum allowable injection volume must also be decreased. A combination of Eqs. 5 and 7 provides an estimate of the maximum allowable value of σ_{inj}^2,

$$\sigma_{(inj)m}^2 = \frac{V_{sm}^2}{K^2} = 0.0324\, Ld_p d_c^4 \qquad (8)$$

b. Detector Variance. Another contribution to the extracolumn variance is from the detector variance, which is the sum of the variances due to the detector cell volume and the detector time variance (4),

$$\sigma_{det}^2 = \sigma_{dv}^2 + \sigma_{dt}^2\, F^2 \qquad (9)$$

Guiochon and Colin (3) have shown that

$$\sigma_{dt}^2 = \frac{V_d^2}{F^2} \qquad (10)$$

and that the maximum allowable detector cell volume is

$$V_{dm} = \frac{V_{sm}}{K} \qquad (11)$$

If K has a value of 2 (as was assumed for Eq. 5), the maximum allowable detector cell volume is one-half the maximum sample injection volume.

A combination of Eqs. 7 and 11 expresses V_{dm} in terms of column length, packing material diameter, and column diameter,

$$V_{dm} = 0.18\sqrt{Ld_p}\, d_c^2 \qquad (12)$$

Thus as the column length and the diameter of the packing material are decreased, the volume of the detector cell must also be decreased (3). Kirkland et al. (5) have shown that the detector cell volume should be no greater than 0.1 of the peak band volume (4σ). The effect of flow cell volume on the system efficiency is illustrated in Fig. 9.1. The reduction in efficiency is more significant for early eluting peaks, which generally have narrower band volumes and are more susceptible to extracolumn variance effects than are later eluting peaks with larger band volumes. In the example shown, the early-eluting peak obtained with a 12-μL flow cell exhibits only 65% of the efficiency for the same peak obtained with the use of a 2.4-μL flow cell. The late eluting peak obtained with the use of a 12-μL cell displays 74% of the efficiency noted for the same peak in a system employing a 2.4-μL flow cell.

Since the solute band passes through the detector cell very quickly, the detector time constant must be small enough to enable the detector to respond rapidly to the brief presence of the analyte in the detector cell. Kucera (6) has shown that the detector time constant should be less than one-third of the time standard deviation of a nonretained peak,

$$t_d < \frac{\sigma_t^0}{3} \tag{13}$$

where the time standard deviation of a nonretained peak is defined as

$$\sigma_t^0 = \frac{t_m}{\sqrt{N}} \tag{14}$$

When dealing with retained peaks, the maximum allowable detector time constant can be calculated (3) by

$$t_{dm} = \frac{\theta t_r}{\sqrt{N}} \tag{15}$$

Thus the use of shorter columns (which reduces t_r) and smaller diameter packings (which increases N) requires the use of faster detector time constants to monitor accurately the separation actually achieved on the chromatographic column. Figure 9.2 clearly illustrates the effect of the detector time constant on the system efficiency. Note that the larger time constant results in a loss of sensitivity, resolution, and efficiency.

Assuming $\sigma_{dv}^2 = V_d^2$ combination of Eqs. 10 through 12 provides a means of calculating the maximum allowable value of σ_{det}^2 as

$$\sigma_{(det)m}^2 = \frac{\Theta^2 V_R^2}{N} + 0.0324 \, L d_p d_c^4 \tag{16}$$

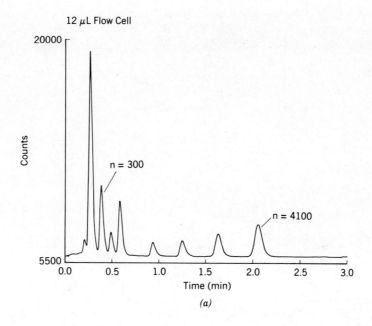

(a)

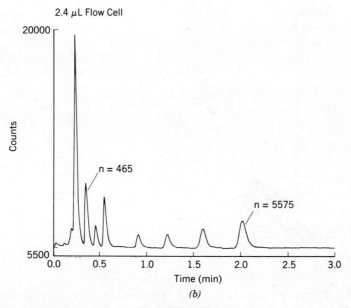

(b)

Figure 9.1. Effect of detector flow cell volume on system efficiency. (*a*) 12-μL flow cell. (*b*) 2.4-μL flow cell.

381

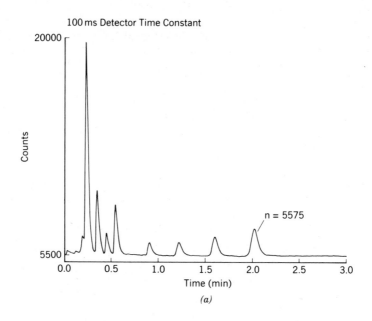

(a)

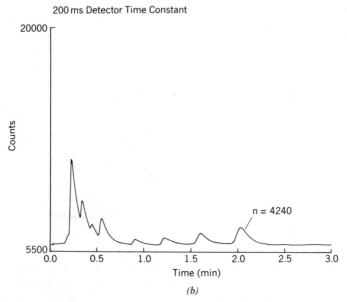

(b)

Figure 9.2. Effect of detector time constant on resolution, system efficiency, and sensitivity. Flow cell volume 2.4 μL. Both chromatograms recorded at same sensitivity. (*a*) 100-ms time constant. (*b*) 200-ms time constant.

where the first and second terms correspond to the maximum allowable values of $\sigma_{dt}^2 F^2$ and σ_{dv}^2 respectively. Equation 16 indicates that $\sigma_{(det)m}^2$ is directly proportional to both L and d_p.

c. Tubing Variance. The variance contribution from the connecting tubing used in the system is defined (3, 7) as

$$\sigma_t^2 = \frac{\pi r^4 l F}{24 D_m} \tag{17}$$

Scott and Kucera (8) have shown that although both the tubing's internal radius and its length contribute to σ_{det}^2, the radius is the major contributing factor. Thus the use of longer lengths of narrow bore tubing is preferred over the use of shorter lengths of wider bore tubing.

When the chromatographer must balance the tubing radius against the length, Eq. 18 has been shown (8, 9) to be a useful guide,

$$r^4 l \leq 6\theta^2 D_m \epsilon (1 + k') d_c^2 L \frac{t_r}{N} \tag{18}$$

Equation 18 indicates that as the column length is decreased, the tubing radius and/or length must also decrease. Studies (10, 11) have shown that in practice, the amount of band broadening achieved with high flow rates through short lengths of narrow bore tubing is less than expected; thus Eq. 18 provides a more stringent guideline for tubing length than is absolutely necessary.

9.2.2. Pressure Drop and Flow Rate

The general relationship between column length, efficiency, and packing material diameter can be expressed as

$$N = \frac{L}{H} \tag{19}$$

A reduction in d_p results in a decrease in H, while an increase in L results in an increase in N. Ideally both d_p (and thus H) and L could be balanced against each other to provide the optimum column efficiency for a given application. The result would be complete resolution of the analytes in a minimum time span. Unfortunately commercial manufacturers do not provide columns covering a continuous range or combination of H and L.

Rather, d_p is generally limited to 3, 5, or 10 μm and L is typically 3, 5, 10, or 25 cm. Thus the chromatographer may be forced to use a high speed column (L = 3 or 5 cm and d_p = 3 μm), which provides an excessive value of N and requires more than the minimal time to complete the entire separation.

To offset partially the resulting longer than optimum analysis time, the mobile phase flow rate may be increased. Relative to larger diameter packings, small d_p material exhibits less of an increase in H as the mobile phase linear velocity (and thus flow rate) is increased. As a result, flow rates much higher than those used in conventional HPLC may be used before H increases to an unacceptable level. The separation time required by the use of elevated flow rates is subsequently decreased due to two factors, (1) an increase in H to match more closely the optimum value for a given application, and (2) more rapid solute transit through the column as expressed by.

$$t_r = \frac{V_r}{F} \tag{20}$$

In practice, the maximum flow rate used is limited by the maximum pumping capacity (typically 10 mL/min) or the maximum operating pressure (typically 6000 psi) of the chromatographic system. The pressure drop across the column can be related (12) to column length, packing material diameter, mobile phase linear velocity, and viscosity by

$$\Delta P = \frac{\phi \eta L u}{d_p^2} \tag{21}$$

Smaller packing particles result in a higher pressure drop. This pressure drop is offset by the use of very short column lengths in HSLC; thus elevated flow rates can be used routinely. For example, a 3-cm × 4.6-mm column packed with 3-μm material typically produces a pressure drop of 2000–3000 psi with an aqueous-based mobile phase flow rate of 3 mL/min.

9.2.3. Sensitivity

One of the most important aspects of a separation is the sensitivity, or detector response, obtainable. Since chromatography is essentially a dilution process, the goal is to minimize postinjection dilution of the solute to maintain maximum sensitivity. The concentration of the analyte at the chromatographic band maximum may be described (3) by

$$C_m = \frac{m\sqrt{N}}{V_r 2\pi}$$ (22)

The dilution ratio of the analyte, that is, the on-column dilution factor, may be similarly described (13) by

$$\frac{C_m}{C_0} = \frac{V_s\sqrt{N}}{V_r\sqrt{2\pi}}$$ (23)

The use of small-diameter packing material will increase N and thus the analyte concentration at the band maximum. In addition, using short columns will result in increased values of C_m and thus dilution ratio. Therefore with short columns packed with small-diameter material there is less sample dilution, and a higher detector response (that is, increased sensitivity) results. Studies have shown that a 33×4.6-mm column packed with 3-μm material provides a three- to sixfold increase in detector response relative to that obtained with a 250×4.6-mm column packed with 10-μm material (14, 15).

Another key factor in achieving good sensitivity in high speed systems is the proper design and construction of the detector flow cell. The requirements for minimizing the flow cell volume have been discussed previously in this chapter. The path length of the flow cell should not be sacrificed as a means of decreasing the cell volume. Use of a longer path length will aid in achieving good sensitivities.

9.2.4. Data Acquisition

To collect accurately and process the detector output signal, the data acquisition system must operate at a time constant comparable to that of the detector. The maximum allowable time constant for the data acquisition system is described (7) by

$$t_a \leq \frac{\theta t_m (1 + k')}{\sqrt{N}}$$ (24)

Digital processing equipment will generally satisfy the requirement expressed by Eq. 24. Simple analog recorders are typically not capable of responding fast enough to the detector output, resulting in a decrease in the apparent operating efficiency of the high speed chromatographic system.

9.3. PRACTICAL CONSIDERATIONS

9.3.1. Instrumental Bandwidth

The preceding discussion has dealt extensively with the system variance. Often the chromatographer may find the measurement of extracolumn variance quite tedious. An easier parameter to measure and evaluate extracolumn band broadening is the instrumental bandwidth IBW. The instrumental bandwidth is equal to 4σ, or the baseline width of a triangulated Gaussian peak. The experimental measurement of the instrumental bandwidth is obtained by replacing the column with a zero dead volume union and measuring the peak width at half-height of an injected solute. Equation 25 converts the peak width to the instrumental bandwidth,

$$IBW = 1.7 \left(\frac{W}{CS}\right) F \tag{25}$$

IBW is expressed in microliters since W is in terms of centimeters, CS is in centimeters per minute, and F has units of microliters per minute. If IBW is much smaller than the column bandwidth, the loss in effective column efficiency is negligible. However, if IBW approaches or exceeds the column bandwidth, a significant reduction will occur in the effective column efficiency.

Short columns packed with small-diameter material produce very narrow column bandwidths. Thus high speed HPLC systems present more demanding requirements for minimizing IBW than do conventional HPLC systems. Dong and Gant (14) have stated that a high speed HPLC system should have an IBW value no greater than 20 μL. In comparison, a conventional HPLC system has a typical IBW value in the range of 75–150 μL (16).

Reduction of the instrumental bandwidth to an acceptable value may be accomplished by following a few simple guidelines. The use of minimal lengths of connecting tubing is mandatory. The ends should be square cut and deburred to obtain a zero dead volume connection. An easy way to ensure square-cut ends is to use short (that is, 5-cm) lengths of commercially precut tubing. The internal diameter of the tubing should also be minimized. Since an exponential relationship exists between the internal radius and volume, reduction of the tubing internal diameter by a factor of 2 will result in a fourfold decrease in internal volume. Thus the use of 0.005 or 0.007-in id tubing is required, with the smaller diameter tubing being preferred.

Another step to aid in minimizing the instrumental bandwidth is to custom swage each compression screw–ferrule set to the specific fitting with which it is intended to be used. Due to variations in casting and machining, all individual fittings are not necessarily identical. For example, minor variations in the swage depth are often encountered, creating small dead volumes which will contribute to the instrumental bandwidth. Fittings from different manufacturers are not necessarily interchangeable. One fitting that varies the most in swage depth is the column end fitting. The designs of various manufacturers possess different swage depths. Therefore column end fittings are not universal, and a mismatch of fittings will cause an increased instrumental bandwidth.

A convenient method of eliminating the requirement of a section of permanently swaged tubing for each column manufacturer is through the use of commercially available finger-tightened fittings which utilize reusable polymeric ferrules. The use of finger-tightened systems for the column inlet and outlet will permit a custom-swaged connection without exchanging sections of connecting tubing.

Sample loops should also be custom swaged with the actual injection valve to be used. In addition, the sample loop should always be reinstalled in the same configuration as it was originally swaged. For example, the side of the loop originally swaged into port 4 of the injection valve should always be reconnected to port 4 of that same injection valve.

9.3.2. Mobile Phase Filtration

The use of narrow bore connecting tubing mandates filtration of the mobile phase to prevent blockage of the tubing. Filtration of the mobile phase, which is especially important if buffers are involved, can be accomplished by using a 0.5 μm filter. Even the use of HPLC grade buffer salts does not eliminate the need for proper filtration procedures.

Use of a 0.5-μm in-line filter between the pump and the injector is also a prudent precaution. The in-line filter serves to trap particulates originating from within the pumping system. These particulates may be from a variety of sources but primarily consist of small particles from worn piston seals.

9.3.3. Guard Columns

The use of guard columns in conjunction with very short (that is, 3- or 5-cm) analytical columns may present a series of problems. A conventional length guard column (2.5–5 cm) results in a significant increase in the overall length of the combined column system. Thus the guard column may become substantially involved in the separation process. Such a situation is not desirable and the increased length of the column system is in direct conflict with the use of minimal-length columns.

Another potential problem is the introduction of additional band broadening. The band broadening may be due to two sources, (1) guard column void volume and (2) connecting tubing. Guard columns which are dry packed with pellicular material generally contain void volumes which are significant relative to the narrow peak widths obtained with short analytical columns. Thus the efficiency generated by the analytical column is not fully realized.

If possible, the use of guard columns with short analytical columns should be avoided. Cartridge column systems reduce the replacement cost of the analytical column significantly. Thus if the lack of a guard column results in decreased analytical column lifetime, the more frequent replacement of the analytical cartridge is not prohibitively expensive. If a guard column is absolutely necessary, the guard column should be narrow bore and as short as possible, thus reducing its void volume and contribution to the separation process. Since injection volumes are small for high speed columns, the capacity of the guard column does not have to be as great as is required in conventional HPLC systems.

9.3.4.　Saturator Columns

High mobile phase flow rates result in the passage of large volumes of mobile phase through the analytical column over relatively short time periods. The result may be a higher rate of dissolution of silica-based packings than is encountered in conventional HPLC systems. The problem may become even more significant when using aqueous-based mobile phases at pH values near or above 7, due to the increased solubility of silica at higher pH values. To reduce the rate of dissolution of the silica, a saturator column may be beneficial. A saturator column is a short column dry packed with large-diameter silica. The column is placed between the pump and the injector. The mobile phase becomes "presaturated" with silica due to passage through the saturator column. Thus the rate of dissolution of silica in the analytical column may be significantly reduced, resulting in extended analytical column lifetime.

9.3.5.　Delay Volume

To fully realize the strength of HSLC during gradient separations, the delay volume should be minimized. The delay volume is simply the total volume between the pump and the column head. Contributions to the delay volume are from the connecting tubing and mixing chambers. Reduction of the delay volume allows the gradient profile to reach the column more rapidly and become involved in the chromatographic process. The result is a faster response of the separation to the gradient program. The delay volume should be reduced to a value that still ensures adequate mixing of multiple-solvent mobile phases. DiCesare et al. (17) suggested that a delay volume of 0.5 mL be employed in an HSLC system.

9.3.6. Column Lifetime

One of the most practical factors to consider is the column lifetime. Various investigations (18–20) have demonstrated that the lifetimes of very short columns are comparable to those of conventional length columns. Precautions that should be taken are proper mobile phase filtration, sample preparation, and use of a saturator column. Applying an in-line filter between pump and injector is also useful for the removal of particulates generated by pump-seal wear.

9.3.7. Precision

The precision of the analytical results may vary in different HPLC assays. However, studies have shown that the analytical precision obtained with HSLC is comparable to that obtained with conventional HPLC (21, 22).

9.3.8. Data Processing

Through the use of HSLC, very large amounts of data may be generated in short time periods. Therefore the use of a rapid data processing system is essential. Generally data acquisition, processing, and output are performed by a digital computer system adapted for chromatographic applications. Such systems generally allow the chromatographer to manipulate the graphics and data processing to obtain the type and amount of information required.

9.4. ADVANTAGES AND DISADVANTAGES

As with every analytical technique, HSLC has several inherent advantages and disadvantages.

9.4.1. Advantages

The primary advantage of HSLC is that high quality separations can be obtained in very short times, often under 2 min. The combination of rapidity and good resolution makes HSLC valuable for use in methods development.

Due to the use of small-diameter packings, the resolution of chromatographic bands is excellent, and the mass sensitivity is generally fourfold better than in conventional HPLC. The increase in mass sensitivity makes HSLC especially useful for applications where the availability of samples and the amount of each sample are quite limited, for example, with pediatric serum samples.

The analytical precision of HSLC is comparable to that of conventional HPLC. Thus HSLC may be used to obtain accurate quantitative measurements rapidly.

An additional advantage is the reduced cost per analysis in HSLC. The savings are due to three factors, (1) reduced solvent consumption, (2) reduced column cost, and (3) reduced analysis time.

9.4.2. Disadvantages

The most severe disadvantage of HSLC is the rigid requirement for a properly designed chromatographic system. Factors requiring consideration include dead volume, injection volume, detector cell volume, and detector response time. Failure to optimize these factors will generally result in a poor-quality chromatographic separation.

Another point to be considered is the limited number of theoretical plates generated by a short column. Thus for extremely complex separations, HSLC may not be appropriate and use of a longer column is mandated.

9.5. CONCLUSIONS

HSLC is being used in an increasing variety of applications, such as pharmaceutical analyses (14, 23), biochemical assays (14, 24), and priority pollutant monitoring (25). At the present time, the major limitations to HSLC are the rigid requirements for small IBW values, rapid detector response times, and high speed data acquisition and processing. As these obstacles are overcome, high speed liquid chromatography will mature and continue to emerge as a powerful analytical technique.

REFERENCES

1. J. F. K. Huber, J. A. R. J. Hulsman, and C. A. M. Meijers, *J. Chromatogr.,* *62,* 79–91 (1972).

2. B. L. Karger, M. Martin, and G. Guiochon, *Anal. Chem.*, *46*, 1640–1647 (1974).

3. G. Guiochon and H. Colin, in P. Kucera, Ed., *Microcolumn High Performance Liquid Chromatography*, *J. Chromatogr. Libr.*, *28*, Elsevier, Amsterdam, 1984, chap. 1.

4. J. L. DiCesare, M. W. Dong, and J. G. Atwood, *J. Chromatogr.*, *217*, 369–386 (1981).

5. J. J. Kirkland, W. W. Yau, H. J. Stocklosa, and C. H. Dilks, Jr., *J. Chromatogr. Sci.*, *15*, 303–316 (1977).

6. P. Kucera, *J. Chromatogr.*, *198*, 93–109 (1980).

7. R. A. Hartwick and D. D. Dezaro, in P. Kucera, Ed., *Microcolumn High Performance Liquid Chromatography*, *J. Chromatogr. Libr.*, *28*, Elsevier, Amsterdam, 1984, chap. 3.

8. R. P. W. Scott and P. Kucera, *J. Chromatogr.*, *125*, 251–263 (1976).

9. M. Martin, C. Eon, and G. Guiochon, *J. Chromatogr.*, *108*, 229–241 (1975).

10. M. J. E. Golay and J. G. Atwood, *J. Chromatogr.*, *186*, 353–370 (1979).

11. J. G. Atwood and M. J. E. Golay, *J. Chromatogr.*, *218*, 97–122 (1981).

12. L. R. Snyder and J. J. Kirkland, *Introduction to Modern Liquid Chromatography*, 2nd ed., Wiley, New York, 1979, chap. 2.

13. L. R. Snyder and J. J. Kirkland, *Introduction to Modern Liquid Chromatography*, 2nd ed., Wiley, New York, 1979, chap. 13.

14. M. W. Dong and J. R. Gant, *LC Mag.*, *2*, 294–302 (1984).

15. R. C. Simpson, P. R. Brown, and M. K. Schwartz, *J. Chromatogr. Sci.*, *23*, 89–94 (1985).

16. J. L. DiCesare, M. W. Dong, and L. S. Ettre, *Chromatographia*, *14*, 257–268 (1981).

17. J. L. DiCesare, M. W. Dong, and L. S. Ettre, *Introduction to High Speed Liquid Chromatograhy*, Perkin-Elmer Corp., Norwalk, CT, 1981, chap. 4.

18. J. L. DiCesare, M. W. Dong, and J. R. Gant, *Chromatograhia*, *15*, 596–598 (1982).

19. R. W. Stout, J. J. DeStefano, and L. R. Snyder, *J. Chromatogr.*, *261*, 189–212 (1983).

20. J. G. Atwood, G. J. Schmidt, and W. Slavin, *J. Chromatogr.*, *171*, 109–115 (1979).

21. J. L. DiCesare, M. W. Dong and L. S. Ettre, *Introduction to High Speed Liquid Chromatography*, Perkin-Elmer Corp., Norwalk, CT, 1981, chap. 3.

22. E. Katz and R. P. W. Scott, *J. Chromatogr.*, *253*, 159–178 (1982).

23. J. Y.-K. Hsieh, B. K. Maglietto, and W. F. Bayne, *J. High Resol. Chromatogr. Chrom. Commun.*, *9*, 392–396 (1986).

24. J. A. Reinhard and R. Perry, *J. Liq. Chromatogr.*, *7*, 1211–1220 (1984).

25. M. W. Dong and J. L. DiCesare, *J. Chromatogr. Sci.*, *20*, 517–522 (1982).

CHAPTER

10

A THEORETICAL APPROACH TO DERIVATIZATIONS FOR HPLC

IRA S. KRULL, STEPHEN T. COLGAN, AND CARL M. SELAVKA

Department of Chemistry and The Barnett Institute
Northeastern University
Boston, Massachusetts

10.1. INTRODUCTION TO PRE- AND POSTCOLUMN DERIVATIZATIONS

Another publication on derivatizations for improved identification and detection in liquid chromatography (LC) is clearly unnecessary in view of the relatively large number of existing texts and reviews on the subject (1–7). Most reviews of derivatizations for gas chromatography (GC) or LC have tended to summarize published reports, including specific reagents, reaction conditions, the nature of the derivatives, percent yields, and so on. Since voluminous texts already exist for the chromatographer, often itemized according to analyte class or derivatizing reagent, it is not our intention or desire to repeat that which is already in the chemical literature. Rather, what appears to be of more interest and need is a theoretical discussion of the general area of derivatizations for LC. How does one theoretically discuss derivatization approaches without becom-

393

ing unnecessarily encumbered with specific examples? We believe that the approach presented here is unique, and that it should prove useful to most chromatographers, whether they use LC or GC. Perhaps the best attempt to do this in the past was that by Lawrence (4).

To begin with, we need to define what is meant by pre- or postcolumn derivatization, and then how on-line or off-line positioning fit into the scheme. Precolumn derivatizations are performed prior to sample introduction, that is, prior to the injection valve of the LC system. While precolumn once meant that derivatizations were performed away from the LC entirely, in an off-line fashion, several precolumn off-line automated approaches have become very popular (8, 9). In addition, derivatizations may be performed on-line, after the injection, before the analytical column. While most off-line or on-line precolumn derivatizations are meant to be completed prior to the analytical column, it is also possible to have the derivatizations occur in situ, that is, upon injection into a mobile phase containing a derivatization reagent. There are some theoretical advantages to in situ type derivatizations, but these methods have not, as yet, been widely used by chromatographers. This area is discussed further below.

Postcolumn on-line reactions have been widely utilized within the past decade, and the basic principles and approaches have been described (1). Such methods may involve the addition of a separate derivatization solution to the eluent, in a low dead volume mixing device after the analytical column, or flow through a solid phase reactor. For homogeneous (solution) reactions, a separate reaction coil or delay device is often inserted after the mixing device in order to effect a semiquantitative or quantitative conversion of the original analytes into final detector-suitable derivatives. Reaction chambers are configured differently for various delay times, and may be subjected to temperature control for optimal reactions. The final derivatives are passed directly into one or more detectors. Alternatively, a liquid–liquid extraction can be performed on the eluent to remove derivatives or ion pairs. On-line extraction displaces the desired derivative from the reaction solution and puts it into a new, usually organic, solvent for improved detector selectivity and sensitivity. A slight variation on this approach has been to introduce derivatization reagents into the organic extraction solvent, thereby extracting and simultaneously derivatizing/chelating the analytes in an organic solvent before final detection. A number of postcolumn on-line reaction configurations are described.

Postcolumn off-line derivatizations in LC are possible, but they require the removal of each analyte from the LC system, followed by an off-line derivatization step and introduction of the final derivative solution into a suitable detector. There are substantial increases in operator involvement, sample manipulation, time, and equipment, the results of which include a decrease in system reproducibility and throughput, as well as an increase in the cost per sample. For these reasons, the off-line postcolumn approach to LC derivatizations has never become widely employed; this probably will not change very much in the future.

Whereas almost all steps in derivatizations for chromatography 20 years ago involved manual intervention, a growing number of these have now been automated. Robotic automation of off-line derivatizations for LC is now routine, as evidenced by the various commercial systems available for this purpose (10). Table 10.1 summarizes the various derivati-

Table 10.1. Summary of Derivatization Approaches in LC

Approach Taken	Description
1. Precolumn, off-line	Derivatization is performed before the injection, using apparatus which is not part of the LC system or its peripherals.
2. Automated, precolumn, off-line	Derivatization occurs off-line, in a continuous fashion, in an LC system peripheral, before the injection step, controlled by an autosampler or robotics.
3. Precolumn, switched "off-line"	Derivatization is performed after injection, before the LC column, using column switching techniques.
4. Precolumn, on-line	Derivatization occurs after the injection, before the analytical column.
5. Precolumn, in situ	Derivatization occurs upon injection into mobile phase containing reagent(s), prior to separation.
6. Postcolumn, in situ	Derivatization occurs postcolumn, usually as a function of increased temperature or addition of reagents or light.
7. Postcolumn, off-line	Derivatization occurs after the analytical column, before detection, away from the LC system.
8. Postcolumn, on-line	Derivatization occurs after the analytical column, before detection, as a part of the LC system.
9. Extraction, on-line	Derivatization occurs using on-line postcolumn liquid–liquid extraction, before final detection.
10. In detector, on-line	Derivatization occurs after the first detector, on-line, followed by another detection step, that is, derivatization occurs between two detectors.

zation approaches possible. Most of these are discussed in greater detail in the remainder of the chapter, along with some specialized reaction detection methods.

10.2. THEORY, PRINCIPLES, AND DESIGN OF PRECOLUMN DERIVATIZATIONS

In this section we describe and discuss the basic theory, principles, and design of off-line and on-line precolumn derivatizations for HPLC. We will not describe any specific reactions or reaction conditions or reagents, but rather the theory of how to design and implement these approaches. There are certain fundamental reasons for performing precolumn derivatizations in LC, as outlined in Table 10.2.

Off-line approaches can utilize almost any solvent, time, temperature, or pressure conditions to approach completion (100% reaction). It is not necessary for any derivatization to provide 100% of the theoretical yields as long as there is good sample-to-sample reproducibility. Quantitative conversion, however, is desirable when the product of the reaction has better detection properties than the original analyte.

Nonautomated approaches to off-line precolumn derivatizations require operator attendance, time, effort, and manual manipulations. However, as robotics utilization increases in off-line derivatizations, these operational limitations will change drastically. Advantages of nonautomated approaches include nearly unlimited flexibility in reaction conditions, and the ability to use harsh reagents or stringent reaction conditions (such as inert atmospheres) that are difficult during automation. The conditions for off-line reactions must be selected to convert as much of the analyte to a single desired product as possible, in as short a time as is

---

Table 10.2. Justifications for Employing Precolumn Derivatization

1. Improved chromatographic performance of a derivative versus the underivatized analyte
2. Improved chromatographic separation of derivatized analyte from potential detector interferents
3. Reduced sample cleanup and preparation
4. Improved sensitivity and lowered detection limits
5. Improved detector selectivity for derivatized analytes as compared to that for sample components
6. Optical isomer separation without chiral columns or mobile phase additives
7. Confirmation of a preliminary identity assignation

---

practical, with a minimum of operator attendance and reagent/solvent consumption. Also, the final reaction solution should contain a minimum amount of unconsumed derivatization reagent, so that it does not contaminate the analytical column or overload the detector. For this reason, a preliminary off-line or on-line sample cleanup step may be necessary to separate underivatized reagent from final derivatives. However, more than a single derivative may, at times, be useful (11, 12). The derivatization solvent should be miscible with the LC mobile phase and, under reversed-phase conditions, should be of equal or higher polarity, so that the derivative will not be injected in a solvent of higher eluotropic strength. With normal phase LC, a less polar solvent for the derivatization is desirable. By observing these solvent precautions, the possibility of generating "solvent effects," which degrade chromatographic efficiency and resolution, is greatly reduced. If the reaction solvents are immiscible with the mobile phase, the solvents needed for the derivatization can be removed, off-line, under a stream of inert gas and mild heat. The residue (derivative) can then be reconstituted in a mobile phase compatible solvent. In addition, some type of sample preconcentration upon injection may be desirable to overcome the dilution effects of the off-line derivatization conditions. This can be accomplished through the use of gradient elution, column switching, or by injecting samples in an eluotropically weak chromatographic solvent.

Automated precolumn derivatizations have the same basic requirements and limitations as nonautomated techniques. However, automation generally increases throughput and decreases operator attendance and analysis cost. In principle, automation should have been widely employed and described for derivatization in LC, but quite the opposite is the case. Perhaps the necessity of extra instrumentation, method codification and sophistication, or method development, has dissuaded many from employing this approach. As the hardware and software become less expensive and easier to use, automated techniques should become more prevalent.

On-line precolumn derivatizations may be accomplished using either homogeneous or heterogeneous approaches. Homogeneous derivatizations occur in a single phase, that is, the derivatization solution is miscible with the HPLC mobile phase. Precolumn homogeneous derivatizations require the mixing of a reagent stream with the injection stream, compatibility of solvents, and compatibility of reagent solution with column packing material; they lead to dilution of the sample, and may require the use of delay times and elevated temperatures for quantitative conversion. Although this approach is possible, it has rarely been used.

Heterogeneous derivatizations, on the other hand, involve reaction of an analyte in solution with a solid supported or solid phase immobilized reagent that does not dissolve in the mobile phase. On-line precolumn solid phase derivatization approaches are currently of great interest, especially those incorporating immobilized enzymes (1). These methods appear to offer several real advantages, while concurrently presenting several limitations on the composition of the mobile phase. In this approach, a small precolumn is packed with the immobilized reagent. Following injection, no dilution with added solvents takes place, very little dead volume is added, no additional pumps or related hardware are required, and the reactor can be switched rapidly into or out of the analytical line, as needed. Microprocessor-controlled automated switching valves for this application are commercially available. Also, a delay time, with valve switching, can be used to improve derivatizations, as well as the use of elevated reaction temperatures. This is an example of a precolumn switched "off-line" derivatization in LC. Specific advantages and disadvantages of solid phase reagents (SPRs) are described in detail later in the chapter.

10.3. THEORY, PRINCIPLES, AND DESIGN OF POSTCOLUMN DERIVATIZATIONS

In the first decade of LC development, a wide variety of derivatizations were performed using off-line precolumn approaches. Disadvantages were numerous and obvious, and though some attempts were made to use postcolumn off-line techniques, such tactics were less than ideal. These disadvantages are summarized in Table 10.3.

Such "failures" naturally led to the formulation of on-line postcolumn derivatizations. Many related publications and commercial systems have

Table 10.3. Major Deficiencies in Off-Line Derivatizations for Liquid Chromatography

1. Dilution of sample with reagent solutions, before or after the analytical column, and loss of chromatographic resolution in the postcolumn mode
2. Additional sample handling and operator intervention and attention, with possible sample loss and/or contamination
3. Inability to readily automate off-line derivatizations until the introduction of sophisticated robotics
4. Need for additional instrumentation and sample manipulation
5. Lack of continuous operation, batch-mode requirements
6. Increased error (decreased precision) in quantitative analytical measurements

been introduced during the past decade, and a number of reviews already exist summarizing this area, including specific reactions and conditions (1, 13–15). In the postcolumn mode, since the analytes have already passed through the analytical column, a derivatization will not change the chromatographic properties of the reactive analytes. Thus postcolumn reactions are usually less specific than their precolumn counterparts. Postcolumn derivatizations are generally engineered to increase the sensitivity of an analysis. Any improvements in specificity are solely a result of the specificity of the reaction chemistry involved and the final detection method. However, it is our opinion that the postcolumn on-line derivatization approach represents the state of the art in derivatization for LC, and is where most of the current research and development work exists. In the following section, the theoretical requirements of on-line derivatization methods are presented.

10.4. DESIGN OF ON-LINE DERIVATIZATIONS FOR LIQUID CHROMATOGRAPHY

In all on-line approaches, band broadening or peak dispersion problems must be a prime consideration. The extent of this band broadening is generally measured as an increase in the variance of the analyte peak (13–18) and results in a loss of chromatographic resolution and a loss in sensitivity when peak height is used for quantitation. Thus an important requirement for the design of all on-line work is the minimization of extracolumn variance. Since extracolumn variance can arise from a number of sources (19), all on-line derivatizations require consideration of several experimental parameters, which are enumerated in Table 10.4.

Table 10.4. Summary of On-Line Derivatization Requirements in Liquid Chromatography

1. Minimization of the length and internal diameter of connective tubing
2. Zero or low dead volume unions, mixing tees/chambers, and reaction chambers
3. Minimization of reagent pump pulsations, with constant reproducible reagent delivery
4. Nondetectability of excess unreacted reagents introduced into the analytical stream
5. Compatibility of derivatization reagents and solvents with mobile phase solvents, pH, ionic strength, buffer composition, and so on
6. Minimization of analyte peak dilution (careful control of solvent to mobile phase ratios)

While Table 10.4 indicates the LC requirements of on-line derivatizations, there are also sample/analyte needs that must be met. A reagent must be chosen that will react selectively with the analytes of interest, under the mobile phase/derivatization solvent conditions used, within reasonable LC time constraints (1–5 min or less), and with a reproducibly high degree of completion. In the precolumn mode, the excess reagent solution must not degrade or modify the analytical packing, and must not change the eluotropic strength of the mobile phase. It is also desirable that no excess reagent solution appear as elevated detector baseline noise or detector spikes, and reactants must be resolved from the analyte's derivative. In the on-line postcolumn mode, excess reagents will often enter the detector along with each analyte, and thus detector selectivity toward the two should be very different. Ideally, the derivatization reagent will not have any detector response. Although a number of useful reagents already exist that fulfill these requirements, the analytes for which they are applicable are limited, making immobilization of the reagent in the on-line mode an advantageous approach (20, 21).

There are different types of reactor configurations available for postcolumn on-line homogeneous derivatizations, and the type chosen is generally determined primarily by the reaction time needed. For relatively fast reactions (up to 1 min), knitted open tubular (KOT) reactors are generally sufficient; reactions of intermediate reactivity (needing from 1 to 5 min) are often performed using packed bed reactors or KOTs, while long reactions (more than 5 min) are generally best performed using air segmented or solvent segmented reactors. The advantages and deficiencies of each reactor type have recently been reviewed (1, 3).

10.5. SOLID PHASE DERIVATIZATIONS IN LIQUID CHROMATOGRAPHY

As mentioned previously, the use of homogeneous derivatization approaches, in either the pre- or the postcolumn mode, places rather stringent constraints on the reaction chemistry and conditions available to the chromatographer. Table 10.5 summarizes some of the limitations of on-line homogeneous derivatizations.

Heterogeneous reactions may be employed to avoid a number of the problems encountered with homogeneous approaches. Table 10.6 lists some of the more obvious advantages of solid phase (supported, covalently bound, or ionically bound) reagents, on-line, pre- or postcolumn (21).

There are certain operational requirements for any on-line, solid phase

Table 10.5. Limitations of On-Line Homogeneous Derivatization in Liquid Chromatography

1. Requires the use of additional instrumentation, including pumps, tubing, connective fittings, mixing chambers, and reaction chamber
2. Introduces additional extracolumn dead volume that increases system variance and may deteriorate chromatographic resolution and sensitivity
3. Increases overall time and cost for each analysis
4. Provides a possible source of contamination via added reagents and solvents
5. Introduces underivatized excess reagents into the analytical stream, which often raise background noise level or generate extraneous large peaks that may obscure the derivative peak
6. Is difficult to use in the precolumn mode without sophisticated instrumentation, so is often restricted to postcolumn applications

or immobilized reactor in HPLC (21). For any on-line solid phase reagent to be practical in HPLC, it must be chemically stable to the mobile phase solvent, salts, pH, and any metal additives. Instability will lead to rapid degradation of the reagent, loss of reactivity, and rapid deactivation of the reactor. Immobilized enzymes are often denatured and inactivated in the presence of high organic solvent concentrations. However, a small amount of an organic modifier can, at times, stabilize an enzyme. An ideal solid phase reagent is stable for long periods of time, over wide temperature ranges, to large volumes of various mobile phase compositions, ranging from purely aqueous to purely organic. Another important, often limiting, requirement is that the mobile phase solvent be suitable for the desired derivatization. It is often the case that optimal separation and derivatization conditions cannot be met easily, with a single solution composition, for all analytes. Though immobilized or solid phase reagents may, at times, be incompatible with the particular mobile phase needed, there have been examples where this limitation was not a problem (22–25). In addition, solid phase reagents must not collapse under pressures that may rise to 6000 psi in the precolumn mode. Experience has shown that reagent bed collapse can be more prevalent with polymeric (rather than silica) supported reagents, but the recent development of highly cross-linked polymeric supports seems to have solved this problem.

Any solid phase or solution reactor system must not contribute excessively to the extracolumn variance. Ideally, heterogeneous reagents should be based on small particle size materials having narrow particle size distributions, which are impervious to the mobile phase and are efficiently packed into short, small internal diameter guard-type columns. With shorter bed lengths, variance and band spreading are minimized,

Table 10.6. Advantages of On-Line Solid Phase Derivatization in Liquid Chromatography

1. Does not require additional instrumentation, hardware, mixing chambers, or reaction chambers; only requires the solid phase reactor, a dummy reactor (support alone, no reagent added), and fittings.
2. Does not introduce any additional extracolumn dead volume other than that normally introduced with a conventional guard column.
3. The increases in overall cost per analysis are usually lower than those incurred with the corresponding homogeneous reactions.
4. Does not increase overall time for each analysis, unless the reaction is displaced from the analytical stream, using column switching techniques (when longer reaction times are required).
5. Introduces a significantly lower amount of impurities and derivatizing reagent into the eluent.
6. Can be used in either the pre- or the postcolumn mode, resulting in differences in chromatography (pre-) or differences in detectability (post-).
7. Does not require any additional sample handling or manipulations; requires only the direct injection into the HPLC.
8. Can be engineered to be compatible with a wide variety of mobile phases for both normal and reversed-phase HPLC.
9. Reactions that occur on a solid support can be more selective and often give fewer side products as a result of a tailor-made microenvironment near the support surface.
10. Reactions can be made to be more "user friendly" through the immobilization of an unsafe or undesirable reagent onto a solid support.
11. Immobilization of reagents can result in a more stable preparation that can retain activity for a longer period of time than homogeneous reaction reagent solutions.
12. Reactions that are only possible in solution at high dilution, due to the low solubility of a reagent, may be carried out in relatively high concentrations on the support.

and with attention to detail, reasonable reaction times (up to 2 min) may be obtained. End fittings, end frits, and tubing connections must always be chosen to minimize unnecessary dead volume.

The lifetime of any solid phase reagent is dependent on the reagent loading (equivalents per gram) of support, the availability of this reagent for reactions, the amount of final solid phase reagent packed into the reactor column, and its stability to mobile phase conditions. Ideally, the final lifetime should be on the order of hundreds or even thousands of reactions/derivatizations so that there is no daily or weekly need to change the reaction column. At the present time, this optimistic goal has not been accomplished. The desire to produce a solid phase reagent which exhibits both high reactivity and long lifetime may continue to lead to compromises of the latter feature. However, the use of column switching, to remove the solid phase reactor from the eluent stream when it is not needed, may bring this dual goal closer to fruition. Finally, changing the reaction column should be as simple and fast as possible to minimize downtime of the HPLC system.

One last point which should be emphasized is that the solid phase reagent and its underlying support must not irreversibly bind, adsorb, or partition the analyte or its derivative. These deleterious properties lead to band spreading, tailing peaks, loss of resolution, loss of overall efficiency, and increase in detection limits. Reactions on the support surface can generate new sites that have adsorptive or ion exchange properties toward the next injection of analyte (26). Methods must be devised to overcome such potential problems, either by reactor and support design, or through mobile phase modification. Solid phase reagents have also been used for off-line derivatizations. In such a configuration, they have several advantages over off-line solution reactions. These advantages are indicated in Table 10.6, items 9–12.

10.6.   MICROCOLUMN DERIVATIZATIONS IN LIQUID CHROMATOGRAPHY

Recently there has been a considerable amount of interest in the area of narrow bore HPLC systems. This has stemmed from the desirability of interfacing LC with gas phase detectors such as mass spectrometers, photoionization detectors, and inductively coupled plasmas. The problems of the LC–detector interface are minimized if microbore columns, with their inherent low flow rates, are used. Often the total column effluent can be directly introduced into the detector. Selectivity and sensitivity also can be improved when using ultraviolet/visible (UV/VIS), fluorescent (FL) or other popular LC detectors. There are other advantages to microbore systems, including savings when using expensive reagents, the compactness of the unit, and their usefulness for sample-limited applications.

Although all of the existing derivatization methodology, in principle, could be applied to microbore systems, the effects of band broadening from reaction detectors are significantly more problematic with narrow bore columns. Reaction times must be kept to a minimum while mixing chambers, heaters, and cell volumes all must be as small as possible to maintain chromatographic performance. With these problems it is not surprising that most of the derivatization methodology to be used in conjunction with microbore columns has involved off-line precolumn reactions. This area has recently been reviewed (27). A major source of band broadening in on-line microbore derivatizations is the mixing chamber incorporated in the instrumental design. Of course this problem is largely eliminated if solid phase reactors, which require no reagent mixing, are used. However, the choice of a packed bed reactor, KOT reactor, or segmented stream reactor for homogeneous derivatizations depends on the chemistry involved, the reaction rate, and the maximum allowable pressure drop. A review of postcolumn reactors for miniaturized HPLC has recently been prepared (28).

10.7. THEORY, PRINCIPLES, AND DESIGN OF IN SITU DERIVATIZATIONS

In situ derivatization has been used far less than the other methods described, seemingly in spite of its intrinsic worth and value. The method involves the addition of the derivatizing reagent to the mobile phase, which is then constantly pumped through the analytical system. Upon injection of a sample, three modes of in situ derivatization may be performed, (1) an instantaneous reaction can occur before the analytical column, (2) a reaction can be forced to take place on injection, before the column, by having a separate heated delay device in place, or (3) the analytes can be chromatographed in their native forms and then forced to react postcolumn, in a reaction coil, to effect better detectability. Metal complexation before reversed-phase separation is a typical example of the first mode of operation. The improved resolving power of modern reversed-phase packings over cation exchangers is used to advantage for trace metal determinations. The second mode is similar to an on-line precolumn reaction, but the reagent is already present in the mobile phase. In general this mode presents more problems with respect to band broadening, and may necessitate the use of column switching and gradient elution to preconcentrate the derivatized forms of the analytes before chromatography. In the third mode, a heated chamber can be placed postcolumn in order to perform a derivatization reaction which is kinetically unfavorable during the separation process (that is, without heating). Actual examples of all of these variations have already been described in the literature. There are some obvious advantages and disadvantages to this approach, as summarized in Table 10.7.

While in situ derivatizations have true potential, they typically introduce more requirements and limitations than most other on-line approaches. This may very well be why they have not been widely used.

Table 10.7. Advantages and Disadvantages of in situ Derivatizations in Liquid Chromatography

Advantages

1. No need for additional pumps, mixing chambers, fittings, connective tubing, etc.
2. Excess reagent is always present in the mobile phase for subsequent injections; reagent is never consumed.
3. There is no need to separate the reagent from derivatives.

Disadvantages

1. Reagent may cause elevation in the baseline of the detector and interfere with trace determinations.
2. Reagent may cause a change in the column packing material, undergo a reaction with the support, irreversible binding/complexation, reaction with metal surfaces, etc.
3. Reagent may cause a change in separation conditions, and thus alter resolution and retention times, in an irreproducible manner.

Also, there are limitations on the analyte classes for which such methods are applicable, due to the stringent chemical requirements involved in these applications. Other than for the reduction in extracolumn variance and freedom from the need for another pump, there are no significant advantages to a postcolumn in situ method when compared to a postcolumn homogeneous (reagent added) approach. Since the effective dead volume of mixing devices and reaction chambers may be decreased to low levels in on-line postcolumn derivatizations, there seems little reason to add a reagent precolumn to the mobile phase, unless the derivative has better chromatographic properties than the starting analyte. More problems may be created than are worth the effort, as seen in Table 10.7.

A variation of this in situ approach has been developed in a postcolumn liquid–liquid extraction system, where the reagent is added to the organic extractant (29). For example, metal cations can be separated by any accepted method and then extracted using a liquid–liquid extraction coil into an organic extractant containing a metal chelating reagent. The chelates are instantaneously formed when the metal ions contact the organic–aqueous interface, and these are then extracted into the organic phase leading to the detector. Derivatization occurs in situ at the liquid–liquid interface. A variation on this approach was described years ago by Frei and Lawrence in the use of an ion-pairing reagent in the mobile phase or added postcolumn. Ion pairs were formed, pre- or postcolumn, and these could then be extracted into an organic phase and detected (30). Another variation of in situ derivatization involves the addition of a reagent to the mobile phase, but the reagent only becomes activated after suitable post-column electro- or photochemistry. The active reagent can then derivatize appropriate analytes (see below).

10.8. DERIVATIZATIONS WITHIN A DETECTOR

In some cases the detector itself can participate in a derivatization scheme. For example, in series dual-electrode electrochemical detectors, the first (upstream) electrode can generate a product that is detectable by the downstream electrode (31, 32). It is also possible to orient two or more electrodes opposite to each other, juxtaposed across the flowing stream within the detector. If an electrochemically active species can be reversibly electrolyzed between its reduced and oxidized forms, a cascading effect occurs, and the detector current is higher than that which could be obtained with a single oxidation or reduction process (33).

An electrochemical detector can also be used for reagent generation. In a typical example, bromine was generated from bromide, which was present in the mobile phase. The bromine could then be reacted with compounds eluting from the column in a postcolumn reactor, and the compounds of interest were quantitated by amperometric measurement of the excess bromine (34). A variation of this approach involved electrochemical generation of luminescence; the luminescence could then be monitored using a photomultiplier tube (35, 36).

The moving belt LC/MS interface can also be used to facilitate derivatizations. In one application, butyl esters of small carboxylic acids were formed by pyrolytic dehydration of the ion pairs (37). Many other pyrolytic reactions could also be accomplished with this reaction scheme. Summarily, such methods exhibit good selectivity and sensitivity, but may be limited overall by the lack of a large number of appropriate analyte–reagent pairs.

10.9. THEORY, PRINCIPLES, AND DESIGN OF ON-LINE PHOTOCHEMICAL DERIVATIZATIONS

Postcolumn on-line photochemical derivatizations have become quite popular within the past several years (38). Precolumn photochemical derivatization is possible, but has not been used as an analytical technique for improved detection. Generally precolumn methods are employed only in photochemical mechanism studies, designed to elucidate photochemical reaction pathways. It should also be noted that the term "derivatization" is used very loosely here. In other words, previous discussions of chemical derivatization techniques (homogeneous, heterogeneous, and in situ) involve the formation of a product which is chemically distinct (by the connectivity of atoms in the molecule) from the starting analyte. In photochemical derivatization it is possible that an excited species, or photogenerated radical, of the analyte may be detected with greater sensitivity and selectivity. Thus although the photochemical product is distinguishable from its precursor in terms of its detection properties, it may not be chemically distinct (the connectivity may not be changed), so conventional derivatization does not occur. Of course, the overall result of chemical and photochemical derivatizations, improved detection, is the same. There are several very good reasons for the recent popularity of photochemical derivatization methods, especially with the introduction of low effective dead volume photochemical reactors. Table 10.8 summarizes some of the advantages of these systems.

Experience has shown that at the same time there are some requirements for an efficient utilization of postcolumn on-line photochemical derivatizations. These would include the areas listed in Table 10.9.

Postcolumn on-line photochemical derivatizations offer some serious advantages to routine chemical reactions. Electronically excited analytes may undergo chemical reactions with efficiencies that are much greater than those of the ground-state species, and photochemical processes tend to offer excellent reproducibility. Also, these methods may be nondestructive, so a photochemical reaction detector may be placed in tandem with another detection device without hampering detection at the second detector. Photochemical methods, however, have some limited applica-

Table 10.8. Advantages of Postcolumn, On-Line Photochemical Derivatizations in Liquid Chromatography

1. Continuous high-intensity on-line irradiation is possible with narrow or broad wavelength sources, offering differences in selectivity.
2. High photoconversion (quantum efficiency) is often possible, depending on the particular analyte, thus offering potentially high sensitivity.
3. There is no need to separate excess derivatization reagents from derivative or analyte.
4. Inexpensive light sources are available, making photochemical derivatization approaches as financially attractive as conventional chemical derivatizations.
5. Photoproducts often have quite different detection properties than original analyte.
6. Automatability.

Table 10.9. Requirements of Postcolumn On-Line Photochemical Derivatizations in Liquid Chromatography

1. Precursor (parent compound) should be activated by light wavelengths not strongly absorbed by photoproducts.
2. Precursor should have detection properties that are distinctly different from those of the photoproducts.
3. Derivatives (photoproducts) should be chemically and thermally stable on the time scale needed. Reactor/detector geometries must be appropriately engineered.
4. Photoconversion processes should be efficient, with high quantum yields, and *must* be reproducible.
5. Photoproduct should have high extinction coefficient in UV, high fluorescence efficiency in FL, or good oxidative/reductive properties in EC detection.

bility, in that the starting analyte must have a suitable chromophore in an easy-to-reach wavelength region. The starting analyte must also undergo an efficient, reproducible, photochemical conversion to, ideally, a single, major photoproduct with good ultraviolet, fluorescent and/or electrochemical properties. The photoproduct must have different detector properties than the starting analyte, some of which may still remain after the photochemical reaction. Photochemical formation of tars or polymers may be a real but undesirable part of postcolumn on-line derivatizations.

This derivatization approach is also not as general as are straight chemical approaches, but when the analyte and the photoproduct meet the requirements of Table 10.9, it can be an ideal method of improving detection sensitivity and specificity in HPLC. Photolytic derivatizations have been performed before, adjacent to, and directly on electrochemical cells (38). In addition to electrochemical detection, photolytic derivatives can be monitored with a conductivity detector, as in the photoconductivity detector by Tracor Corp. (39), or using fluorescence or ultraviolet absorbance detection. Chemical reactions may also be used to monitor for the products of a photochemical process (40). The future of these methods will probably involve greater use of lasers for improved specificity and sensitivity, as well as miniaturized cell geometries.

10.10.   SUMMARY, CONCLUSIONS, AND FUTURE TRENDS

The decision to perform the reaction pre- or postcolumn will depend on the particular analytes of interest. Because precolumn reactions tend to be more selective and can be tailored to yield better chromatography, they should be used whenever possible. If the derivative is unstable, a postcolumn reaction may be the only choice. In all derivatization approaches, one must recognize that nonlinearity of the derivatization efficiency may be caused by the presence of interferences in a sample matrix. In addition, loss of analyte on insufficiently passivated glassware, or in incomplete extraction or cleanup procedures, may cause deviations in accuracy and precision. For these reasons, it is *imperative* that blanks, controls, and calibration standards be subjected to the entire analytical procedure, including dilutions, extractions, cleanups, and derivatizations. It appears that, both theoretically and in practical experience, on-line methods will introduce substantially less variance into an analytical measurement than off-line approaches. The additional benefit of automatability of on-line methods makes them even more attractive. The requirement of subjecting controls, blanks, and samples to the full analytical procedure is most easily met through the use of an on-line, automated system.

In the future, derivatizations should be engineered to be selective as a function of reaction conditions, so that compound classes, or even individual compounds, can be distinguished by their degree of derivatization/reaction. In addition, the derivative should be such that it can be monitored with several different selective detection schemes. In complex matrices, separating coeluting peaks through selective detection may be easier than through different chromatographic conditions. The increased use of chemometrics (the application of mathematics and statistics to chemical analyses) will allow greater yield of meaningful data from multiple detectors in such derivatization/detection approaches.

It is time for derivatizations for HPLC to abandon common procedure wherein the sample mixture is reacted off-line in a Reacti-Vial with a specific concentration of reagent for a specific period of time at a specific temperature. During the performance of such labor-intensive derivatizations, the reagents must be prepared at a known concentration just prior to the reaction, the reaction temperature and time must be monitored carefully, and the reaction must be manually quenched or cooled, centrifuged, filtered, and an aliquot removed for injection into the HPLC. Today most routine precolumn derivatizations for HPLC may be fully automated. A wide variety of postcolumn on-line derivatizations may be performed in real time and at room temperature, and the final derivatives may be passed immediately to the detectors.

Recent advances in derivatization for LC involve replacing solution reagents with immobilized enzymes, solid supported reagents, ionic catalytic reactors, and ionically or covalently attached reagents. Eventually all solution derivatizations, off-line or on-line, may be effectively replaced with on-line postcolumn or precolumn solid phase reactions. Photochemical reactions will never provide as many derivatization options as chemical methods, but they will continue to provide a sensitive and selective approach for suitable analytes. Since all of the advantages discussed will accrue to the analytical HPLC system and the operator, eventually final analyses will be easier, more reproducible, faster, less expensive, less involved, and more accurate and precise as well. In the next few years we hope to be able to report the successful completion of this stage in the evolution of derivatizations for HPLC.

ACKNOWLEDGMENTS

The authors acknowledge the assistance of various members of their research group who read and constructively commented on the contents, including B. D. Karcher and W. R. LaCourse. Some of the work described was supported by various contract research and development programs to Northeastern University. The authors are especially grateful to Waters Chromatography Division (Millipore Corporation), EM Science, Inc., Pfizer, Inc., and Allied Analytical Systems, Inc., for their continued interest and support in these programs. Finally, they acknowledge the contributions made by everyone involved over the past several years in their group in solid phase derivatizations (S. T. Colgan, T.-Y. Chou, C.-X. Gao, K.-H. Xie, M.-Y. Chang, C. Dorschel, C. Stacey, B. Bidlingmeyer, C. Santasania), postcolumn metal chelations (B. Karcher), postcolumn hydride generation (D. S. Bushee), postcolumn liquid–liquid extractions for organometals (D. S. Bushee), photolytic derivatizations for LCEC (C. Selavka), photoionization detection in HPLC (K.-H. Xie and R. Nelson), and photoelectrochemical detection in HPLC (W. R. LaCourse).

Partial support of this work was provided by the Barnett Fund for Innovative Research at Northeastern University and NIH Biomedical Research Support Grant RR07143, Department of Health and Human Services.

REFERENCES

1. I. S. Krull, Ed., *Reaction Detection in Liquid Chromatography*, Dekker, New York, 1986, chaps. 1–7.
2. J. F. Lawrence and R. W. Frei, *Chemical Derivatization in Liquid Chromatography*, Elsevier, Amsterdam and New York, 1976.

3. R. W. Frei and J. F. Lawrence, Eds., *Chemical Derivatization in Analytical Chemistry*, vol. 1, *Chromatography*, Plenum, New York, 1981.

4. J. F. Lawrence, *Organic Trace Analysis by Liquid Chromatography*, Academic, New York, 1981.

5. K. Blau and G. S. King, Eds., *Handbook of Derivatives for Chromatography*, Heyden, 1977.

6. D. R. Knapp, *Handbook of Analytical Derivatization Reactions*, Wiley, New York, 1979.

7. I. S. Krull, C. M. Selavka, W. Jacobs, and C. Duda, "Derivatization and Post-Column Reactions for Improved Detection in Liquid Chromatography/Electrochemistry," *J. Liq. Chromatogr., 8,* 2845 (1985).

8. S. A. Cohen, T. L. Tarvin, and B. A. Bidlingmeyer, "Analysis of Amino Acids Using Precolumn Derivatization with Phenylisothiocyanate," *Am. Lab.* (Aug. 1984).

9. "Waters Technical Bulletin on AutoTag Amino Acid Analysis for HPLC," Waters Chromatography Div., Millipore Corp., Milford, MA, 1982.

10. R. Vivilecchia, "Integration of Laboratory Robotics and HPLC," *LC Mag., 4*(2), 94 (1986).

11. S. T. Colgan, I. S. Krull, U. Neue, A. Newhart, C. Dorschel, C. Stacey, and B. Bidlingmeyer, "Derivatization of Alkyl Halides and Epoxides witih Picric Acid Salts for Improved HPLC Detection," *J. Chromatogr., 333,* 349 (1985).

12. S. T. Colgan, I. S. Krull, C. Dorschel, and B. Bidlingmeyer, "Derivatization of Ethylene Dibromide (EDB) with Silica Supported Silver Picrate for Improved HPLC Detection," *Anal. Chem., 58,* 2366 (1986).

13. R. W. Frei, "Reaction Detectors in Liquid Chromatography," in R. W. Frei and J. F. Lawrence, Eds., *Chemical Derivatization in Analytical Chemistry*, vol. 1, *Chromatography*, Plenum, New York, 1981, chap. 4.

14. L. R. Snyder and J. J. Kirkland, *Introduction to Modern Liquid Chromatography*, 2nd ed., Wiley, New York, 1979, chap. 17.

15. B. Lillig and H. Engelhardt, "Fundamentals of Reaction Detection Systems," in I. S. Krull, Ed., *Reaction Detection in Liquid Chromatography*, Dekker, New York, 1986, chap. 1.

16. C. Horvath and W. R. Melander, "Theory of Chromatography," in E. Heftmann, Ed., *Chromatography: Fundamentals and Applications of Chromatographic and Electrophoretic Methods, Part A: Fundamentals and Techniques*, Elsevier, Amsterdam, 1983, chap. 3.

17. L. R. Snyder and J. J. Kirkland, *Introduction to Modern Liquid Chromatography*, 2nd ed., Wiley, New York, 1979, chap. 2.

18. J. P. Foley and J. G. Dorsey, "Equations for Calculation of Chromatographic Figures of Merit for Ideal and Skewed Peaks, *Anal. Chem., 55,* 730 (1983).

19. L. R. Taylor, "Effect of a Low Dead-Volume Deoxygenator on Peak Broadening Using Reductive Electrochemical Detection," *LC Mag., 4*(1), 34 (1986).

20. I. S. Krull and E. P. Lankmayr, "Derivatization Reaction Detectors in HPLC," *Am. Lab.*, 18 (May 1982).
21. S. T. Colgan and I. S. Krull, "Solid Phase Reaction Detectors for LC," in I. S. Krull, Ed., *Reaction Detection in Liquid Chromatography*, Dekker, New York, 1986, chap. 5.
22. K.-H. Xie, S. Colgan, and I. S. Krull, "Solid Phase Derivatization Approaches in HPLC," *J. Liq. Chromatogr.*, 6(S-2), 125 (1983).
23. K.-H. Xie, C. Santasania, I. S. Krull, U. Neue, B. Bidlingmeyer, and A. Newhart, "Solid Phase Derivatization Reactions in HPLC: Polymeric Permanganate Oxidations of Alcohols and Aldehydes," *J. Liq. Chromatogr.*, 6, 2109 (1983).
24. I. S. Krull, K.-H. Xie, S. Colgan, T. Izod, U. Neue, R. King, and B. Bidlingmeyer, "Solid Phase Derivatization Reactions in HPLC: Polymeric Reductions for Carbonyl Compounds," *J. Liq. Chromatogr.*, 6, 605 (1983).
25. I. S. Krull, S. Colgan, K.-H. Xie, U. Neue, R. King, and B. Bidlingmeyer, "Solid Phase Derivatization Reactions in HPLC: Borohydride/Silica Reductions for Carbonyl Compounds," *J. Liq. Chromatogr.*, 6, 1015 (1983).
26. I. S. Krull, X. Gao, S. T. Colgan, and C. Dorschel, unpublished results, 1985–1986.
27. P. Kucera and H. Umagat, "Chemical Derivatization Techniques Using Microcolumns," in P. Kucera, Ed., *Microcolumn High Performance Liquid Chromatography*, Elsevier, Amsterdam and New York, 1984, chap. 5.
28. H. Jansen, U. A. T. Brinkman, and R. W. Frei, "Post-Column Reaction and Extraction Detectors for Narrow-Bore High Performance Liquid Chromatography," *J. Chromatogr. Sci.*, 23, 279 (1985).
29. I. S. Krull and B. D. Karcher, unpublished results, 1985–1986.
30. J. F. Lawrence, U. A. T. Brinkman, and R. W. Frei, "Ion-Pairing as a Means of Detection in Liquid Chromatography," in I. S. Krull, Ed., *Reaction Detection in Liquid Chromatography*, Dekker, New York, 1986, chap. 6.
31. R. W. Andrews, C. Schubert, J. Morrison, E. W. Zink, and W. R. Matson, *Am. Lab.* (Oct. 1982).
32. W. A. MacCrehan and R. A. Durst, *Anal. Chem.*, 53, 1700 (1981).
33. S. G. Weber and R. A. Durst, *Anal. Chem.*, 54, 1757 (1982).
34. W. T. Kok, J. J. Halvax, W. H. Voogt, U. A. T. Brinkman, and R. W. Frei, *Anal. Chem.*, 57, 2580 (1985).
35. A. J. Bard and L. R. Faulkner, *Electrochemical Methods: Fundamentals and Applications*, Wiley, New York, 1980, chap. 14.
36. H. Tachikawa and L. R. Faulkner, in P. T. Kissinger and W. R. Heineman, Ed., *Laboratory Techniques in Electroanalytical Chemistry*, Dekker, New York, 1984, chap. 23.
37. C.-P. Tsai, A. Sahil, J. M. McGuire, B. L. Karger, and P. Vouros, *Anal. Chem.*, 58, 2 (1986).

38. W. R. LaCourse and I. S. Krull, "Post-Column Photochemical Derivatizations for Improved Detection in HPLC," in I. S. Krull, Ed., *Reaction Detection in Liquid Chromatography*, Dekker, New York, 1986, chap. 7.

39. I. S. Krull, "Recent Advances in New and Potentially Novel Detectors in High Performance Liquid Chromatography and Flow Injection Analysis, in S. Ahuja, Ed., *Chromatography and Separation Chemistry, Advances and Developments*, ACS Symp. Ser. 297, American Chemical Soc., Washington, DC, 1986, chap. 9.

40. W. Iwaoka and S. R. Tannenbaum, IARC Sci. Publ. 14, International Agency for Research on Cancer, Lyon, France, 1976, pp. 51–56.

41. "Solid Phase Catalytic Reactor for Carbamate Analysis in HPLC," Tech. Bull., Kratos Corp., Ramsey, NJ, 1985.

42. R. Weinberger, "Advances in Carbamate Pesticide Analysis by Liquid Chromatography," presented at the Association of Official Analytical Chemists (AOAC) Spring Workshop/Meeting, Dallas, TX, Apr. 1985.

CHAPTER

11

PREPARATIVE LIQUID CHROMATOGRAPHY

HENRI COLIN

PROCHROM
54250 Champigneulles, France

11.1. INTRODUCTION

Preparative liquid chromatography (PLC) is the earliest form of chromatography. The first experiments were made at the beginning of the century (1), when chromatography was used primarily for the isolation of substances rather than their analyses. Actually, the first 40 years of existence of chromatography were dominated by preparative chromatography (2). Displacement chromatography was recognized early as a powerful technique for preparative purposes and was slowly replaced by elution chromatography, primarily because of the development of analytical chromatography (3). For many years PLC has been considered a low-efficiency technique, using low-grade silica gel and low-pressure equipment. Accordingly, the results obtained were not very exciting, particularly in view of the fantastic development experienced by analytical liquid chromatography during the same period. The situation is changing now, and it has been realized that modern PLC can be a very powerful tool, especially for the purification of complex mixtures requiring high column efficiencies (2). The present trend in PLC is to use small particles with a narrow size distribution (10–15 μm or 15–25 μm) rather than the large particles (50–200 μm) popular several years ago. The interest in modern PLC is growing quickly. There are now two series of meetings (one in the United States and one in Europe) specifically dedicated to PLC. The popularity of the technique will certainly continue to grow in the future, particularly with the explosion of the pharmaceutical and biotechnology fields (4).

The goal of chromatography is to analyze or isolate pure substances. Based on the amount to be purified and the type of equipment used, it is possible to classify several types of liquid chromatography (5, 6). Table 11.1 is a tentative classification of PLC. As it can be seen, the range covered in terms of load, column dimensions, and flow rate requirements is large. Modern PLC can be made on analytical columns (7) as well as on columns with a diameter of up to 45 cm (8) with the same quality of separation (9). This is a key point which explains the recent increase in use of modern PLC.

PLC can be performed either continuously or discontinuously. In continuous chromatography, the feed is introduced at a constant rate during the purification process and the feed components are collected at different places of the system (10–12). Most often the stationary phase is actually moving (moving bed technique) and there can be multiple injection ports. Continuous chromatography is used mostly for large and very large scale process applications and requires a complex and thorough optimization. This form of chromatography is highly sophisticated and specialized, and it is beyond the scope of this chapter to discuss it. Discontinuous chromatography is the classical mode of chromatography (individual injections) and can be practiced under many different forms. To clarify the situation, it is convenient to distinguish two categories, "old", and "modern" PLC.

Old PLC is characterized by the use of large particles with a large size distribution. This is a low-pressure and low-efficiency technique. Solvents are forced through the columns simply by gravity (open-column chromatography with glass equipment) or with a low-pressure pumping system

Table 11.1 Tentative Classification of Liquid Chromatography Columns

	Column Diameter				
	100 μm	2 mm	7 mm	2 cm	>15 cm
Length	Up to 50 m	10 cm–10 m	5–30 cm	10–50 cm	20 cm–2 m
Particle size	Empty	5–20 μm	3–10 μm	5–60 μm	10–150 μm
Typical flow rate	<20 μL/min	50 μL/min–2 mL/min	1–5 mL/min	5–100 mL/min	100 mL/min–100 L/min
Load	<100 ng	100 ng–10 μg	10 μg–10 mg	50 mg–1 g	>100 g
Type	Analytical microbore capillary	Analytical narrow bore, small bore	Analytical classic microprepa-ration	Semipreparative	Preparative process
Typical application	Analysis	Analysis	Analysis identification micro-preparation	Synthesis lab-scale production	Medium-size production process

417

(peristaltic type). Old PLC is still used, particularly for many biochemical purifications based on affinity chromatography (13), for instance. With some packing materials such as soft gels, low-pressure chromatography is actually the only technique that can be used. This type of PLC is not discussed in this chapter because there is abundant literature on this topic and there have not been many new developments in this area during the past years. Other "old" techniques such as flash chromatography, vacuum chromatography, or preparative thin-layer chromatography are not addressed because of their limited use.

Modern PLC is in some respects a direct scale-up of analytical chromatography. High-quality packing materials are used along with high-flow rate/high-pressure pumping systems. Plate counts achieved are typically in the range of 10,000–50,000 per meter, whereas the figures are 500–5,000 for old PLC. Modern PLC is growing very rapidly for the simple reason that more difficult separation problems have to be solved at the preparative level in many fields, pharmaceuticals and biotechnology being the most important ones. Whereas old PLC was mainly based on silica gel, modern PLC makes increasing use of reversed-phase supports (which parallels the extensive use of these materials for analytical applications) and more generally sophisticated (and expensive) packing materials.

This chapter discusses some critical aspects of modern PLC. Column operation is examined in the cases of volume and concentration overload (linear and nonlinear elution), and the concept of column capacity is addressed. A limited form of optimization is then discussed in both cases. Finally, the hardware is examined with particular emphasis on the packing materials and the column technology.

Several terms specific to PLC are used in this chapter. It may be useful to restate their meanings (14). *Feed* is the mixture injected in the column; *throughput* is the quantity of feed injected per unit time (averaged over a number of cycles); and *purity* is the amount, in percent, of the component of interest in the collected fractions. Purity is generally based on analytical HPLC analysis of the collected material and is often calculated from peak area values. Most often this does not provide information on the actual concentration composition since different solutes may have different response factors. The *production rate* is the amount of pure material (based on the purity criterium required) collected per unit time. *Recovery* or *yield* is the ratio of the amount of a given component injected to the amount of pure component collected. The throughput characterizes the feed as a whole, whereas production rate and recovery are related to the composition of the feed and also the degree of separation.

Finally, it should be indicated that literature is available on PLC. Several review articles (4–6, 9–32) as well as a practical guide (2) have been published recently.

11.2. COLUMN OPERATION CONDITIONS

The migration of a zone of solute in a column is controlled by two types of effects, kinetic and thermodynamic. Kinetic effects concern dispersion of the solute zones, whereas thermodynamic effects control solute retention. Thermodynamic effects are related to the distribution equilibrium of the solute between the mobile and the stationary phases. Of particular interest for preparative chromatography is the distribution isotherm, which represents the variation of the solute concentration in the stationary phase with that in the mobile phase. The slope of the tangent of the isotherm is the distribution constant K. The retention volume of a sample, V_R, is related to K and the phase ratio ϕ through the capacity ratio k' according to the known equations

$$k' = K\phi \tag{1}$$

$$k' = \frac{V_R - V_0}{V_0} \tag{2}$$

where V_0 is the column dead volume.

The output profile resulting from an injection in a chromatographic column depends on several contributions: the column, the injection (quantity injected and quality of the injector), and other sources such as the detector and the connecting tubings. In the following discussion it is assumed that only the two first contributions are important. The other ones can be neglected or considered as part of the column contribution. With this assumption, the output profile is the result of the combination of the injection function and the column function (defined as the output profile resulting from an infinitely small injection). The first and second centered moments can be used to describe the output profile (33). The first moment, m_1, is the gravity center (or the apex for a symmetrical peak) and the second moment, m_2, is the variance. These moments can be calculated from the output signal $c(V)$ according to

$$m_1 = \frac{1}{m_0} \int_0^V c(V)V \, dV \tag{3}$$

$$m_2 = \frac{1}{m_0} \int_0^V c(V)(V - m_1)^2 \, dV \tag{4}$$

where m_0 is the zero-order moment (peak area).

The first and second moments can be used to evaluate the mode of operation of a column (34). There are two possibilities. In the first case the moments of the output profile are the sums of the corresponding moments of the injection function and the column function. This is *linear chromatography*. In this case, the concentration of the sample is sufficiently low to be in the linear part of the distribution isotherm. This corresponds to analytical conditions or preparative conditions with volume overload (see below). When the sample concentration is too large, the isotherm is no longer linear and the moments of the output profile cannot be calculated from the injection and the column functions. This is *nonlinear chromatography*. It is typical of preparative applications when large amounts of sample are injected. Before discussing in more detail linear and nonlinear chromatography, it is worth mentioning the concept of infinite diameter (35).

In order to achieve infinite diameter conditions, it is necessary to inject the solute as a very small droplet at the center of the column. Such conditions are not recommended in PLC because a fraction only of the column packing is used (about 50%) (36). The design of the column and in particular of the column ends is critical as far as the wall effect is concerned. It has been shown that with a proper design of the column extremities and well-packed columns identical results are obtained when the solute is injected at the center of the column or distributed over the entire column cross section (8, 9). More information is given below.

11.2.1. Linear Chromatography

Linear PLC is associated with the injection of large volumes of dilute sample, V_{INJ}. The first and second centered moments of the injection function are related to V_{INJ} and the quality of the injector (shape of the injection profile). The best injection is achieved when the sample is delivered as a rectangular plug. Under these conditions, m_1 and m_2 are given by

$$m_1 = \frac{V_{INJ}}{2} \qquad m_2 = \frac{V_{INJ}^2}{f} \tag{5}$$

where f is equal to 12 (37). When the injection function is not rectangular but presents some tailing (this is very often the case in practice), f is smaller than 12. A value between 6 and 8 is typical of many commercially available injectors (38).

It is important to be able to calculate the output profile for optimization purposes (see below). A numerical expression has been derived (39–42),

$$c(V) = \frac{c_{INJ}}{2}\left[\text{erf}\left(\frac{V - V_R}{\sqrt{2}\,\sigma_C}\right) - \text{erf}\left(\frac{V - V_R - V_{INJ}}{\sqrt{2}\,\sigma_C}\right)\right] \tag{6}$$

where σ_C is the standard deviation of the column function (in volume units) and c_{INJ} is the feed concentration. Some experimental data are shown in Fig. 11.1 (43). As long as V_{INJ} is small enough ($<\sigma_C$) there is no significant contribution of the injection to the output profile. For larger injected volumes the column is volume overloaded. The peak keeps a nearly Gaussian shape until V_{INJ} becomes larger than about $5\sigma_C$. After that, a plateau is reached. The solute concentration on the plateau is equal to c_{INJ} (no dilution). The retention time of the peak is apparently shifted by $V_{INJ}/2$ compared to an infinitely small injection. This is because time zero is taken at the beginning of the injection.

The effect of the injected volume on the column efficiency is simply related to V_{INJ} according to

$$\Delta N = \frac{N_C \theta^2}{1 + \theta^2} \tag{7}$$

where ΔN is the decrease in column plate number, N_C is the intrinsic column efficiency, and θ^2 is given by (44)

$$\theta^2 = \left(\frac{V_{INJ}}{V_R}\right)^2 \frac{1}{fN_C} \tag{8}$$

For a given θ, the larger the column volume and the capacity ratio (thus V_R), the larger is V_{INJ}. Columns with large plate numbers are overloaded more rapidly than those with smaller efficiencies. A similar situation will be observed in the case of nonlinear chromatography. This is an important point which explains why columns packed with small particles are usually considered to be inadequate for preparative applications. It is important to insist on the fact that everything else being kept constant, the particle size has no effect. Only the plate number is important.

11.2. Nonlinear Chromatography

Nonlinear chromatography is very complex. Because of the nonlinearity of the distribution isotherm it is impossible to find a solution to the differential equation describing the mass balance in the column. Accordingly, it is very difficult to calculate the output profile. Composite isotherm phenomena make the situation even more complicated. Three approaches have been proposed to describe the output profile in nonlinear elution. The first is based on a modelization of the distribution isotherm. It was developed by Haarhoff and Van der Linde (45) and Houghton (46) and was used recently by Cretier and Rocca (47) Poppe and Kraak (48), and Knox and Pyper (48[1]). It is valid only when the departure from lin-

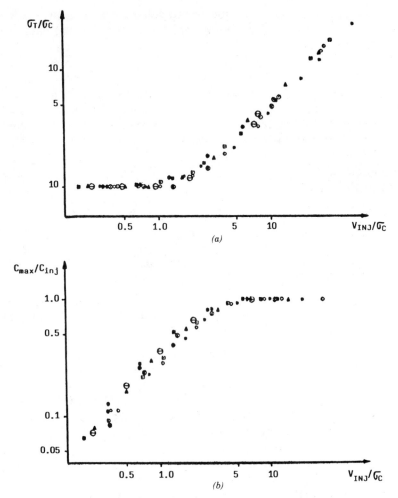

Figure 11.1. Influence of feed volume on (*a*) peak broadening and (*b*) peak concentration. ○—phenol; ⊗—anisole; ⊖—benzophenone (stationary phase: LiChrosorb RP-8; mobile phase: acetonitrile–water–triethylamine, 447:547:6); ▲—benzyl alcohol; ■—5-phenyl-1-pentanol; ●—benzophenone (stationary phase; LiChrosorb RP-8; mobile phase: acetonitrile–water–triethylamine, 348:646:6); □—3-nitroaniline; ⊡—3-nitrophenacetin (stationary phase: LiChrosorb SI-100; mobile phase: dichloromethane–methanol–water, 989:10:1). [From Ref. (43) with permission.]

earity of the isotherm is not too pronounced. The second approach is semiempirical (29, 30, 49, 50). It does not provide an analytical form for the output profile but gives some general trends at high loads. The third approach is to make a direct numerical integration of the mass balance equation (51). It is beyond the scope of this chapter to discuss this method, but it is likely that it could offer the ultimate solution to the optimization in PLC. We only discuss below the principal features of the two first approaches.

11.2.2.1. Haarhoff–Van der Linde Model

This model is based on a quadratic modelization of the distribution isotherm. Some typical isotherms are shown in Fig. 11.2. In practice, the shape is often of type a (Langmuir) and sometimes b, such as in reversed-phase chromatography with nonpolar solutes. Such isotherms can be described by a power series according to

$$\phi c_S = a_1 c_M + a_2 c_M^2 + a_3 c_M^3 + \cdots \tag{9}$$

where c_S and c_M are the solute concentrations in the stationary and mobile phases, respectively. The coefficient a_1 is the capacity ratio ($c_M \to 0$) in analytical conditions (small injected quantity).

Based on this model and certain additional assumptions, Haarhoff and Van der Linde have solved the mass balance equation in the column (47). The equation of the output profile is

$$c(V) = c_{\text{INJ}} \left[1 + \frac{Z^*(T) + [1 - Z^*(T + V^*)] \exp(-M^*)}{Z^*(T + V^* + C^*) - Z^*(T + C^*)} \right.$$
$$\left. \times \exp\left(-C^*T - \frac{C^{*2}}{2} \right) \right] - 1 \tag{10}$$

(10)

T, C^*, V^*, and M^* are dimensionless parameters given by

$$T = \frac{V - V_R - V_{\text{INJ}}}{\sigma_C} \tag{11}$$

$$C^* = -2V_0 a_2 c_{\text{INJ}} V_{\text{INJ}} \tag{12}$$

$$V^* = \frac{V_{\text{INJ}}}{\sigma_C} \tag{13}$$

$$M^* = C^* V^* \tag{14}$$

The function $Z^*(x)$ is defined by

$$Z^*(x) = \frac{1}{2} \left[1 + \text{erf}\left(\frac{x}{\sqrt{2}} \right) \right] \tag{15}$$

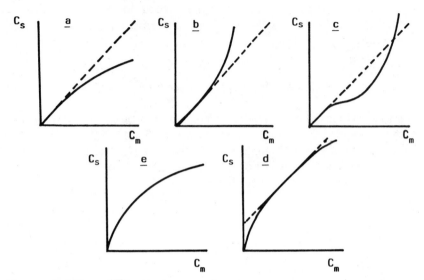

Figure 11.2. Different types of isotherms. Isotherms a (Langmuir), b, and c can be described by a power series. Isotherms d and e (Freundlich) cannot because derivatives at $c_M = 0$ are infinite. [From Ref. (48) with permission.]

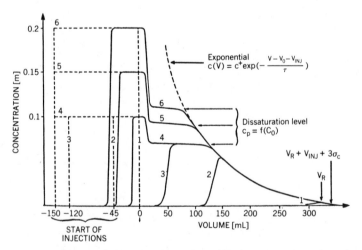

Figure 11.3. Evolution of column output profile in linear and nonlinear elution. $- - - - -$ injection profile; $—$ output profile; 1: pulse injection: $V_{INJ} = 1$ mL, $c_{INJ} = 0.1M$; 2: $V_{INJ} = 45$ mL, $c_{INJ} = 0.1M$. 3: $V_{INJ} = 120$ mL, $c_{INJ} = 0.1M$. 4: $V_{INJ} = 150$ mL, $c_{INJ} = 0.1M$. 5: $V_{INJ} = 150$ mL, $c_{INJ} = 0.15M$. 6: $V_{INJ} = 150$ mL, $c_{INJ} = 0.2M$. [From Ref. (5) with permission.]

424

Cretier and Rocca (47) have reported good correlations between experimental and calculated profiles for various combinations of injected volume and mass. They have also demonstrated that isotherms can be determined fairly accurately from a single asymmetrical elution profile. It thus seems feasible to use Eqs. 10 and 15 to optimize PLC when overload is limited.

11.2.2.2. Gareil Model

Experiments have shown that under widely varying chromatographic conditions (adsorption, ion exchange, partition) the output profile has a unique shape and the following properties [see Fig. 11.3 and Refs. (34, 49, 50)]:

1. When the injected quantity increases (provided the injected volume stays smaller than a certain limit), the front part of the peak becomes more and more vertical and the ratio Q_0/Q_0' decreases and plateaus at 1.2 (Q_0 is the total peak area and Q_0' the area of the second part of the peak, see Fig. 11.4).

2. The peak "ends" at an elution volume $V_R + V_{INJ} + 3\,\sigma_C$, as in linear chromatography.

3. If time zero is taken at the end of the injection, the peaks corresponding to a given injected quantity under different injection conditions are superimposable.

4. The peaks corresponding to different injected quantities have their rear profile on a common curve; this equation is

$$c(V) = C^+ \exp\left(- \frac{V - V_0 - V_{INJ}}{\tau} \right) \qquad (16)$$

C^+ is equivalent to a concentration. It decreases slowly with increasing k' and a typical value is $0.5M$. τ is related to the retention volume and is close to $0.2V_R$.

5. When the injected quantity is larger than about 1% of the maximum column capacity (maximum quantity of solute in the column at saturation of the stationary phase), the output profiles are independent of the particle size (and thus are C^+ and τ).

Based on these rules, Gareil and coworkers proposed a simple optimization of the injected quantity in nonlinear PLC. This is discussed below.

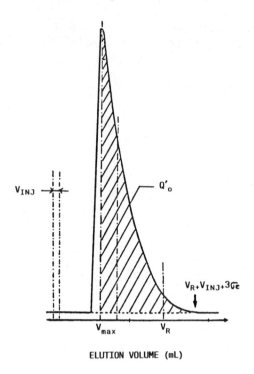

ELUTION VOLUME (mL)

Figure 11.4. Peak profile under concentration overload conditions. [From Ref. (29) with permission.]

11.2.3. Column Loadability

To determine how much sample (volume and mass) can be injected into a column without deteriorating the separation, it is convenient to define the column capacity. Two capacities should be defined, the volume capacity and the concentration (or mass) capacity. The volume capacity can be calculated using Eqs. 7 and 8. Most of the data published in the literature concern the mass capacity [such as (52, 53, 54)]. It is usually defined *for a given column* as the mass of sample per unit mass of packing (specific load) which produces a certain change in the plate number or k' value (5 or 10%, for instance) (52).

The *volume capacity* of a column depends on its size. It also depends on the efficiency and the solute capacity factor. This can be demonstrated using Eq. 11 to calculate the value of the injection volume (linear elution conditions) which produces a relative variance increase θ^2 relative to the column variance,

$$V_{\text{INJ}}^2 = \frac{\theta^2 V_0^2 (1 + k')^2 N}{f} \tag{17}$$

The concentration or mass capacity (also called *linear capacity*) is more difficult to calculate since it corresponds to nonlinear conditions. It is probably better to define a mass capacity rather than a concentration capacity because the practical parameter in PLC is how much sample can be injected. Actually, the two concepts are somewhat similar. The critical parameter is the solute concentration at the top of the column. This concentration determines the isotherm nonlinearity contribution to band broadening. Provided the injected volume is not too large, the concentration achieved in the first plates at the top of the column does not depend on the injection conditions (with the limit on V_{INJ}) but only on the injected quantity. A given mass injected spreads rapidly over a certain number of theoretical plates at the top of the column. When c_{INJ} is large and V_{INJ} small, the resulting strong isotherm nonlinearity is produced by a rapid spreading of the solute zone at the top of the column. The concentration decreases then rapidly and the broadening process stabilizes after a certain concentration is achieved. Conversely, when c_{INJ} is small and V_{INJ} large, the broadening due to isotherm nonlinearity is not as important, but there is an additional contribution of the injected volume. It has been shown that the final situation is very similar in both cases (34, 48) and the output profile is independent of the injection conditions.

Although widely accepted, the usual definition of linear capacity is misleading because it characterizes not only the thermodynamic nature of the chromatographic system, but also the *kinetic conditions* (in terms of column efficiency) and thus the column design (although the linear capacity is expressed in mass per mass), similar to the volume capacity. Accordingly, it makes no sense to compare stationary phases on the basis of linear capacity data only. This point has been clearly discussed by Poppe and Kraak (48), who have shown that the critical point is not the absolute isotherm nonlinearity effect but its contribution *relative* to dispersion in the column. A given isotherm nonlinearity effect is more critical for a column with low dispersion (high efficiency) than with large dispersion. The column plate number is a critical parameter. This is illustrated in Fig. 11.5, which shows plots of efficiency versus specific load in logarithmic scale for several columns. A comparison of the curves corresponding to 10-μm particles indicates that the longer column is more rapidly saturated than the shorter one. The two columns have different linear capacities although they are packed with the same material. This explains why small .particles are often reported to be less suitable for preparative applications than larger ones because they are allegedly more rapidly saturated. The point is that small particles give more efficient columns, which are consequently more sensitive to overload. The particle size itself is not directly involved, only the plate number is important. As it can be seen in Fig. 11.5, the column packed with 5-μm particles is two times shorter than that packed with 10-μm particles, and both columns have the same plate count. They also have identical linear capacities.

Poppe and Kraak (48) have used the model of Haarhoff and Van der Linde to investigate the change in the output profile first and second centered moments (gravity center and variance) with the column load.

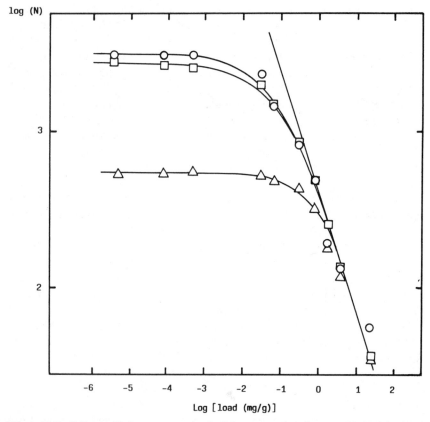

Figure 11.5. Column efficiency versus load. Solvent—methanol–water 70:30 (v/v); stationary phase—LiChrosorb RP 18; sample—benzene. $\bigcirc = L$ (column length) = 30 cm; d_p (particle size) = 10 μm; $\square = L = 10$ cm, $d_p = 5$ μm; $\triangle = L = 10$ cm, $d_p = 10$ μm.

Their treatment leads to a dimensionless definition of the loadability according to

$$\psi = \frac{2M}{SV_{MP}} \left(\frac{k'}{1 + k'} \right)^2 \tag{18}$$

where V_{MP} is the volume of mobile phase in one theoretical plate (V_0/N), M the injected mass, and S a parameter that characterizes the saturation of the stationary phase per unit volume by the solute. With such a definition of loadability, it has been shown that the relative importance of thermodynamic and dispersion broadening does not depend on the length of the column and thus its efficiency. It is determined only by the amount of solute per unit mass of stationary phase (specific load q_0) in one theoretical plate in addition to some physical properties of the stationary phase and some intensive thermodynamic parameters. This clearly appears when Eq. 18 is rewritten,

$$\psi = \frac{2q_0}{N} \frac{\rho}{S} \left(\frac{k'}{1 + k'} \right)^2 \tag{19}$$

where ρ is the packing density (mass of stationary phase per column unit volume). The variations of the first and second moments are shown in Fig. 11.6. From these curves it is easy to calculate the mass injected that generates, for a given column, a certain change in peak gravity center or variance. For instance, a 10% increase in peak variance (or plate number, assuming a negligible change in first moment) corresponds to $\psi = 4$. This corresponds to a $1/\sqrt{N}$ % change in first moment and $1.5/\sqrt{N}$ % in peak maximum. Incidentally, it must be noted that for a given change in efficiency, the change in retention or capacity ratio depends on the column efficiency. Once the criterion of linear capacity has been defined (ψ value), it is easy to calculate the corresponding injected mass.

In conclusion, the only way to characterize the mass loadability of a given chromatographic system without considering the column size is to use the specific load per plate. If only the specific load is given, the data are meaningless (55).

11.3. SEPARATION OPTIMIZATION

Optimization in PLC usually means achieving the maximum throughput for given purity and recovery conditions. Other optimization strategies can be foreseen as well. For instance, the production cost (56, 57) or the

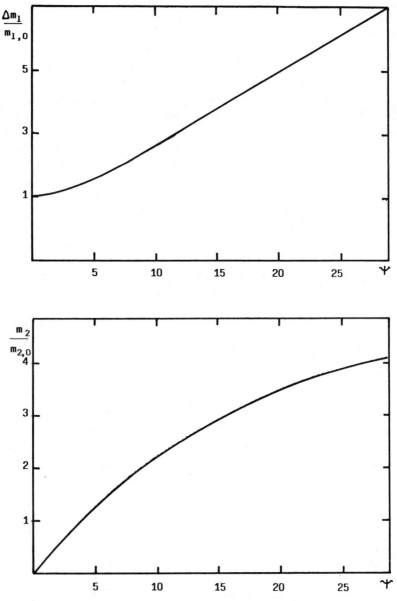

Figure 11.6. Relative change in first moment $(-\Delta m_1/m_{1,0})$ and second moment $(m_2/m_{2,0})$ according to the Haarhoff and Van der Linde expression. [From Ref. (48) with permission.]

solvent consumption (29) can be optimized. The optimization of a preparative separation is a difficult task because of the large number of parameters that can be included in the optimization scheme. Indeed, in addition to the thermodynamic variables (choice of the mobile and stationary phases and possibly the temperature), additional parameters have to be considered, such as sample solubility, sample and solvent recycling, sample recovery and purity, as well as economic factors. The equipment is also a critical aspect. Before discussing the different optimization strategies, it is important to point out that optimizing a separation in view of large-scale production and optimizing an occasional separation are two completely different problems. It is generally a necessity to fully optimize process chromatography based on the expected production rate, cost, purity, recovery, and equipment constraints. In the case of an occasional separation, the optimization is less critical and factors such as cost or solvent availability, for instance, can often be ignored.

There is no general-purpose "absolute" optimization scheme available, although some approaches have been proposed (36, 49, 58–61). They are valid in certain conditions and have certain limitations. They can be divided into three categories. The first is oriented toward a general "hydrodynamic" optimization with particular emphasis on column design and solvent flow rate. Studies in the second category deal with volume overload, and the last group is related to mass overload. For the sake of simplicity, optimization in the following discussion is limited to achieving the maximum throughput with a given equipment, taking into consideration purity and yield but not cost.

No discussion will be made on the thermodynamic aspect of the optimization. The reason is that the criteria involved in the choice of the system in PLC are often very different from those in analytical chromatography and therefore cannot be discussed because each case is a special one. It is clear, however, that up to a certain limit it is of utmost importance to have the largest possible selectivity. It is also clear that there is a price to pay for this selectivity. This can be the solvent, the stationary phase cost, the sample solubility, or the time of analysis.

11.3.1. Scalability

For obvious reasons the optimization is generally not studied on the preparative column but on a column of much smaller size (analytical column). This raises the question of scalability, which is a very important matter, particularly for industrial applications. When scaling up a separation, it is necessary that both the analytical and the preparative columns have identical retention characteristics. Ideally, the columns should be packed with the very same material. If the particle sizes are not the same for the two columns, the materials should originate from the same batch. It makes no sense to optimize a separation on an analytical column packed with a given material and transfer it to a preparative column packed with a different support, even of the same type.

There are two aspects to consider in scaling up a separation. The first is theoretical: how to extrapolate to a large column the results obtained on a small one? In other words, how to calculate the dimensions of the preparative column and the solvent flow rate to achieve a certain production rate, yield, and purity based on analytical data. The second aspect is practical: is it possible to build a large column which has the same hydrodynamic characteristics as a small one (in particular, the same plate count and peak shape)? The first aspect of scaleup has to be discussed in the case of linear chromatography only. Indeed, in nonlinear chromatography the separation must be optimized on an analytical column having the same efficiency as the preparative column since the mass loadability depends on the efficiency, as previously mentioned.

Unexpected experimental problems may occur when scaling up a separation. One problem concerns the column permeability with semirigid and compressible packings. When increasing the column diameter, the benefit of bed support by friction along the column wall is lost (9). Accordingly, if the mechanical properties of the packing material are not adequate, the bed collapses. This becomes more and more critical with increasing column length because of the associated increase in pressure. Another problem with increasing the column diameter is the thermal effect (62). Fluid energy is transformed into heat in the column. When the column diameter is small enough, the heat is dissipated by the wall and the column temperature is radially homogeneous. There can be a longitudinal temperature gradient, but this is not critical as far as column efficiency and peak shape are concerned. With increasing column diameter, the column becomes warmer at the center than near the wall because usually the packing material has very poor heat conduction properties. This radial temperature profile produces a radial velocity gradient which can be the source of intense peak broadening and deformation (63) column thermostating may become necessary.

11.3.2. General Considerations on Column Hydrodynamics

In optimizing the design of a preparative column, it is essential to take into account the instrument limitations. These limitations are mostly dictated by the pumping system and concern the maximum flow rate and pressure. Often there is also a pressure limitation imposed by the column.

The parameters to be optimized are the column length and diameter, the particle size, and the flow rate of solvent. These parameters determine the efficiency of the column and its volume. The volume of the column is an important point. It makes no sense to use a column much bigger than necessary to accommodate the quantity of sample to be purified, even if the column has a larger throughput than another one of smaller volume. On the other hand, it is not practical (and often not economical) to use small columns for which a large number of injections is required to purify a given quantity of feed.

A difficult question is what should be the column plate number. A minimum plate number is required to obtain a given purity (36). It is often considered that large plate numbers result in high throughputs. Two extreme situations can be foreseen in linear PLC, assuming 100% recovery yield. In the first case, the column efficiency is equal to the minimum required to achieve the expected resolution. The injected volume has to be infinitely small, otherwise it would increase the peak variance and then decrease the resolution. The throughput is thus almost zero at 100% recovery yield. It increases with decreasing recovery yield. The other case is a very efficient column. The column contribution to the peak variance can be neglected compared to the injection contribution. In this case, the injection volume is maximum and equal to the difference in retention volumes of the two solutes to separate. Increasing the column length and thus column efficiency results in a proportional increase of the injected volume, but also a proportional increase of the analysis time. The throughput is then constant and maximum. An "optimum" design has been proposed that compromises between throughput and column length. Under "optimal" conditions the contribution of the injection to the output profile variance is equal to the column contribution. This definition is arbitrary, however, and does not correspond to a true optimum. The corresponding "optimum" length is twice the length required to make the separation with the desired resolution in analytical conditions (58).

The situation is more complicated in nonlinear PLC because the column loadability decreases with the column efficiency, as previously seen. Assuming that a given plate number N is required, it can be shown that the throughput (Th) is given by (64)

$$\text{Th} = F q_0 \epsilon_T \frac{\rho}{1 + k'} \tag{20}$$

where F is the eluent flow rate, q_0 the specific load corresponding to the required plate number, ϵ_T the column porosity, and k' the capacity factor of the last eluted peak. q_0 is a function of N and decreases with increasing

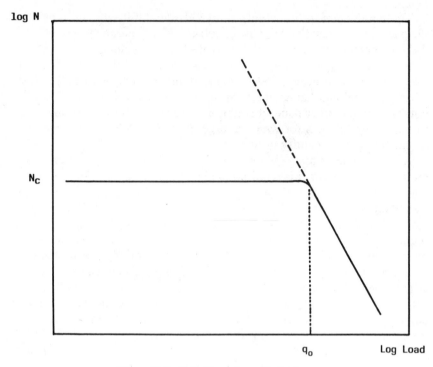

log N

N_C

q_0

Log Load

Figure 11.7. Definition of specific load q_0.

N (see Fig. 11.7). A similar equation was proposed by De Jong et al. (36) in which q_0 does not appear, but rather is a term related to N. A straightforward conclusion from Eq. 20 is that the throughput is proportional to the flow rate and independent of the column design, provided the column generates the required plate number. Some general considerations on column design can be made from Eq. 20. Several authors have reached different conclusions based on different assumptions about system limitation.

De Jong et al. (36) have considered a pressure limitation and discussed the role of the particle size and the column diameter. They have shown that the throughput is proportional to v/h for columns of the same diameter generating the same pressure (h is the reduced plate height and v the reduced velocity). They have concluded that it is preferable to work in the ascending part of the plate height curve and to use large particles packed in long columns compared to small particles and short columns because the large particles generate larger v/h values at a given pressure. These conclusions have to be tempered by the fact that the column packed with small particles is not operated at the maximum flow rate available and the volume of the column is smaller—if not much smaller—than that of the column packed with large particles since both columns have the same diameter.

Different conclusions have been reached by Colin based on a flow limitation (64). This choice is justified *a posteriori*. Since short and bulky columns are more convenient in practice than long and skinny ones, operating pressures are generally not very large. Results of calculations indicate that in order to take full advantage of the flow capability of the pumping system, large particles should be used with very bulky columns. The curves shown in Fig. 11.8 correspond to columns giving the same throughput and operated at the same flow rate. Each curve is drawn for a given column length. For practical reasons, the column length is kept between 10 and 200 cm and the particle size between 10 and 100 μm. It can be seen in Fig. 11.8 that there is a maximum particle size for a given length. This particle size corresponds to operation of the column at the minimum of the plate height curve where the efficiency is maximum. Larger particles would not achieve the required efficiency. The diameter

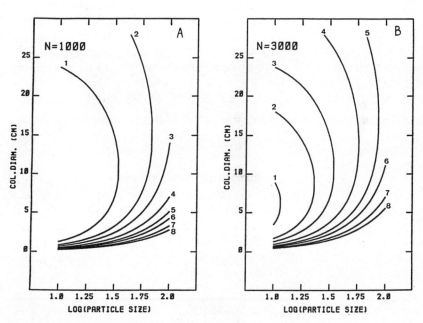

Figure 11.8. Variation of column diameter with particle size (μm) for columns of various lengths giving 1000 (A) or 3000 (B) theoretical plates. Flow rate = 100 mL/min; column length (cm): 1:10; 2:20; 3:30; 4:50; 5:75; 6:100; 7:150; 8:200.

of the particle must decrease when the plate number increases in order to keep the column diameter within reasonable limits. The previous results do not consider the pressure drop nor the column volume, however. These parameters are taken into account in Fig. 11.9, where the variations of the column diameter and pressure with the particle size are shown for several column volumes. The results reveal that, for a given column volume, column efficiency, and solvent flow rate, the pressure increases initially very slowly with increasing particle size and then extremely rapidly when the particle size becomes larger than a critical value. This critical value decreases with increasing flow rate. This suggests that small particles are preferable for large-scale operation. Cost considerations should obviously be made at a certain point.

11.3.3. Optimization in Linear Elution

Two optimization approaches have been described. The first consists in determining the maximum injection volume to achieve a given resolution with a given column. It is quite limited, but nevertheless useful for occasional purifications on a given column (29, 49, 59). The second approach is more powerful since it permits optimizing the column design in order to achieve the maximum throughput with the available equipment for a given purity and yield. This method also provides a means of calculating the cut points. The use of the purity and recovery ratio along with a parameter describing the feed composition is better than using the resolution only (29).

Using the resolution equation, it is possible to calculate the maximum sample volume to achieve a given resolution (40),

$$V_{\text{INJ,max}} = V_{R,2} - V_{R,1} - 2(\sigma_1 + \sigma_2) = 2(\sigma_1 + \sigma_2)(R_s - 1) \quad (21)$$

where R_s is the resolution in analytical conditions and σ_i is the standard deviation of peak i in volume units. Equation 21 is not very accurate, and more appropriate expressions have been derived (40). Using purity and yield as the separation constraints rather than the resolution is more difficult but more powerful. In the following it is assumed that the feed is a binary mixture and the first peak is to be collected. The purity p and the recovery ratio r are defined according to

$$p = \frac{c_{0,1}V_{\text{INJ}} - a}{c_{0,1}V_{\text{INJ}} - a - a'} \quad (22)$$

$$r = 1 - \frac{a}{c_{0,1}V_{\text{INJ}}} \quad (23)$$

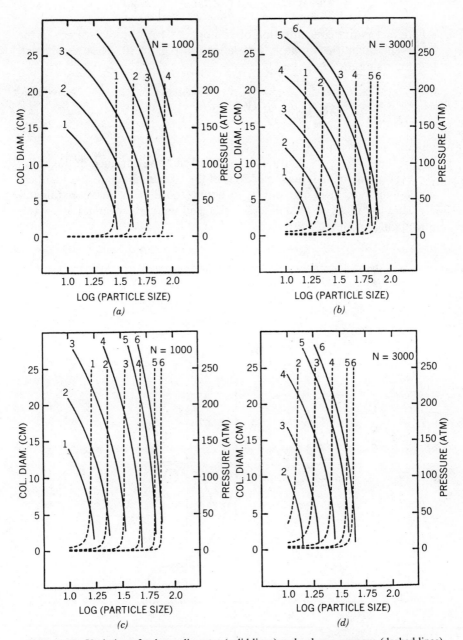

Figure 11.9. Variation of column diameter (solid lines) and column pressure (dashed lines) for columns of various volumes, giving 1000 and 3000 theoretical plates. (*a*) and (*b*) flow rate = 500 mL/min; (*c*) and (*d*) flow rate = 1500 mL/min; volume (mL): curve 1, 500; curve 2, 1000; curve 3, 2000; curve 4, 4000; curve 5, 7500; curve 6, 10,000.

437

where $c_{0,1}$ is the concentration of component 1 (to be recovered) in the feed. The quantities a and a' are defined in Fig. 11.10. They can be calculated using the equations

$$a = \int_{V_C}^{\infty} c_1 \, dV \qquad a' = \int_{0}^{V_C} c_2 \, dV \qquad (24)$$

where V_C is the cut volume.

For the sake of simplicity, it is assumed that the variance of the peaks corresponding to analytical injections of the two components are identical and equal to σ. It is convenient to define the dimensionless parameter $K_{1/2}$ and the reduced parameters x (reduced injection volume) and y (reduced cut volume) according to

$$K_{1/2} = \frac{c_{0,1}}{c_{0,2}} \qquad x = \frac{V_{INJ}}{\sigma} \qquad y = \frac{V_C - V_{R,1} - V_{INJ}}{\sigma} \qquad (25)$$

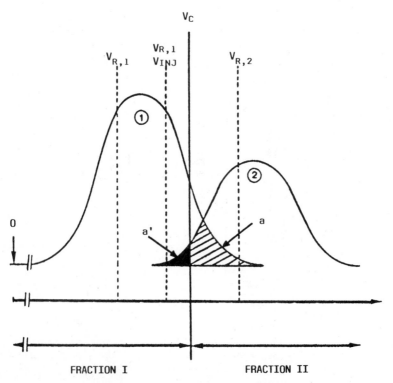

Figure 11.10. Definition of a and a'. [From Ref. (39) with permission.]

$K_{1/2}$ characterizes the composition of the feed. It must be noted that when the component to be collected is the last one, y is given by

$$y = \frac{V_{R,2} - V_C}{\sigma} \tag{26}$$

Coq et al. (39) have shown that it is only possible to calculate x and y using an iterative numerical method. Some results are shown in Figs. 11.11 and 11.12. Data in Fig. 11.11 indicate the critical role of the resolution. The value of x increases from 0 to almost 10 when the resolution increases from 1 to 2. As is well known, the best way to increase the resolution is to improve the selectivity of the phase system. For a given resolution, the injection volume decreases very rapidly with increasing purity and recovery (see Fig. 11.12). It is interesting to look at the variation of the recovered amount of solute with yield. This amount is proportional to the product xr. The curves in Fig. 11.12a show the variations of xr with r for several purities. The existence of an optimum yield r_{opt} should be noted. When the recovery is less than the optimum, the re-

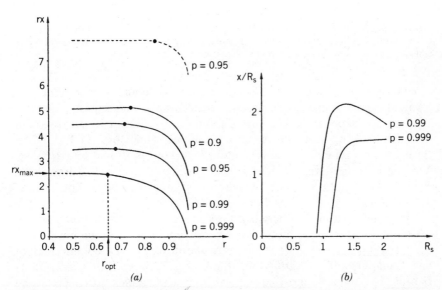

Figure 11.11. (a) Plots of recovered solute amount rx versus recovery ratio r for equimolar sample composition ($K_{1/2} = 1$). —— $R_s = 1.25$; --- $R_s = 2$; ●—position of optimal recovery ratio r_{opt} on curves. (b) Plots of throughput characterized by x/R_s^2 versus analytical resolution R_s for a 0.90 recovery ratio and an equimolar sample composition. [From Ref. (39) with permission.]

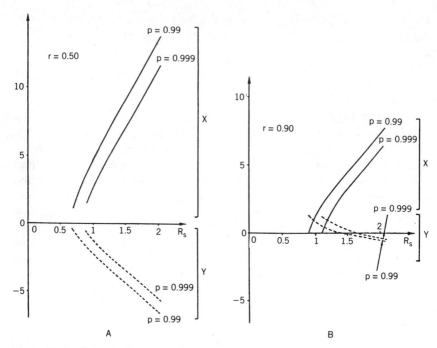

Figure 11.12. Plots of reduced parameters x and y versus analytical resolution R_s. (a) 0.50 recovery ratio and equimolar sample composition ($K_{1/2} = 1$). (b) 0.90 recovery ratio and equimolar sample composition. [From Ref. (39) with permission.]

covered amount is almost independent of the yield and is maximum. For r values larger than r_{opt} the product xr decreases very rapidly. The optimum yield depends on the purity and the resolution. It decreases with increasing purity and decreasing resolution (see Fig. 11.12a). The throughput is related to the resolution and is proportional to x/R_s^2 (39). Variations of the ratio x/R_s^2 with R_s are shown in Fig. 11.12b. There exists a critical resolution above which the throughput does not change significantly.

Once the relationship between the reduced injection volume and the resolution is known for the particular problem (amount to be purified, r, p, and $K_{1/2}$), it is possible to optimize the design of the column (length and diameter) and operating conditions (d_p, F, *number of operation n*) in order to obtain the maximum throughput while taking into account the instrument limitations (flow and pressure). There are 11 variables (L, D_c, d_p, F, Th, u, H, R_s, x, P, and n) related by seven equations. These equations are the five classical hydrodynamic equations of liquid chromatography (plate height–velocity, Darcy's law, resolution, plate number, and diameter–velocity–flow), the definition of the throughput, and the relationship between x and R_s. There are consequently four independent variables. Different optimization schemes can be proposed depending on the choice of the independent variables.

Cretier and Rocca (59) have selected the particle size, column diameter, and number of operation (d_p, D_C, and n) as three of the independent variables. This choice is reasonable considering the practical limitations on the column diameter and the particle size. The authors consider that it is generally a good choice to start with $n = 1$. The last variable to be chosen is either the column length if the problem has to be solved with the available columns, or the solvent flow rate if the user has to adjust the length of the column. This second choice is probably more realistic because the main limitation comes from the pumping system (maximum flow rate and pressure rating). Adjusting the length of the column is very simple with the axial compression technique (see Section 11.5).

Once these four variables are set, all the other ones can be calculated using the proper equations. The only experimental work required at this point consists of determining the plate height curve. This approach has been illustrated by the separation of a mixture of diethyl and dibutyl phthalates using two types of silica (5–20 μm and 25–40 μm) and a 2-cm id axial compression column. The flow rate was limited to 20 mL/min, the pressure to 1.2 MPa (instrument limitations), and the quantity of sample to be recovered was 10 mg. Some results are shown in Table 11.2. They indicate that the largest throughput is obtained with a 10-cm-long column operated at the maximum flow rate. The pressure is the maximum permitted. Three runs are necessary to produce the required quantity of sample ($n = 3$). It must be pointed out that these conditions are not necessarily the absolute best. They simply give the best result within the range of variation selected for the independent variables.

11.3.4. Optimization in Nonlinear Elution

The semiempirical approach of Gareil et al. offers a simple way to optimize the injected quantity in certain cases (34). Another method has been proposed by De Jong et al. (65). It is based on the assumption that the nonlinearity contribution to the peak variance should be half of the total variance. This is quite arbitrary. Moreover, the injected quantity is calculated without taking into account the presence of other components in the feed. This is a limiting assumption.

The optimization method proposed by Gareil et al. (34) is valid when the isotherms are convex, the first peak is eluted in linear conditions and the second one in nonlinear conditions. This corresponds to the most

difficult situation because most of the isotherms are convex and consequently the retention times decrease with increasing injected quantity.
The model can be used *a fortiori* when both peaks are nonlinear since in
this case the retention time of the first component also decreases with
increasing load.

As indicated previously, the end of the first peak is obtained at $V_E = V_{R,1} + 3\sigma + V_{INJ}$ (see Fig. 11.3). The injected quantity of the second
component $Q_{2,INJ}$ must be such that the apex of this peak is eluted at V_E.
The area defined by the exponential of the second peak and V_E is equal
to $Q_{2,INJ}$ and is given by

$$Q_{2,INJ} = C_{2,max}\tau \tag{27}$$

Table 11.2. Working Conditions Requisite to Produce 10 mg of DBP with Available Chromatograph

d_p (μm)	D_C (cm)	n	F (cm³/ min)	u (cm/min)	H (μm)	R_m	Th (mg/min)	L (cm)	ΔP (bar)
5–20	2	1	5	1.9	56	2.7	0.6	24.9	7.1
			10	3.8	62	2.6	1.2	24.9	14.2
			15	5.7	64	2.55	1.9	24.9	21.2
			20	7.6	65	2.5	2.5	24.9	28.3
		2	5	1.9	56	1.95	0.6	13.2	3.8
			10	3.8	62	1.87	1.15	13.5	7.6
			15	5.7	64	1.84	1.7	13.7	11.7
			20	7.6	65	1.83	2.2	13.8	15.6
		3	5	1.9	56	1.65	0.5	9.6	2.7
			10	3.8	62	1.59	1	10	5.7
			15	5.7	64	1.57	1.55	10	8.5
			20	7.6	65	1.56	2.1	10	11.3
25–40	2	1	5	1.9	110	2.1	0.7	22.9	3.2
			10	3.9	139	1.9	1.25	24.7	6.9
			15	5.8	167	1.75	1.8	25.7	10.8
			20	7.8	197	1.64	2.3	26.6	14.9
		2	5	1.9	110	1.57	0.6	13.8	1.9
			10	3.9	139	1.44	1.1	14.4	4
			15	5.8	167	1.35	1.5	15.4	6.4
			20	7.8	197	1.29	1.9	16	9

SOURCE: From Ref. (59) with permission.

$C_{2,\max}$ is calculated using Eq. 16 with $V = V_{2,\max} = V_E$. Equation 27 then becomes

$$Q_{2,\text{INJ}} = C_m \exp \left(\frac{V_{R,1} + 3\sigma - V_M}{\tau} \right) \tag{28}$$

It is remembered that the injected volume must be less than σ for the model to be valid (otherwise the second peak would present a plateau and be larger than predicted).

11.3.5. Comparison of Throughput in Linear and Nonlinear Elution

It is usually recognized that much larger quantities can be injected in nonlinear elution and this mode should be used whenever possible in preparative chromatography. It is difficult to give absolute figures comparing the two types of elution since this would require, among other things, the knowledge of the distribution isotherms. It is nevertheless possible to give some general figures (49).

The throughput depends on the injection frequency (66). The time between two injections can be defined as

$$\Delta t = \frac{V_{R,2} + 3\sigma_2 + V_{\text{INJ}}}{F} \tag{29}$$

This definition corresponds to the case when there are other peaks eluted before the peaks of interest. It prevents the next injection from being made before the solutes of the actual injection have been eluted. If the feed contains only two components, the injection frequency can be increased and the separation time is the time between the beginning of the first peak and the end of the last one (unless gradient elution is used in which case the column regeneration time has to be included). Equation 29 then becomes

$$\Delta t = \frac{V_{R,2} + 3\sigma_2 + V_{\text{INJ}} - V_{R,1} - 3\sigma_1}{F} \tag{30}$$

Combining Eqs. 21, 28, and 29 or 30 gives the throughput in linear (assuming $R_s > 1.3$) and nonlinear elutions (assuming $\tau = 0.2 V_{R,2}$),

$$\text{Th}_{\text{LIN}} = F c_{0,\text{LIN}} f_1(k', \alpha) \tag{31}$$

$$\text{Th}_{\text{NLIN}} = 0.2 F c_M f_2(k', \alpha) \tag{32}$$

Table 11.3. Variation of $f_1(k', \alpha)$ and $f_2(k', \alpha)$ with k' and α (see Eqs. 43 and 44)

	α	k'				
		1	2	5	10	∞
f_1	1.1	0.045	0.06	0.07	0.08	0.08
	1.5	0.17	0.20	0.23	0.24	0.25
	2	0.25	0.29	0.31	0.32	0.33
	5	0.40	0.42	0.43	0.44	0.44
		0.50	0.50	0.50	0.50	0.50
f_2	1.1	0.09	0.04	0.02	0.01	0.01
	1.5	0.11	0.07	0.04	0.03	0.03
	2	0.14	0.10	0.07	0.06	0.05
	5	0.26	0.23	0.22	0.21	0.20
		0.50	0.50	0.50	0.50	0.50

SOURCE: From Ref. (49) with permission.

where the functions f_1 and f_2 are given by

$$f_1 = \frac{k'(\alpha - 1)}{1 + k'(2\alpha - 1)} \tag{33}$$

$$f_2 = \frac{1 + \alpha k'}{1 + k'(2\alpha - 1)} \exp\left(\frac{-5k'}{1 + \alpha k'}\right) \tag{34}$$

Both $c_{0,\text{LIN}}$ and c_M are slightly decreasing functions of k'. To the first approximation it can be assumed that the dependence is similar and can

Table 11.4. Effect of Mobile Phase Composition (R_s, α) and Optimization Procedure on Production Rate

R_S	Optimization	V_{INJ} (mL)	C_{INJ} (M)	Inj. Quantity (mM)	P_1 (%)	P_2 (%)	r_1 (%)	r_2 (%)	Production Rate (mM/h)
6.5	L	25.8	10^{-2}	0.26	0.25	0.4	99.1	99.1	0.88
3.9	NL	3.15	8×10^{-1}	2.52	0.2	0.4	99.4	98.8	12.0
4.6	L	7.15	1.26×10^{-2}	0.09	0.2	0.35	99.4	99.4	0.85
4	NL	1.7	6×10^{-1}	1.02	0.1	0.6	99.2	99.1	12.0
2.2	L	1.35	3.5×10^{-2}	0.05	0.3	1.1	98.5	98.4	0.87
3.1	NL	0.6	8×10^{-1}	0.48	0.25	2.2	97.3	97.3	8.8

SOURCE: From Ref. (49) with permission.
P = purity; r = recovery; L = linear; NL = nonlinear.

be neglected for comparison purposes. Some typical values of f_1 and f_2 are given in Table 11.3. They show that Th does not depend much on k' in linear elution (f_1 increases slowly with k' but $c_{0,\text{LIN}}$ decreases slowly with k'), whereas it decreases quite rapidly in nonlinear elution. As far as the effect of selectivity is concerned, both f_1 and f_2 increase with α, but the rate is faster for f_1 at low k' and faster for f_2 at high k'.

Assuming typical values for $c_{0,\text{LIN}}$ and c_M (10^{-2} and $0.5M$, respectively), Gareil et al. have shown that the throughput is about 10 times larger in nonlinear compared to linear elution (49). Some experimental results are given in Fig. 11.13 and Table 11.4.

11.4. ANCILLARY TECHNIQUES

11.4.1. Displacement Chromatography

Displacement chromatography is no longer used for analytical purposes. It has some interesting features, however, and can be a powerful preparative technique. This has been discussed recently by Horvath and co-workers (3, 67–70).

In displacement chromatography, the column is first equilibrated with a carrier which has little affinity for the stationary phase and is able to solubilize the feed, preferably at high concentrations. After equilibration has been achieved, the feed mixture is introduced into the carrier stream and preconcentrates at the column inlet. The components saturate the stationary phase and a frontal chromatographic process occurs. The next step consists in pumping the *displacer*, which sequentially desorbs the feed components. The displacer must have a stronger affinity to the stationary phase than any of the feed components. A competitive adsorption process causes the sample components to move down the column at speeds determined by the displacer front velocity. Strongly adsorbed components displace those which have less affinity to the stationary phase. The mixture separates into a *displacement train*. Once fully developed, the train is composed of adjacent square bands of near uniform concentration, all moving at the same velocity (*isotachic* elution). After all the feed components are eluted, the column has to be regenerated and reequilibrated with the carrier before another injection can be made. The procedure is summarized in Fig. 11.14.

The nature of the displacer, its concentration, and the injected quantity are critical and determine the time and the column length required for the separation. The velocity of the displacer front depends on its adsorption isotherm and its concentration in the carrier. This is illustrated in Fig.

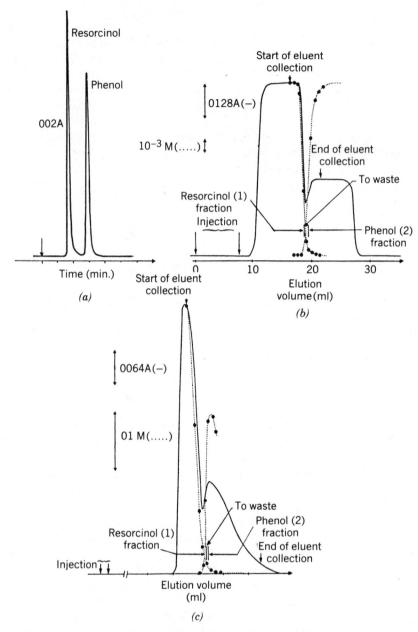

Figure 11.13. (*a*) Analytical chromatogram of equimolar resorcinol–phenol mixture. 15 cm × 0.48 cm id column; stationary phase—Lichroprep RP 8 (25–40 μm); mobile phase—methanol–water (45:55); flow rate—100 mL/h; injection—V_{INJ} = 2 μL; c_{INJ} = 0.228M; spectrophotometric detection—246 nm. (*b*) Linear preparative chromatogram of equimolar resorcinol–phenol mixture. 25 cm × 0.76 cm id column. Stationary and mobile phases as in (*a*); F = 275 L/h; injection—V_{INJ} = 7.75 mL; c_{INJ} = 1.26 × 10^{-2}M; solid line—recorder

446

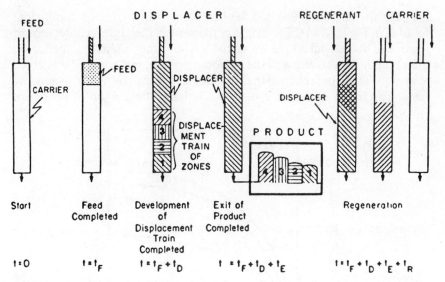

Figure 11.14. Stages of operation in displacement chromatography. Initially the column is equilibrated with the carrier. The mixture to be separated is fed into the column and thereafter the displacer solution is introduced. As the displacer front moves down the column, the displacer train containing adjacent zones of the separated feed components is developed. After the product zones progress down the column it is regenerated by removing the displacer and reequilibrating with the carrier. Time requirement for individual steps: t_F, feed time; t_D, development time of displacement train; t_E, exit time of product; t_R, time of column regeneration. [From Ref. (3) with permission.]

11.15 for a four-component mixture. Figure 11.15 shows the adsorption isotherms of the various species involved. The curve corresponding to the displacer is the highest because the displacer has the strongest affinity to the stationary phase. The velocity of a concentration step is related to the carrier velocity and the slope of the cord of the adsorption isotherm at that concentration. The cord corresponding to the displacer at the concentration c_D (Fig. 11.15a) is called the *operating line*. The slope of the operating line is proportional to the net retention volume of the displacer front. Since all the zones in the fully developed train move at the same velocity, the concentration in each zone corresponds to the intersection

---

trace of absorbance at 237 nm; flat upper portions correspond to concentration injected ($V_{INJ} > 4\sigma_C$); ●—concentrations calculated from quantitative analysis of effluent fractions. (c) Nonlinear preparative chromatogram of equimolar resorcinol–phenol mixture. Operating conditions: see (a), except mobile phase is methanol–water (62:38) and injections are V_{INJ} = 0.585 mL; c_{INJ} = 0.79M. [From Ref. (49) with permission.]

between the operating line and the isotherm of the particular component (c_2 to c_4 in Fig. 11.15a). The height of the zone is therefore a characteristic feature of the chemical nature of the system (the isotherm depends on the sample, the solvent, and the stationary phase, but *not* on the injected quantity). The more retained the component, the higher the corresponding zone. The length of the zone is proportional to the amount of substance

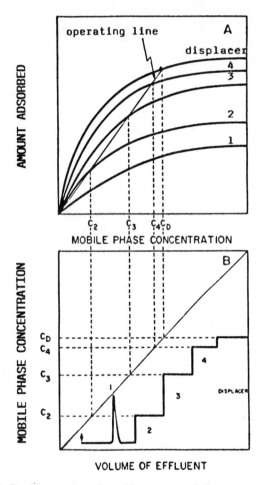

Figure 11.15. (a) Graphic representation of isotherms of feed components and operating line. (b) Corresponding fully developed displacement train. Concentrations of component zones, determined by intersections of operating line and adsorption isotherms of components, are projected from (a) to (b) with the aid of 45° line. Isotherm of first component lies below operating line at all concentrations, hence it elutes from column. [From Ref. (3) with permission.]

injected since the height does not depend on the feed concentration. For the displacement development to be complete, it is necessary to have convex isotherms (Langmuir type) because of the need for self-sharpening boundaries required for minimum zone cross-contamination. It is also necessary that the operating line intersect the isotherms of all feed components. If this does not occur, the components elute as a usual elution peak. The column efficiency required for complete development increases when the isotherms of the feed components become more similar. Similar to elution chromatography, selectivity is an important parameter. It is controlled by the chemical nature of the stationary phase (and the carrier to some extent).

The choice of the displacer is probably the most critical aspect in the separation optimization. For correct development to occur, the adsorption isotherm of the displacer must overlie those of the feed components. According to Fig. 11.15, different displacers should give identical results provided the operating line does not change. More strongly retained displacers require higher concentrations to work with the same operating line. The concentration of the displacer is a key parameter. Large concentrations provide more concentrated sample zones. The separation time is thus decreased, but the yield may also decrease because the relative magnitude of the boundary regions becomes larger. In addition, large concentrations sometimes generate problems with column reequilibration. It is possible to define a maximum and a minimum displacer concentration. The maximum concentration corresponds to the saturation of the most retained component in the carrier. If the displacer concentration is above the limit, the component precipitates in the solvent. At the minimum concentration the operating line just touches the isotherm of the least retarded compound. If the displacer concentration is below this limit, the least retained solute is eluted in the elution mode and the displacement train is distorted. There is an optimum value of the concentration, which is a compromise between speed and yield.

The displacer must not interact with the feed components and it must be sufficiently soluble in the carrier, easy to remove, not toxic, and so on. A substance that has functional groups similar to those of the feed components and a stronger affinity to the stationary phase is a good selection. For reversed-phase applications, the displacer must have both hydrophilic and hydrophobic groups to be soluble in the carrier (usually an aqueous solvent) and have at the same time a strong affinity to the surface. Salts are often used in reversed-phase applications (for example, ammonium salts), but solvents like butanol can also be selected.

The selection of the carrier is less critical. It must be a good solvent for the displacer and the feed, and must provide at the same time high k'

values. In the case of reversed-phase chromatography, such properties as ionic strength and pH are important.

Although there are no theoretical models which take into account non-ideal flow rate effects, it has been experimentally verified several times that yield decreases as the flow rate increases. There is an optimum flow rate which compromises between the time of separation and the yield. Typical flow rate values are very low. For instance, using analytical columns (4.6-mm id) packed with 5-μm particles, it is necessary to use flow rates in the range of 0.1–0.4 mL/min. Accordingly, separation times are usually quite long and although the load put on the column can be much larger than in elution chromatography (10–100 times more), the throughput is not much larger (usually less than one order of magnitude). Low flow rates are necessary to limit axial dispersion effects. If large particles are used, it is necessary to use even lower flow rates than indicated.

The possibility of using narrower columns in displacement compared to elution chromatography is an attractive feature which has several advantages (less costly equipment). The column length is critical. The column must be long enough for the development to be complete. For a given separation there is a minimum column length. If the column is too short, the contribution of the frontal effect associated with the injection process may become very significant, resulting in abnormal peak shape. It is a very peculiar feature of displacement chromatography that the column length has to be increased when the amount of feed injected is increased (whereas in the elution mode, the column diameter has to be increased). This is because the length of the zones in the displacement train increases with the amount of sample, the zone height being only related to the chemical nature of the system. On the other hand, the column must not be longer than required. Increasing the length above the optimum has, at least, two negative consequences, (1) throughput decreases because separation time increases and (2) yield (or purity) also decreases for a given recovery because additional band broadening in the column decreases the ratio of the length of the plateau to the length of the boundary region. ·

The most significant advantage of displacement compared to elution is that much larger loads can be put on the column. The separation times are much longer, however, and consequently the throughput increase is not as large as what would be expected. It is necessary to regenerate the column after each separation. From the examples given in the literature, it appears that the regeneration procedure is often quite complicated and time consuming (more than 1 h). When the regeneration time and the injection time are included in the calculation of the throughput, the gain compared to elution chromatography is not always very large.

Finally, it must be noted that displacement chromatography requires the use of an on-line analyzer because the detectors usually used do not provide sufficient information on the boundary regions. An analytical chromatograph equipped with short columns for very fast analyses is the best choice.

11.4.2.   Recycling

The technique of recycling is a powerful method for increasing the column efficiency and consequently improving the separation, in both preparative and analytical chromatography (71–73). This technique is much more popular in PLC than in analytical chromatography because it is easier to implement on preparative columns than on analytical ones due to the larger ratio of column volume to system volume.

Two types of operation are possible (71). The first is the usual *closed-loop* recycling using one column as described in Fig. 11.16*a*. It requires

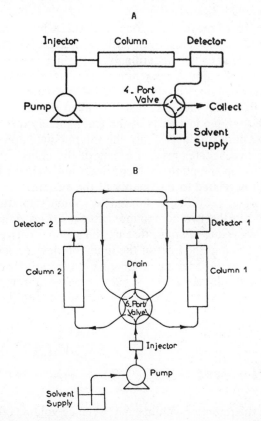

Figure 11.16. (*a*) Closed-loop recycling. (*b*) Alternate pumping recycling. [From Ref. (71) with permission.]

one four-way valve located after the detector to collect the eluate or to direct it to the pump where it is pressurized and reintroduced to the column. Dilution and additional spreading of the sample zone (and then remixing) occur between the column outlet and inlet, particularly in the pumping system. The second type of recycling is called *alternate pumping* recycling and is shown in Fig. 11.16b. It requires two columns (preferably identical), two detectors (or one dual-cell detector), and one six-port two-position switching valve. The position of the valve is changed after the zone to be recycled has been transferred from one column to the other. Because this approach requires two columns, the operating pressure is twice as great as in the closed-loop operation (although it should be noted that the columns do not have to be in series all the time). However, since the sample does not have to go through the pumping system at each cycle, less dilution and spreading occur. The alternate pumping approach is nevertheless not well suited for PLC because of the cost requirements. It will not be further discussed.

A critical aspect of recycling is zone remixing, which occurs when the early eluted bands of the actual cycle start to interfere with the late eluted bands of the previous cycle. Accordingly, recycling of complex mixtures without fraction collection is very limited (71). It is preferable to carry out recycling only on binary mixtures, with intermediate sample collection whenever possible. This method is usually called *shave* recycling (74). At each cycle, the first and last parts of the incompletely resolved zone are collected giving pure products. Only the intermediate zone is recycled. This technique necessitates precise timing of the collection valves. Remixing also exists in the pump, solvent lines, and valves. The extent of zone dispersion is related to the quality of the system.

In the following, recycling is limited to a binary mixture. R_n is the resolution between the two compounds after n cycles (initial resolution R_1). V_A is the *recycling volume*, that is, the volume between the outlet of the detector cell and the inlet of the injection device, and the volume variance of the peak broadening originating in this volume is denoted by σ_A^2. The retention volumes after n cycles (n passes through the column and $n - 1$ passes through the extracolumn volume) and the volume variances are given by

$$V_{1,n} = nV_{R1} + (n - 1)V_A \qquad V_{2,n} = nV_{R2} + (n - 1)V_A \qquad (35)$$

$$\sigma_{1,n}^2 = n\sigma_{C1}^2 + (n - 1)\sigma_A^2 \qquad \sigma_{2,n}^2 = n\sigma_{C2}^2 + (n - 1)\sigma_A^2 \qquad (36)$$

V_{Ri} and σ_{Ci}^2 are the retention volume and variance volume of solute i after one pass through the column. The resolution between solutes 1 and 2 after

n cycles is given by the usual definition

$$R_n = \frac{V_{1,n} - V_{2,n}}{2(\sigma_{1,n}^2 - \sigma_{2,n}^2)} \tag{37}$$

Since recycling is usually made in the case of closely eluted peaks, it can be reasonably assumed that σ_{C2} and σ_{C1} are not very different and equal to σ_C. Combining Eqs. 35–37 gives

$$R_n = R_1 \frac{\sqrt{n}}{\sqrt{1 + (n - 1)\lambda/n}} \tag{38}$$

where λ is the ratio σ_A^2/σ_C^2.

The increase in resolution appears when the number of cycles becomes larger than λ, where λ characterizes the quality of the recycling hardware (the smaller λ, the better the system). It can be seen that the separation is improved after the first cycle only if λ is smaller than approximately 0.5. It is not possible to give an average value for λ because it depends on the column, the hardware (particularly the pump heads), and the solvent flow rate. Most often in PLC, however, λ is smaller than 1.

There is an optimum cycle number for which the remixing resolution (between the first peak of cycle n and the second peak of cycle $n - 1$) is equal to the direct resolution (72). It is given by

$$n_{\text{opt}} = \frac{4(1 + \beta) + (5\alpha - 1)k'}{8k'(\alpha - 1)} \tag{39}$$

where β is the ratio V_A/V_0, V_0 being the column dead volume. Some results are shown in Fig. 11.17 for an ideal system (β and $\lambda = 0$).

When the column is volume and not concentration overloaded, it has been shown (76) that the maximal injection volume with recycling under optimal conditions ($n = n_{\text{opt}}$) is larger than n_{opt} times the maximal injection volume with a single passage. This combined with the fact that the operation time for n_{opt} cycles is roughly equivalent to the time for n_{opt} injections with no recycling indicates that the recycling technique gives higher throughputs than successful injections.

The theoretical optimization of shave recycling is very difficult because the volume of sample recycled is not necessarily the same at each cycle. Most often the optimum number of cycles is smaller than without shaving and better resolution can be obtained.

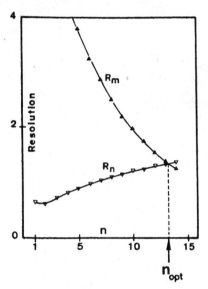

Figure 11.17. Resolutions R_n and R_m versus cycle number n. $\lambda = 2.09$; $\beta = 0.23$; $k_1 = 1.23$; $\alpha = 1.08$. [From Ref. (72) with permission.]

11.5. INSTRUMENTAL ASPECTS OF PREPARATIVE CHROMATOGRAPHY

The most significant differences between a preparative and an analytical chromatograph concern the injection technique and the column technology. These points are discussed in this section. The properties of stationary phases relevant to PLC are also examined. The rest of the equipment is not discussed since it is similar to that used for analytical applications, except for size and flow rate.

11.5.1. Packing Materials

11.5.1.1. Physical Properties

The size of the particles to be used in PLC is a subject of controversy. The actual trend is nevertheless to use smaller particles. There seems now to be an agreement that the "optimum" particle size is in the range of 10–25 μm [see (2, 30, 64)]. Actually, the proper choice depends on how much the column is going to be overloaded (30). If the degree of overloading is so large that the peak profile is independent of the particle size, there is no reason to use small particles. Often these conditions correspond to easy separations for which high efficiencies are not required. If overloading is moderate or the column operated in linear conditions, it is preferable to use small (10–20-μm) particles, which always give more plates than larger particles, whatever the load. The particle size distribution is also a very important parameter which influences both column efficiency and permeability. It should be mentioned that it is not very clear how the particle distribution should be described: weight, size, or number distribution. The situation is even more complex in the case of irregular particles because the concept of size itself is not clear. Nevertheless it is possible to calculate an "average" particle size from the size distribution. The chromatographic meaning of this parameter is questionable, however. It is probably better to define a "hydrodynamic" size which is calculated from Darcy's law, assuming a given value for the permeability constant ϕ,

$$d_P = \sqrt{\frac{\phi \eta L u}{\Delta P}} \tag{40}$$

ϕ is usually 600 for spherical particles and 1000 for irregular ones, η is the solvent viscosity, u the solvent velocity, and L the column length.

When the size distribution is quite large, it is usually claimed that the smallest particles determine the permeability and the largest ones the efficiency. The situation is probably more complicated, as suggested by Dewaele and Verzele (75), who have shown that a large distribution does not impair the efficiency provided the solvent speed is close to the optimum (minimum of the plate height curve) but has a negative effect on the permeability. Some results are given in Fig. 11.18. They show the variation of plate count and pressure drop for columns packed with mixtures of 3- and 8-μm particles in different proportions (weight). For a given mixture, the "average" particle size (X axis) is proportional to the composition. This particle size is *not* the hydrodynamic size, however. The distributions are not symmetrical, except for the mixture 50%–50%. Such an asymmetry is often the case for silicas (75). The column efficiency appears to be linearly related to the average particle size, whereas the pressure increases much more rapidly than expected. The lower part of Fig. 11.18 shows the relative pressure increase, which is almost 200% for the largest distribution. The presence of small particles results in a very rapid increase in column pressure. This suggests that very fine dust should absolutely be avoided (76). It has been proposed that the ratio d_{P90}/d_{P10} be used to characterize the size distribution (d_{P90} and d_{P10} being the sizes

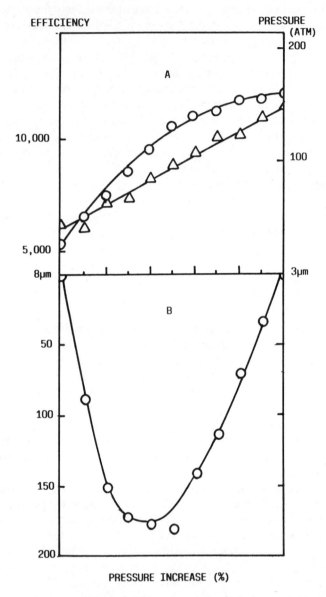

Figure 11.18. (*a*) Plate numbers (Δ) and column pressure (○) of 10- × 0.46-cm id columns packed with 8- and 3-μm ROSIL-C18-D packing materials mixed in 10 proportions as shown. (*b*) Relative pressure increase. Theoretical pressure is calculated from average particle size (see text). [From Ref. (78) with permission.]

456

limiting 90 and 10% of the total weight of particles, respectively) (75). For a good material, this ratio should be less than 1.5.

The shape of the particles is critical, at least as far as price is concerned. Spherical materials are often reported to give larger plate numbers and smaller or similar pressures than irregular ones, but contradictory opinions can be found (77). Column reproducibility has also been said to be related to the shape of the particles. Whatever the truth is, spherical materials are (much) more expensive. Actually it is not simple to make a fair comparison between both types since it is not clear how to define the size of an irregular particle. Moreover, there are several degrees of "irregularity," and disklike particles probably do not behave the same way as rodlike ones. Irregular particles often contain fines (because of the manufacturing process), thus this is probably one of the reasons (if not the only one) why irregular particles often given larger pressures than spherical ones. It has been reported (77) that spherical materials produce 1.5–2 times more plates than irregular ones when the packing technique is optimized. This may be the case, although it seems difficult to know when a packing method is fully optimized.

There are other properties of the silica which are critical for preparative applications. The specific surface area S_{SP} is one of them. It is preferable to use materials with large S_{SP} values in order to obtain large sample capacities. Everything else being constant, the loading capacity is proportional to S_{SP}. It is not a good strategy, however, to use materials with values of S_{SP} that are too large because of the presence of micropores with slow desorption kinetics and possible exclusion effects. A good choice is 300–400 m^2/g. The specific surface area of chemically bonded silicas is necessarily smaller, in the range of 150–250 m^2/g. Other properties such as pH, pore size and size distribution, porous volume, apparent density, and mechanical properties are also important, but cannot be discussed here [see (2, 78) for more details].

The situation is more complex with soft and semirigid materials. Shrinking or swelling and compressibility phenomena are often the source of column instability, in particular at large column diameters (9). Soft gels cannot be used at elevated or even moderate flow rates. This is unfortunate for preparative applications since throughput increases with flow rate. Semirigid and rigid materials recently introduced (79) eliminate the compressibility problem. These materials still potentially suffer from sensitivity to solvent changes, which is almost the rule with polymeric materials. Intense efforts are being made to develop rigid polymer-based packings (79), although at the same time many phases traditionally available as soft gels are now prepared from a silica matrix (80).

11.5.1.2. Chemical Properties

Most of the preparative separations published in the literature are based on straight or nonpolar chemically modified silicas. The chemical properties of these phases are not discussed in this section since abundant information is available [see, for instance, (2)]. Only some specific points are mentioned.

Trace elements should be avoided since they can generate mixed retention mechanisms (such as charge transfer) with associated asymmetrical peaks. In the particular case of bonded silicas, it is very important that the material be *extensively* rinsed after the bonding process. HCl is generated during the chemical reaction and can be trapped inside the pores of the silica. Since these phases are often used with aqueous eluents, irreversible column damage may occur if the column is stored while containing water and traces of acid. This is actually a very critical problem with the use of bonded silicas for process applications.

It is known that silica is soluble at basic pH. Attention must be paid to the solvent pH to avoid dissolution of the silica (particularly with nonpolar bonded materials). This is important in PLC since not only are the columns very expensive, but the collected fractions may be contaminated by shaved bristles. In this respect the use of unbonded silica with aqueous eluents has been reported to give more reproducible results than bonded phases and to be easily scaled up for preparative procedures (81).

The type of moieties grafted is an additional variable for bonded silicas. Most of the work published so far has been done with C_{18} silicas, probably because until recently they were the only materials available for preparative applications. It is difficult to comment on the best chain length, although the situation should be similar to analytical chromatography. It has been mentioned that long chains (C_{18}) allow larger loads (3). On the other hand, a report indicates that for protein purification, short chains (C_4, C_8) give better results (82). Besides nonpolar materials, the use of other types of bonded phases has also been described [see (83)]. Silica coated with silver (argentation chromatography) has been used for preparative purposes as well (84–87).

Polymeric materials are potentially very good candidates for preparative applications because they do not suffer the same chemical limitations as adsorbents based on silica. Moreover, they are inert and eliminate the sample degradation problems sometimes encountered with silica. Some reports describe the use of such stationary phases as polyethylene powder (88), styrene-divinylbenzene copolymers (89–91), polystyrene gels (92), polyamides (93), triacetylcellulose (94), agarose (95), and Sephadex (96), to name just a few. Soft gels do not allow the achievement of high efficiencies and are used for preliminary cleanup rather than for fine separation. The chromatographic characteristics of modern and rigid polymer-based high performance packings have been discussed recently (79), and these phases will probably become popular in modern PLC.

It is very likely that affinity chromatography will become more and more popular in the future. This form of chromatography makes use of highly selective materials which allow high column loadings. The column efficiency (and thus the particle size) becomes less critical. Therefore it is possible to use low-pressure equiment which might be less expensive than high-pressure equipment. On the other hand, it is likely that the price of the packing materials will be significantly larger than that of conventional materials (silica or bonded silica). Intense research efforts are being made in the development of these highly specific adsorbents.

11.5.2. Injection Techniques

The proper injection method depends on the volume to be injected. There are four basic injection procedures: (1) syringe, (2) sample loop, (3) solvent pump (pumping the sample into the column), and (4) "dry" injection. Syringe injections with a septum-type injection port are now seldom used and will not be discussed further. The technique of split-flow injection is related more to the column design and is discussed in the corresponding section.

Injection with a loop, as in analytical chromatography, is only applicable when the injected volume is sufficiently small. The design of the loop must be a compromise between volume (it is typical to inject several milliliters in a 1-inch column, particularly if on-column preconcentration is made), back pressure, and shape of the injection profile. If V_L is the loop volume and D_L its diameter, the pressure drop in the loop is given by

$$\Delta P = \frac{256}{\pi^2} \frac{F V_L \eta}{D_L^6} \tag{41}$$

The pressure increases with the inverse of the sixth power of the loop internal diameter. For instance, at a flow rate of only 10 mL/min, the pressure drop in a 5-mL loop made of 0.5-mm id tubing is more than 8.4 MPa (1200 psi), assuming a solvent viscosity of 1 cP. It drops to 20 psi if 1-mm id tubing is used. This suggests the use of large bore tubing.

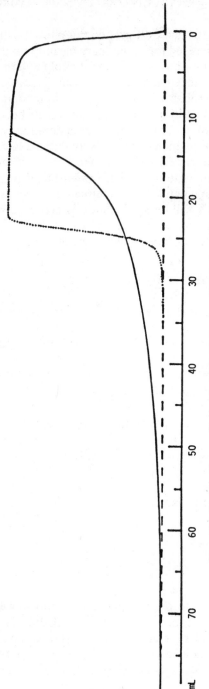

Figure 11.19. Injection profile from sample loop. ——: Empty loop (V = 25 mL, 4.6 mm id). ·····: Packed loop (V = 25 mL, 4.6 mm id, packed with 120-μm beads). Flow rate: 30 mL/min. [From Ref. (99) with permission.]

However, band dispersion inside the loop increases with the square of the diameter according to (44, 98).

$$\sigma_L^2 = \frac{D_L^2 V_L F}{96 D_M} \tag{42}$$

D_M is the diffusion coefficient of the solute in the mobile phase. Equation 42 is based on the assumption of a laminar flow. The peak profile is more deformed (with a strong tailing) when the tube diameter increases. Coiling the loop to produce secondary flow does not greatly help (97). There are two possibilities of compromising between pressure drop and dispersion: (1) pack the loop with an inert material (silanized glass, for instance) using large particles to avoid excessive pressures and (2) use the temporary injection technique (98). The packed loop concept permits the construction of large loops (up to 100 mL) using wide bore tubing (0.5 or 1 inch). Almost perfect rectangular injection profiles can be obtained (see Fig. 11.19) (99). The packing material is used to generate radial mixing and achieve plug flow conditions. Details on the design of the loop are given in Ref. (98). The volume of the loop must be larger than the desired injection volume (about double if a nonporous material is used to pack the loop).

The temporary technique consists of switching the loop on-line during such a time that the tail of the profile is not transferred into the column. A significant part of the sample may be left in the loop. Precise timing of the operation is necessary to avoid excessive sample loss. The loss of sample does not make this technique suitable for PLC.

With a loop injector, at least one valve is required to load the loop and switch it on-line. During valve rotation it is very common to experience a temporary pressure increase (more than several thousand pounds per square inch) because solvent flow is momentarily interrupted. This generates a sudden burst of solvent at the column inlet when the flow is resumed. This has been reported to have a very detrimental effect on analytical columns (100). The situation can be even worse with preparative columns because of the higher flow rates used.

When the injected volume is larger than 50 mL, it is usually preferable to inject with the solvent pump. This technique is almost the only way to inject in large-scale applications where injection volumes can exceed several liters. There must not exist poorly swept zones inside the pump; otherwise strongly tailing peaks are obtained. On some equipment, a special pump is dedicated to sample injection. Almost perfectly rectangular injection profiles can be obtained this way (97). Other devices have been proposed to replace an expensive additional pump. One is to use a pressurized reservoir. The flow rate is controlled by the pressure. Although this technique is very simple, it has a major drawback: the injected volume depends on the column permeability. If for any reason (such as partial clogging) the column permeability decreases, so does the injection volume. This method is not recommended.

The last injection technique to be mentioned is sometimes called *dry injection*. It consists in coating some packing material (either glass beads or preferably an adsorbant similar to the one in the column) with the sample [see, for example, (100, 101)]. The column is opened and the first layers of packing are carefully removed. The coated material is then placed in the void created and the column closed. This method is not very practical, but it is sometimes the only method of injection (in the case of solubility problems, for instance).

11.5.3.   Column Technologies

There are two categories of preparative columns, ''regular'' type and compression type (axial, radial, or both). Glass columns are not included in the following discussion. Besides technological considerations, the major difference between the two categories is that the packing is kept under mechanical compression in the compression columns, whereas it is "free" in regular columns. The two types of columns are discussed in this section and some general considerations on the design of preparative columns are also made.

11.5.3.1.   *General Considerations on Column Design*

An impressive number of papers have been published on the optimum design of preparative columns, with often contradictory conclusions (5, 9, 28, 36, 41, 61, 102–111). The most critical aspect of column design seems to be the column extremities. The reason is that intense band distortion can occur at the column inlet because the flow profile is distorted when the eluent (and the sample) is transferred from the inlet line into the column (112). The change in flow cross section can be as large as several thousandfold. The same phenomenon takes place at the column outlet, where it may actually play an even more critical role. Connected to the design of the column is the problem of transferring the sample onto the packing. In addition the wall effect can also produce intense zone distorsion (35). The so-called point injection with the split-flow technique has been suggested for analytical and preparative columns (41, 113). The flow coming from the pump is divided into two parts. One is used to transfer the sample from the injector to the center of the column and the other one is distributed over the entire cross section of the column. This permits plug flow conditions at the column inlet and prevents sample zone deformation. This technique has been used in PLC with apparently good success [see, for example, (41)].

It is not clear yet what is the best column end design. However, based on the fact that it is now possible to find on the market large packed columns (up to 3 inches in diameter), giving highly symmetrical peaks and plate numbers corresponding to reduced plate heights between 2 and 3 with simple on-line injection (no split flow), it is possible to draw the following conclusions.

1. The design of a column extremity as shown in Fig. 11.20 gives satisfactory results. The end fitting should be machined to provide an empty chamber on top of the frit for proper distribution of liquid

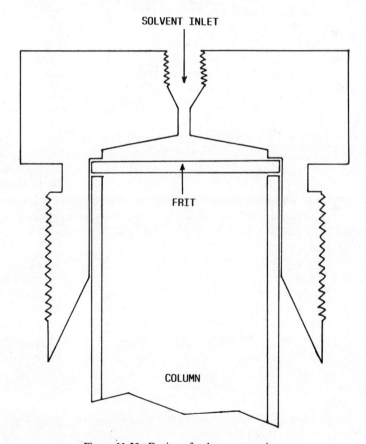

Figure 11.20. Design of column extremity.

over the entire column cross section. The frit must be carefully selected (112). The volume of the chamber should be kept reasonably small to avoid excessive band spreading.

2. For a column with a diameter larger than 1 inch, it is recommended to have a flow distributor to help spread the solvent. Several types of distributors have been tested (9, 41, 111). With the right design it is possible to achieve excellent eluent distribution over the entire cross section of very large columns, even greater than 20 inches (8, 9).

The design shown in Fig. 11.20 usually solves the problem of flow profile distortion but not the problem of wall effect. Experience indicates that this has a lot to do with the packing technique. The packing procedure is particularly critical with regular columns.

11.5.3.2. Regular Columns

It is not necessary to discuss the design of regular columns because they are basically identical to analytical columns, except for the size. With up to 1-inch diameter, normal compression fittings can be used, preferably of the inverted type. Above 1 inch, these fittings are not very practical (too much torque is required to seal the column) and it is better to use flanges.

Packing a preparative column is not a simple matter. The packing procedure is apparently very critical and changing certain parameters may result in columns of completely different qualities (37). Actually, it seems that very often poor results have been justified on the basis of inappropriate column design (or injection technique), whereas they were mostly due to incorrect packing procedures. There are two categories of packing techniques: dry packing and slurry packing, and there are several alternatives in each category (114).

Dry packing is used with large particles (>15–20 μm). It is important to vibrate the column during the filling process, but lateral tapping should be avoided unless it is done very uniformly over the whole circumference of the column. Mechanical devices to dry pack columns have been described (2, 104, 105) as well as the influence of the operating parameters studied (104). Very often the mechanical stability of dry packed columns is not very satisfactory and bed compaction occurs rapidly. This becomes more critical with decreasing particle size. Columns packed this way should not be used at elevated pressures and are not really suitable for modern PLC.

The slurry packing technique is more appropriate. Packing a column is still an art, particularly large-diameter columns. Several methods can be used [see (115) for a review]. The choice of the operating parameters (such as slurry solvent and concentration, packing pressure, and nature of the solvent) is more limited for preparative columns than for analytical ones because of the large volumes involved. There is not much data available on the proper recipe to pack large columns since the information is most often proprietary. Among others, a critical factor seems to be the duration of the packing process. Pumping the solvent too long may generate too much heat in the column and subsequently cause trouble because the expansion coefficients of stainless steel and silica are different.

11.5.3.3.   Compression Columns

There are two types of compression, axial and radial. A new technology combining both aspects has been introduced recently. The various types of systems are shown in Fig. 11.21. Radial compression necessitates the use of columns with flexible walls. The column is placed in a container filled with an appropriate fluid, which is then pressurized (111). The column is consequently compressed and the packing particles are pushed against each other. This technique should eliminate the most common problem with liquid chromatographic columns: the formation of a cavity at the top of the column. The column should then be very stable. Another advantage is the simplicity of the packing procedure. Because of the radial

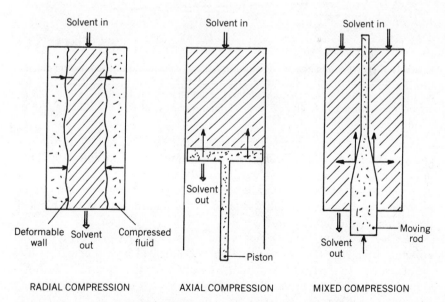

Figure 11.21. Different types of compression techniques.

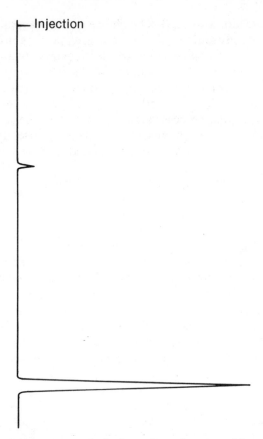

Figure 11.22. Test chromatogram obtained on a large diameter axial compression column packed with 10 μm spherical particles. Column diameter = 30 cm. Column length = 48 cm (bed height). Aromatic test mixture contains uracil and methyl benzoate. Mobile phase: acetonitrile–water 60:40 (v/v); flow rate: 2 L/min. [Courtesy of Eli Lilly and Company.]

compression, it is not necessary to achieve a very dense bed during the packing operation. Radial compression columns are apparently available only with large particles and consequently offer limited efficiencies. It has been reported that the number of compression/decompression cycles is somewhat limited, probably due to a degradation of the mechanical properties of the flexible wall.

The technique of axial compression uses a moving piston inside the column (109, 110). The piston motion is controlled by either gas, a hydraulic jack, or the solvent of the separation itself ("floating" piston). Gas compression (the original method) suffers from a pressure limitation. It is not possible to exceed 400 psi. The hydraulic jack and the floating piston can be used at much higher pressures (up to 1500 psi). Columns up to 20 inches in diameter have been made with a hydraulic jack and up to 6 inches with a floating piston (8). Axial compression makes it very simple to pack and unpack the column. Remarkable results have been obtained using spherical 10-μm particles in a preparative scale axial compression system (Fig. 11.22). Operation of the columns is simple and the columns are very well suited for large-scale process applications.

As indicated earlier, the rest of the equipment will not discussed here. There is abundant literature dealing with various subjects related to instrumentation in PLC. References (116–127) are typical examples.

11.6. CONCLUSIONS

The interest in preparative liquid chromatography is increasing rapidly because of the growing demand for compounds with very high degrees of purity. Modern PLC does not closely resemble the old technique with low-pressure equipment, poor-quality packing materials, and long separation times. Recent technological developments make it possible to pack very large columns with small particles and obtain remarkable efficiencies. This opens news doors in many fields, particularly the pharmaceutical and biotechnology industries.

There are hundreds of publications dealing with applications of PLC, and it is not possible to list them in this chapter. Several review articles are available in which many references can be found (2, 4–6, 16–18, 128–147).

GLOSSARY

C^+	Coefficient equivalent to a concentration (Gareil's model)
C^*	Parameter of Van der Linde model
D_c	Column internal diameter
D_M	Sample diffusion coefficient in mobile phase
D_L	Internal diameter of injection loop
F	Solvent flow rate
H	Theoretical plate height
K	Distribution isotherm coefficient
$K_{1/2}$	Ratio of concentrations of components 1 and 2 in feed
L	Column length
M	Injected mass
M^*	Parameter of Van der Linde model
N	Actual theoretical plate number
N_C	Number of theoretical plates in column

Q_0	Total peak area (Gareil model)
Q_0'	Peak area delimited by the rear part of the profile (see Fig. 11.4)
$Q_{2,\text{IN}}$	Injected quantity of second component (Gareil model)
R_1	Initial resolution (recycle operation)
R_S	Resolution
R_n	Direct resolution after n recycle operations
R_m	Remixing resolution = (second peak cycle $n - 1$)/(first peak cycle n)
S	Parameter characterizing saturation of stationary phase
S_{SP}	Specific surface area
Th	Throughput
Th$_{\text{LIN}}$	Throughput in linear elution
Th$_{\text{NLIN}}$	Throughput in nonlinear elution
V	Solvent volume actually pumped through column
V^*	Parameter of Van der Linde model
V_A	Recycle volume
V_C	Cut volume
V_{INJ}	Injection volume
V_E	Volume corresponding to end of first peak
V_{MP}	Volume of mobile phase in one theoretical plate
V_0	Column dead volume
V_L	Injection loop volume
V_R	Solute retention volume
Z^*	Parameter of Van der Linde model
a, a'	Parameters characterizing peak overlap
a_i	Coefficients of power series describing isotherm
c_M	Solute concentration in mobile phase
$c_{0,i}$	Concentration of component i in feed
$c_{0,\text{LIN}}$	Maximum concentration in linear elution
c_S	Solute concentration in stationary phase
$c(V)$	Output profile
d_P	Particle size
f	Parameter characterizing injection device
f_1, f_2	Functions of k' and α
h	Reduced plate height
k'	Capacity ratio
m_i	Centered moment of order i
n	Number of operation
n_{opt}	Optimum number of recycle operations
p	Purity
q_0	Specific load
r	Recovery

u	Solvent velocity
x	Reduced injection volume
y	Reduced cut volume
ΔP	Column pressure
ΔN	Variation of plate number resulting from injection
Δt	Time interval between two injections
α	Selectivity
β	Relative recycle volume
η	Solvent viscosity
ϵ_T	Total column porosity
θ	Parameter characterizing injection
λ	Relative recycle variance
ν	Reduced velocity
ρ	Packing density
σ_i	Standard deviation of peak i
σ_A	Recycle standard deviation
σ_C	Column standard deviation
τ	Parameter of Gareil model
ϕ	Permeability constant
ψ	Dimensionless loadability

REFERENCES

1. S. Claesson, *Ark. Kem., Mineral. Geol., 23A,* 1 (1947).

2. M. Verzele and C. Dewaele, *Preparative High Performance Liquid Chromatography—A Practical Guideline*, Gent, Belgium, 1986.

3. C. Horvath, A. Nahum, and J. F. Frenz, "High Performance Displacement Chromatography," *J. Chromatogr., 218,* 365–393 (1981).

4. R. Sitrin, P. DePhilips, J. Dingerdissen, K. Erhard, and J. Filan, "Recent Advances in the Preparative Chromatography of Low Molecular Weight Substances," in I. W. Wainer, Ed., *Liquid Chromatography in Pharmaceutical Development: An Introduction,* Aster, Springfield, 1985, pp. 265–303.

5. M. Verzele and E. Geeraert, "Preparative Liquid Chromatography," *J. Chromatogr. Sci., 18,* 559–570 (1980).

6. R. Rosset, "Modern Preparative Liquid Chromatography," *Analysis, 5,* 253–264 (1977).

7. M. Verzele, C. Dewaele, J. Van Dijck, and D. Van Haver, "Preparative Scale High Performance Liquid Chromatography on Analytical Columns," *J. Chromatogr., 249,* 231–238 (1982).

8. PROCHROM, Internal Report (PROCHROM Co., Champígneulles, France).

9. P. Rahn, W. Joyce, and P. Schratter, "Scalability: The Challange of Chromatography," *Ann. Biotech. Lab.*, 34–43 (July/Aug. 1986).

10. P. C. Wankat, "Improved Preparative Chromatography," *Ing. Eng., 23,* 256–260 (1984).

11. M. V. Sussman and R. N. S. Rathore, "Continuous Modes of Chromatography," *Chromatographia, 8,* 55–59 (1975).

12. P. E. Barker, B. W. Hatt, and A. N. Williams, "Fractionation of a Polymer Using a Preparative-Scale Continuous Chromatograph," *Chromatographia, 11,* 487–493 (1978).

13. V. R. Meyer, "Preparative High Performance Liquid Chromatography as an Aid in Organic Synthesis," *J. Chromatogr., 316,* 113–124 (1984).

14. G. Guiochon and H. Colin, "Theoretical Concepts and Optimization in Preparative Scale Liquid Chromatography," *Forum,* 21–28 (Sept./Oct. 1986).

15. K. Nakamura and Y. Kato, "Preparative High Performance Ion Exchange Chromatography," *J. Chromatogr., 333,* 29–40 (1985).

16. R. Sitrin, P. DePhilipps, J. Dingerdissen, K. Erhard, and J. Filan, "Preparative Liquid Chromatography, a Strategic Approach," *LC/GC Mag., 4,* 530–550 (1986).

17. J. Lesec, "Preparative Gel Permeation Chromatography," *J. Liq. Chromatogr., 8,* 875–923 (1985).

18. D. A. Small and T. Atkinson, "Preparative High Performance Liquid Affinity Chromatography," *J. Chromatogr., 266,* 151–196 (1983).

19. R. E. Majors, "Practical Aspects of Preparative Liquid Chromatography," *LC/GC Mag., 3,* 862–866 (1985).

20. R. P. W. Scott and P. Kucera, "Some Aspects of Preparative-Scale Liquid Chromatography," *J. Chromatogr., 119,* 467–482 (1976).

21. J. L. Tayot, "Industrial Development of Chromatography in Biologic Productions," *Biofutur,* 69–77 (Nov. 1986).

22. S. A. Bormann, "Preparative and Process Liquid Chromatography," *Anal. Chem., 57,* 998A–1000A (1985).

23. D. E. Nettleton, Jr., "Preparative Liquid Chromatography. I. Approaches Utilizing Highly Compressed Beds," *J. Liq. Chromatogr., 4(Suppl. 1),* 141–173 (1981).

24. D. E. Nettleton, Jr., "Preparative Liquid Chromatography. II. Approaches on Non-Compressed Beds," *J. Liq. Chromatogr., 4(Suppl. 2),* 359–398 (1981).

25. F. M. Rabel, "Practical Procedures in Preparative LC," *Int. Lab.,* 91–98 (Nov./Dec. 1980).

26. J. J. De Stefano and J. J. Kirkland, "Preparative High Performance Liquid Chromatography, Pt. I.," *Anal. Chem., 47,* 1103–1108 (1975).

27. J. J. De Stefano and J. J. Kirkland, "Preparative High Performance Liquid Chromatography, Pt. II," *Anal. Chem., 47,* 1109–1115 (1975).

28. M. Verzele and C. Dewaele, "Preparative Liquid Chromatography," *LC/ GC Mag., 3,* 22–28 (1985).

29. P. Gareil and R. Rosset, "Preparative and Liquid Elution Chromatography, I. The Basic Concepts, Linear and Non-Linear Chromatography Process," *Analysis, 10,* 397–408 (1982).

30. P. Gareil and R. Rosset, "Preparative and Liquid Elution Chromatography, II. Practical Aspects, Scale-up Strategy and Apparatus," *Analysis, 10,* 445–459 (1982).

31. D. R. Bundle, T. Iversen, and S. Josephson, "Preparative Medium and High Pressure Chromatography," *Int. Lab.,* 27–32 (Nov./Dec. 1980).

32. L. R. Snyder and J. J. Kirkland, in *Introduction to Modern Liquid Chromatography,* 2nd ed., Wiley, New York, 1979, chap. 15.

33. E. Grushka, "Characterization of Exponential Modified Gaussian Peaks in Chromatography," *Anal. Chem., 44,* 1733–1738 (1972).

34. P. Gareil, L. Personnaz, J. P. Feraud, and M. Caude, "Study of Non-Linear Elution in Preparative and Liquid Chromatography," *J. Chromatogr., 192,* 53–74 (1980).

35. J. H. Knox, G. R. Laird, and P. A. Raven, "Interaction of Radial and Axial Dispersion in Liquid Chromatography in Relation to the 'Infinite Diameter Effect,'" *J. Chromatogr., 122,* 129–145 (1976).

36. A. W. J. De Jong, H. Poppe, and J. C. Kraak, "Contribution to the Choice of Optimal Geometric Conditions for Preparative Liquid Chromatography," *J. Chromatogr., 148,* 127–141 (1978).

37. J. C. Sternberg, in J. G. Giddings and R. A. Keller, Eds., *Advances in Chromatography,* vol. 2, Dekker, New York, 1986.

38. H. Colin, unpublished results.

39. B. Coq, G. Cretier, and J. L. Rocca, "Injection Volume and Cut Elution Volume in Preparative Liquid Chromatography," *Anal. Chem., 54,* 2271–2277 (1982).

40. L. Personnaz and P. Gareil, "Effect of Sample Volume in Linear Preparative Chromatography: A More Rigorous Treatment," *Sep. Sci., 16,* 135–146 (1981).

41. B. Coq, G. Cretier, and J. L. Rocca, "Preparative Liquid Chromatography: Sample Volume Overload," *J. Chromatogr., 186,* 485–502 (1979).

42. R. A. Barford, R. McGraw, and H. L. Rothbart, "Large Sample Volumes in Preparative Chromatography," *J. Chromatogr., 166,* 365–372 (1978).

43. A. Wehrli, U. Hermann, and J. F. K. Huber, "Effect of Phase Selectivity in Preparative Column Liquid Chromatography," *J. Chromatogr., 125,* 59–70 (1976).

44. M. Martin, C. Eon, and G. Guiochon, "Pertinency of Pressure in Liquid Chromatography, Part I. Problem in Equipment Design," *J. Chromatogr., 108,* 229–241 (1975).

45. A. Haarhoff and H. J. Van der Linde, "Concentration Dependence of Elution Curves in Non-Ideal Gas Chromatography," *Anal. Chem., 38,* 573–584 (1966).

46. J. Houghton, *J. Phys. Chem., 67,* 84–95 (1963).

47. G. Cretier and J. L. Rocca, "Mass Overload in Liquid Chromatography, Experimental Validity of the Haarhoff–Van der Linde Treatment," *Chromatographia, 18,* 623–627 (1984).

48. H. Poppe and J. C. Kraak, "Mass Loadability of Chromatography Columns," *J. Chromatogr., 255,* 395–414 (1983).

48[1]. J. H. Knox and H. M. Pyper

49. P. Gareil, C. Durieux, and R. Rosset, "Optimization of Production Rate and Recovered Amount in Linear and Non-linear Preparative Elution Liquid Chromatography," *Sep. Sci., 18,* 441–459 (1983).

50. P. Gareil, L. Personnaz, and M. Claude, "Peak Shape Influence on Preparative Chromatography of a Binary Mixture," *Analysis, 7,* 401–407 (1979).

51. G. Guiochon, to be published.

52. L. R. Snyder, *Principles of Adsorption Chromatography,* Dekker, New York, 1968.

53. J. N. Done, "Sample Loading and Efficiency in Adsorption, Partition and Bonded-Phase High-Speed Liquid Chromatography," *J. Chromatogr., 125,* 43–57 (1976).

54. R. A. Wall, "Some Sampling Effects in Liquid Chromatography," *J. Liq. Chromatogr., 2,* 775–798 (1979).

55. A. W. J. De Jong, H. Poppe, and J. Kraak, "Column Loadability and Particle Size in Preparative Liquid Chromatography," *J. Chromatogr., 209,* 432–436 (1981).

56. P. Legoff and N. Midoux, "Energetics and Cost Optimization of Preparative Chromatography Columns in Percolation Processes: Theory and Applications," The Netherlands, 1981.

57. I. Mazsaroff and F. E. Regnier, "An Economic Analysis of Performance in Preparative Chromatography of Proteins," *J. Liq. Chromatogr., 9,* 2563–2583 (1986).

58. K. P. Hupe and H. H. Lauer, "Selection of Optimal Conditions in Preparative Liquid Chromatography," *J. Chromatogr., 203,* 41–52 (1981).

59. G. Cretier and J. L. Rocca, "Preparative Liquid Chromatography: Choice of Column Dimensions and Working Conditions for Occasional Preparative Separations," *Chromatographia, 16,* 32–38 (1982).

60. B. Coq, G. Cretier, C. Gonnet, and J. L. Rocca, "How to Approach Preparative Liquid Chromatography," *Chromatographia, 12,* 139–146 (1979).

61. A. W. J. de Jong, J. C. Smit, H. Poppe, and J. C. Kraak, "Optimum Conditions for High Performance Liquid Chromatography on the Preparative Scale," *Anal. Proc., 12,* 508–513 (1980).

62. H. Poppe, J. C. Kraak, J. F. K. Hube, and J. H. M. Van den Berg, "Temperature Gradients in HPLC Columns Due to Viscous Heat Dissipation," *Chromatographia, 14,* 515–523 (1981).

63. P. Achener, R. Simpson, and F. Flink, "Effect of Radial Thermal Gradients in Elevated Temperature High Performance Liquid Chromatography," *J. Chromatogr., 218,* 123–135 (1981).

64. H. Colin, "Simple Considerations on Column Design in Preparative Scale Liquid Chromatography," *Sep. Sci., 22,* 1851–1869 (1988).

65. A. W. J. de Jong, J. C. Kraak, H. Hoppe, and F. Nooitgedacht, "Isotherm Linearity and Sample Capacity in Liquid Chromatography," *J. Chromatogr., 193,* 181–195 (1980).

66. J. R. Conder and M. K. Shingari, "Throughput and Band Overlap in Production and Preparative Chromatography," *J. Chromatogr. Sci., 11,* 525–534 (1973).

67. J. Frenz, P. Van der Schrieck, and C. Horvath, "Investigation of Operating Parameters in High Performance Displacement Chromatography," *J. Chromatogr., 330,* 1–17 (1985).

68. H. Kalasz and C. Horvath, "Preparative Scale Separation of Polylixins with an Analytical High Performance Liquid Chromatography System by Using Displacement Chromatography," *J. Chromatogr., 215,* 295–302 (1981).

69. C. Horvath, J. Frenz, and Z. El Rassi, "Operation Parameters in High Performance Displacement Chromatography," *J. Chromatogr., 255,* 273–293 (1983).

70. H. Kalasz and C. Horvath, "High Performance Displacement Chromatography of Cortiscosteroids. Scouting for Displacer and Analysis of the Effluent by TLC," *J. Chromatogr., 239,* 423–438 (1982).

71. M. Martin, F. Verillon, C. Eon, and G. Guiochon, "Theoretical and Experimental Study of Recycling in High Performance Liquid Chromatography," *J. Chromatogr., 125,* 17–41 (1976).

72. B. Coq, G. Cretier, J. L. Rocca, and J. Vialle, "Recycling Technique in Preparative Liquid Chromatography," *J. Liq. Chromatogr., 4,* 237–249 (1981).

73. K. E. Conroe, "Recycling in Preparative Liquid Chromatography," *Chromatographia, 8,* 119–120 (1985).

74. S. Mohanraj and W. Herw, "Use of the Peak Shaving-Recycle Technique for Separation of Labdadiene and Labdatriene Isomers by HPLC," *J. Liq. Chromatogr., 4,* 525–532 (1981).

75. C. Dewaele and M. Verzele, "Influence of the Particle Size Distribution of the Packing Material in Reversed-Phase High Performance Liquid Chromatography," *J. Chromatogr., 260,* 13–21 (1983).

76. P. A. Bristow, "Packing Performance and Permeability of Larger and Wider Liquid Chromatography Columns and Their Use in Preparing Samples for Identification," *J. Chromatogr., 149,* 13–29 (1978).

77. M. Verzele, J. Van Dijck, P. Mussche, and C. Dewaele, "Spherical versus Irregular-Shaped Silica Gel Particles in HPLC," *J. Liq. Chromatogr., 5,* 1431–1448 (1982).

78. M. Verzele, C. Dewaele, and D. Duquet, "Quality Criteria and Structure of Silica Gel Column Packing Material," *J. Chromatogr., 329,* 351–357 (1985).

79. J. V. Dawkins, L. L. Llyod, and F. P. Warner, "Chromatographic Characteristics of Polymer-Based High Performance Liquid Chromatography Packings," *J. Chromatogr., 352,* 157–167 (1986).

80. D. R. Nau, "Lecture on Preparative Scale Purification of Protein," presented at the Liquid Chromatography–Biotechnology Interface Seminar, Boca Raton, FL, 1986.

81. J. Adamovics and S. Unger, "Preparative Liquid Chromatography of Pharmaceuticals Using Silica Gel with Aqueous Eluents," *J. Liq. Chromatogr., 9,* 141–155 (1986).

82. J. D. Pearson and F. E. Regnier, "The Influence of Reversed-Phase N-Alkyl Chain Length on Protein Retention, Resolution, and Recovery: Implications for Preparative HPLC," *J. Liq. Chromatogr., 6,* 497–510 (1983).

83. M. Zief, L. J. Crane, and J. Horvath, "Low Pressure Preparative Liquid Chromatography, Amine Bonded Phases," *Int. Lab., 5,* 72–80 (1982).

84. S. Hara, A. Ohsawa, J. Endo, Y. Sashida, and H. Itokawa, "Liquid Chromatographic Resolution of the Unsaturated Sesquiterpene Alcohol Isomers Using Silica Gel–Binary Solvent Systems," *Anal. Chem., 52,* 428–430 (1980).

85. R. R. Heath, J. H. Tumlinson, and R. E. Doolittle, "Analytical and Preparative Separation of Geometrical Isomers by High Efficiency Silver Nitrate Liquid Chromatography," *J. Chromatogr. Sci., 15,* 10–13 (1977).

86. R. R. Heath and P. E. Sonnet, "Technique for in Situ Coating of Ag⁺ onto Silica Gel in HPLC Columns for the Separation of Geometrical Isomers," *J. Liq. Chromatogr., 3,* 1129–1135 (1980).

87. M. Morita, S. Mihashi, H. Itokawa, and S. Hara, "Silver Nitrate Impregnation of Preparative Silica Gel Columns for Liquid Chromatography," *Anal. Chem., 55,* 412–414 (1983).

88. H. Chow, M. B. Caple, and C. E. Strouse, "Polyethylene Powder as a Stationary Phase for Preparative Scale Reversed-Phase High Performance Liquid Chromatography," *J. Chromatogr., 151,* 357–362 (1978).

89. D. J. Pietrzyk and J. D. Stodola, "Characterization and Applications of Amberlite XAD-4 in Preparative Liquid Chromatography," *Anal. Chem., 53,* 1822–1828 (1981).

90. D. J. Pietrzyk, W. J. Cahill, Jr., and J. D. Stodola, "Preparative Liquid Chromatography Separation of Amino Acids and Peptides on Amberlite XAD-4," *J. Liq. Chromatogr., 5,* 443–461 (1982).

91. D. J. Pietrzyk and W. J. Cahill, Jr., "Amberlite XAD-4 as a Stationary Phase

for Preparative Liquid Chromatography in a Radially Compressed Column,'' *J. Liq. Chromatogr., 5,* 781–795 (1982).

92. Y. Kato, T. Kitamura, and T. Hashimoto, ''Preparative High Performance Hydrophibic Interaction Chromatographie of Proteins on TSK Gel Pheny-5PW,'' *J. Chromatogr., 333,* 202–210 (1985).

93. D. W. Lamson, A. F. W. Coulson, and T. Yonetani, *Anal. Chem., 45,* 2273–2278 (1973).

94. K. H. Rimbock, F. Kastner, and A. Mannschreck, ''Liquid Chromatography on Triacetylcellulose. Preparative Separation of Enantiomers on an Axially Compressed Column,'' *J. Chromatogr., 329,* 307–310 (1985).

95. T. Anderson, M. Carlsson, L. Hagel, P. A. Pernemalm, and J. C. Janson, ''Agarose-Based Media for High-Resolution Gel Filtration of Biopolymers,'' *J. Chromatogr., 326,* 33–44 (1985).

96. J. V. Jizba, V. Prikrylova, and H. Lipavska, ''Preparative Chromatography of Epimers and Anomers of Daunomycin Derivatives on Sephadex LH-20,'' *J. Chromatogr., 329,* 193–195 (1985).

97. M. Kaminski and J. Reusch, ''Comparison of Methods of Sample Introduction during Scale-up of Liquid Chromatography,'' *J. Chromatogr., 356,* 47–58 (1986).

98. B. Coq, C. Cretier, J. L. Rocca, and M. Porthault, ''Open or Packed Column Sampling Loops in Liquid Chromatography,'' *J. Chromatogr. Sci., 19,* 1–12 (1981).

99. VAREX, Tech. Note. (Varex Corp., Rockville, MD).

100. J. L. Dicesare, M. W. Dong, and J. R. Gant, ''Influence of Injector Bypass on Lifetime of Small-Particle Liquid Chromatography Columns,'' *Chromatographia, 15,* 1–12 (1982).

101. J. Kriz, M. Brezina, and L. Vodicka, ''Solid Sample Introduction in Preparative High Performance Liquid Chromatography: Separation of Diamantanols,'' *J. Chromatogr., 248,* 303–307 (1982).

102. L. A. Anderson, N. S. Doggett, and M. S. F. Ross, ''Preparative HPLC of the Lipid Fraction of Teucrium Canadense 1,'' *J. Liq. Chromatogr., 2,* 455–461 (1979).

103. D. R. Braker, R. A. Henry, R. C. Williams, and D. R. Hudson, ''Preparative Columns in High-Speed Liquid Chromatography,'' *J. Chromatogr., 83,* 233–243 (1973).

104. J. Klawiter, M. Kaminski, and J. S. Kowalczyk, ''Investigation of the Relationship Between Packing Methods and Efficiency of Preparative Columns. Part 1: Characteristics of the Tamping Method for Packing Preparative Columns,'' *J. Chromatogr., 243,* 207–224 (1982).

105. M. Kaminski, J. Klawitter, and J. S. Kowalczyk, ''Investigation of the Relationship Between Packing Methods and Efficiency of Preparative Columns. Part 2: Characteristics of the Slurry Method of Packing Chromatographic Columns,'' *J. Chromatogr., 243,* 225–244 (1982).

106. K. Prusiewicz, M. Kaminski, and J. Klawiter, "Devices for Packing Preparative Chromatographic Columns by 'Dry Packing' Techniques," *J. Chromatogr., 238,* 232–236 (1982).

107. G. A. Fisher and J. J. Kabara, "Simple, Multibore Columns for Superior Fractionation of Lipids," *Anal. Biochem., 2,* 303–309 (1964).

108. P. A. Bristow, "Packing Performance and Permeability of Larger and Wider Liquid Chromatography Columns and Their Use in Preparing Samples for Identification," *J. Chromatogr., 149,* 13–29 (1978).

109. E. Godbille and P. Devaux, "Use of an 18-mm I.D. Column for Analytical and Semi-Preparative-Scale High-Pressure Liquid Chromatography," *J. Chromatogr., 122,* 317–329 (1976).

110. E. Godbille and P. Devaux, "Description and Performance of an 8 cm i.d. Column for Preparative Scale High Pressure Liquid–Solid Chromatography," *J. Chromatogr. Sci., 12,* 564–569 (1974).

111. J. N. Little, R. L. Cotter, J. A. Prendergast, and P. D. McDonald, "Preparative Liquid Chromatography Using Radially Compressed Columns," *J. Chromatogr., 126,* 439–445 (1976).

112. S. Fujine, K. Saito, and K. Shiba, "Liquid Mixing in a Large-Sized Column of Ion Exchange," *Solv. Extr. Ion Exch., 1,* 113–126 (1983).

113. B. Coq, G. Cretier, J. L. Rocca, and R. Kastner, "End-Effects and Band Spreading in Liquid Chromatography," *J. Chromatogr., 178,* 41–61 (1979).

114. T. N. Webber and E. H. McKerrel, "Optimization of Liquid Chromatographic Performance on Columns Packed with Microparticulate Silicas," *J. Chromatogr., 122,* 243–258 (1976).

115. M. Martin and G. Guiochon, "Review and Discussion of the Various Techniques of Column Packing for High Performance Liquid Chromatography," *Chromatographia, 10,* 194–204 (1977).

116. D. Berger and B. Gilliard, "Design and Applications of a Microprocessor Controlled System to Optimize Preparative Liquid Chromatography," *J. Chromatogr., 210,* 33–44 (1981).

117. F. R. Sugnaux and C. Djerassi, "Complete Computer Automation of Preparative Liquid Chromatography Through Intelligent Fraction Collection, with Unlimited Injection Volume and Repetitive Collection of Separated Solute Peaks," *J. Chromatogr., 248,* 373–389 (1982).

118. A. F. Hadfield, R. N. Dreyer, and A. C. Sartorelli, "Conversion of an Analytical High Performance Liquid Chromatographic System into an Automated Semi-Preparative Unit and Its Application to the Separation of a Mixture of Benzyl and B-D-Glucofuranosides and Glucopyranosides," *J. Chromatogr., 257,* 1–11 (1983).

119. H. Colin, G. Lowy, and J. Cazes, "Design and Performance of a Preparative-Scale HPLC," *Am. Lab., 3,* 55–63 (1985).

120. C. J. Little and O. Stahel, "Role of Column Switching in Semipreparative Liquid Chromatography, Isolation of the Sweetener Stevioside," *J. Chromatogr., 316,* 105–111 (1984).

121. S. A. Martin and L. Chan, "Preparative HPLC. Part I: A Comparison of Three Types of Equipment for the Purification of Steroids," *HRC CC*, 570–576 (1984).

122. J. Miller and R. Strusz, "A UV Detector for Preparative LC," *Int. Lab.*, *3*, 87–93 (1980).

123. E. von Arx, P. Richert, R. Stoll, K. Wagner, and K. H. Wuest, "New Aids for High Performance Liquid Chromatography," *J. Chromatogr.*, *238*, 419–425 (1982).

124. W. H. Pirkle and R. W. Anderson, "An Automated Preparative Liquid Chromatography System," *J. Org. Chem.*, *39*, 3901–3903 (1974).

125. P. A. Bristow, "Computer-Controlled Preparative Liquid Chromatograph," *J. Chromatogr.*, *122*, 277–295 (1976).

126. H. Loibner and G. Seidl, "A Low-Cost Medium Pressure Liquid Chromatography System for Preparative Separations," *Chromatographia*, *12*, 54–67 (1979).

127. L. Garpe, H. Lundin, and J. Sjodahl, "Automated Peak Collection," *Sci. Tools*, *29*, 13–18 (1982).

128. M. Rubinstein, "Preparative High Performance Liquid Partition Chromatography of Proteins," *Anal. Biochem.*, *98*, 1–7 (1979).

129. W. Kullmann, "Liquid Adsorption Chromatography on Preparative Scale of Protected Synthetic Peptides," *J. Chromatogr.*, *2*, 1017–1029 (1979).

130. J. Jumanotani, R. Oshima, Y. Yamauchi, N. Takai, and Y. Kurosu, "Preparative High Performance Gel Chromatography for Acidic and Neutral Saccharides," *J. Chromatogr.*, *176*, 462–464 (1979).

131. J. Rivier, R. McClintock, R. Galyean, and H. Anderson, "Reversed-Phase High Performance Liquid Chromatography: Preparative Purification of Synthetic Peptides," *J. Chromatogr.*, *288*, 303–328 (1984).

132. R. Westwood and P. W. Hairsine, "Use of Routine Preparative High Performance Liquid Chromatography in the Separation of Isomers," *J. Chromatogr.*, *219*, 140–147 (1981).

133. E. Soczewinski and T. Wawrzynowciz, "Thin-Layer Chromatography as a Pilot Technique for the Optimization of Preparative Column Chromatography," *J. Chromatogr.*, *218*, 729–732 (1981).

134. W. H. Pirkle and J. M. Finn, "Preparative Solution of Racemates on a Chiral Liquid Chromatography Column," *J. Org. Chem.*, *47*, 4037–4040 (1982).

135. D. R. Knighton, D. R. K. Harding, J. R. Napier, and W. S. Hancock, "Facile, Semi-Preparative, High Performance Liquid Chromatographic Separation of Synthetic Peptides Using Ammonium Bicarbonate Buffers," *J. Chromatogr.*, *249*, 193–198 (1982).

136. M. R. Kilbourn, D. D. Dischino, C. S. Dence, and M. J. Welch, "Sep-pack Preparative Chromatography: Use in Radiopharmaceutical Synthesis," *J. Liq. Chromatogr.*, *5*, 2005–2016 (1982).

137. P. Pei, J. Britton, Jr., and S. Hsu, "Hydrocarbon Type Separation of Lubricating Base Oil in Multigram Quantity by Preparative HPLC," *J. Liq. Chromatogr., 6,* 627–645 (1983).

138. R. Viville, A. Scarso, J. P. Durieux, and A. Loffet, "Side Reaction of Synthetic Peptides During their Purification by Preparative High Performance Liquid Chromatography Using Formate Buffers," *J. Chromatogr., 262,* 411–414 (1983).

139. S. P. Djordjevic, M. Batley, and W. Redmond, "Preparative Gel Chromatography of Acidic Oligosaccharides Using a Volatile Buffer," *J. Chromatogr., 354,* 507–510 (1986).

140. H. A. Chase, "Prediction of the Performance of Preparative Affinity Chromatography," *J. Chromatogr., 297,* 179–202 (1984).

141. J. Rivier, R. MacClintock, R. Galyean, and H. Anderson, "Reversed Phase High Performance Liquid Chromatography: Preparative Purification of Synthetic Peptides," *J. Chromatogr., 288,* 303–328 (1984).

142. C. C. Ku, S. C. Hwang, and T. A. Jacob, "Semi-Preparative High Performance Liquid Chromatographic Separation of Carbon-14 Labeled Avermectin B_1a from a Mixture of Avermectins," *J. Liq. Chromatogr., 7,* 2905–2914 (1984).

143. M. A. Adams and K. Nakanishi, "Selected Uses of HPLC for the Separation of Natural Products," *J. Liq. Chromatogr., 2,* 1097–1136 (1979).

144. M. P. Strickler and M. J. Gemski, "Protein Purification on a New Preparative Ion Exchanger," *J. Chromatogr., 9,* 1655–1677 (1986).

145. E. H. Cooper, R. Turner, J. R. Webb, H. Lindblom, and L. Fagerstam, "Fast Protein Liquid Chromatography Scale-Up Procedures for the Preparation of Low-Molecular-Weight Proteins from Urine," *J. Chromatogr., 327,* 269–277 (1985).

146. A. M. Cantwell, R. Calderone, and M. Sienko, "Process Scale-Up of a B-Lactam Antibiotic Purification by High Performance Liquid Chromatography," *J. Chromatogr., 316,* 133–149 (1984).

147. M. N. Schmuck, K. M. Gooding, and D. L. Gooding, "Preparative Chromatography of Proteins," *J. Liq. Chromatogr., 7,* 2863–2873 (1983).

CHAPTER

12

PROCESS HIGH PERFORMANCE LIQUID CHROMATOGRAPHY

WILLIAM M. SKEA

Millipore Corporation, Systems Division
Bedford, Massachusetts

12.1. INTRODUCTION

Large-scale open-column chromatography has been used for the purification of chemical and biological compounds for a number of years. Today industrial chromatography can be found in pharmaceutical, biotechnology, and chemical pilot and production facilities throughout the world. Large-scale chromatography is used in the purification of antibiotics, therapeutic drugs, and a variety of therapeutic and diagnostic peptides and proteins. The fractionation and the purification of proteins, especially blood products, are well-known examples of incorporating large-scale chromatography into the separation scheme (1–7). The bulk of the world's supply of insulin has been produced using ion exchange and gel filtration chromatography. Texts have been written which describe the engineering

479

and chemical principles of large-scale chromatographic unit operations (8, 9), and it is common to hear of new stationary phase chemistries and the development of chromatographic techniques that will be used to purify the new generation of biologically active products. Although there is a considerable amount of information available in the literature that deals with large-scale open-column chromatography, the scope of this chapter will focus specifically on the emerging technique of process-scale high performance liquid chromatography (HPLC). This form of chromatography is a recent addition to the field of HPLC and is really an extension of preparative HPLC. The purpose of this chapter is to describe the characteristics of process HPLC, current applications, and a practical step-by-step scaleup procedure that can be used to make the transition from the analytical laboratory to the pilot or the manufacturing plant. Since there are many excellent texts that present a through review of chromatographic principles (10, 11), chromatographic theory will not be reviewed in detail in this discussion.

Process HPLC is a relatively new separation technique for the pilot plant and production areas, but over the last three years it has gained recognition and became increasingly utilized. Although the chemical engineer, biochemist, and chemist working in these areas has experience with open-column chromatography, unfortunately he or she may be unfamiliar with HPLC. This situation is beginning to change as chemical engineering curricula include HPLC in undergraduate and graduate courses. Over the past two years, individuals working in the pilot plant and production areas have seen process HPLC emerge as a complementary technique, and in selected cases a replacement, for the more traditional open-column methods of purification. It is also being incorporated as a unit operation within an isolation train to complement other commonly used isolation procedures such as microporous or ultrafiltration, precipitation, crystallization, and reverse osmosis. As new potent biologically active compounds are developed through synthetic or recombinant DNA methodologies, the criteria being established for purity, yield, and yearly production estimates can be met with currently available commercial process-scale HPLC equipment. Figure 12.1 shows a simplified purification flow chart of a recombinant product, highlighting where process HPLC would be used. Although this technique can be incorporated at any stage within the purification scheme, once the desired product is in solution, it is more commonly found toward the end of the isolation train to yield a high-purity product. Obviously the nature of the feed stream and the desired purification objective will dictate where HPLC can be used most effectively.

The inception and design principles of process HPLC came from the

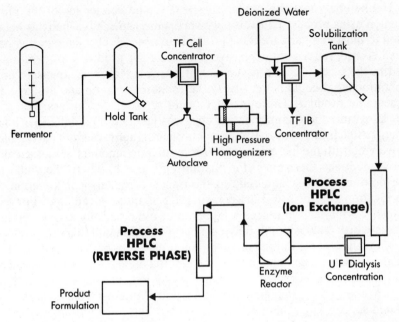

Figure 12.1. Purification flowchart for recombinant product, showing where process HPLC could be utilized. TF, tangential flow; IB, inclusion body; UF, ultrafiltration. (Courtesy of Creative Biomolecules, Hopkinton, MA.)

need to produce larger quantities of pharmaceutical grade compounds than could be purified efficiently by analytical and preparative methodologies. The design principles have originated as extensions of preparative HPLC. The impetus for moving in the direction of process HPLC came first from experimental results on the analytical and preparative scale. Second, it was realized that it was indeed possible to meet the purity and recovery criteria that were demanded for the new high-value-added biological products. Good examples of this are human insulin and luteinizing-hormone-releasing hormone. To meet these requirements, process HPLC systems were needed that would provide relatively fast separations and a degree of purity and yield that was difficult to achieve with existing large-scale techniques. The third reason for the development of process-scale HPLC was the fact that this type of system could be used to purify enough material in a pilot plant, for example, a newly discovered antibiotic, for further testing to determine whether it had the potential of becoming a viable product before optimizing fermentation or synthesis conditions. In essence, it was realized that process HPLC was a way to shorten the develement time for a new biopharmaceutical product.

Unlike analytical and preparative HPLC, the economics of the purification using process HPLC is very important, especially when it is being used to purify a pharmaceutical product. In general, chromatography has been considered a very expensive procedure for production operations, and in the past it has been avoided, if possible, for less expensive unit operations that could better justify the economic criteria and produce an acceptable product. These products were often valued at less than $50 per kilogram, and the purification cost was justified on a dollar per kilogram basis. In most of these cases chromatography was not justifiable. However, with the increasing need for multigram amounts of biologically active peptides (hormones) for pharmaceutical use, where the value of the product is often hundreds to thousands of dollars per kilogram, a purification cost of $40–$200 per kilogram is quite acceptable. For example, Carlbiotech produces a kilogram of a biologically active peptide that is worth $400,000, but only costs a few thousand dollars to produce (12). Further momentum to move in the direction of process-scale HPLC for these types of compounds was the fact that analytical and preparative HPLC were well accepted and often the methods of choice, due to their ability to separate completely the compound of interest from very closely related species in a relatively short period of time. In addition, analytical and preparative HPLC were already proven techniques that had been used for a number of years to separate and purify a wide range of compounds in the research, development, and quality control laboratories.

12.2. COMPARISON BETWEEN ANALYTICAL AND PROCESS-SCALE HPLC

Analytical and process HPLC separations are similar in that the overall objective is to separate a specific compound from its nearest eluting neighbor. To gain a better understanding of the differences between an analytical separation and a process-scale isolation, it is important to understand the characteristics of each scale. Implicit to this discussion is the interrelationship between sample load, speed of separation, and the degree of resolution (Fig. 12.2). Theory and practical experience have shown that one of these attributes can be improved at the expense of the other two, or two of these can be optimized by decreasing the third. For example, in an analytical separation, the ultimate goal is baseline resolution between the compounds of interest and the nearest eluting sample component. The amount of sample injected onto the column is small, on the order of micrograms or nanograms. In this situation one can expect good resolution in a reasonable amount of time. Resolution will decrease as

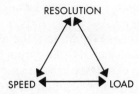

Figure 12.2. Interrelationship of load, speed, and resolution.

the sample load is increased or an attempt is made to shorten the analysis time. At the other extreme is the process purification where the chromatographic goal is expressed in terms of production rate or throughput, which is defined as a specified amount of usable product isolated per unit time. The amount of material that meets specific purity and recovery criteria and time are important considerations. Knowing that resolution will decrease as load is increased, and that at high mass loadings resolution will decrease as separation time is decreased, it is obvious that baseline resolution, as observed on the analytical scale, will not always be achieved at the process scale. The fact that baseline resolution cannot always be achieved does not mean that pure compounds with excellent recoveries' are not produced at the process scale.

There are specific parameters common to HPLC that describe the separation. The primary factors that determine the degree of separation between two compounds include column efficiency N, column length L, mobile phase velocity μ, the particle diameter of the chromatographic packing d_p, the capacity factor of the component of interest k', and selectivity α. Other parameters which will influence the separation include temperature T, column permeability ϵ, the sample size and concentration of the solute C_i, and the diffusion coefficient of the solutes K. However, for practical analytical purposes, only three main parameters are measured and then used in the well-known resolution equation, Eq. 1, to describe the separation mathematically,

$$R_s = \frac{1}{4} \frac{\alpha - 1}{\alpha} \frac{k'}{1 + k'} N^{1/2} \tag{1}$$

Although this is a good starting point and applicable in an analytical situation, once the preparative/process scale is reached, the situation becomes more complicated and the chromatography less well characterized. In other words, at low mass loadings, when distribution isotherms are linear, elution profiles are well defined and provide narrow Gaussian bands. This is characteristic of analytical analysis, and the influence of the various chromatographic factors has been described in detail elsewhere [see (10, 11)].

As one approaches a process separation, the mass and volume loading are greatly increased to achieve a desired throughput or production rate. Hupe and Lauer (13) have expressed mathematically the production rate (the amount of solute i that can be separated per unit time by a specified number of theoretical plates to achieve the required resolution), using the relationship

$$\text{production rate} = A\epsilon_t \mu c_i D \left(\frac{1}{N} - \frac{H}{L} \right)^{1/2} \tag{2}$$

where A is the column cross section, ϵ_t the porosity of the packing, μ the linear velocity, c_i the concentration of the solute, D the ratio of the injection volume to its volume standard deviation at the column inlet, N the plate number, L the column length, and H the theoretical plate height of the column. For a detailed description of how this expression was derived see (13). More recently, Knox and Pyper (14) have described their approach to maximizing throughput. The system is run in what is considered an overload condition, with eluting peaks having widths generally greater than 20% of a corresponding analytical run where very small amounts and volume of sample are injected. One attempts to achieve and maximize the optimum system load, which is feed concentration dependent. Unger and Janzen (15) have described system loadability, at least preparatively, as the maximum sample input $Q_i(\text{max})$ by the relationship

$$Q_i(\text{max}) = 2\pi^{1/2} A \epsilon_m L (C_{i,m})^{\text{max}} (1 + k_i') N^{-1/2} \tag{3}$$

where A is the column cross section, ϵ_m the porosity of the chromatographic bed, L the column length, $(C_{i,m})^{\text{max}}$ the maximum elution profile at the column outlet, k_i' the capacity factor of solute i, and N the column plate number. The maximum load (mass) that can be placed on a chromatographic column has been described by several teams of investigators (16–18). Scott and Kucera (16) used the following relationship to describe the maximum load:

$$M = a\pi r^2 L K d A_s \frac{b d_p^2}{L} c^{1/2} \tag{4}$$

where a and b are constants, L is the column length, K the solute distribution coefficient, d the packing density of the adsorbent (grams per milliliter), and A_s the surface area of the adsorbent. In more practical terms, the maximum load that a process HPLC system can tolerate is determined by the mass that can be handled per cycle and the maximum volume that can be accepted before a predefined loss in resolution is found or there is product breakthrough in the void volume. As larger amounts of solute enter the column, a point is reached where the distribution isotherm becomes nonlinear. The resulting elution profiles are characterized by broad peaks due to isotherm nonlinearity and the fact that the sample feed may be diluted over a large extent of the column because of column dispersion. Column dispersion becomes an insignificant factor at extremely high loads, and distribution nonlinearity is the predominant contributor to the broad elution profile (19). The amount of feed volume that can be loaded onto the column per cycle has been reviewed preparatively (16, 20–23) and is dictated by the mobile phase–solute–stationary phase adsorption kinetics (distribution equilibrium), column dimensions, stationary phase characteristics (particle size, particle type, available surface area, and packing density), and overall production demands. The maximum feed volume that can be processed is also directly proportional to the concentration of components in the feed mixture, which is a result of the solubility characteristics of the components in the mixture, product yield limits, and prior processing steps. Another way to view the differences between analytical and process HPLC is to look at what effect the stationary phase particle size has on overall resolution.

As is known analytically, a primary factory in the degree of resolution R_s is column efficiency N, and particle size is directly proportional to efficiency; the smaller the particle, the higher the efficiency. At the analytical scale, where the goal is optimum resolution, column efficiency is very important. Columns are packed with small particles in the range of 5–10 μm in diameter. As a separation is scaled, load will have a pronounced effect on column efficiency, as has been discussed. Figure 12.3 shows the effect of load on the efficiency of a 10 μm silica column. With increasing load there is an accompanying decrease in efficiency (plates). Figure 12.4 shows the effect of increasing load on column efficiency by comparing two different size silica particles with varying sample injection devices, while Fig. 12.5 shows a comparison between three different C18 particle sizes. As expected, as the load increases, the column efficiency decreases. Once the mass load reaches a point where the sample load is generally greater than 10 mg of sample per gram of packing (this will vary depending on the compound), another facet of process HPLC becomes evident which distinguishes it from analytical chromatography. At low loads, small particles provide more efficient columns than do large particles. At very high loads, generally greater than 10 μg of sample per gram of packing, the efficiency of a nominal 80 μm particle becomes similar to

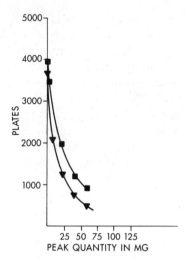

Figure 12.3. Decrease in efficiency as sample load is increased. [From Rausch and Heckendorf (38); reproduced with permission from the authors and Pergamon Press.]

the efficiency of a 10 μm particle under the same operating conditions. Therefore, at extremely high loads it is possible to operate with particles of relatively large diameter. This not only decreases the system back pressure, it also reduces the cost of the packing, which is an important consideration in process HPLC. It is possible to tailor the packing material

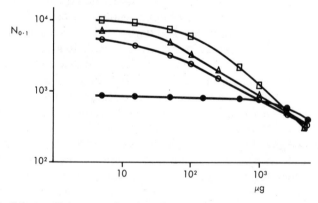

Figure 12.4. Column efficiency as a function of load. Varying amounts of an injected solute, 2,4-dimethylphenol. N was calculated with σ_t measurements at 0.1 of peak height ($N_{0.1}$). Three different injection systems were used: ○,●, central injection; △, conical disperser; □, flow surrounding conical disperser. Particle sizes: ○, △, □, 5–8 μm; ●, 20–25 μm. Column, 10 mm id × 250 mm; packing, silica gel SI 60 (Merck); mobile phase, 2,2,4-trimethylpentane-butanol-1 (99:1, v/v); flow rate, 85 μL/s; detector, UV. [From De Jong, Poppe and Kraak (39); reproduced with permission from the authors and the *Journal of Chromatography*.]

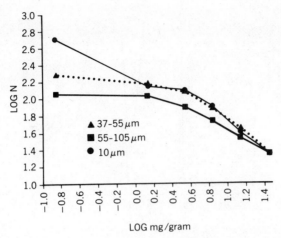

Figure 12.5. Column efficiency as a function of load between three different C18 particle sizes (33).

to the approximate loading and efficiency needed to accomplish a specific separation. Simply stated, if increased resolution is required, the system load is decreased and smaller particles should be used. It is still possible to obtain fairly high production rates under these operating conditions, although the operational cost will increase. The optimum sized packing material then becomes a function of the desired process separation.

From the above studies and relationships it follows that the type of elution profiles will differ between process and analytical HPLC. The elution profile will differ depending on whether the system is run in volume or mass overload. These situations have been characterized and practical examples are shown in Figs. 12.6 and 12.7 [taken from the study of Eisenbeiss et al. (23) on column overloading in preparative HPLC]. Figure 12.6 shows that in volume overload, the elution profile displays broad rectangular symmetrical peaks. Although the front of the peak has the same retention time as an analytical analysis, the peak tends to spread toward a greater retention time. Figure 12.7 is an example of mass overload which shows that as the mass is increased, the retention time decreases and the eluted peaks become asymmetric and characterized by sharp fronts and excessive tailing.

In summary, the overall objective of analytical, preparative, and process chromatography is to separate the compounds of interest from other components in the mixture. The mechanism for separation is essentially the same, although the chromatographic profiles between analytical and preparative/process separations are dissimilar due to nonlinear effects. The reason for these differences are the result of the differences in the

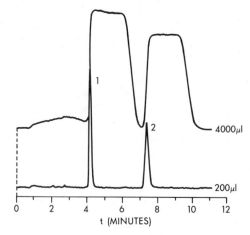

Figure 12.6. Volume overload. Lower profile represents a 200-μL injection of 2.5 μg each of pentylbenzene (1) and octylbenzene (2). Upper profile represents a 4000-μL injection of solutes 1 and 2 at 50 μg each. Column, LiChroCART^R 250-4; packing, 7 μm LiChrosorb RP-8; mobile phase, acetonitrile–water (75:25, v/v); flow rate, 1.5 mL/min; detector, UV, 254 nm, 0.04 AUFS. [From Eisenbeiss et al. (23); reproduced with permission from the authors and *Chromatographia*.]

mass and volume loadings characteristic of each type of chromatography. Furthermore, the particle size usually differs between analytical and process separations, and the size used will depend on the required resolution, the desired load, and the economic considerations for the specific separation.

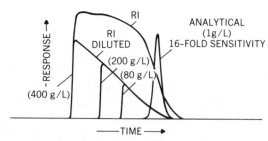

Figure 12.7. Mass overload. 30-μL injections of solutions containing 1, 80, 200, and 400 g/L diethylphthalate. Column, Hibar^R RT, 250-4; packing, LiChrosorb Si 60, 5 μm; mobile phase, n-heptane–ethylacetate (90:10, v/v); flow rate, 1 mL/min. Elution monitored by refractive index; for 1-g/L concentration detection was at full flow at 16-fold sensitivity; for concentrations of 80, 200, and 400 g/L the eluent was diluted fourfold prior to detection. [From Eisenbeiss et al. (23); reproduced with permission from the authors and *Chromatographia*.]

12.3. DESIGN CHARACTERISTICS OF PROCESS HPLC

Although several process HPLC systems have been described in the literature (24–26), this section is general in nature and revolves around systems and components that are provided by commercial manufactures. A process HPLC, in isocratic or gradient configuration, has the same type of components as are found in an analytical or preparative HPLC—pump, column, detector, and recording device. The difference is in the size and electrical coding of the components to meet production rates and explosion-proof requirements. Typically, electric equipment used in process HPLC systems must meet Class I, Division 1, Group D requirements of the National Fire Protection Association Code 493. Class I, Division 2 also seems to be acceptable in certain situations, although the exact code requirement will usually depend on the specific installation safety policies. Figures 12.8–12.11 show several commercially available process HPLC systems. Not shown in these figures is the requirement for accessory items

Figure 12.8. Millipore Kiloprep^R 1000 process HPLC showing self-contained radial compression chamber with pumping and detector modules attached. (Courtesy of Millipore Systems Division, Millipore Corp., Bedford, MA.)

Figure 12.9. YMC process HPLC. (Courtesy of YMC, Inc., Morris Plains, NJ.)

such as vessels for solvent storage, feed handling, fraction collection, and waste storage. An example of such a facility is shown in Fig. 12.12. The type and extent of these accessory vessels will depend on the facility, the products being purified, and the type of solvents being handled.

A schematic of a process HPLC installation is shown in Fig. 12.13. Solvent, feed, collection, and waste vessels are often hard-plumbed to the process HPLC with appropriate valves to direct flow along the desired fluid path during the separation. Waste solvents are recycled by using distillation apparatus and storage tanks to hold redistilled solvents prior to mobile phase makeup. Off-line prefiltration of crude feed materials is used to prevent particulates from building up on the head of the column, thus preventing back-pressure buildup and prolonging column life. It is also possible to install in-line filtration between the mobile phase tanks and the main pumps, although there is the risk of starving the pump if flow is decreased below a level that keeps the pump heads flooded.

A typical pump used in a process HPLC system is shown in Fig. 12.14. Diaphragm pumps that can operate at back pressures of 1000–2000 psi are used as mobile phase pumps in most commercially available systems, although piston-type pumps that can operate at 3000–4000 psi back pressures are also available for use with small-diameter (10 μm) packing materials. The choice of a diaphragm or piston driven pump will depend on the expected back pressures encountered and on the hazardous nature of the material being processed. The primary reason that diaphragm pumps

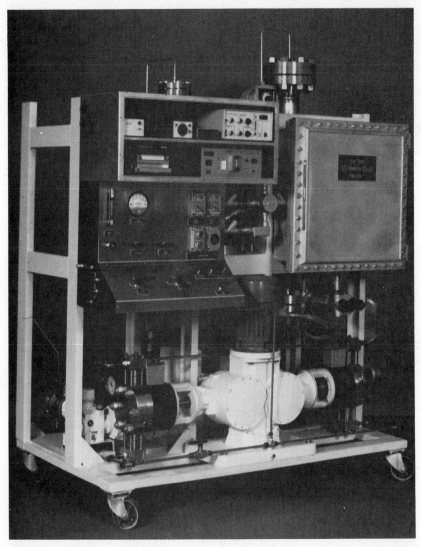

Figure 12.10. Separations Technology ST/process 2000 HPLC. (Courtesy of Separations Technology, Wakefield, RI.)

Figure 12.11. Prochrom VE300 series process HPLC showing axial compression column. Module at right is slurry reservoir and pumping unit for hydraulic jack. (Courtesy of VAREX Corp., Rockville, MD.)

are chosen is the fact that there are no piston seals to replace, and with hazardous materials worker exposure is minimized.

The introduction of sample feed to the column is accomplished via the main pump or by a separate dedicated sample pump. Since the sample feed is usually in a large volume due to solubility constraints, inherent low feed concentration initially, or the result of a prior processing step, it is easier to pump the sample directly onto the column, as opposed to trying to operate a valve/loop injector.

The most important component of the system is the column, since this is where the separation takes place. As already discussed, the mode of chromatography and the mathematical approaches to determining theoretical production and feed input rates have been described for preparative/process-scale chromatography. There are a number of packing chemistries that can be used for process HPLC. These include normal phase, reversed phase, ion exchange, size exclusion, and affinity chemistries (15, 27). Again, these packings are the same as those used in the preparative area and range in size from 15–20 μm to 200 μm. An excellent review of the currently available packing materials is provided in (15). A variety of pore sizes are also available and range from 60 Å to 500 Å. For many HPLC process separations particles having a size range of 40–80 μm can be used. With the new biopharmaceutical products, especially peptides, columns having particle sizes from 15 to 30 μm are more commonly utilized. Since the load is lower for these products, a smaller particle is used to increase the resolution needed to meet purity and recovery

Figure 12.12. Process HPLC facility with accessory storage, feed handling, fraction collection, and waste storage tanks.

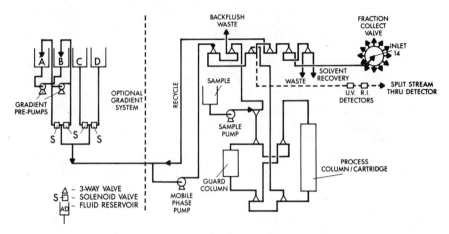

Figure 12.13. Schematic of process HPLC installation.

requirements. Assuming one has optimized the separation with regard to the stationary phase chemistry, two critical factors that will greatly affect the separation on the process scale are fluid distribution and the uniformity/stability of the chromatographic bed.

To achieve the resolution necessary for process HPLC separations, the moving liquid phase must be evenly distributed across the entire diameter of the column, and its frontal movement along the column must be uniform to minimize sample dispersion. This requires that the column have fluid distributor plates at the inlet and outlet ends to uniformly spread and collect the moving eluent. It is important that the fluid be evenly distributed and collected to allow feed components to interact with the entire chromatographic bed and to allow individual components, or poorly resolved compounds, to be either collected in small volumes or recycled back through the column to achieve better separation. Figure 12.15 shows several fluid distribution patterns. In Fig. 12.15a it can be observed that a point source (no distribution) will disrupt a uniform flow pattern, and will result in poor resolution of the feed mixture. If nothing else, the desired compound will elute in a large volume, often requiring additional concentration steps. Figure 12.16 illustrates a typical column fluid distribution system, although there are a number of fluid distributor designs, all having the same primary aim: to distribute fluid evenly across the head of the column.

A second important consideration in a successful process HPLC separation is the chromatographic bed density and homogeneity. The column must be well packed and free of voids, and the stationary bed structure must remain stable with time. If the bed structure changes due to energy

Figure 12.14. Process HPLC diaphragm pump. Patented double diaphragm design with no intermediate fluid between diaphragms prevents contamination of process liquid. (Courtesy of Bran & Luebbe, Buffalo Grove, IL.)

fluctuations caused by varying mobile phase conditions, the bed density and uniformity will become heterogeneous. This will affect flow distribution and impact the adsorption kinetics of the sample feed with the packing. The disruption of flow distribution is depicted in Fig. 12.15b. The end result is poor resolution of the feed components and dilution of the desired compound.

A third variable within the packed column that will decrease resolution

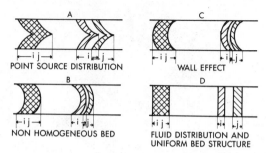

Figure 12.15. Influence of column bed structure and inlet/outlet fluid distribution on chromatographic profiles.

is the extent of the wall effect. Although this point has been discussed and mathematically shown to be insignificant in large-scale open-column chromatography (6), the effect in process HPLC is still poorly understood. The contribution, if any, will distort flow profiles, as can be observed in Fig. 12.15c. Again, this will result in poor resolution and dilution due to band broadening. All of these effects on the column packing density and homogeneity result in the degradation of the overall resolution and dilution of eluting components. These negative effects of poor flow distribution and bed uniformity must be minimized to achieve a successful separation. Commercial manufacturers of process-scale HPLC systems have attacked this problem by using well-known bed stabilization techniques, such as radial compression (Millipore Corporation) or axial compression (Prochrom, formerly Elf Aquatine). There are other packing techniques used in addition to these, and all have the same intent: bed uniformity and homogeneity. Essentially each technique utilizes forces to keep the stationary bed structure intact and uniform for extended periods of time. These techniques are employed to help maintain reproducibility from run to run and to ensure longer column life. The effect of good fluid distribution and bed uniformity will help produce a more concentrated product and will have a positive effect on resolution. This is depicted in Fig. 12.15d.

Monitoring the separation can be accomplished with a refractive index detector, a fixed or variable ultraviolet/visible (UV/VIS) detector, or a light-scattering detector (Fig. 12.17). Other monitoring techniques such as conductivity can also be used. The detector should be designed to meet explosion-proof requirements if it is being used in an explosion-proof area, such as a pilot plant or production facility, unless other safety precautions have been designed into the plant. This means that the electrical components must be inside an approved air purged structure and the fluid stream should be isolated from the electrical components. The sensitivity

Figure 12.16. Typical inlet/outlet column/cartridge fluid distributors. (Courtesy of YMC, Inc., Morris Plains, NJ.)

of analytical detectors is not needed in the process environment, since the mass of eluting material is usually large and the detector will be easily overloaded. Detectors that are designed and commercially available for process HPLC are of two types, full flow and split flow. At present the most common detector is the split-flow version where a small portion of the main stream is diverted through the detection device. Although this stream could be blended into the main fluid path upon exiting the detector, it is collected individually or discarded, because it is difficult to correct and then match for the differences in the fluid velocities of the two streams. Ideally, the full-flow detector is the detector of choice, since

Figure 12.17. Model 404 and model 481 explosion-proof refractive index and variable wavelength UV/VIS detectors. (Courtesy of Millipore Systems Division, Millipore Corp., Bedford, MA.)

one can cut fractions without being concerned about delay times and reblending of a split stream. At this stage in the development of process HPLC, there are few detectors that can accept full flow. Most of the flow cells being used cannot tolerate flows greater than 1–1.5 L/min. However, this is rapidly changing, and several variable-wavelength detectors that have flow cells capable of accommodating more than 6 L/min are appearing on the market. Figure 12.18 is a schematic of an adjustable path length cell that can handle flow streams of more than 6 L/min. Industrial explosion-proof monitors do exist that can accept full-flow streams, but

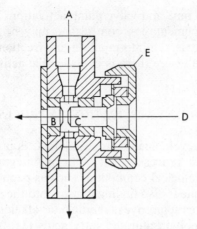

Figure 12.18. Adjustable path length flow cell. Mobile phase enters cell along path outlined by *A*. Light travels along axis *D* through two sapphire rods (*B* and *C*). Rod *B* is fixed in its position while position of rod *C* varies along axis *D* by turning adjustment knob *E*. Fluid flows between and around rods to minimize back pressure. (Courtesy of Linear Instruments Corp., Reno, NV.)

these are usually limited to a single visible wavelength or a narrow wavelength spectrum, and are inadequate for process HPLC systems.

An advanced detection system that is also commercially available is the split-flow detector where the diverted stream, after passing through the detector, goes into a multiport sample valve connected to an in-line analytical HPLC system. This system is useful when the process column split-flow detector becomes overloaded (saturated), since the in-line analytical HPLC system will analyze a fraction of the fluid stream. At this point the system will not provide an instantaneous analysis of the stream and will lag behind the eluting material from the process column. This is an area where there is an opportunity for investigation to improve the speed of analysis and to provide information in a very short period of time to help make accurate decisions relative to sample purity and the blending of fractions.

Commercial process HPLC systems can also be automated to control gradient elution, activate valves, perform data acquisition, and run the system unattended. Automation packages use personal computers or a specially designed system controller. These devices are typically not explosion-proof and are located in a control room. They may have a peak sensing capability (slope sensitivity) to collect specific fractions containing the desired product, the ability to perform recycle and fraction collection through the automation of valves, and automated loading of crude feed based on cycle time and valve/pump activation. Data acquisition is mainly used for documentation control and may be tied into a central computer. As with research into rapid on-line fraction analysis, the area of automation is one where there is considerable activity.

12.4. CURRENT APPLICATIONS IN PROCESS HPLC

At this point in its development, process HPLC is primarily being used in biopharmaceutical processing. It is being integrated into isolation schemes to purify biological compounds such as peptides and antibiotics (28–30). The technique is also finding its way into the speciality chemical area to purify photographic dyes, therapeutic alkaloids, prostaglandins and intermediates, polyunsaturated fatty acids (31), essential oils, and natural products. Unlike analytical and preparative chromatography, process HPLC separations often do not find their way into the scientific literature. For obvious reasons, the scale of separation and the pre- and postprocessing steps are proprietary. Due to the proprietary nature of the pharmaceutical and chemical industries, and the fact that this technique is still relatively new, a review of the literature shows very few references as to the diverse applications that process HPLC is actually being used for.

To date the isolation and purification of peptides has been the primary application where process HPLC has been the natural choice. This stems from the fact that HPLC is a well-accepted technique for peptide purification on the analytical and preparative scale. An example of this is the purification of the analogues of luteinizing-hormone-releasing hormone (LH-RH). LH-RH is a decapeptide, and an analogue is produced by making a modification in the sixth position with a D-amino acid. These analogues are being investigated for the treatment of malignancies, reproductive disorders, and in the control of fertility. Carlsberg Biotechnology (Copenhagen, Denmark) manufactures these analogues by linking two fragments, either LH-RH(1-5) and LH-RH(6-10) or LH-RH(6-9)NHEt together. Process HPLC is used in the purification scheme to isolate the enzymatic peptide synthesis products, LH-RH(1-3)NH_2 and LH-RH(1-5)NH_2 and the final decapeptide. Figure 12.19 shows the separation of the 1-3 fragment on the process scale, and Fig. 12.20 shows the separation of the 1-5 fragment on the process scale. The sample load, 100–200 g, has been reduced to maximize purity. The purity of a fraction shows that it is possible to achieve acceptable purity. Figure 12.21 is the purification of the final decapeptide. In this separation, the packing chemistry has been changed and the particle size has been reduced from 80 μm to 15–

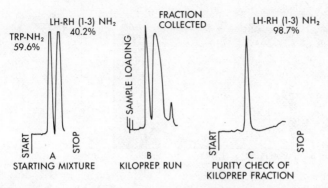

Figure 12.19. Process-scale separation of LH-RH (1-3)NH$_2$ (PyrogluhistrpNH$_2$). Sample feed—200 g in 2.5-L reaction mixtures; column—Kiloprep^R 20 cm id × 60 cm; packing—PrepPAK^R C18, 80 μm; mobile phase—50mM acetic acid–ethanol (9:1, v/v); flow rate—1100 mL/min; detection—UV, 300 nm, 2.0 AUFS. (With permission from Carlsberg Biotechnology, Copenhagen, Denmark.)

20 μm to help maximize resolution. In addition, the separation time has been increased to over 3 h to allow for higher purity. The pooled fraction of the desired product is shown in Fig. 12.21*b*. The overall purity of the product and the recovery are greater than 98% (32).

Another application for which process HPLC has been used is the isolation and purification of antibiotics from fermentation broths and chemical syntheses. An example of this type of application can be illus-

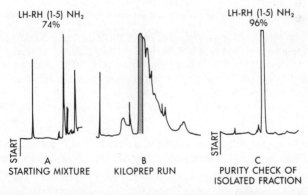

Figure 12.20. Process-scale separation of LH-RH (1-5)NH$_2$ (PyrogluhistrpsertyrNH$_2$). Sample feed, 100 g in 13-L water; column—Kiloprep^R 20 cm id × 60 cm; packing, PrepPAK^R C18, 80 μm; mobile phase, A = 50mM acetic acid, B = 99% ethanol; gradient conditions, linear gradient from 5 to 20% B over 45 min; flow rate, 2000 mL/min; detection, UV, 280 nm, 2.0 AUFS. (With permission from Carlsberg Biotechnology, Copenhagen, Denmark.)

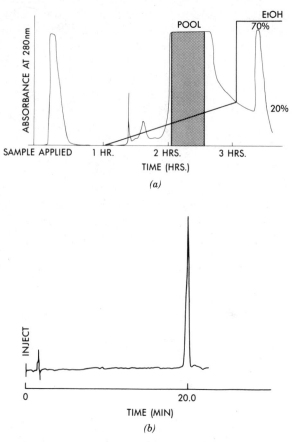

Figure 12.21. (*a*) Process-scale separation of final LH-RH analogue. Sample feed, 152 g of synthetic reaction mixture in 4 L of NH_4 acetate, pH 5.0. Column, Kiloprep^R 20 cm id × 60 cm; packing, Vydac™, 15–20 μm; mobile phase, A = 50mM acetic acid, B = 99% ethanol; gradient conditions, 1 h at 0% B, linear gradient from 0 to 20% B over 2 h, step to 70% B at 3 h; flow rate, 300 mL/min; detection, UV, 280 nm, 2.0 AUFS. (*b*) Analytical separation of pooled LH-RH fraction from process-scale isolation. Column, Radial-PAK™ 8 mm id × 10 cm; packing, NOVA-PAK™ C18, 4 μm; mobile phase, A = 0.1M NH_4 acetate, pH 3.0, B = 0.1M NH_4 acetate (pH 3.0)–acetonitrile (50:50, v/v); gradient conditions, linear from 30 to 80% B over 20 min; flow rate, 2 mL/min; detection, UV, 220 nm, 0.05 AUFS. (With permission from Carlsberg Biotechnology, Copenhagen, Denmark.)

trated with the scaleup of the synthetic β-lactam antibiotic cefonicid (developed by SmithKline Beckman Corporation) (28). Figure 12.22 shows the analytical separation of crude cefonicid (peak eluting at 10 min) from other impurities derived from the synthesis reaction or generated during storage. Figure 12.23 is a large-scale HPLC purification of the crude cefonicid showing contaminants well resolved from the component of interest. However, many of the antibiotic purifications currently being performed with process HPLC involve investigational antibiotics. In these cases, the compound of interest is usually in concentrations less than 0.1% of the total dissolved solids, and the aim is to isolate enough material for more detailed testing with regard to the antibiotic's efficacy, structural studies, and/or toxicology before deciding whether it is feasible to optimize the fermentation or synthesis.

12.5.   DEVELOPING A PROCESS-SCALE HPLC SEPARATION

The transition from an analytical HPLC separation to a large process-scale HPLC isolation is a systematic process. However, the transition can be confusing if a systematic approach is not followed, and therefore a general description of scaleup is helpful to review. Let us begin with

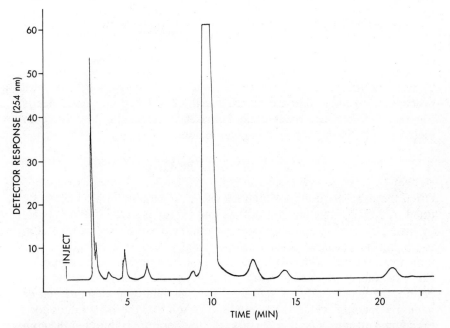

Figure 12.22. Analytical separation of crude cefonicid. Sample, 20 μL of 20 mg/mL solution. Column, 3.9 mm id × 30 cm; packing, μBondapak™ C18, 10 μm; mobile phase, methanol–0.1M ammonium phosphate–H_2O (12.5:10:77.5, v/v), pH 5.3; flow rate, 1 mL/min; detection, UV, 254 nm. (From Cantwell, Calderone and Sienko (28); reproduced with permission from the authors and the *Journal of Chromatography*.)

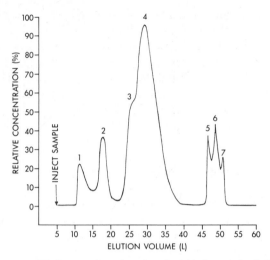

Figure 12.23. Process HPLC separation of crude cefonicid. Sample feed, 270 g (10% solids, 81% purity). Column, 20 cm id × 60 cm; packing, PrepPAK^R C18, 80 μm; mobile phase, H_2O; detection, conductivity (peaks 1–4) and refractive index (peaks 5–7). (From Cantwell, Calderone and Sienko (28); reproduced with permission from the authors and the *Journal of Chromatography*.)

an analytical separation and scale it to a pilot/production purification. This can be used as a guide for scaleup, although one may modify these procedures to those that work best for the individual's specific needs. Figure 12.24 outlines a systematic approach to scaling, and a more complete description of each step is given below.

12.5.1. Define the Problem

Often the exact nature or perception of the separation is not clearly defined. A considerable amount of time and energy is saved in the long run by taking the time to define the specifics of the desired purification before beginning any chromatographic development study. At this stage, existing data about the compound of interest and the sample matrix are compiled. For example, it is helpful to know the molecular structure of the compound; the wavelengths, if any, for absorbing light; the aqueous and organic reagents in which the compound is soluble; the pK_a values for a precolumn cleanup step if necessary; the temperature stability of the compound; the compound's stability in common chromatographic solvents; and whether there are special handling precautions due to toxicity. For the feed matrix, it is important to know the concentration of the compound of interest; where the sample comes from, whether from a fermentation,

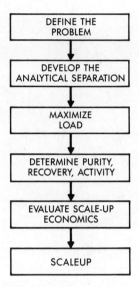

Figure 12.24. Systematic approach to scaling.

an organic or an enzymatic synthesis reaction, or from plant or animal tissue; and whether there are any metabolites or isomers of the compound of interest present in the sample feed.

After the above information has been obtained on the compound of interest and the sample matrix, the next step is to determine realistic criteria for purity and recovery. Ideally one would like to achieve 100% purity and 100% recovery. In fact, it will be possible to obtain 100% purity on selected fractions, and to achieve 100% overall recovery. However, what should be defined is the "usable" purity and recovery. This point is best illustrated by referring to Fig. 12.25 and Tables 12.1 and 12.2. Figure 12.25 is a chromatogram of a large-scale separation of plant alkaloids used for cancer chemotherapy, showing where fractions were taken during the purification step. Table 12.1 shows the results of analytical HPLC analyses of these fractions. As can be seen, fractions 1 through 11 contain the product of interest, which varies from 4 to 99.8% purity. Fractions 7 through 9 contain product that is greater than 99% pure. Assuming the defined usable purity requirement was equal to or greater than 99.0%, fraction 6 could be combined with fractions 7, 8, and 9 (Table 12.2). Although this would lower the overall usable purity, the purity of the combined fractions would still be greater that 99.3%. The usable recovery in this case would be 75.4%. Note that fractions 1 through 5 and 10 and 11 would not be combined, although they contain 23% of

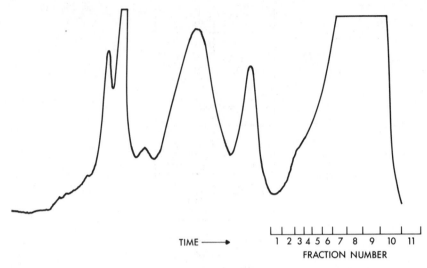

TIME ⟶   1 2 3 4 5 6 7 8 9 10 11
              FRACTION NUMBER

Figure 12.25. Process-scale separation of therapeutic plant alkaloids. Sample feed, 100 g extract in 100 L of water; column, Kiloprep^R, 20 cm id × 60 cm; packing, DeltaPAK^R C18, 15–30 μm; mobile phase, H_2O–95% ethanol-diethylamine–H_3PO_4 (48.9:50:0.8:0.3, v/v); flow rate, 1000 mL/min; detection, 280 nm, 2.0 AUFS.

the product. These fractions could be pooled and reprocessed, or they could be combined with more crude feed. Figure 12.26 shows the relationship between purity and usable recovery. As can be seen for this particular separation, the usable recovery is lowered considerably as the requirement for greater than 99% product purity is reached.

Table 12.1. Analysis of Fractions from Process HPLC Purification

Fraction Number	Volume (mL)	Concentration (mg/mL)	Product Recovered (grams)	Product Purity (%)	Total Mass of Fraction (grams)
1	1500	0.11	0.17	5.08	3.37
2	2000	0.55	1.10	25.65	4.30
3	1000	1.00	1.00	69.28	1.44
4	1000	1.47	1.47	85.36	1.72
5	1500	2.25	3.37	92.52	3.65
6	1500	3.28	4.92	98.24	5.01
7	2000	2.39	4.77	99.83	4.78
8	2000	5.83	11.67	99.58	11.72
9	2500	6.14	15.34	99.20	15.46
10	3000	1.62	4.85	92.03	5.27
11	2500	0.02	0.04	4.30	0.93

Table 12.2. Effect of Combining Fractions on Usable Purity and Recovery

Combined Fractions	Product Purity (%)	Product Recovery (%)
7, 8	99.65	33.76
7–9	99.43	65.26
6–9	99.27	75.36
5–9	98.67	82.29
5–10	97.90	92.24
4–10	97.45	95.26
3–10	96.62	97.31
2–10	90.90	99.58
1–10	85.79	99.93
1–11	84.48	100.00

12.5.2. Develop the Analytical Separation

Now that information has been obtained about the compounds of interest and the sample matrix, and purity and recovery criteria have been considered, it is time to develop an analytical separation. This point in the scaleup procedure is the most critical, since the chromatographic conditions that will be established will be used in the process-scale chromatographic step. These same conditions may also be used in the analytical analysis of crude material and fractions from the large-scale separation. It is important, therefore, to take the time to optimize the parameters for Eq. 1, that is, establish a retention value k', optimize selectivity α, and be aware of the solvents that will be used and their compatibility with other downstream processes and recovery systems.

The first task is to define the mode of chromatography. This will depend on the sample, its solubility characteristics, and the sample matrix. At this stage, analytical columns or cartridges containing 5 to 10 μm packing materials will be used. Several modes may be attempted before selecting the one that serves the chromatographic purpose best. Since the chemistry chosen will be the one that is scaled with, it is helpful to chose a chemistry that is available in various particle sizes.

Once the mode of chromatography has been determined that will be best suited for the sample, a mobile phase must be selected that is compatible with the compound being separated and with the mode of chromatography being employed. There are several key factors to keep in mind when selecting a chromatographic mobile phase. To minimize run time and solvent consumption, the mobile phase should be modified to

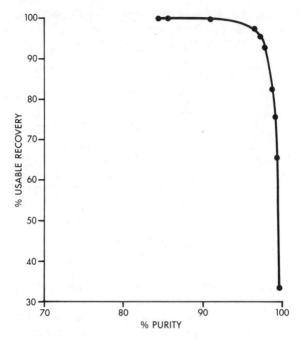

Figure 12.26. Product purity versus usable product recovery.

obtain a k' value of the component of interest between 3 and 6. A k' value in this region allows one to obtain a reasonably high load, it will keep the volume of mobile phase within economic limits, and it permits the separation to be completed in a sufficient amount of time. Second, time must be devoted to optimize selectivity α and obtain as large a value as possible. The larger the selectivity, the greater the load. Time spent optimizng these chromatographic parameters will provide for higher loads durng the process separation. The third factor to consider is the compatibility of the mobile phase with other downstream processes and recovery systems. It will save time, effort, money, and sample if solvents that will not fit into the overall process scheme are determined early in the development process, and not when the process HPLC step is being implemented.

12.5.3.   Maximize Load

When the mode of chromatography and the optimum mobile phase for the separation have been established, the next step is to maximize the sample load and determine the operating parameters that can be expected during the process HPLC purification. The maximum load can be determined using analytical or semipreparative columns packed with the same chemistry and particle size packing that will be used on the large scale. Using this type of column permits the use of analytical equipment to establish maximum load and allows one to simulate a process separation. The objective at this stage is to load as much material as possible before reaching a predetermined R_s value or the compound of interest appears in the void volume (breakthrough).

Typically, a stainless steel column that is in the range of 4.0–8.0-mm id × 25–30 cm in length is used as the scaling column. It is important to emphasize that the column should be packed with the same media that will be used in the large-scale chromatographic system. This column is used to make the transition from the analytical separation (nanograms or milligrams) to the large-scale separation (grams to kilograms). Using a small amount of sample, it represents a convenient and economical way to confirm that the chemistry chosen for the isolation is the most appropriate. In addition, this column can be used with analytical equipment to estimate packing life, define column regeneration procedures, and investigate the overall economics of the process HPLC separation.

The first step in maximizing load is to equilibrate the scaling column with mobile phase and inject an equivalent amount of material that was run on the analytical column. This will give an indication of the difference in resolution between the analytical and the scaling columns. The resolution will decrease, as has already been shown, since the particle size is different between the two columns and, therefore, the column efficiency will be different.

Incrementally increasing the load (sample mass per gram of packing) is used to optimize the system capacity. It is also possible to determine system loadability theoretically, assuming certain parameters are known, as evidenced in (14–16, 23). Usually a more empirical approach is taken. Before beginning these experiments the type of detection system that is best suited should be considered. If an analytical ultraviolet/visible (UV/ VIS) wavelength detector is being used, increasing mass will rapidly exceed the linear range. In other words, the detector will become overloaded very early in the loading study. Therefore, if possible, it may be easier to use a refractive index (RI) detector when doing an isocratic separation to enable the operator to monitor the progress during the loading study. The RI detector monitors refractive index changes (mass) rather than light absorption or transmission, and will not be as sensitive as a UV/VIS detector. On the other hand, for separations requiring a gradient or for component mixtures (such as peptides) that must be monitored by UV/ VIS, it will be easier to conduct loading studies by operating at a wavelength different from the optimum wavelength used for analytical work.

It should be noted that by operating in this fashion the absorption response may be nonlinear. Another consideration to keep in mind is the detector that will be used with the large-scale separation. The analytical and process detection systems should be kept the same. For example, if an RI detector will be used on the process system, then an RI detector should be used during the loading studies. The consistency will make it easier to compare quickly and interpret results from a scaling column with those from a process column before doing any analytical workup.

In order to shorten the time for determining the maximum load to be used on a large scale, the loading study should begin by injecting a fairly large amount of sample onto the column. For example, the study might begin with 10 mg of sample per gram of packing load. It is important to collect fractions and run analytical analyses to determine that the system is not oversaturated, that is, the product is not eluting in the void volume (breakthrough). Assuming breakthrough is not observed, and that the resolution is still adequate to provide the purity and recovery required, the

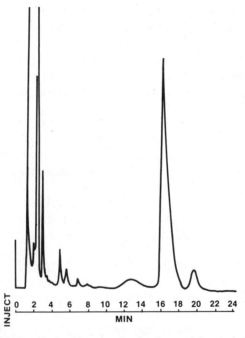

Figure 12.27. Analytical analysis of fermentation broth containing cephalosporin c. Sample amount, 25 μL injection of clarified broth containing 2.4 mg/mL cephalosporin c; column, 3.9 mm id $\times$ 30 cm; packing, μBondapak^{TM}, 10 μm; mobile phase, 20mM NH$_4$ acetate, pH 5.2; flow rate, 2.0 mL/min; detection—254 nm, 0.5 AUFS.

load could be doubled to 20 mg of sample per gram of packing. The loading study is conducted by continuing to double the load until one of the following situations exists: a predetermined R_s is reached or the product appears in the void volume. At this point the load would be decreased until the desired resolution criterion has been achieved. Once the mass loadability is achieved, the maximum volume that can be loaded onto the column must be determined. This is accomplished by increasing the volume at the maximum solute concentration until the desired resolution is no longer achieved or material elutes in the void volume. The scaling progress is monitored (purity and recovery criteria) by analyzing collected fractions via analytical HPLC techniques or by another acceptable analytical method.

An example of what has been discussed can be illustrated with a comparison of chromatographic profiles between an analytical analysis, a scaling study, a preparative separation, and a process purification using a generic antibiotic. In this example, a fermentation broth containing cephalosporin C was initially clarified by removing cells and other particulate matter with microporous flow filtration prior to HPLC analysis and scaleup (29). Figure 12.27 shows an optimized analytical analysis of 0.5 mg of crude feed per gram of packing. The cephalosporin C is well separated from other components and metabolites in the mixture. Figure 12.28 is a chromatogram of 120 mg of crude feed per gram of packing on a scaling column. This represents the maximum load (mass and volume)

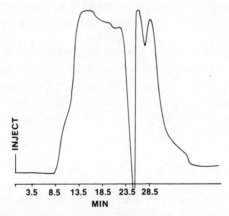

Figure 12.28. Loading study of fermentation broth containing cephalosporin c. Sample amount, 21 mL of clarified broth at a solids concentration of 45.7 mg/mL; column, 7.8 mm id × 30 cm; packing PrepPAK^R C18, 80 μm; mobile phase, 20mM NH$_4$ acetate (pH 5.2)–acetonitrile (95.5:2.5, v/v); flow rate, 2 mL/min; detection, refractive index, 2×; sample load, 120 mg/g packing.

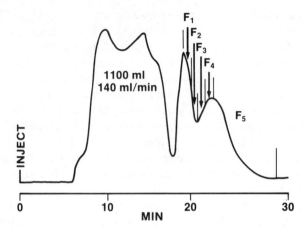

Figure 12.29. Preparative isolation of cephalosporin c from clarified fermentation broth. Sample feed—1100 mL of clarified broth at a solids concentration of 39.3 mg/mL, 3.6 mg/mL ceph C; column, 5.7 cm id × 30 cm PrepPAK^R cartridge; packing, PrepPAK^R C18, 80 μm; mobile phase, 20mM NH$_4$ acetate (pH 5.2)–acetonitrile (97.5:2.5, v/v); flow rate, 140 mL/min; detection, refractive index, 2×; sample load, 120 mg/g packing.

that could be run through the column without adversely affecting the separation. In Fig. 12.29 the same load of crude feed is used per gram of packing separated at the preparative scale, and Fig. 12.30 shows a separation at the process scale. The chromatography between the simulated, preparative, and process scale is essentially identical. Although the chromatography between the scaling column study and the process isolation is not as well defined as the chromatography displayed by the analytical separation, the analysis of defined fractions (Table 12.3) shows that it is possible to achieve 93% purity in one pass with 100% recovery of the desired product. The final purity of the product will depend on where the product fractions are taken and on the desired usable recovery. As discussed previously, as the requirement for product purity increases, the usable product recovery decreases.

12.5.4. Determination of Purity, Recovery, and Activity

There are several methods that may be used to measure purity and recovery. These methods involve either the direct analytical measurement of an eluting fraction or the analysis of a column fraction that has been taken to dryness and reconstituted in a small volume before it is analyzed. With an analytical HPLC measurement of an eluting fraction, or the "solution assay method," the peaks of interest will be quantitated by peak

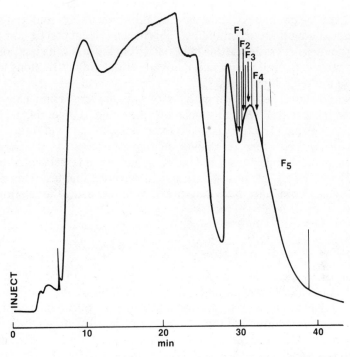

Figure 12.30. Process-scale purification of cephalosporin C from clarified fermentation broth. Sample feed, 1600 mL of clarified broth at a solids concentration of 37.5 mg/mL, 4.5 mg/mL ceph C; column, Kiloprep^R 15 cm id × 60 cm; packing—PrepPAK^R C18, 80 μm; mobile phase, 20mM NH$_4$ acetate (pH 5.2)–acetonitrile (97.5:2.5, v/v); flow rate, 1.0 L/min; detection, refractive index, 8×; sample load, 120 mg/g packing.

Table 12.3 Summary of Cephalosporin-C Purification Using Process HPLC

Fraction Number	Volume (mL)	Cephalosporin-C Concentration (mg/mL)	Cephalosporin-C Recovered (grams)
1	450	2.98	1.34
2	800	8.94	7.15
3	900	14.23	12.81
4	1200	16.27	19.52
5	5800	5.70	33.06
Total	9150		73.88

Feed volume: 16 L; feed concentration, 5.76 mg/mL; purity of combined fractions 1–5 = 93%.

513

area or peak height using standards and calibration curves and by following common quantitative techniques. This measurement can be compared with an analysis of crude starting material. From these results it is possible to estimate the overall purity and recovery of individual fractions and to calculate the overall purification.

A second method of determining purity and recovery is one which uses a "dry weight assay." With this technique the liquid sample, that is, crude sample or collected fraction, is taken to dryness, usually via rotary evaporation or freeze-drying. A known amount of sample is weighed and dissolved in the mobile phase, and the component of interest is quantitated by analytical HPLC using standard quantitative techniques. This procedure provides a more accurate measure of purity by using the relationship

$$\% \text{ purity} = \frac{\text{area}_{\text{sample}}/\text{weight}_{\text{sample}}}{\text{area}_{\text{standard}}/\text{weight}_{\text{standard}}} \tag{5}$$

With this assay technique a comparison is made between the peak heights or peak areas of a known standard concentration of the compound of interest and an unknown concentration of crude or collected fraction containing a certain amount of the compound of interest in a solution that

Table 12.4. Scaling Data Input

Variable	Input
1. Scaling load	45 mg/g
2. Column/cartridge cost	$7500/column/cartridge
3. Feed purity	55%
4. Recovery of product	98.5%
5. Mobile phase flow rate	3 L/min
6. Cycle time	35 min/cycle
7. Column/cartridge life	250 cycles
8. Operation time	3720 h/yr
9. Mobile phase cost	$0.20/L
10. Labor rate	$30/h
11. System cost	$85,000
12. Ancillary equipment cost	$40,000
13. System useful life	5 yr
14. Value of crude feed	$120/kg
15. Value of pure product	$550/kg
16. Rejuvenation solvent cost	$0.25/L
17. Rejuvenation volume	75 L
18. Cycles/rejuvenation	10

also contains dissolved buffer salts and minor contaminants, which have added to the dry weight. The solution assay method does not account for extraneous materials such as buffer salts. Although the dry weight assay method will give a lower purity calculation, it is a more accurate measure of overall purity.

Another important factor is the measurement of the compound's con-

Table 12.5. Economic Analysis of Process HPLC Derived from Scaling Column Studies

Process Operating Conditions	
Column load/cycle	450 g/cycle
Feed purity	55%
Product recovery	98.5%
System flow rate	3 L/min
Cycle time	35 min/cycle
Column/cartridge life	250 cycles
Operation time	3720 h/yr
Mobile phase cost	$0.20/L
Labor rate	$30/h
System useful life	5 yr
Value crude product	$120/kg
Value pure product	$550/kg
Rejuvenation solvent cost	$0.25/L
Rejuvenation volume	75 L
Cycles/rejuvenation	10
System Output	
System feed rate	0.77 kg/h crude
System product rate	0.42 kg/h
System product rate	1555 kg/yr
System Economics	
Value of purified product	$550/kg product
Value of production	
Labor	$ 72/kg product
Column/cartridge cost	$123/kg product
Solvent cost	$ 94/kg product
Depreciation	$ 16/kg product
Operating cost	$305/kg product
Value of crude product	$222/kg product
Total value of production	$527/kg product
Contribution	$ 23/kg product
Contribution (annual)	$35,765

tinued suitability to do the job it was intended to do. One must determine, using an appropriate assay, that the molecule has not been altered during the chromatographic step. For example, if an antibiotic was being purified, it would be important to determine that the molecule was still biologically active. Further structural studies using such techniques as nuclear magnetic resonance or mass spectroscopy would be conducted to verify that the molecules had not been altered during the chromatographic step. Similarly, it would also be important to determine the activity of an isolated enzyme or peptide using a biological activity measurement, and if feasible, to perform an amino acid analysis to determine structural integrity.

12.5.5. Evaluate Scaleup Economics

Once the optimum load (mass) and volume have been determined for the large-scale system using a simulated process with a scaling column, it is possible to investigate the economics of the process to determine throughput and overall chromatographic costs. This type of analysis is accomplished to determine whether the chromatographic purification is economically viable. There are a number of variables that can be considered. The variables listed in Table 12.4 are fairly comprehensive, although there are other variables, such as energy costs, that could be included. Using these parameters and data generated from loading studies, column life, and rejuvenation experiments, it is possible to estimate the cost of the chromatographic step. It is also possible to determine the effect of changing certain variables on the overall cost. Table 12.4 lists the input data, which include certain assumptions such as labor cost, ancillary equipment cost, and the depreciation of equipment. For this particular analysis a very conservative figure of 250 cycles was used for column life. Table 12.5 is a completed analysis showing the value of production and the contribution that a process HPLC system could offer to the product. To calculate the system feed and production rates, the following equations were used:

$$\text{system feed rate (kg/h)} = \text{column load} \times \frac{60}{\text{cycle time}} \qquad (6)$$

$$\text{system production rate (kg/h)} = \text{system feed rate}$$
$$\times \% \text{ feed purity} \times \% \text{ recovery} \qquad (7)$$

$$\text{system production rate (kg/yr)} = \text{system feed rate}$$
$$\times \% \text{ feed purity} \times \% \text{ recovery} \times \text{operation time} \qquad (8)$$

The value of production was calculated from the following equations:

$$\text{labor (\$/kg)} = \frac{\text{labor rate}}{\text{system production rate (kg/h)}} \tag{9}$$

kiloprep cartridge ($/kg)

$$= \frac{\text{number of segments} \times \text{cartridge cost}}{\text{cartridge life} \times \text{column load} \times \% \text{ feed purity} \times \% \text{ recovery}} \tag{10}$$

$$\text{solvent cost (\$/kg)} = \frac{\text{mobile phase cost} \times \text{ flow rate} \times 60}{\text{system production rate (kg/h)}}$$

$$+ \frac{\text{rejuvenation solvent cost} \times \text{rejuvenation volume}}{\text{cycles per rejuvenation} \times \text{column load} \times \% \text{feed purity} \times \% \text{recovery}} \tag{11}$$

$$\text{depreciation (\$/kg)} = \frac{\text{ancillary equipment cost} + \text{kiloprep cost}}{\text{system useful life} \times \text{system production rate}}$$

$$\tag{12}$$

This economic analysis shows that with column loads of 450 grams per cycle, a 35-min cycle time, and the ability to operate 3720 hours per year (approximately two shifts per day with 10% downtime), it will be possible to produce 1555 kg of product per year. The analysis also shows that it will cost $305 to produce a kilogram of product that has a purified value of $550. Since it will take 1.85 kg of the crude material (55% pure) to produce 1 kg of pure material, the value of the crude material is estimated to be $222. Taking this into account, the total value of production becomes $527/kg of product. Therefore the contribution that process HPLC would provide in this case would only be $23/kg, or $35,765 per year. From the results of this analysis, process HPLC would probably not meet an economic justification. However, let us look at three important variables and why optimizing these variables can impact the overall chromatographic processing cost.

Table 12.5 shows that the cost of purifying this particular compound is estimated to be $305/kg. This assumes that the loading is 450 grams of crude material, the column lasts 250 cycles before it has to be replaced, and eluting solvent is discarded after every cycle. Let us look first at how increasing column life can affect the overall operating cost. Again, a scaling column can be used to determine column life. This type of study can provide information using a small amount of solvent and a limited amount of sample compared to the quantities of solvent and labor that would be required to perform this analysis on a large scale. An example of this kind of investigation has been described (33) using a partially purified heat labile preparation of a proprietary antibiotic. The crude antibiotic feed was repeatedly run through the scaling column at the same operating conditions, column load (milligram of feed per gram of packing) and linear velocity that would be used at the process scale. Figure 12.31 summarizes the results from this study. The elution of the product was monitored by plotting the retention time k' against the injection number. As the retention time decreased, due to the nonspecific adsorption of material in the crude feed mixture, the resolution (purity) of the primary product decreased. When the k' value reached 2.5–3.0, the column was rejuvenated with 1.8 column volumes of ethanol. The type of solvent and the volume required to clean the column were determined during the scaleup experiments. Although solvents such as methanol and acetonitrile were also able to rejuvenate the column, they were considered not to be as pharmaceutically acceptable as was the ethanol. The study was allowed to continue for 400 injection cycles before being terminated. From these results, and from data collected during other column life experiments, it is possible to predict that the packing material would last substantially longer than 250 cycles before it would have to be replaced. By assuming that column life can be increased to 800 cycles before the column is replaced, it is possible to reduce the processing cost from \$305 to \$220 (Table 12.6) for a 28% reduction in overall operating cost per kilogram of material produced.

A second important variable focuses on load. Again, maximum load is determined by using the scaling column. In the initial example, an estimate of 450 grams of feed material was chosen for each cycle. If it is possible to increase the load from 450 grams to 900 grams by making modifications to the chromatographic mobile phase or by simply pushing

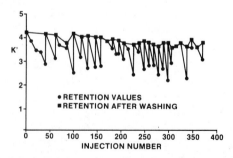

Figure 12.31. Injection number versus k'. Sample feed, 8 mL of proprietary antibiotic solution; column, 7.8 mm id × 30 cm; packing PrepPAK^R C18, 80 μm; mobile phase—H$_2$O; flow rate, 2.5 mL/min; detection, refractive index, 4×.

Table 12.6. Effect on Process Economics by Improving Column Life

Column/cartridge life	From 250 to 800 cycles

System Output

System feed rate	0.77 kg/h crude
System product rate	0.42 kg/h
System product rate	1555 kg/yr

System Economics

Value of purified product	$550/kg product
Value of production	
Labor	$ 72/kg product
Column/cartridge cost	$ 38/kg product
Solvent cost	$ 94/kg product
Depreciation	$ 16/kg product
Operating cost	$220/kg product
Value of crude product	$222/kg product
Total value of production	$442/kg product
Contribution	$108/kg product
Contribution (annual)	$167,940

Table 12.7. Effect on Process Economics by Improving Column Load

Column load	From 450 to 900 grams

System Output

System feed rate	1.54 kg/h crude
System product rate	0.84 kg/h
System product rate	3109 kg/yr

System Economics

Value of purified product	$550/kg product
Value of production	
Labor	$ 36/kg product
Column/cartridge cost	$ 62/kg product
Solvent cost	$ 47/kg product
Depreciation	$ 8/kg product
Operating cost	$153/kg product
Value of crude product	$222/kg product
Total value of production	$375/kg product
Contribution	$175/kg product
Contribution (annual)	$544,075

the system to higher loadings, it will be possible to drop the cost from $305 to $153 (Table 12.7) for a 50% reduction in overall operating cost per kilogram of material produced. Therefore, the amount of material that can be loaded onto the column will also impact the overall cost.

Solvent costs can be compared with and without recovery to determine the impact on operating cost. Table 12.4 shows that it will cost 20¢ for each liter of mobile phase (includes disposal). If solvent is reclaimed, the cost could be decreased to approximately 5¢/L. The impact on chromatographic cost would be a decrease in cost per kilogram of 21% (Table 12.8), from $305 to $240.

The impact on the overall economics using all of these improvements in the process is shown in Table 12.9. This analysis of the operating costs uses a column life of 800 cycles, a load of 900 grams, and a solvent estimate of 5¢/L for a total cost per kilogram of $78 and an annual contribution of $777,250. This figure represents a reduction of 74% of the operating cost from our original estimate and a 22-fold increase in the annual contribution. By using a scaling column it was possible to investigate these improvements and estimate the process performance without using full-scale equipment and large quantities of possibly expensive and limited supplies of valuable compound. One should be aware that other factors such as cycle time and overall recovery are also important and will have

Table 12.8. Effect on Process Economics by Improving Solvent Cost

Solvent cost	From $0.20/L to $0.05/L
System Output	
System feed rate	0.77 kg/h crude
System product rate	0.42 kg/h
System product rate	1555 kg/yr
System Economics	
Value of purified product	$550/kg product
Value of production	
Labor	$ 72/kg product
Column/cartridge cost	$123/kg product
Solvent cost	$ 29/kg product
Depreciation	$ 16/kg product
Operating cost	$240/kg product
Value of crude product	$222/kg product
Total value of production	$462/kg product
Contribution	$ 88/kg product
Contribution (annual)	$136,840

Table 12.9. Effect on Process Economics by Improving Column Life, Column Load, and Solvent Cost

Process Operating Conditions

Column load/cycle	900 g/cycle
Feed purity	55%
Product recovery	98.5%
System flow rate	3 L/min
Cycle time	35 min/cycle
Column/cartridge life	800 cycles
Operation time	3720 h/yr
Mobile phase cost	$0.05/L
Labor rate	$30/h
System useful life	5 yr
Value crude product	$120/kg
Value pure product	$550/kg
Rejuvenation solvent cost	$0.25/L
Rejuvenation volume	75 L
Cycles/rejuvenation	10

System Output

System feed rate	1.54 kg/h crude
System product rate	0.84 kg/h
System product rate	3109 kg/yr

System Economics

Value of purified product	$550/kg product
Value of production	
Labor	$ 36/kg product
Column/cartridge cost	$ 19/kg product
Solvent cost	$ 15/kg product
Depreciation	$ 8/kg product
Operating cost	$ 78/kg product
Value of crude product	$222/kg product
Total value of production	$300/kg product
Contribution	$250/kg product
Contribution (annual)	$777,250

an impact on the economics. The more these factors can be improved, the lower the overall processing cost will be.

12.5.6. Scaleup

Having completed the economic analysis, it is now time to scale the separation. Assuming that the same packing material that was used during the scaling study is being used in the process-scale purification, and the maximum load per gram of packing has been determined, the following equation, based on bed volume, can be used to calculate the load:

$$LOAD_2 = LOAD_1 \frac{R_2^2}{R_1^2} \frac{L_2}{L_1} \tag{13}$$

where $LOAD_2$ = amount of material to be loaded on larger column
   $LOAD_1$ = amount loaded on scaling column
    L_2 = length of larger column
    L_1 = length of scaling (smaller) column
    R_2 = radius of larger column
    R_1 = radius of scaling (smaller) column

For example, if one determined that the load on the scaling column (7.8 mm id × 30 cm) was 800 mg, the theoretical load on a 20-cm id × 60-cm column/cartridge would be 1067 grams.

Another method that has been used to calculate load is to base the estimate on milligrams of sample per gram of packing. For example, if one found that it was possible to load 100 mg of sample per gram of packing on a scaling column, one could then use the data to estimate the load on the larger column by knowing the amount of packing in the larger column.

Although column flow rate (linear velocity) has not been specifically discussed, it is important to the scaling process. In order to maintain consistency and simplify the scaling process, the linear velocities should be kept constant. The following equation can be used:

$$V_2 = V_1 \frac{D_2^2}{D_1^2} \tag{14}$$

where V_2 = flow rate of large column
   V_1 = flow rate run on scaling column
   D_2 = diameter of larger column (cm)
   D_1 = diameter of scaling column (cm)

For example, if the flow rate on a 7.8-mm id scaling column was 3 mL/min, then the flow rate on the 20-cm id column/cartridge would be 1967 mL/min. Realistically, one would run at approximately 2 L/min.

12.5.7. Scaleup Summary

Using the preceding approach, that is, simulating the process HPLC sep-aration, allows the chemist or the chemical engineer to transition from the milligram scale to the kilogram scale. A logical approach to the scaling process can be accomplished by using analytical instrumentation and scaling columns packed with the same stationary phases that will be used during process-scale purification. The primary consideration in the simulated scaleup process is to keep the chemistry consistent between scales. Once the chemistry and the particle size have been determined, it is important to maximize the sample load that the packing material can handle while still providing the specified purity and recovery of product that is required. Using such a system, it is possible to investigate column life and rejuvenation procedures. Data collected from these studies can be used to generate estimated production rates on a yearly basis. These data can then be used to determine the economic feasibility of employing process HPLC in the overall isolation scheme by using simple equations and a standard basic computer program. The economic information can then be used as a basis of comparison with other purification techniques.

The main advantage of using a simulated large-scale HPLC system with valuable product is the ability to investigate large-scale conditions with a relatively small amount of sample. Using this approach, chromatographic conditions can be optimized and the economics can be predicted at the bench level before resources are committed to a specific pilot or product purification process.

12.6. FUTURE TRENDS

Process HPLC is still in its infancy and there is much to be learned about the optimization of production rates through more efficient utilization of the column and the use of multiple columns and column sequencing to meet production needs. Within the next five years we will see the emergence of new equipment and techniques to improve and extend the capability of industrial HPLC. One possible technique, yet to be proven in the process HPLC area, is displacement chromatography. Another device that may find applicability in certain applications is the continuous annular chromatograph (25, 26). Examples of displacement chromatography, although known for a number of years, have only recently appeared in the literature. These have been concerned with peptides and nucleic acids (34, 35). In displacement chromatography (Fig. 12.32), the feed components are in concentrations high enough to be in the nonlinear region of the distribution isotherm. A mobile phase is chosen to support the feed components in high concentration and provide suitable retention. A "displacer" compound is chosen which when pumped through the column will have a higher affinity for the stationary phase than the feed components. As the displacer molecule travels through the packed bed, feed

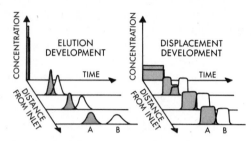

Figure 12.32. Elution chromatography versus displacement chromatography. (From Helf- ferich (40); reproduced with permission from the author and the *Journal of Chroma- tography*.]

components move ahead of this molecule in the order of their relative retention characteristics, the weaker retained molecule moving ahead of the more strongly retained compounds. In this fashion compounds are displaced from the column as discrete bands with almost no tailing. This type of chromatography allows for a very high sample load, on the order of 50–80 mg of sample per injection on an analytical column. The key to this technique is finding a displacer molecule that will perform its desired function and also be acceptable as a chromatographic agent, especially in a situation where the products of interest will be used in a therapeutic mode. Work in this area is now being undertaken.

A second type of chromatography that has been investigated is the continuous annular chromatograph. This system, developed at Oak Ridge National Laboratory in 1976 (36), has the potential of increasing through- put by continuously feeding in a multicomponent mixture and continu- ously removing purified components. This system is a cross-flow moving bed type chromatograph and separates components by allowing the bed to rotate with feed introduced at a fixed point and eluent introduced at ports around the entire annulus of the bed. As the sample components migrate, they begin to separate vertically due to the typical solute–sta- tionary phase distribution. In this system the rotating bed also causes the feed components to separate circumferentially. The result is that solute molecules begin to elute as helical bands. Most of the studies to date have been on system design and the separation of metals (37). Further work in this area with regard to biological molecules would be helpful, espe- cially investigations concerned with overall resolution and throughput.

Process HPLC is a powerful separation technique, and we have only witnessed the beginning of this technology being utilized in the pilot and production environment. Although it is too early to predict accurately how well this technology will be accepted, the future is full of promise.

APPENDIX 12.1

The following is a list of U. S. suppliers of process HPLC equipment:

Millipore Corporation, Waters Chromatography Division, 34 Maple Street, Milford, MA 01757

Separations Technology, 2 Columbia Street, P. O. Box 63, Wakefield, RI 02879

YMC Incorporated, 51 Gibraltar Drive, Morris Plains, NJ 07950

Varex Corporation, 12221 Parklawn Drive, Rockville, MD 20852

REFERENCES

1. J. M. Curling, "Albumin Purification by Ion Exchange Chromatography," in J. M. Curling, Ed., *Methods of Plasma Protein Fractionation,* Academic, London, 1980, pp. 77–91.
2. H. G. J. Brummelhuis, "Preparation of the Prothrombin Complex," in J. M. Curling, Ed., *Methods of Plasma Protein Fractionation,* Academic, London, 1980, pp. 117–128.
3. R. A. M. Delaney, "Industrial Gel Filtration of Proteins," in R. A. Grant, Ed., *Applied Protein Chemistry,* Appl. Sci. Publ., London, 1980, pp. 233–280.
4. A. J. Milkulski, J. W. Heine, H. V. Le, and E. Sulkowski, *Prep. Biochem., 10,* 103–119 (1980).
5. J. M. Curling and J. M. Cooney, *J. Parent. Sci. Technol., 36,* 59–64 (1982).
6. J. C. Janson and P. Hedman, "Large-Scale Chromatography of Proteins," in A. Fiechter, Ed., *Advances in Biochemical Engineering,* Springer, Berlin, 1983, pp. 43–99.
7. G. Cueille and J. L. Tayot, *World Biotech. Rep., 1,* 141–160 (1985).
8. P. C. Wankat, *Large-Scale Adsorption and Chromatography,* vols. I and II, CRC Press, Boca Raton, FL, 1986.
9. R. E. Anderson, "Ion-Exchange Separations," in P. A. Schweitzer, Ed., *Handbook of Separation Techniques for Chemical Engineers,* McGraw-Hill, New York, 1979, pp. 1-359-1-414.
10. L. R. Snyder and J. J. Kirkland, *Introduction to Modern Liquid Chromatography,* 2nd ed., Wiley, New York, 1979.
11. B. L. Karger, L. R. Snyder, and C. Horvath, *An Introduction to Separation Science,* Wiley, New York, 1973.
12. *IMS Int. Marketletter, 13*(30), 7 (1986).

13. K. P. Hupe and H. H. Lauer, *J. Chromatogr., 203,* 41–52 (1981).

14. J. H. Knox and H. M. Pyper, *J. Chromatogr., 363,* 1–30 (1986).

15. K. K. Unger and R. Janzen, *J. Chromatogr., 373,* 227–264 (1986).

16. R. P. W. Scott and P. Kucera, *J. Chromatogr., 119,* 467–482 (1976).

17. A. W. J. De Jong, J. C. Kraak, H. Poppe, and F. Nooitgedacht, *J. Chromatogr., 193,* 181–195 (1980).

18. H. Poppe and J. C. Kraak, *J. Chromatogr., 255,* 395–414 (1983).

19. A. W. J. De Jong, H. Poppe, and J. C. Kraak, *J. Chromatogr., 209,* 432–436 (1981).

20. B. Coq, G. Cretier, C. Gonnet, and J. L. Rocca, *Chromatographia, 12,* 139–146 (1979).

21. L. Personnaz and P. Gareil, *Sep. Sci. Technol., 16,* 135–146 (1981).

22. P. Gareil, C. Durieux, and R. Rosset, *Sep. Sci. Technol., 18,* 441–459 (1983).

23. F. Eisenbeiss, S. Ehlerding, A. Wehrli, and J. F. K. Huber, *Chromatographia, 20,* 657–663 (1985).

24. R. S. Timmins, L. Mir, and J. M. Ryan, *Chem. Eng.,* 170–178 (May 19, 1969).

25. C. K. Shih, C. M. Snavely, T. E. Molnar, J. L. Meyer, W. B. Caldwell, and E. L. Paul, *Chem. Eng. Prog., 6,* 53–57 (1983).

26. A. H. Heckendorf, E. Ashare, and C. Rausch, "Process Scale Chromatography: The New Frontier in High Performance Liquid Chromatography," in D. LeRoith, J. Shiloach, and T. J. Leahy, Eds., *Purification of Fermentation Products,* (ACS Symp. Ser. 271, American Chemical Soc., Washington, DC, 1985, pp. 91–103.

27. Y. D. Clonis, K. Jones, and C. R. Lowe, *J. Chromatogr., 363,* 31–36 (1986).

28. A. M. Cantwell, R. Calderone, and M. Sienko, *J. Chromatogr., 316,* 133–149 (1984).

29. M. Kalyanpur, W. Skea, and M. Siwak, "Isolation of Cephalosporin C from Fermentation Broths Using Membrane Systems and High-Performance Liquid Chromatography," in L. A. Underkofler, Ed., *Developments in Industrial Microbiology,* vol. 26, Soc. for Industrial Microbiology, Arlington, VA, 1985, pp. 455–470.

30. J. L. Dwyer, "Chromatographic Purification of Semisynthetic Beta-Lactam Antibiotics," in G. T. Tsao, Ed., *Annual Reports Fermentation Proc.,* vol. 8, 1985, pp. 93–110.

31. H. J. Wille, H. Traitler, and M. Kelly, *Rev. Franç. Corps Gras,* vol. *34,* 69–74 (Feb. 1987).

32. Carlsberg Biotechnology Ltd., Copenhagen, Denmark, private communication.

33. W. Skea, "Process HPLC Scale-Up Techniques and Process Economics Prediction," presented at the Preparative Liquid Chromatography Symposium, Washington, DC, May 15–16, 1985.

34. C. Horvath, J. Frenz, and Z. El Rassi, *J. Chromatogr., 255,* 273–293 (1983).

35. J. Frenz and C. Horvath, *J. Am. Inst. Chem. Eng.*, *31*, 400–404 (1985).

36. C. D. Scott, R. D. Spence, and W. G. Sisson, *J. Chromatogr.*, *126*, 381–400 (1976).

37. J. M. Begovich, C. H. Byers, and W. G. Sisson, *Sep. Sci. Technol.*, *18*, 1167–1191 (1983).

38. C. W. Rausch and A. H. Heckendorf, "High Performance Liquid Chromatography," in M. Moo-Young, C. L. Cooney, and A. E. Humphrey, Eds., *Comprehensive Biotechnology: The Principles, Applications and Regulations of Biotechnology in Industry, Agriculture and Medicine*, Pergamon, New York, 1985, pp. 537–555.

39. A. W. J. De Jong, H. Poppe, and J. C. Kraak, *J. Chromatogr.*, *148*, 127–141 (1978).

40. F. G. Helfferich, *J. Chromatogr.*, *373*, 45–60 (1986).

CHAPTER

13

PRECISION IN HPLC

ELI GRUSHKA and ILANA ZAMIR

Department of Inorganic and Analytical Chemistry
The Hebrew University
Jerusalem, Israel

13.1. INTRODUCTION

Liquid chromatography, with all of its variations, is probably the most prevalent method of analysis in the modern analytical laboratory. Since chromatography by itself is not a unique method for the identification of analytes, the key to the widespread use of the technique is in its applicability to quantitative analysis, its ease of operation, and its relatively low cost. Moreover, it is a good tool to prove that a certain component is not present in a mixture. A cursory examination of all the current method handbooks, such as the official methods of the AOAC and the USP, will attest to the popularity of the liquid chromatographic technique. One of the reasons for the success of high performance liquid chromatography (HPLC) is that it is a powerful separation technique. Thus impurities can be readily separated from the analytes. In fact, this technique can be used to characterize simultaneously both the major components and the impurities. The availability of a wide range of detectors, some selective and others universal, and of accurate integrators is also responsible for the exponential increase in the popularity of chromatography as a quantitative instrument.

Since HPLC has become such a popular analytical technique, one might expect the publication of a large number of studies dealing with the accuracy and precision of chromatographic data. However, relatively few papers can be found in the literature describing and discussing the factors that influence this precision and accuracy. A recent book (1), devoted solely to the topic of quantitation in chromatographic analysis, is one of the few sources containing information on the precision of chromatographic data. Scott and Reese (2) examined the precision involved in several operating conditions. With these notable exceptions, it would be difficult for a novice to find any tangible information on factors controlling the validity of the data. As a consequence, analysts do not know what kind of errors they can expect and accept from their data. Topics such as reproducibility and accuracy of equipment are frequently neglected, and time is spent in attempting to obtain performance that the equipment cannot support. No attempts are made to analyze the effects of propagation of error. Since many reports can be found in the literature where corroborative studies are analyzed, there is a great deal of discussion on inter- and intralaboratory agreement of the results. However, little or no discussion is presented concerning the validity of the results in light of instrumental limitations. As a result, there is a general misunderstanding of the capability of the chromatographic technique to perform adequately in quantitative analysis. Based solely on their experience, some analysts are aware of the kind of precision that they can expect from measurements such as retention times or peak areas. However, very few of them know the effect of the various components of the chromatograph on the precision.

In this chapter we deal with errors and their propagation in liquid chromatography. The influence of the various components of the chromatograph on the precision and reproducibility of the results is analyzed using basic statistical concepts. First the effects of individual sources of error on the precision and reproducibility of retention times and of peak areas are examined. The effects of the interaction of all these sources of errors are investigated, and then the limitation is examined that the errors place on the comparison of inter- and intralaboratory studies.

13.2. BASIC CONCEPTS

In this section we introduce, very briefly, elementary concepts that will aid us later on in our analysis of the possible errors and their effects. The first concept is that of average. In an ideal world, an experimental measurement would always give the true value of the quantity that is sought. Moreover, all repeated measurements would always give identical results. In the real world, the results of the measurements are corrupted by errors of various sources and types. Thus we have no guarantee that the answer obtained is correct. Even worse, repeated measurements of identical experiments do not generally yield identical values. Therefore the true value must be estimated from the experimental results. We also need to calculate the spread of the experimental data, since the spread will have a bearing on the validity of the estimation.

An estimate of the true value of the result as obtained from a series of measurements is given by the average. The average of a set of N measurements is defined as

$$X = \frac{\sum x_i}{N} \tag{1}$$

where X represents the average value, x_i is the result obtained during the ith measurement, and N is the number of repeated measurements. The spread in the results of individual experiments can be described in several ways. We will be using the standard deviation, which is defined as

$$s = \left[\frac{\sum (X - x_i)^2}{N - 1} \right]^{1/2} \tag{2}$$

The numerator is a summation of the differences between the individual results and the average value. The denominator takes into account the number of repetitions and the fact that we lost one degree of freedom by using the average value as the reference point. The square of the standard deviation, s^2, is the variance. In a later section it will be shown that the variance is the important parameter when analyzing the effects of errors on the precision.

The error of each individual measurement is defined as the difference between the real value and the experimental one,

$$\epsilon = |x_i - \mu| \tag{3}$$

where ϵ is the error and μ is the true value. Difficulties arise when the true value is not known. In such cases we need an estimate of the true value based on the measured observation. The estimate can be best obtained with the relationship

$$\mu = \overline{X} \pm \frac{t \cdot s}{\sqrt{N}} \tag{4}$$

In this equation t is the Student's t value. The value of t, which is available in tables, is a function of the number of experiments n and the degree of probability (confidence level) that we wish to assign to the set of measurements. A 0.9 confidence level means that 90% of the time μ will fall within the range $\mu = X \pm ts/\sqrt{N}$. For a given number of measurements, the value of t increases as the confidence level increases to unity. For a given confidence level, t decreases as the number of experimental points increases.

In this discussion we assumed that the measurements followed a Gaussian distribution, that is, a symmetrical and narrow spread of the results around the mean. However, the distribution of the data does not always follow Gaussian statistics. Under such conditions, the individual data points must be weighted in order to ensure that far-off points do not contribute disproportionately to the average value. Weighting is not dealt with in this chapter, and the interested reader is referred to the excellent review on the topic by Kafadar and Eberhardt (3).

13.3. ERRORS AND THEIR PROPAGATION

The reasons for the deviation of the experimentally obtained results from the true value are the various errors that are omnipresent in every step of the measurement. The errors are the cause of both the deviation from the expected value and the spread in the results. Errors can be divided into two general types. The first type will be called systematic (or determinant) errors while the second type will be called random errors.

13.3.1. Determinant Errors

Determinant errors result from sources that can be traced and identified. They are usually due to systematic errors in one or more of the following three sources: the methodology, the instrumentation, or the operator. Typical examples are (i) misreading of the absorbance scale, in instruments such as the Spectronic 21, due to parallax (operator error), (ii) incorrect flow rate delivery by the HPLC pump (instrument-related error), and (iii) determination of chloride ions in an acidic medium by Mohr titration (wrong method). These errors are characterized by the fact that they can be traced and eliminated, or at least greatly reduced, by suitable calibration techniques. Systematic errors affect mainly the accuracy but not the precision of the measurements.

The propagation of systematic errors and their contributions to the total error of the measurement are rather simple to compute, at least in the first approximation. Let us say that the measured result Z is the sum of individual contributions u, v, and w,

$$Z = u + v - w \tag{5}$$

If each of the contributions has an error associated with its determination, δu, δv, and δw, respectively, then the error in Z is

$$Z + \delta Z = (u + \delta u) + (v + \delta v) - (w + \delta w) \tag{6}$$

or

$$\delta Z = \delta u + \delta v - \delta w \tag{7}$$

One very important point must be made here. Since these errors are systematic, the sign associated with them is uniquely defined, that is, the error does not carry the sign $\pm$ in front of the numerical value.

When the result of the measurement is a product of several individual determinations,

$$Z = \frac{uv}{w} \tag{8}$$

the errors are more conveniently discussed in terms of relative quantities,

$$\frac{\delta Z}{Z} = \frac{\delta u}{u} + \frac{\delta v}{v} - \frac{\delta w}{w} \tag{9}$$

We wish to stress again the fact that the individual errors have a sign associated with them which must be taken into account.

13.3.2. Random Errors

Random errors are more difficult to control and to correct. They are the results of random fluctuations in the measured signal due to effects such as the cycling of thermostats, thermal noises in electric components, and mechanical imprecision of moving parts. As their name indicates, their magnitude *and sign* are random. Thus unlike systematic errors, they do not offset the outcome of the measurements. Instead, they contribute to the spread of the results of the measurements around a mean value. Because of their random nature, it is difficult to eliminate their effects by making calibration curves or by running blanks. Assuming a Gaussian behavior, the effect of random errors can be reduced by repeating the measurements many times.

As a direct result of the random nature of these errors, the propogation of random errors and their effects on the total error of the measured quantities are more complicated than in the case of systematic errors. When the result of the analysis is a linear combination of several steps, as shown in Eq. 5, the error δZ is given by

$$\delta Z = (\delta u^2 + \delta v^2 + \delta w^2)^{1/2} \tag{10}$$

Equation 10 is a result of the fact that variances, and not standard deviations, are additive. Note that the sign of the error is not a factor here since it is not known.

In cases where the measured quantity is a result of multiplicative steps, such as Eq. 8, it is easier to characterize the error in the final result in terms of relative errors of the individual steps,

$$\frac{\delta Z}{Z} = \left[\left(\frac{\delta u}{u}\right)^2 + \left(\frac{\delta v}{v}\right)^2 + \left(\frac{\delta w}{w}\right)^2\right]^{1/2} \tag{11}$$

The meaning of Eq. 11 is very important in the context of this chapter. According to Eq. 11 the expected deviation δZ (or the standard deviation s) in the result of an experimental determination is estimated by the value of the result multiplied by the square root of the sum of the squares of the relative errors of the individual steps in the determination,

$$s_z = \delta Z = Z\left[\left(\frac{\delta u}{u}\right)^2 + \left(\frac{\delta v}{v}\right)^2 + \left(\frac{\delta w}{w}\right)^2\right]^{1/2} \tag{12}$$

This expression will be used in order to evaluate the expected errors in the chromatographic data.

At times the dependence of the final result on the analysis steps is logarithmic in nature,

$$Z = \ln w \qquad (13)$$

In this case the error in the final outcome of the determination is given by

$$\delta Z = \frac{\delta w}{w} \qquad (14)$$

The error in Z in this case is a function of the *relative error* in w. This type of interrelation is important in chromatography since the retention time has an exponential dependence on the temperature and, possibly, on the amount of the modifier in the mobile phase.

13.4. PRECISION, REPRODUCIBILITY, AND REPEATABILITY

Before proceeding with an analysis of the errors in chromatographic measurements, some definitions must be made. For the purposes of this chapter, precision will be defined as the repeatability of the result in a series of experiments run during a *single session, with identical reagents and equipment by a single operator*. Reproducibility will defined as the agreement in the results of identical experiments made by the same operator *on two different occasions* or the results of identical experiments run in *two different laboratories*. These definitions are similar to the conventional definitions in which precision and repeatability are considered equivalent (4, 5).

With these statistical tools an analysis can be made with respect to the expected precision of typical chromatographic data. Initially the analysis will center on the expected precision in retention time measurements. After that, attention will be directed toward examining the precision involved in area measurements.

13.5. EXPECTED PRECISION IN RETENTION TIMES

This section studies the effects of individual experimental parameters affecting the retention time. The cumulative effects of these parameters are also examined.

13.5.1. Effects of Flow Rate Variations

The equation that relates the retention time t_R to the chromatographic parameters is

$$t_R = \frac{L}{u}(1 + k')$$ (15)

where L is the column length, u the velocity of the mobile phase, and k' the capacity ratio. The length of the column can be measured rather accurately, and thus its influence on the precision of the retention time is insignificant. Moreover, since the column length does not vary in a random fashion between runs, any error introduced by the column will be a systematic error. The effect of the velocity of the mobile phase on the precision of the retention time is another matter. If all else were constant, Eq. 11 shows that the relative error in t_R is directly related to the relative error in the velocity,

$$\frac{\delta t_R}{t_R} = \frac{\delta u}{u}$$ (16)

The precision in the velocity is, to a great extent, controlled by the precision of the pump delivering the mobile phase. The velocity is a function of the flow rate,

$$u = \frac{F}{A}$$ (17)

where A is the available column cross section, which has been corrected for the column porosity, and F is the flow rate as delivered by the pump. Hence the relative error in retention time is equal to the relative error in flow rate. Manufacturers usually report the relative error in the flow rate of the pumps; typical values are around 0.3%. Thus by Eq. 16 the relative error in retention, due to flow rate fluctuations, is also around 0.3%. The absolute error in the retention time, therefore, is

$$\delta t_R = 0.003 t_R$$ (18)

For example, if the retention time is 300 s, the error in that time is approximately ± 0.9 s.

13.5.2.   Effects of Temperature Variations

The relative errors in the retention time, as shown by Eq. 18, are correct only if the fluctuation in flow rate is the only factor contributing to the error in the measurements. In reality, however, other factors, such as the temperature, must be also considered.

The capacity factor in Eq. 15 is an exponential function of the temperature,

$$k' = \phi \exp\left(\frac{-\Delta\mu}{RT}\right) \tag{19}$$

where ϕ is the phase ratio, R is the gas constant, T is the temperature and $\Delta\mu$ is the chemical potential change. The relative error in k', which results from the relative error in the temperature control of the column, is

$$\frac{\delta k'}{k'} = \frac{\Delta\mu\,\delta T}{RT^2} \tag{20}$$

Equation 20 shows that the relative error in the capacity ratio is a function not only of the relative error in the temperature, but also of the absolute temperature and the free energy change $\Delta\mu$ involved in the partition process.

When fluctuations in both temperature and flow rate are considered, then the relative error in the retention time is given by

$$\frac{\delta t_R}{t_R} = \left[\left(\frac{\delta F}{F}\right)^2 + \left(\frac{\Delta\mu\,\delta T}{RT^2}\right)^2\right]^{1/2} \tag{21}$$

When the column is not thermostated, the temperature fluctuations can be as high as $\pm 3°C$. Given typical values of μ, and assuming a mean temperature of 25°C, the associated relative error in the capacity factor is about 1–2%. The resultant error in the retention time is

$$\frac{\delta t_R}{t_R} = [(0.003)^2 + (0.015)^2]^{1/2}$$

$$= 0.0153 \tag{22}$$

In this calculation it is assumed that the error in k' is 1.5%. The relative error is just over 1.5%, and assuming the retention time to be 300 s, the precision is ± 4.6 s.

There are several important implications in this example: (1) A seemingly small temperature variation can have a large effect on the precision of the measured retention times. (2) Due to the addition of the variances, the major influence on the precision of the retention time is due to the uncertainties in the temperature. The effect of the variation in the flow rate is rather small. (3) A corollary of statements (1) and (2) is that in an unthermostated operation, the effect of temperature fluctuation may be the limiting factor controlling the precision of the retention data. This point can be further demonstrated by analyzing the error introduced by a $\pm 0.1°C$ temperature fluctuation, which is the case in many thermostated columns, especially those heated in block heaters or in air ovens. A typical variation in the capacity ratio due to this $\pm 0.1°C$ change in temperature is around 0.1–0.3% in relative units. Assuming a value of 0.2%, the relative error in the time is

$$\frac{\delta t_R}{t_R} = [(0.003)^2 + (0.002)^2]^{1/2}$$

$$= 0.0036 \tag{23}$$

or around 0.36%. Again, in the case of 300-s retention time, the precision is about ± 1.1 s. Thus the conclusion to be drawn is that to achieve high precision in retention data the column must be maintained in ovens in which the temperature control is better than $\pm 0.05°C$. This result is in agreement with the findings of Scott and Reese (2), who observed that for a precision of 0.1% in the capacity factor, the temperature control should be better than 0.05°C.

It was mentioned in a previous section that the effect of the variation in the temperature is a strong function of the temperature itself and of the change in the chemical potential involved in the transfer from one phase to another. Figure 13.1 shows the influence of the change in the chemical potential on the precision in the retention data. In this figure the dependence of the error in the retention time is plotted versus the variation in temperature. Each line represents a different enthalpy of transfer. In each case it was assumed that the error in the flow rate is similar, 0.3%. From the figure it is clear that when the error in the temperature is small, the dominating effect is the flow rate variations. When the fluctuations in the temperature increase, the control of the temperature becomes the dominating factor that influences the precision of the determination of the retention times. As the enthalpy increases, that is,

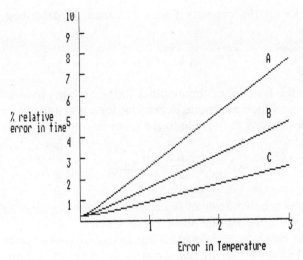

Figure 13.1. Effect of errors in temperature on relative error in retention time for different enthalpies. *A*—4750 cal; *B*—2500 cal; *C*—1750 cal; temperature—25°C.

as the temperature dependence of the capacity factor increases, the demand for controlling the temperature is more stringent.

The column temperature by itself can also determine the variation in the retention times. This is clear from Eq. 21. Interestingly enough, the error is less as the column temperature increases.

13.5.3. Effects of Mobile Phase Composition

Up to this point it was assumed that the chromatographic system was operated in an isocratic mode, using either a pure solvent as the mobile phase or a mixture pumped from a single reservoir. Frequently, however, the isocratic mixture is pumped from two or more different reservoirs. In such cases, variations in the mobile phase composition must be anticipated. The relationship of the makeup of the mobile phase with retention is similar to that of temperature with retention; the capacity factor varies in an exponential manner with changes in the mobile phase composition. The dependence is rather strong, and a relatively small change in the composition of the solvent can change the retention time drastically. Generally speaking, a quadratic equation describes the variation of the capacity ratio over a wide range of the modifier content of the mobile phase. However, in the context of the present work we are interested in small variations in the mobile phase composition around an average value.

The behavior of the capacity factor can then be described using the expression

$$\ln k' = aX + b \tag{24}$$

where X is the fraction of the modifier in the mobile phase and a and b are constants, which are solute dependent for a given mobile phase. The error in the capacity factor due to the error in composition is

$$\frac{\delta k'}{k'} = a \, \delta X \tag{25}$$

Therefore the relative error in the capacity factor is proportional to the absolute error in the mobile phase composition.

When the only source of error is due to mobile phase variations, then Eqs. 15 and 25 show that the relative error in the retention time is directly related to the absolute error in the mobile phase composition. As anticipated, the precision in the retention time is a direct function of the slope that indicates the steepness of the dependence of $\ln k'$ on the modifier content in the mobile phase. Thus solutes that are greatly affected by the concentration of the modifier will show large errors in the precision of retention data associated with relatively minor variations in the composition of the mobile phase. Typical values of the slope are between -4 and -7. Due to the logarithmic nature of the dependence shown in Eq. 24, this represents an extremely wide range of possible variations in capacity factors. If the precision in the mobile phase composition is around $\pm 1\%$, the relative error in the retention is between ± 4 and $\pm 7\%$. Using the 300-s retention time example, the variation in the retention time will then be between ± 12 and ± 21 s, clearly an unacceptable situation. Fortunately the modern instruments can control the precision of the mobile phase composition to within $\pm 0.1\%$. In this case the precision in the retention, for the same 300-s analysis time, is between ± 1.2 and ± 2.1 s. Therefore high precision in the generation of multicomponent mobile phases is obviously needed to prevent large errors in retention data.

Trying to account for all the sources of errors discussed so far is complicated by the fact that the capacity factor is a function of both the temperature and the mobile phase composition. To take into account the effects of the temperature and the mobile phase composition, the capacity factor can be written as (6)

$$\ln k' = -\frac{\alpha X + \beta}{RT} + \frac{\tau X + \epsilon}{R} \tag{26}$$

where α, β, τ, and ϵ are solute-related constants. According to Eq. 11, the relative error in the capacity factor due to errors in the mobile phase composition and the temperature is given by

$$\frac{\delta k'}{k'} = \left\{ \left[\left(-\frac{\alpha}{RT} + \frac{\tau}{R} \right) \delta X \right]^2 + \left[\frac{(\alpha X + \beta)\, \delta T}{RT^2} \right]^2 \right\}^{1/2} \tag{27}$$

The error in the retention time, then, is equal to

$$\frac{\delta t_R}{t_R} = \left\{ \left(\frac{\delta F}{F} \right)^2 + \left[\left(-\frac{\alpha}{RT} + \frac{\tau}{R} \right) \delta X \right]^2 + \left[\frac{(\alpha X + \beta)\, \delta T}{RT^2} \right]^2 \right\}^{1/2} \tag{28}$$

Equation 28 shows that the precision of the retention time is a complicated function of the temperature, the mobile phase composition, and the nature of the solute (via constants α, β, and τ). As an example, Table 13.1 shows typical values of the solute-related constants for the solute phenetol. These values, based on the data of Gant et al. (7), were obtained on a reversed column using a water–methanol mobile phase. With the aid of these data, the combined effects of all the parameters discussed so far can be demonstrated. Table 13.2 shows the expected magnitude of the errors in the retention time of phenetol due to errors in temperature and mobile phase composition. In the calculation it was assumed that the temperature was 25°C. The table shows two cases, each with a different methanol content in the mobile phase—0.8 and 0.5 mass fraction of methanol. For each composition four cases are given, describing two different magnitudes in the errors in the temperature and two in the mobile phase composition. In each case the error in the flow rate was taken to be ±0.3%.

Several important conclusions can be reached from the data in Table 13.2. When the precision in the mobile phase composition is poor, the effect of the variation in the temperature is less pronounced. On the other hand, if the control of the mobile phase composition is fairly tight, then the temperature fluctuations become the dominating factor. The relative errors in the retention times, at least in the case of phenetol, are fairly insensitive to the amount of modifier in the mobile phase. However, precise retention data require a chromatographic system with an excellent

Table 13.1. α, β, τ, and ϵ Values for Phenetol

Solute	α	β	τ	ϵ
Phenetol	5885	−6607	4.12	−7.12

**Table 13.2. Precision Involved in Measuring
Retention Time of Phenetol Due to Errors in
Temperature and Mobile Phase Composition**

δX (%)	δT (°C)	δt_R (%)	± seconds*
Methanol Mass Fraction 0.8			
1.0	3.0	7.10	15.3
0.1	3.0	3.31	7.1
1.0	0.5	6.34	13.7
0.1	0.5	0.88	1.9
Methanol Mass Fraction 0.5			
1.0	3.0	7.4	66.6
0.1	3.0	6.27	56.4
1.0	0.5	4.09	36.8
0.1	0.5	1.15	10.4

* Based on retention times of 216 and 900 s for the first
and second parts of the table, respectively.

mobile phase controller and a thermostated column. These findings concur, again, with the work of Scott and Reese (2).

The discussion so far has centered on isocratic runs in which the contents of two reservoirs are mixed to yield a mobile phase whose composition is constant throughout the entire chromatographic run. Gradient runs, in which the mobile phase composition changes as a function of time, are subject to greater errors than in isochratic operation. The errors are due to the inability of the pumping system to reproduce exactly the time-based changes in the mobile phase composition. Typical errors in the precision of the mobile phase composition during a gradient run are about 1% in *absolute* terms, irrespective of the actual composition. Thus the error in the composition is a strong function of the actual composition. A complete analysis of the expected precision in the retention data is, indeed, complicated by the fact that the precision is a strong function of both the actual mobile phase composition during the gradient and the range of the gradient. Therefore a detailed analysis is not presented here, and the precision values presented in the discussion on isocratic analysis can serve as lower limit estimators for the expected precision in gradient runs.

13.5.4. Effects of Integrator

The retention times are usually determined with the aid of digital integrators. The use of integrators can add to the error of determination due

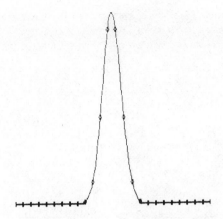

Figure 13.2. Simulated peak with digitized data points indicated as small circles.

to the actual process of integration. In the first place, digital integrators sample the signal periodically and not continuously. Therefore the peak maximum can be missed by a factor which, at most, is equal to half the time span between two sequential samplings. Figure 13.2 illustrates the problem. This effect can be greatly reduced with curve fitting techniques; however, most integrators do not use such an approach.

In addition to the sampling period problem, noise in the data can also have deleterious effects on the precision of the measured retention time since the integrator can identify the highest signal value as the retention time of the solute. Figure 13.3 illustrates the difficulty. This problem can be diminished using averaging techniques, and most integrators have an averaging method built into the data collection procedures. Often the averaging is done by "bunching" procedures in which several sampling points are summed together to give one acquired point. However, in cases of noisy data the problem can persist. Scott and Reese (2) have shown that the error in the determination of the retention times, due to noisy data, can be significant.

A quantitative analysis of the contribution of the integration process to the precision of the retention data is beyond the scope of this chapter; see, for example, Rijswick (8). In order to illustrate the magnitude of the effect, however, the problem will be discussed from a qualitative point of view.

When the sampling rate of the integrator is 100 ms per data point, the maximum error in the exact placing of the peak maximum will be ± 0.05 s. To a first approximation, this error represents a spread of ± 2 standard deviations in the location of the retention time. The error, which is ± 0.025

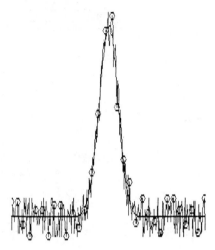

Figure 13.3. Simulated peak with SNR of about 5. Small circles indicate points of digitization; superimposed is smooth peak.

s, is rather insignificant in all but superfast chromatography, where the retention times are on the order of 10–60 s. In such cases the sampling rate of 10 points per second is too slow and the operator should use a much faster digitizing rate, which will decrease the error in the location of the peak maximum. Even with longer sampling intervals in conventional chromatography, such as 400 ms, the error in the retention time is ±0.1 s. This is still an insignificant contribution to the error in the retention time, and the effect of flow rate and temperature fluctuation will overshadow those due to digitization.

The effect of a noisy signal is a function of several parameters, (1) the signal-to-noise ratio (SNR), (2) the frequency of the noise, (3) the sampling rate, (4) the width of the peak, and (5) the retention time of the solute. It is obvious that the error should be larger when the SNR is small. For example, the SNR of a particular Gaussian peak is 3, the peak width (four standard deviations, 4σ) is 10 s, and the sampling interval of the integrator is 250 ms. To a first approximation the noise in the range of $\pm\sigma$ (one standard deviation, or ± 2.5 s) around the true maximum can be equal to or higher than the true peak height. In terms of time, the noise in the range of ± 2.5 s can cause a faulty determination of the retention time. Since this range represents a $\pm 2\sigma$ range in the error in the retention time, the precision involved is ± 1.25 s. This error is significant and cannot be neglected. Moreover, since this error is absolute, the situation can be worse for fast eluting peaks.

The example indicates that the effect of this contribution is a function of the peak width; for equal SNRs narrower peaks suffer less. As anticipated, when the signal is clean (high SNR values), the present contribution can be ignored, with the exception of very broad peaks. Moreover, broad peaks require a much higher SNR in order not to be affected. Thus good chromatography is essential to ensure good precision. Another corollary is that high-speed chromatography requires high SNRs in order to yield reproducible results.

13.5.5. Effects of Operator Initiation of the Integrator

Frequently the initiation of the integrator is done by the operator at the injection point. This procedure can introduce errors of two kinds. A systematic error can be caused by a sequential rather than a parallel operation: the injection is made and then the operator presses the start button of the integrator. The second error is due to random fluctuation in reaching and actually pressing the integrator initiation button. It is difficult to assess the exact contribution of these errors since, to a great extent, they are operator dependent. We will assume that the systematic part of the error is corrected by calibration, and that the random error is ±0.3 s. When the precision in the measurement is controlled by mobile phase and temperature effects, the influence of the error in starting the integrator is minuscule. On the other hand, in isocratic systems which use only one reservoir, and which have good temperature control, the effect of integration initiation can be important. This type of system was described in Eq. 23. It was found there that the relative error in the retention time, due only to errors in the flow rate and temperature, is 0.36%. If the retention time of that system is 150 s, then the error due to these two sources is ±0.54 s. In this case the contribution of the integrator initiation, which is ±0.3 s, is significant, and the overall error in retention time is

$$\delta t_R = [(0.54)^2 + (0.3)^2]^{1/2}$$

$$= 0.62 \text{ s} \tag{29}$$

13.5.6. Experimental Verification

The theoretical material discussed up to this point can be verified experimentally. For the purpose of the verification, retention data were collected using a Spectra-Physics model 8700 liquid chromatograph equipped with a 254-nm ultraviolet detector and a Rheodyne injection valve. The dimensions of the reversed-phase (ODS) column were 125 × 4.1 mm. The column was thermostated in a water bath whose temperature

Table 13.3. Precision in Retention Time, Delivery from Single Reservoir

Number of Runs	Benzene		n-Propylbenzene		n-Butylbenzene	
	t_R	%RE	t_R	%RE	t_R	%RE
6	2.71	0.15	6.22	0.21	13.15	0.29
5	2.72	0	6.10	0.13	13.07	0.53
7	2.67	0.25	6.09	0.08	12.83	0.41
6	2.73	0.14	6.29	0.12	13.36	0.25
7	2.71	0.55	6.23	0.73	13.20	0.76

%RE—percent relative error in retention time.

was kept at 31.6 ± 0.3°C with a Haake E12 thermostat. A Hitachi D-2000 integrator wa used to obtain the retention times and peak areas. The mobile phase was a methanol–water mixture (75:25).

Table 13.3 shows the retention times of several solutes. The mobile phase was premixed and delivered from one solvent reservoir.

The data in Table 13.3 were collected on several occasions. Each of the five sets of experiments was obtained on a different day. In each case, the mobile phase was prepared afresh by mixing 75:25 methanol–water in a single bottle and using it as the reservoir. The data in the table are representative of many experiments which were carried out over a period of 2 months. A later section in this chapter examines the precision involved in the day-to-day operation of the liquid chromatograph.

Table 13.3 describes a situation similar to that discussed in Eqs. 23 and 29. The major sources of imprecision are fluctuation in the flow rate and temperature and the random errors involved in the manual starting of the integration. The relative error in the temperature (in kelvins) is ±0.1%. Thus to a good approximation, the contribution of fluctuations in the flow rate and temperature to the relative error in the retention time is about ±0.36%, as shown by Eq. 23. Assuming an uncertainty of ±0.3 s in starting the integrator, the contribution of this source to the error in the retention time is about ±0.2% in the case of benzene, ±0.1% in the case of n-propylbenzene, and ±0.04% in the case of n-butylbenzene. Therefore the overall precision in the retention should be about ±0.4% for each solute. The relative errors shown in Table 13.3 are in good agreement with the predicted values.

The data in Table 13.3 indicate that the precision is somewhat worse at the longer retention. Two factors may contribute to the poorer precision in the longer retained compound. One factor may be related to the fact that the error in retention time is a function of the enthalpy of transfer of the solutes. Longer retentions are usually associated with greater absolute values of the enthalpy. The second factor responsible for the poorer precision may be due to integrator difficulties: the peak top of the longer retained peak is broad, and the integrator may have difficulties in determining the exact position of the maximum of that peak. This source of error was not discussed at all in the present communication. It is felt that its importance is minimal as compared with the importance of fluctuations in temperature and flow rate.

The mobile phase used to obtain the results in Table 13.3 was delivered from a single reservoir. Table 13.4 shows typical results obtained with a mobile phase that was mixed, by the chromatographic pump, from two reservoirs.

In addition to the fluctuations in temperature and flow rate, an imprecision in the mixing of the solvents by the pump contributes here to the error in retention time. As expected, the precision of the retention time is worse in this case than when only one reservoir is used for the mobile phase. The data in Table 13.4 are commensurate with an (absolute) error of about 0.1% in the mobile phase composition (see the discussion related to Table 13.2). The data in Table 13.4 comprise only a small representative sample of the many experiments performed during the course of the study. In every case the precision in the retention data was poorer when the mobile phase components were mixed by the pump from two reservoirs.

As was observed with Table 13.3, the relative error in the retention time is generally greater for longer retained compounds. The same explanations given in conjunction with Table 13.3 can be used here, namely, temperature fluctuation and integrator difficulties are responsible for the poorer precision of the more retained solutes. However, in the present case the retention dependence of the solute on the mobile phase composition will also affect the precision of the retention times (see Eqs. 25 and 28.)

Table 13.4. Precision in Retention Time, Delivery from Two Reservoirs

Number of Runs	Benzene		n-Propylbenzene		n-Butylbenzene	
	t_R	%RE	t_R	%RE	t_R	%RE
9	2.79	0.40	6.53	0.58	14.03	0.50
5	2.70	0.00	6.20	0.52	13.17	0.88
7	2.72	0.70	6.18	1.18	13.10	1.62
5	2.68	0.52	6.11	0.24	12.87	0.48
6	2.71	0.50	6.48	1.00	14.05	0.65

%RE—percent relative error in retention time.

13.5.7. Summary

This section showed that precise retention data call for good habits in the development of the chromatographic system. The concept of precision is subjective; one chromatographer's idea of precision is another's nightmare. No limit of precision can be set, and frequently the requirement of the analysis dictates the acceptable limits. However, in order to minimize the spread in the retention data, certain guidelines should be followed:

1. Good constant flow rate pumps should be used.
2. The column should be thermostated to within $\pm 0.5°C$.
3. Isocratic runs should be made using a single solvent reservoir.
4. Noisy signals should be avoided.

In addition, the analysis presented shows that high-speed chromatography puts extra demands on the system since many of the sources of imprecision are absolute and not relative. However, with proper precaution, the analyst can expect to acquire retention data with less then 1% relative error.

13.6. EXPECTED PRECISION IN PEAK AREAS AND IN QUANTITATIVE CALCULATIONS

Although there are several methods of measuring peak areas, currently the predominant technique is by electronic integrators. Consequently the present treatment will assume that the integration is accomplished with these systems. Those who are interested in errors involved with other methods of integration should read Grob [9, chap. 8]. That chapter deals with the integration of gas chromatographic signals. However, the underlying principles, and problems, are the same in HPLC as in gas chromatography.

The analysis of the factors controlling the precision of integrated peak areas differs, in some respects, from that studying retention data. The factors controlling the retention are primarily thermodynamic in nature, whereas the factors controlling the areas are related to the amounts injected and to the kinetic processes occurring in the column during the separation. Hence the magnitude of similar effects may be very different. For example, the influence of the integration process is more pronounced in area calculation than in time determination. Temperature variation, on the other hand, should have a stronger effect in retention calculations than in measuring peak areas.

In this section the effects of the experimental parameters on the expected precision in the determination of peak areas are examined. When possible, the influence of the appropriate parameter on both the area and the time will be compared.

13.6.1. Effects of Flow Rate

The great majority of liquid chromatographic detectors are concentration-sensitive devices. For such detectors it can be shown that the detector signal is equal to

$$A = \frac{S_c w}{F} \tag{30}$$

where A is the peak area, S_c a response factor of the detector for the solute, w the weight of solute injected, and F the flow rate of the mobile phase. It is clear from Eq. 30 that fluctuations in the flow rate will introduce errors in the measured areas. The expression for the error in the area is similar to that given by Eq. 18,

$$\frac{\delta A}{A} = \frac{\delta F}{F} \tag{31}$$

Thus a relative error of 0.3% in the flow rate will cause the same relative error in the area and, therefore, in the concentration of the unknown. The effect of flow rate fluctuation on the area uncertainty is similar in magnitude to the effect of the same parameter on the error in time measurements.

13.6.2. Effects of Temperature and Mobile Phase Composition

The temperature affects the retention time and peak width through variations in the diffusion coefficients of the solutes. Variations in the peak width will influence the precision of the measured areas. However, fluctuation of $\pm 3°C$ will not have a great effect on the diffusion coefficients and, hence, on the peak width. To a good approximation, this factor can be neglected.

In the context of errors in area measurements, the effect of the temperature on the retention can be considered as similar to that on flow rate fluctuations. A detailed analysis of the effect of temperature fluctuations on the retention time was presented in the previous section. The results of that analysis can be utilized here.

In cases where the temperature control is $\pm 3°C$, Eq. 22 shows that the relative error in retention time, due to flow rate and temperature variation, is about 1.5%. The relative error in the area of the peak will be equivalent. If the temperature of the column is controlled at $\pm 0.2°C$, the relative error in the area is around 0.36%, which is similar to the effect on the retention time, as shown by Eq. 23.

The effects of fluctuations in the mobile phase composition are very similar to those discussed previously. The retention time and peak width are affected most by variations in the mobile phase composition. The peak width depends on the mobile phase composition via viscosity and diffusion effects, which are a function of the composition. Similar to the case of temperature variations, the contribution of fluctuations in the mobile phase composition to the precision of the measured areas can be neglected. The effects of retention on the precision of the area were already discussed, and the orders of magnitude involved in the effect were analyzed in the previous section. In conclusion, good precision in area measurements requires good precision in the mobile phase composition during the chromatographic run. Thus for systems in which the temperature is well controlled and the mobile phase is maintained constant to within $\pm 0.1\%$, the error contribution in the areas due to these two factors can be maintained below 1%. When the mobile phase, in isocratic runs, is pumped from a single reservoir, the resulting precision is much better, with typical values of about 0.4% in relative units.

13.6.3. Effects of Integration

Probably the biggest source of error in area determination is found in the process of integration. Within the integration process there are several sources of potential difficulties which will affect the precision of the area measurements. Two examples are noise in the signal and difficulties in the recognition of the beginning and end of peaks. These problems will have a minimal effect on the precision of the retention time, but they could have a major effect on the reliability of area determination.

13.6.3.1. Noise Effects

Area calculations can be accomplished by many methods such as Simpson's rule, trapezoid rule, and moment analysis. Perhaps the simplest is the moment analysis, for which the definition of the area is

$$A = \sum y_i \Delta x \tag{32}$$

where y_i is the height of the signal at the ith sampling point and x is the sampling time. The error due to noise is manifested in an error δy in the y_i values. The relative error in the area then is

$$\frac{\delta A}{A} = \Sigma\left(\frac{\delta y}{y_i}\right) \tag{33}$$

where the summation is taken over the width of the peak. Since δy is constant, this equation can be rewritten as

$$\frac{\delta A}{A} = \frac{n\,\delta y}{\Sigma y_i} \tag{34}$$

To a first approximation, δy can be related to the noise N by

$$\delta y = N\,\Delta x\sqrt{w} \tag{35}$$

where w is the peak width. Thus,

$$\frac{\delta A}{A} = n\,\Delta x\,N\,\frac{\sqrt{w}}{\Sigma y_i} \tag{36}$$

where n is the number of data samplings over the width of the peak. Division and multiplication by the peak height H yields

$$\frac{\delta A}{A} = n\,\Delta x\left(\frac{S}{N}\right)^{-1}\frac{H\sqrt{w}}{\Sigma y_i} \tag{37}$$

The ratio $H/\Sigma y_i$ is related to the peak width by a functionality which depends on the model assumed for the peak shape. In the case of a Gaussian peak, Eq. 37 reduces to

$$\frac{\delta A}{A} = \frac{4}{\sqrt{2\pi}}\left(\frac{S}{N}\right)^{-1}\frac{w^{3/2}}{n} \tag{38}$$

As expected, the relative error in the area is inversely proportional to the SNR; the higher the noise (low SNR), the poorer the precision of the measured area. In addition, the relative error in the area due to signal noise is proportional to the width of the peak. This makes sense, since the wider the peak, the more noise is included in the integration process. The frequency of data sampling is also a factor in the magnitude of the error. Equation 38 shows that the error in the area is inversely proportional to the number of digitizations during the width of the peak.

Equation 38 has some important implications. The first obvious implication was already mentioned. High SNRs are desirable if the error in the peak area is to be small. Another ramification of Eq. 38 is that narrow peaks are desirable not only for good efficiencies, but also for minimal error in the peak area calculation. The third implication is that a suitable number of data points across the peak is needed for minimal noise influence as well as for good peak shape reproduction. Since peak widths in chromatography tend to increase with the retention times of the solutes, the integrator should be able to vary the sample collecting rate as a function of time. More important, the operators should be aware that they may have to adjust the integrators to sample at different rates as the analysis times increase.

Figure 13.4 shows the effect of S/N on the precision of the area. Poor S/N values were chosen intentionally to show the effect of noise. Each line in the figure represents a different sampling rate. The peak width (4σ) was taken as 4 s. For a peak eluting at 300 s this width represents 5600 plates. Thus the very strong influence that the SNR can have on the precision of the area measurements is evident. An S/N value of 10 can result in about 40% error in the measurement of the area when there are

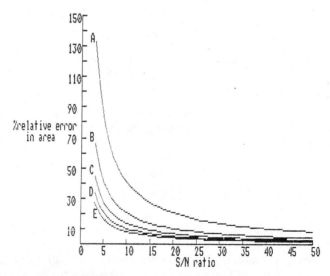

Figure 13.4. Effect of SNR on relative error in area for different number of digitizations across peak. Error here is due to area contribution of noise. A—$n = 10$; B—$n = 20$; C—$n = 30$; D—$n = 40$; E—$n = 50$.

only 10 data points per peak width. When the digitization rate is 50, the error is still 8%. Only when S/N values increase to 100 (a range not shown in the figure), does the error decrease to acceptable error levels: 4% at 10 samples per peak width and 0.8% at $n = 50$. Again, it is seen that "clean" signals can aid in increasing the overall precision of the chromatographic measurements.

13.6.3.2. Effects of Initiation and Termination of Integration

The decision to start and stop the integration can also affect the precision of the measured areas. The error in this case can be random or systematic. Systematic errors occur when the integrator is set wrong and the integration begins or stops well inside or outside the peaks. The error involved is easily calculable, especially if Gaussian peaks are assumed, by evaluating error functions which are well known and are tabulated in many sources. More importantly, the errors can be greatly diminished by the correct use of the integrator. Modern integrators allow the user to indicate on the chromatogram the integration limits and the baseline used. It is recommended that these parameters be checked routinely in order to ascertain whether or not this type of error is biasing the results. In any event, this error affects the accuracy but not the precision of the data.

Noise can also modify the integration limits. Here, however, the effect is random and the precision of the area is altered. Generally speaking, integrators employ two means of recognizing the beginning and the end of a peak. In a simplified manner, one approach is based on a voltage threshold. When this threshold is exceeded successively by several signal readings, a peak is recognized. In the second method, a slope limit is set and when it is exceeded, a peak is recognized. Some integrators use a combination of these two techniques. These two methods of integration initiation can be adversely affected by noise, as well as by baseline drift.

If the integrator fails to recognize the start of the peak, the integration will start late by a factor which is a function of S/N and of the sampling rate. For example, it is assumed that the integrator opened the integration process one digitization period too late as a result of the noise in the signal. The area lost in this process can be estimated by further assuming that, in the sampling period missed, the signal can be approximated by a straight line. The area lost is therefore the area of the triangle,

$$\delta A = 0.5 \Delta x \, N \qquad (39)$$

where N is the root mean square magnitude of the noise and Δx is the sampling interval. The sampling interval is the peak width w divided by the number n of samplings during that width. Equation 39 can then be written as

$$\delta A = 0.5H\left(\frac{w}{n}\right)\left(\frac{S}{N}\right)^{-1} \tag{40}$$

where H is the peak height. In the case of Gaussian peaks, the factor Hw is the peak area divided by $\sqrt{2\pi}$. Therefore Eq. 40 can be rewritten in terms of the relative error of the area,

$$\frac{\delta A}{A} = \left(\frac{2}{\pi}\right)^{1/2}\left(\frac{S}{N}\right)^{-1}\left(\frac{1}{n}\right) \tag{41}$$

Since the same error can occur at the termination of the integration, the relative error in the area can be twice that shown by Eq. 41.

This example is probably a good approximation for cases where the peak is symmetrical and the frequency of digitization is relatively fast. For asymmetrical peaks the error in the approximation is in the constant $(2/\pi)^{1/2}$. If the sampling frequency is too slow, the approximation of the triangular area, made in connection with Eq. 39, is probably wrong. However, Eq. 41 can be used to estimate the contribution to the precision of the area due to errors in starting and ending the integration.

Figure 13.5 shows the relative error in the area as a function of S/N for several sampling rates. Some of the conclusions drawn from this figure are similar to those discussed in connection with Fig. 13.4, namely, clean signals are relatively immune to the integration errors. On the other hand, the influence of the peak width is different in the two integration-related errors in area measurements. While in the case of noise the error in the area is directly dependent on the width, in the case of integration initiation and termination the error in the area is only an implicit function of width.

13.6.4. Experimental Verification

The concepts discussed in this section on precision in area measurements are examined experimentally in Tables 13.5 and 13.6. The chromatographic system used to collect the data was described in the preceding section that deals with errors in retention times.

Table 13.5 shows the areas of the peaks obtained for five sets of experiments, each carried out on a different day. These data correspond to the data in Table 13.3. They were collected without making any attempts to improve the precision of the system. Conventional equipment, including the integrator, was used for the duration of the study. However, low

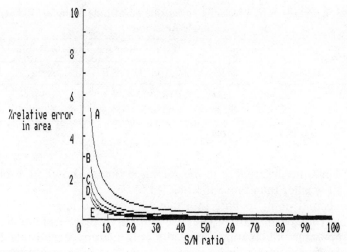

Figure 13.5. Effect of faulty start and stop of integrator, due to noise, on relative error in area for different number of digitizations across peak. $A—n = 10; B—n = 20; C—n = 30; D—n = 40; E—n = 50.$

concentrations of the solutes were employed intentionally in order to obtain chromatograms with fairly small SNRs, typically between 20 and 50.

The first point to note in Table 13.5 are the large variances in the averages of the areas from one set of experiments to another. Chiefly responsible for these variances is the fact that while within a set of experiments the sample was the same, different sets used different samples. The retention time is not affected by differences in the amount of the components in the samples provided that the adsorption isotherms are

Table 13.5. Precision in Area, Delivery from Single Reservoir

Number of Runs	Benzene		n-Propylbenzene		n-Butylbenzene	
	Area	%RE	Area	%RE	Area	%RE
6	51,893	2.69	60,825	5.71	49,263	11.1
5	53,400	4.75	61,687	3.13	49,628	12.2
7	32,323	4.99	66,923	2.35	120,641	3.16
6	57,171	8.50	63,533	2.46	50,106	9.33
7	52,512	2.54	64,462	3.62	49,837	8.92

%RE—percent relative error in area.

Table 13.6. Precision in Areas of Concentrate and Dilute Samples, Delivery from Single Reservoir

	Benzene		n-Propylbenzene		n-Butylbenzene	
	Area	%RE	Area	%RE	Area	%RE
Concentrate	282,460	1.34	318,791	1.41	2,633,047	1.50
Dilute	52,774	3.26	63,155	4.49	486,300	6.57

%RE—percent relative error in area.

linear. Thus the differences in the retention times, shown in Table 13.3 are much smaller than those in Table 13.5.

The second point to note in Table 13.5 are the large values of the relative errors, particularly in the case of n-butylbenzene. If mobile phase and temperature fluctuations were the only factors responsible for the imprecision, then the magnitude of the relative errors would have been much smaller, as was the case with the precision in the retention data. The data in Table 13.5 were collected using a single reservoir of the mobile phase. When two reservoirs were used to generate the mobile phase, the errors in the measured areas were, in general, of the same order of magnitude as those in Table 13.5. The fact that the magnitude of the errors in the area measurements is independent of the method of mixing the mobile phase lends credence to the statement that phase fluctuations do not control the precision of the areas. The main source of error in the present case can be traced to integration problems. The detector signals were noisy, and the integrator had difficulties in detecting the start and end of the peaks. This problem was particularly noticeable in the case of butylbenzene. This compound eluted, after about 13 min, as a relatively broad peak with relatively gentle slopes at the beginning and end of the peak. Due to the noisy baseline, the integrator could not sense properly the extremities of the peak. As an example, Fig. 13.6 shows a chromatogram in which the integrator was instructed to draw markers indicating peak beginning and end, and to indicate the baseline used. From the figure it is clear that the integrator grossly misjudged the end of the peak. Moreover, the drawn baseline shows that a large fraction of the peak area was excluded from the integration.

The third point to note in Table 13.5 are the larger relative errors of the last eluting solute, n-butylbenzene. It is not immediately clear whether this observation is real, and whether it occurs with all late eluting peaks.

The last point to observe in Table 13.5 are the large day-to-day variations in the relative errors of the areas. It is felt that this variation is a function of the low SNR observed with the data.

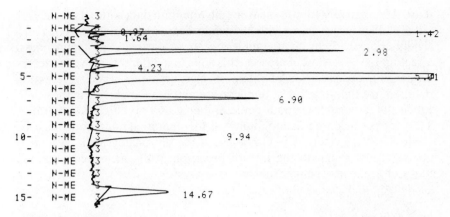

Figure 13.6. Chromatogram showing errors made by integrator in deciding of peak start and end.

Table 13.6 demonstrates the effect of the SNR on the precision of area measurements. The data were generated by chromatographing a concentrated sample solution and a fivefold dilution of that solution. Each sample was chromatographed six times.

The superiority of the data obtained with the concentrated solution is obvious; the consistency of the integrator in recognizing the beginning and the end of the peaks of the concentrated sample is far better than with the dilute sample. This improved consistency is the result of the higher SNR in the case of the more concentrated sample.

The preceding discussion indicated that noise in the signal can have two effects, (1) contribution from the area of the noise to the true area of the peak and (2) faulty triggering of the limits of the integration. When the SNR is low, the more important of the two effects is the faulty start or stop of the integration process. Frequently the misjudgment in the initiation or termination of the peak can be greater than the one digitization period that was analyzed in a previous example. In such cases, area measurements can be quite imprecise, as shown in Tables 13.5 and 13.6. Signal filtering and/or smoothing may be required when the detector signal is noisy. This method of handling noisy signals is outside the scope of the present work. Those interested in the topic are referred to the monograph by Sharaf et al. (10).

13.6.5. Summary

The discussion on integration errors indicates that it is more difficult to obtain precise area measurements than it is to obtain precise retention data. It is relatively easy to measure the retention data with relative errors >0.5%. Measuring areas with the same degree of precision requires a rigidly controlled chromatographic system. Moreover, the factors most affecting the precision in area calculations are different from the factors influencing the precision of retention measurements. Precise retention data require tight control of temperature and mobile phase composition. Precise areas dictate smooth detector signal and an efficient integrator. Thus for precision chromatography all the operating conditions must be controlled very rigorously. Otherwise if the chromatographic system is not carefully designed, the lack of precision in the measured retention times or areas may render the results useless.

13.7. DAY-TO-DAY AND LABORATORY-TO-LABORATORY REPRODUCIBILITY

The preceding sections discuss the precision with which chromatographic data can be acquired in a series of runs, all done sequentially by the same operator, on the same instrument and column. Frequently, however, a chromatographic method of analysis is used in a routine manner day after day, over a long period of time. The analysis may be carried out in a single laboratory, or transferred from one laboratory to another. It is of great importance to establish the expected reproducibility of the measured retentions or areas.

13.7.1. Variation in Retention Data

All the souces contributing to the imprecision in the retention data within a run are present between runs as well. However, the magnitude of the contribution to the imprecision differs in the two cases. For analyses on the same instrument, the same column, and by the same operator, the greatest source of error can probably be traced to the preparation of the mobile phase. It is simply difficult to prepare consistently, either by manual or by pump mixing, mobile phases with identical composition. The errors can be related to wrong proportions of the mobile phase components, as well as to varying sources of the components. A second source of error, causing the day-to-day variability in the retention data, lies in temperature fluctuations, which can occur in unthermostated columns.

The magnitude of the errors in the retention data can be calculated using the arguments and equations discussed in the section dealing with precision in the retention data. If we estimate the day-to-day reproducibility in the mobile phase composition to be about 0.5%, then from Table

Table 13.7. Reproducibility of Retention Times

Number of Days	Benzene		n-Propylbenzene		n-Butylbenzene	
	t_R	%RE	t_R	%RE	t_R	%RE
1. 16	2.68	2.05	6.19	2.30	13.16	2.58
2. 11	2.71	1.65	6.34	2.59	13.64	3.74

%RE—percent relative error in retention time; 1—using a single mobile phase reservoir; 2—using two reservoirs.

13.2 we can estimate the relative error in the retention time to be between 1–4%, depending on the actual mobile phase composition and on the magnitude of the error in the temperature. Table 13.7 shows the averages of the retention times of three solutes, accumulated over a period of over two months. The retention times were collected using a premixed mobile phase pumped from a single reservoir and the data were collected using pump mixing from two reservoirs.

The data in Table 13.7 were collected with a conventional liquid chromatograph without any modifications. The sample concentration was such that the S/N was 50 or less. The relative errors shown are in good agreement with the values calculated in Table 13.2.

The reproducibility of the retention data from laboratory to laboratory poses new difficulties. In such cases, variations in the equipment, including the chromatographic column, introduce additional sources of errors, which affect the precision of the data. Unfortunately these contributions to the relative errors in the retention times are not always easy to quantify. In general, however, the discussion concerning day-to-day reproducibility can serve as a rough guide to the reproducibility that can be expected in interlaboratory studies.

It should be stressed that the present discussion deals only with the expected precision in the data. No attempts were made to perform significance tests in order to check the validity of a set of measurements accumulated over a period of time. Tests such as the t test, the F test, and ANOVA should be performed in order to recognize the set of results that need to be examined more critically. These tests are outside the scope of this chapter, and the reader is referred to Kafadar and Eberhardt (3) and Miller and Miller (5).

13.7.2. Variation in Area Measurements

The discussion concerning the precision in the areas indicated that integration problems dictate the quality of the data obtained. The noise in the signal is a major factor affecting the precision of the measured areas. The noise in a chromatographic system can vary from day to day as a result of many causes. For example, differences in ambient temperature can affect the performance of the electronic components of the detector, thus influencing the quality of the signal. Another possible cause of the variability in the noise lies in the makeup of the mobile phase. Different batches of the mobile phase may have different impurities, which will affect the detector baseline in an unpredictable manner. Therefore the magnitude of the resulting noise will vary from batch to batch of the mobile phase. Similarly, interlaboratory measurements will be affected by different noise characteristics, which may cause large differences in the measured areas.

The day-to-day and laboratory-to-laboratory variations in the noise are difficult to predict and to quantify. For the purpose of this chapter we can estimate the variation in the areas measured in several laboratories from the work of McCoy et al. (11, 12). These two papers reported on the precision involved in measuring peak heights and areas in many laboratories. One study was published in 1984, the second in 1986. Although the later work shows more precise results, the relative errors in the measured areas were still rather large, covering the range of 2–5.6% for uncorrected data. When results grossly deviating from the average (outliers) were eliminated, the precision varied between 1.8 and 4%. McCoy and his coworkers found (12) that the errors were larger with more dilute samples. This larger error found in cases of dilute samples is in agreement with the discussions in previous sections concerning the effects of smaller S/N values on the precision of the measurements.

The many collaborative studies which can be found in the literature allow us to gauge the differences in the precision of intralaboratory and interlaboratory measurements [see, for example, Slahck (13) and Carter and Overton (14)]. As expected, interlaboratory measurements are, in general, less precise. Frequently the relative error observed in interlaboratory studies is twice as large as that in intralaboratory determinations. In the section dealing with area measurements during a single set of experiments, it was found that the expected relative error in the area is around 1–3%. Therefore a relative error of 2–5% in interlaboratory measurements seems to be a reasonable expectation.

13.8. CONCLUSION

The purpose of this chapter was to call attention to the expected precision of liquid chromatographic data. The discussion attempted to give reasonable limits for the precision with which retention data and peak areas can be measured using conventional chromatographic equipment without rigorously controlling the operating parameters. It was shown that retention times can be measured quite precisely. Area measurements, in general, are less precise than retention times. One reason for the difference in the precision is related to the fact that different parameters control the repeatability of retention times and area.

In conclusion, to obtain highly precise data, the chromatographic equipment must be adjusted specially and controlled very stringently. This conclusion is in agreement with the findings of Scott and Reese (2). The flow rate delivered by the pump, the composition of the mobile phase, the temperature, the quality of the detector signal, and the operation mode of the integrator must all be controlled and regulated very tightly. The more rigid this control, the more precise will be the measured quantities. Good chromatographic habits are required not only for good efficiencies, but also for highly precise chromatographic data.

REFERENCES

1. E. Katz, *Quantitative Analysis Using Chromatographic Techniques*, Wiley, Chichester, UK, 1987.
2. R. P. W. Scott and C. E. Reese, *J. Chromatogr., 138,* 283–307 (1977).
3. K. Kafadar and K. R. Eberhardt, in J. C. Giddings, E. Grushka, J. Cazes, and P. R. Brown, Eds., *Advances in Chromatography*, vol. 24, Dekker, New York, 1984, pp. 1–34.
4. H. C. Hamaker, *J. Assoc. Off. Anal. Chem., 69,* 417–428 (1986).
5. J. C. Miller and J. N. Miller, *Statistics for Analytical Chemistry*, Ellis Horwood, UK, 1984.
6. I. Atamna and E. Grushka, *J. Chromatogr., 355,* 41–56 (1986).
7. J. R. Gant, J. W. Dolan, and L. R. Snyder, *J. Chromatogr., 185,* 153 (1979).
8. M. H. J. van Rijswick, *Chromatographia, 7,* 491–501 (1974).
9. R. L. Grob, Ed., *Modern Practice of Gas Chromatography*, 2nd ed., Wiley, New York, 1985.
10. M. Sharaf, D. L. Illman, and B. R. Kowalski, *Chemometrics*, Wiley, New York, 1986.
11. R. W. McCoy, R. L. Aiken, R. E. Pauls, E. R. Ziegel, T. Wolf, G. T. Fritz, and D. M. Marmion, *J. Chromatogr. Sci., 22,* 425–431 (1984).
12. R. E. Pauls, R. W. McCoy, R. E. Ziegel, T. Wolf, G. T. Fritz, and D. M. Marmion, *J. Chromatogr. Sci., 24,* 273–277 (1986).
13. S. C. Slahck, *J. Assoc. Off. Anal. Chem., 69,* 490–492 (1986).
14. P. L. Carter and K. C. Overton, *J. Assoc. Off. Anal. Chem., 69,* 908–911 (1986).

CHAPTER

14

HPLC AS A SOURCE OF INFORMATION ABOUT CHEMICAL STRUCTURE OF SOLUTES

ROMAN KALISZAN

Medical Academy of Gdańsk
Gdańsk, Poland

14.1. INTRODUCTION

Much effort has been directed to describing chromatographic retention in terms of classical thermodynamics. However, as Prausnitz has remarked (1), "classical thermodynamics is revered, honored and admired, but in practice it is inadequate." Although in chromatography the thermodynamic data can give some insight into the nature of solute distribution between the mobile and the stationary zones, it is unfortunately knowledge acquired after the fact (2). Thus based on thermodynamics alone, one cannot predict the retention of an individual solute nor any other property encoded in its structural formula. This is because the thermodynamic properties of a given system are bulk properties reflecting

563

only the net interactive effects in the system. The magnitude of thermodynamic parameters represents the combination of individual interactions that may take place at the molecular level. In effect, the thermodynamic analysis of chromatographic processes provides information of a physical rather than chemical nature.

It is obvious that the chromatographic behavior of solutes depends on their structural features. Thus it should also be possible to apply chromatography to extract and quantify information about the structure and properties of chemical compounds of interest. This would be of utmost importance not only for our understanding of the basis of chromatographic separations, but also for an explanation of other processes in which no permanent chemical changes are made to the chemical compounds involved. In fact, chromatographic interactions may be viewed as generally similar to other physicochemical interactions determining such vital phenomena as drug action in a living system, the activity of hormones and pheromones, taste and olfactory effects of chemicals, the physical properties of crystals and solutions, dissolution rates and solubilities, and even the hazards of environmental pollution by pesticides.

The chemical discipline that is to provide maximum information through analysis of chemical data is called chemometrics. When computers became more widely available in the 1960s, chemometrical approaches to problem solving in chemistry became possible, and new research areas were created within the science, one of these being quantitative structure–retention relationships (QSRR) (3).

Chromatography offers a unique model system for studying structure–property relationships resulting from intermolecular interactions. In a chromatographic experiment all the conditions may be kept constant or controlled, thus permitting solute structure to be studied as the single independent variable in the system. Contrary to complex biological systems, chromatography readily yields large amounts of unequivocal data with a high degree of precision.

Relationships between chromatographic retention data and the quantities related to the solute structure cannot be solved in strict thermodynamic terms. Such relationships are of the so-called extrathermodynamic type, where the term *extrathermodynamic* means that the science lies outside the formal structure of thermodynamics, although the approach resembles that of thermodynamics in that detailed microscopic mechanisms do not need to be explicitly identified during use (4). Extrathermodynamic approaches are combinations of detailed models with the concepts of thermodynamics. Since it involves model building, this kind of approach lacks the rigor of thermodynamics, but it can provide information not otherwise accessible. The manifestations of extrathermodynamic relationships are the linear free-energy relationships (LFER). Although LFERs are not a necessary consequence of thermodynamics, their occurrence suggests the presence of a real connection between the correlated quantities, and the nature of this connection can be explored (5, 6).

Having the QSRRs established one can attempt to use them for:

1. Explanation of the molecular mechanism of chromatographic separations
2. Prediction of the retention behavior of individual substances at specific separation conditions
3. Determination and selection of structural data of importance for activity (both chemical and biological) of the compounds of interest

14.2.   INTERMOLECULAR INTERACTIONS IN CHROMATOGRAPHY

When attempting to extract structural information from chromatographic data, it seems reasonable to first ask what may be the nature of the intermolecular interactions governing the chromatographic separation. Certainly these are not the interactions leading to definite chemical alterations of the solute molecules through protonation, oxidation, reduction, complex formation, or other stoichiometric chemical processes. Only in ion-exchange chromatography, where the separation determining forces are ionic in nature, can discrete chemical alterations be said to occur. In other chromatographic techniques and modes only the forces that can occur between closed-shell molecules are involved. The intermolecular interactions known to appear in chromatography are listed in Table 14.1.

14.2.1.   Physicochemical Forces Determining Intermolecular Interactions

14.2.1.1.   Ion–Dipole Interactions

The ion–dipole interactions do not belong to the intermolecular forces in the narrower sense. These coulombic forces, however, may play an important role in specific chromatographic processes and should be considered for the sake of completeness. When placed in the electric field resulting from an ion, the dipole will orient itself so that the end with charge opposite to that of the ion will be directed toward the ion, and the other,

Table 14.1. Intermolecular Interactions

Physical Interactions

Ion–dipole
Dipole–dipole
Dipole–induced dipole
Instantaneous dipole–induced dipole

Chemical Interactions

Hydrogen bonding
Electron pair donor–electron pair acceptor (EPD–EPA)

Solvophobic Interactions

General term, incorporates both physical and
   chemical interactions

---

repulsive, end directed away. The potential energy of an ion–dipole interaction E_{i-d} is given by

$$E_{i-d} = -W^2 Z\mu \, \frac{\cos \alpha'}{\epsilon r^2} \qquad (1)$$

where the coefficient W depends on the unit system applied (for the SI system $W = 1/4\pi\epsilon_0$), Z is the charge on the ion, μ the dipole moment of the neutral molecule, r the distance from the ion to the center of the dipole, and α' the dipole angle relative to the line joining the ion and the center of the dipole; ϵ is the relative electric permittivity of the medium (eluent). It can be noted that the lower ϵ is, the lower is the energy of attractive ion–dipole interactions. Thus a decrease of the solvent permittivity by adding ethanol ($\epsilon = 27$) to water ($\epsilon = 81$) should increase electrostatic interactions between solutes and the mobile phase and decrease retention.

14.2.1.2. Orientation Interactions

The forces resulting from the electric field generated by the molecules (un-ionized) are commonly called van der Waals interactions. These interactions are divided into three groups, orientation interactions, inductive interactions, and dispersive interactions.

The potential energy E_{d-d} of orientation interactions (Keesom effect) between molecules of dipole moments μ_1 and μ_2 is given by

$$E_{d-d} = -W^2 \frac{2}{3kT} \frac{\mu_1^2 \mu_2^2}{\epsilon r^6} \tag{2}$$

14.2.1.3. Inductive Interactions

The electric dipole in one molecule can induce a dipole moment in a neighboring molecule. The induced moment always lies in the direction of the inducing dipole. Thus attraction always exists between the two entities, which is independent of temperature. The induced dipole moment will be larger, the higher the polarizability α of the apolar molecule. The potential energy E_{d-id} of dipole–induced dipole interactions (Debay effect) equals

$$E_{d-id} = -W^2 \frac{\alpha_1 \mu_2^2 + \alpha_2 \mu_1^2}{\epsilon r^6} \tag{3}$$

The dipole–induced dipole interactions do not depend on temperature and are not additive.

14.2.1.4. Dispersive Interactions

Even in atoms and molecules without permanent dipole moment, the continuous electron density fluctuations result, at any instant, in a small dipole moment which can fluctuatingly polarize the electron system of the neighboring atoms or molecules. An approximate expression for the calculation of the potential energy E_d of dispersion interactions (London effect) is

$$E_d = -W^2 \, 3I_1 I_2 \frac{\alpha_1 \alpha_2}{2(I_1 + I_2)} \frac{1}{(\epsilon r^6)} \tag{4}$$

where I_1 and I_2 are the ionization energies of the interacting molecules 1 and 2. In the London equation molecular polarizabilities are given as scalar quantities. Thus the dispersion interactions cannot be saturated and are additive. These types of interactions usually prevail among the three kinds of van der Waals interactions. It should be added here that the van der Waals interactions increase with pressure as the intermolecular distances decrease.

14.2.1.5. Hydrophobic Interactions

The distribution of solutes between the mobile and the stationary zones in liquid chromatography, especially in the so-called reversed-phase systems, is often interpreted in terms of hydrophobic (or, more generally, solvophobic) interactions (7). The phenomenon of hydrophobic interactions is still the subject of interest and controversy (8). However complex, the hydrophobic interactions are the net effect of the same well-known physiochemical interactions that determine the state of all matter.

14.2.1.6. Hydrogen Bonding

The intermolecular interactions discussed so far are classified as chemically nonspecific. There are also intermolecular interactions in chromatography which result from characteristic structural features of the molecules involved. That category comprises hydrogen bonding and electron pair donor–electron pair acceptor (EPD–EPA) forces. Whereas the van der Waals interactions are more "physical" in nature, the latter group of interactions has a more "chemical" character.

As is commonly known, when a covalently bound hydrogen atom forms a second bond to another atom, the second bond is referred to as a *hydrogen bond*. The best known proton donor groups are —O—H; —S—H; Cl—H; Br—H. The most important hydrogen bond acceptors are the oxygen atoms in alcohols, ethers, and carbonyl compounds, halogen atoms and ions, nitrogen atoms in amines, and the π-electron systems of aromatic compounds.

14.2.1.7. Electron Pair Donor–Electron Pair Acceptor Interactions

The formation of an additional bonding interaction between two valency-saturated molecules is possible only if an occupied molecular orbital of sufficiently high energy is present in the electron pair donor molecule, and a sufficiently low unoccupied orbital is available in the electron pair acceptor molecule. Whereas in a normal chemical bond each atom supplies one electron to the bond, in EPD–EPA bonding one molecule supplies the pair of electrons, while the second molecule provides the vacant molecular orbital. The energetically highest orbitals are the lone pair of the n electrons (such as amines, ethers, sulfoxides, phosphines); the electron pair of σ bond (such as cyclopropane, halogen derivatives); and the pair of π electrons of unsaturated and aromatic compounds (such as polyaromatic compounds). Similarly, electron pair acceptor compounds are those possessing a vacant valency orbital of a metal atom (such as organometallic compounds); a nonbonding σ orbital (such as halogen molecules); and those having a system of π bonds (such as aromatic polynitro compounds, tetracyanoethylene).

14.2.2. Interpretation of HPLC Retention in Terms of Fundamental Intermolecular Interactions

Although the intermolecular interactions that take place during a chromatographic process seem to be fairly simple, the details of the mechanism governing retention are not completely understood. Certainly, the distribution of a solute between a mobile phase and a stationary phase during the chromatographic separation results from the forces that operate between solute molecules and the molecules of each phase. If one considers, for example, reversed-phase (RP) HPLC on chemically bonded hydrocarbon silicas, then one has to consider the following interacting entities: the solute, the hydrocarbon bonded to silica, the mobile phase components preferentially adsorbed on the stationary phase material, the free silanol groups of the silica support, and the components of the eluent. Evidently the situation is quite complex and no satisfactory model exists that would permit precise quantitative retention prediction. However, a rational explanation of the observed retention changes in terms of intermolecular interactions is possible. Thus a search for some chemometric model of retention seems to be justified.

For the sake of illustration let us now consider processes in RP HPLC with hydrocarbon bonded stationary phases in terms of the fundamental intermolecular interactions mentioned.

The characteristic feature of RP HPLC is an increase of log k' with an increasing number of carbon atoms for homologous series (Fig. 14.1). As far as interactions of solutes with the hydrocarbonaceous moiety are concerned, the forces of importance for the separation of homologues are the dispersion forces. The orientation interactions of solutes with stationary phase are negligible as the polarity (dipole moment) of the hydrocarbon moiety is practically zero. The dipole–induced dipole interactions should be similar for all homologues because the dipole moments within homologous series are similar. The magnitude of dispersion interactions increases with the polarizability of the solutes, which in turn reflects the molecular size. In the case of solute–mobile phase interactions the orientation forces and the dipole–induced dipole forces are stronger than for the case of solute–hydrocarbon stationary phase interactions. The orientation interactions with mobile phase are similar for homologues due to the similarity of their dipole moments. Attraction by a mobile phase resulting from the dispersive and the dipole–induced dipole interactions

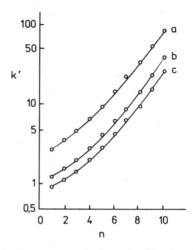

Figure 14.1. Relationships between capacity factor log k' and carbon number of alkyl chain n for homologous series of a 3-Cl, b 2-OCH$_3$, and c 2-NO$_2$-substituted alkylbenzoates. [After E. Tomlinson, H. Poppe, and J. C. Kraak, Thermodynamics of functional groups in reversed-phase high performance liquid–solid chromatography, *Int. J. Pharm.*, *7* (1981) 225. With permission of Elsevier Science Publishers B.V., Biomedical Division, Amsterdam, The Netherlands].

is affected by the solute polarizability. However, because the dispersive interactions are usually prevailing among intermolecular interactions and the polarizability of octadecyl chains of the stationary phase is greater than the polarizability of the smaller molecules of the eluent, the net effect of intermolecular interactions will be the increased retention of bigger homologues. Of course, the interactions of solutes with the components of a chromatographic system can be further complicated by the weak chemical interactions previously mentioned.

What possible effect on RP HPLC retention of a given solute can be expected when the organic modifier content in aqueous mobile phase is increased? From experiment it is known that for a given solute, retention usually decreases with increasing content of organic solvent (Fig. 14.2). Certainly, due slightly higher polarizability of the eluent (as compared to water) the dispersive interactions of a solute with the mobile phase increase. Thus one could expect some decrease of retention. The main reason for the observed retention decrease with increasing organic mod-ifier content in the mobile phase, however, seems to be the increase of the strength of all electrostatic interactions between solutes and the mo-bile phase resulting from a decrease of the relative electric permittivity ϵ of aqueous eluent after diluting water with organic solvent (see Eqs. 1–4).

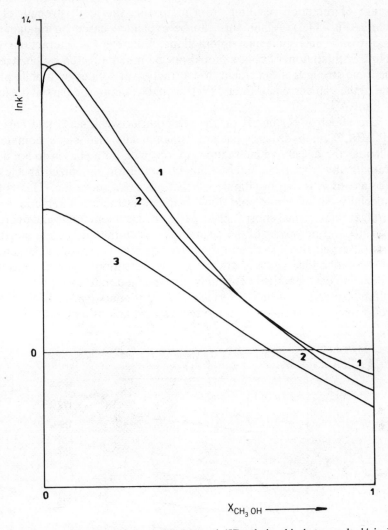

Figure 14.2. Example of experimentally observed (57) relationship between ln k' in RP HPLC and volume fraction of methanol in methanol–water eluent. Solutes: 1—naphthalene; 2—benzophenone; 3—p-chlorophenol. [After P. J. Schoenmakers, H. A. H. Billiet, and L. de Galan, Description of solute retention over the full range of mobile phase compositions in reversed-phase liquid chromatography, *J. Chromatogr.*, *282* (1983) 107. With permission of Elsevier Science Publishers B.V., Amsterdam, The Netherlands].

The next experimentally observed RP HPLC phenomenon is the increase of retention resulting from increasing the ionic strength of the eluent (Fig. 14.3). Again, this can be explained based on fundamental electrostatic intermolecular interactions. Whereas the interactions of a solute with stationary phase should not be much affected by changes in the ionic strength of the eluent, the attraction of solutes by mobile phase due to the van der Waals interactions would be diminished in the presence of ions.

The effect of eluent pH on the retention of weak acids and bases in RP HPLC on hydrocarbonaceous silica needs also some comments. Whereas the dispersive interactions of ions with both phases do not differ dramatically from these interactions observed for un-ionized species, in the case of ions the ion–dipole interactions are dominating. This is true especially since ion–dipole long-distance attractive interactions are stronger than normal short-range van der Waals interactions and decrease with the second power of the distance between the interacting moieties. It is observed that the retention of the completely ionized acids on hydrocarbon-bonded silica is lower than the retention of ionized bases of similar molecular weight. This may be explained in terms of interactions with free silanol sites on the surface of the stationary phase. These silanol groups produce some exclusion effect toward anions while the retention

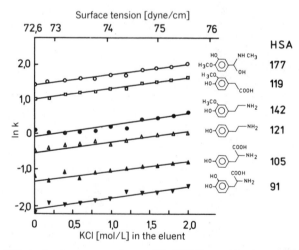

Figure 14.3. Effect of salt concentration on RP HPLC capacity factor. HSA—so-called hydrocarbonaceous surface area. [After Cs. Horvath, W. Melander, and J. Molnar, Solvophobic interactions in liquid chromatography with nonpolar stationary phases, *J. Chromatogr., 125* (1976) 129. With permission of Elsevier Science Publishers B.V., Amsterdam, The Netherlands.]

of cations is increased. This type of electrostatic (coulombic) interactions is diminished at higher ionic strength (above 0.1) of the eluent.

It is observed in RP HPLC that retention decreases with temperature (Fig. 14.4). This fact can again be explained based on the theory of intermolecular interactions. With increasing temperature the dielectric constant of water decreases significantly, from 88.0 at 0°C to 63.7 at 70°C. Since the dispersion interactions with the stationary phase are basically unaffected by temperature changes, the decrease in the dielectric constant increases the affinity of the solutes to the solvent, thus decreasing retention.

14.3. PRINCIPLES OF CHEMOMETRICAL ANALYSIS OF RETENTION DATA

Although a rationalization of the observed HPLC behavior of solutes at specific separation conditions is possible in terms of the fundamental intermolecular interactions, no strict quantitative retention model exists, permitting a precise prediction of the chromatographic parameters for individual solutes analyzed at specific conditions. One reason for this is

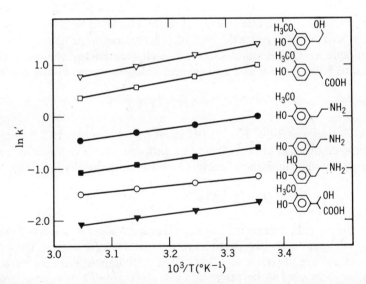

Figure 14.4. Temperature dependence of RP HPLC capacity factors. [After Cs. Horvath, W. Melander, and J. Molnar, Solvophobic interactions in liquid chromatography with nonpolar stationary phases, *J. Chromatogr., 125* (1976) 129. With permission of Elsevier Science Publishers B.V., Amsterdam, The Netherlands.]

certainly the complexity of the chromatographic systems resulting from mutual interactions of several molecular species. The other difficulty is caused by our inability to describe structural properties of the interacting chemical species in a precise, reliable, unequivocal manner.

In spite of the lack of a good model of chromatographic processes, a great deal of chemical information can be extracted from the HPLC data by statistical analysis. Advanced statistical techniques are now in routine use since efficient personal computers have been the common research tool (9). Thus the chemometrical analysis of retention data can be done by researchers looking for specific chemical information to be applied next in various areas of chemical and biological sciences. However, such an analysis requires some working knowledge of chromatography and structural chemistry in order to avoid the folly of developing statistically valid correlations to chemically invalid principles.

14.3.1. Free-Energy Relationships in HPLC

The term *linear free-energy relationships* (LFER) denotes the linear relationships between the logarithms of the rate or equilibrium constants for one reaction series and those for a second reaction series subjected to the same variation in reactant structure or reaction conditions (10, 11). The logarithm of capacity factor log k' in HPLC is normally assumed to be proportional to the free-energy change associated with the chromatographic distribution process. Not all chromatographic data, however, are suitable for QSRR studies. As is known, free-energy changes ΔG are related to enthalpy ΔH and entropy ΔS changes by the Gibbs equation

$$\Delta G = \Delta H - T \Delta S \tag{5}$$

where T is temperature. For LFER to be found between real and model systems, changes in either entropy or enthalpy must be constant, or the enthalpy changes must be linearly related to entropy changes (12),

$$\Delta H = \beta \Delta S + \Delta G_\beta \qquad \text{at } T = \beta \tag{6}$$

When the so-called enthalpy–entropy compensation is observed with a family of compounds in a particular chemical transformation, the values of β and ΔG are invariant and β is called the compensation temperature. Then, as shown by Melander et al. (12), plots of ln k'_T of various solutes measured at a given temperature T under different conditions against the corresponding enthalpy change are linear. From the slope of the compensation plot of ln k'_T versus ΔH the compensation temperature β may be obtained. It is recommended that the reference temperature T be near the harmonic mean of the experimental temperatures used for the evaluation of the enthalpies.

The majority of QSRR reported concern retention parameters as obtained in routine chromatographic measurements. Enthalpy–entropy compensation is checked only occasionally and LFER are assumed a priori. Reservations are justified as far as the HPLC retention parameters are defined or determined for individual solutes at specified experimental conditions.

14.3.2. Determination of HPLC Retention Parameters

The measure of the degree of retention of a solute in HPLC is the *phase capacity ratio* or *capacity factor k'*. This parameter is defined as

$$k' = \frac{t_R - t_M}{t_M} = \frac{V_R - V_M}{V_M} \tag{7}$$

where t_R and V_R are the retention time and the retention volume, respectively, of the chromatographed solute, t_M and V_M are the dead time and the dead volume of the column used.

As it has been shown by Knox and coworkers (13, 14), the determination of the dead volume in HPLC presents both theoretical and practical problems. From the thermodynamic point of view, under the term dead volume we understand the total volume of all eluent components within the column bed, that is, not only the eluent in the interparticle space but also an additional volume of eluent within the pores of the particles of column packing. However, any bonded stationary phase will preferentially adsorb certain components from eluents. Are these adsorbed eluent components to be considered as part of the mobile phase or of the stationary phase?

An unambiguous definition of what constitutes mobile phase and what constitutes stationary phase in the context of HPLC using porous packing materials is impossible. Nevertheless, if one is to make any thermodynamic measurements, which include simple measurements of column capacity ratios, one must provide a clear-cut definition of dead volume so that the above formula for the capacity ratio can be used and data compared between laboratories. It should be stressed that the value used for V_M and its definition should ideally be the same for a single column for all eluents and should be readily determined with adequate precision. Unfortunately the popular recipes to determine dead volume (such as solvent disturbance peak and elution volume of salt or ion) do not provide correct V_M data, except by chance (14). Details concerning the theory and practice of determination of the dead volumes are given elsewhere (14). From a practical point of view it will be simplest to determine V_M by flushing the column with any one-component eluent and determining the elution volume of the isotopically labeled eluent sample. (Deuterated eluent can be detected by a refractive index detector.)

It is obvious that to calculate reliable capacity factors in HPLC, not only the column dead volume must be known, but the retention volumes V_R of solutes should also be determined precisely. The experimental conditions that provide good retention measurements are described in every chromatographic textbook. It may be reminded that a precise control of the pH and the ionic strength of the mobile phase is important and that ion-pair formation with buffer components should be taken into consideration. The linear Soczewinski–Wachtmeister relationship (15) between log k' and the volume fraction of water in binary water–organic eluent is not obeyed at larger composition ranges, and the isocratic data are pre-

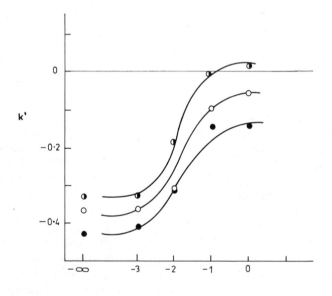

$$\log \left([\, \text{Na NO}_3] / \text{mol dm}^{-3} \right)$$

Figure 14.5. Dependence on ionic strength of capacity factors of benzoic acid (◑), salicylic acid (○), and sulfanilic acid (●) chromatographed at the same conditions. [After J. H. Knox, R. Kaliszan, and G. F. Kennedy, Enthalpic exclusion chromatography, *J. Chem. Soc., Faraday Symp., 15* (1980) 113. With permission of The Royal Society of Chemistry, Burlington House, London, UK.]

ferred for QSRR analysis. When dealing with HPLC data determined on hydrocarbonaceous silica stationary phases, the possibility of enthalpic exclusion of anionic solutes due to the presence of residual silanol groups within the matrix of the packing material has to be considered (Fig. 14.5).

14.3.3. Numerical Descriptors of Solute Structures

Analyzing the intermolecular interactions involved in the chromatographic process one can conclude that generally two main structural features of the solutes affect their retention. One type would be related to molecular size and would reflect the ability of a solute to undergo interactions which are usually termed nonpolar or nonspecific. The other type of structural parameters providing an input to retention are the so-called polar, or chemically specific, structural descriptors. Thus a free-energy related retention parameter R could be described by the regression equation.

$$R = af(P) + bf(NP) + c \qquad (8)$$

where $f(P)$ is a function of the ability of a solute to undergo polar interactions, $f(NP)$ is a function reflecting the tendency of a solute to participate in molecular interactions of a nonpolar type, and a, b, and c are constants.

Data supporting the general validity of Eq. 8 can be found, for example, in the paper by Karger et al. (16), published in 1976. These authors obtained linear plots of a solubility parameter versus the molar volume of a solute for the homologous series of a given functional group. The lines are parallel for different monofunctional homologous series except n-alcohols (Fig. 14.6). The linearity of the plots in Fig. 14.6 is the result of practically equal polarity among the members of any homologous series, whereas the nonpolar (bulky) properties depend largely on the size of a given member of the series.

On the other hand, if one considers a series of solutes of the same molar volume but possessing different functional groups, then one can expect a linear relationship of the solubility parameter as a function of some measure of polarity of the solutes. Unfortunately the quantitation of the structural features of the solutes determining their polar character is difficult. For monofunctional derivatives considered by Karger et al. (16) such a polarity measure can be taken as the dipole moment of a solute (Fig. 14.7). However, in general to characterize intermolecular interactions in HPLC one cannot rely on the net structural descriptors valid for a molecule as a whole, but one has to consider contributions of individual

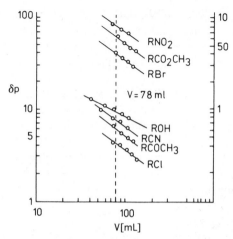

Figure 14.6. Logarithmic plots of polar solubility parameter against molar volume for homologous series of monofunctional group compounds. [After B. L. Karger, L. R. Snyder, and C. Eon, An expanded solubility parameter treatment for classification and use of chromatographic solvents and adsorbents. Parameters for dispersion, dipole and hydrogen bonding interactions, *J. Chromatogr., 125* (1976) 71. With permission of Elsevier Science Publishers B.V., Amsterdam, The Netherlands.]

submolecular fragments of the solute to intermolecular interactions. To find reliable substructural descriptors is an extremely difficult task, which often requires a great deal of what we call chemical intuition.

Studies of the quantitative relationships between the structure of solutes and their retention can be divided into two groups. In one type of QSRR studies the experimentally determined physicochemical properties of solutes are employed as structural parameters, for example, boiling point, refraction index, and especially *n*-octanol–water partition coefficient. Certainly, the QSRR relating retention to any arduously determinable physicochemical property can be of practical analytical value. An example may be the utilization of RP HPLC for the evaluation of partitioning properties of biologically active agents. From the point of view of fundamental chemical studies, the second group of QSRR is probably the more informative one. Here retention is related to the molecular structural parameters, which can be determined a priori based on the structural formulas of the compounds of interest.

The molecular descriptors most often applied in QSRR studies, which can be extracted from information encoded in structural formulas, are summarized in Table 14.2.

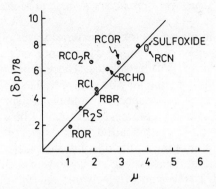

Figure 14.7. Plot of polar solubility parameter normalized to molar volume of 78 mL against dipole moment for monofunctional group compounds. [After B. L. Karger, L. R. Snyder, and C. Eon, An expanded solubility parameter treatment for classification and use of chromatographic solvents and adsorbents. Parameters for dispersion, dipole and hydrogen bonding interactions, *J. Chromatogr., 125* (1976) 71. With permission of Elsevier Science Publishers B.V., Amsterdam, The Netherlands.]

Table 14.2. Structural Descriptors Used in QSRR Analysis

Additive "Bulk" Parameters

Carbon number
Molecular mass
Parachor
Molecular volume
Molecular refractivity
Polarizability
Total energy

Electronic Parameters

Dipole moments
Substituent constants
Various quantum chemical indices

Molecular Shape Parameters

Topological Indices

Employ information from bulk, electronic, and shape categories

579

14.3.3.1. Structural Parameters Related to Molecular Size

When dealing with homologous (or congeneric) series of solutes one usually observes regularities of retention dependence on parameters reflecting the size ("bulkiness") of the solutes. These types of parameters are characterized by additivity and may be interpreted as reflecting the ability of solutes to participate in nonspecific, nonpolar, dispersive intermolecular interactions. The bulky parameters can easily be quantified and are highly intercorrelated. From this author's experience it appears that the most reliable descriptor of the ability of solutes to participate in dispersive chromatographic interactions has been the quantum chemically calculated total energy, which is the sum of electronic energy and the core–core repulsion (3, 6).

14.3.3.2. Molecular Descriptors Reflecting Electron Distribution in Solute Molecules

Unfortunately the quantitation of the structural features of the solutes determining their chemically specific character has been much more difficult than was the case with their "bulkiness." In some individual cases the so-called polarity of the solutes can be represented satisfactorily by one or more electronic, topological, and molecular shape descriptors. Among the structural descriptors reflecting the electron distribution within a molecule, the most promising seem to be quantum chemical data. Dipole moments, for example, appeared to be poor descriptors of the polarities of a set of aromatic derivatives for which RP HPLC data were obtained. When the quantum chemically calculated electron charge distribution on individual atoms was considered instead, it was shown that a submolecular polarity measure Δ (see Fig. 14.8 for definition) was a statistically significant parameter for describing retention (17, 18). This observation is in agreement with early findings by Karger et al. (16) and Scott (19), who noted that molecules like 1,4-dioxane, which have a dipole moment of zero, behave in chromatography as if composed of two polar subunits.

There are a variety of semiempirical electronic substituent constants of the Hammett and Taft types as well as Hansch hydrophobic constants or Rekker fragmental constants which have been widely used in medicinal chemistry (20). The application of these data to determine the molecular fragment contribution to retention is occasionally reported in studies aimed at the prediction of retention and is discussed in detail elsewhere (3).

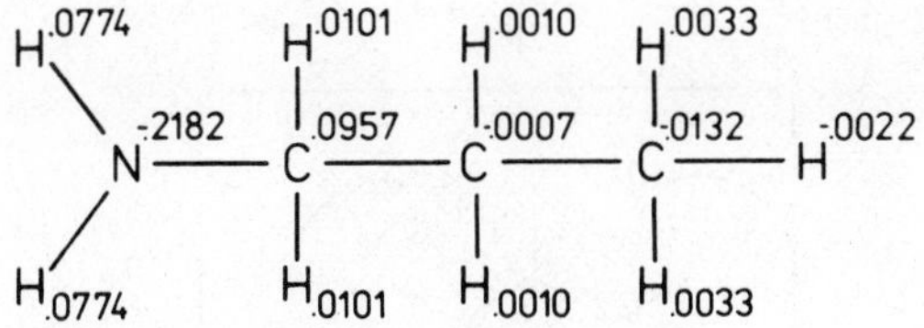

$$\Delta = 0.0957 - [-0.2182] = 0.3139$$

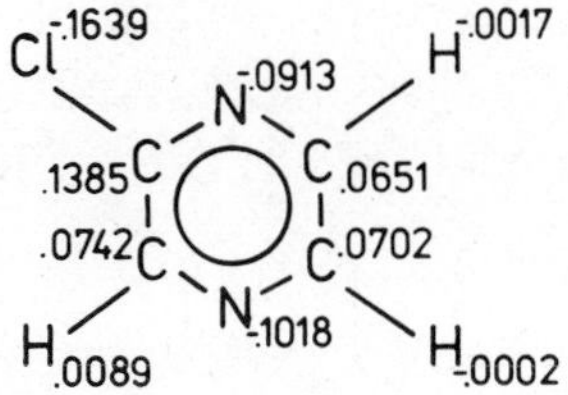

$$\Delta = 0.1385 - [-0.1639] = 0.3024$$

Figure 14.8. Examples of electron excess charge density distribution and determination of submolecular polarity parameter Δ.

14.3.3.3. Quantitation of Molecular Shape

For specific groups of solutes chromatographed at specific chromatographic conditions, the molecular shape may determine separation. It is difficult to describe the molecular shape in quantitative terms. For rigid molecules like polyaromatic hydrocarbons (PAH) the shape parameter was proposed (21) as being the ratio of the longer side to the shorter side of a rectangle having a minimum area, which can envelop the formula of a molecule (Fig. 14.9). Wise et al. (22) applied the shape parameter to the prediction of differences in the liquid chromatographic elution of isomeric methyl-substituted PAH. The correlation between determined and predicted retention data is illustrated in Fig. 14.10.

14.3.3.4. Topological Indices

The translation of molecular structures into unique numerical descriptors is interesting not only to QSRR, but also to many subdisciplines of chemistry and pharmacology. This can be attempted by means of the chemical graph theory, where a chemical structural formula is expressed as a mathematical graph. Such a molecular graph may be represented by either

β

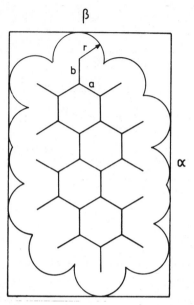

α

Figure 14.9. Determination of shape parameter as exemplified by chrysene (58). *a, b*—approximate lengths of C—C and C—H bonds (1.4 and 1.1 Å, respectively); *r*—van der Waals radius of hydrogen atom, = 1.2 Å. [After A. Radecki, H. Lamparczyk, and R. Kaliszan, A relationship between the retention indices on nematic and isotropic phases and the shape of polycyclic aromatic hydrocarbons, *Chromatographia, 12* (1979) 595. With permission of Friedrick Vieweg and Sohn Verlagsgesellschaft mbH, Wiesbaden, FRG.]

a matrix, a polynomial, a sequence of numbers, or a numerical topological index. The most popular topological index, the molecular connectivity index, was introduced by Randic (23) for the characterization of molecular branching.

An example calculation of simple connectivity indices is given in Fig. 14.11. At first the atom connectivities are assigned to individual atoms in the hydrogen-suppressed molecular graph, then simple arithmetic calculations are done. Atom connectivity is equal to the number of valence electrons minus the number of hydrogen atoms attached to a given atom.

Although the potency of molecular connectivity as a parameter quantitatively reflecting molecular branching is unquestionable, its ability to differentiate solute participation in complex chromatographic interactions should not be overestimated. In fact, Clark and coworkers (24) obtained high correlation (a correlation coefficient above 0.99) between the RP HPLC experimental data and the log k' values calculated by a three-parameter regression equation involving different modifications and arithmetic transformations of connectivity indices. However, there is no in-

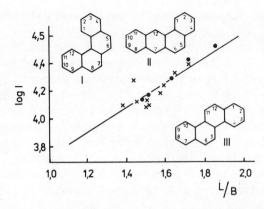

Figure 14.10. Linear correlation of shape parameter L/B versus liquid chromatographic retention parameter for methyl-substituted isomers of benzo(c)phenanthrene (I), benz(a)anthracene (II), and chrysene (III). [After S. A. Wise, W. J. Bonnett, F. R. Guenther, and W. E. May, A relationship between reversed-phase C_{18} liquid chromatographic retention and the shape of polycyclic aromatic hydrocarbons, *J. Chromatogr. Sci., 19* 457 (1981). With permission of Preston Publications, Niles, IL.]

formation concerning the intercorrelation of the variables used, and besides, it is difficult to assign any physical meaning to squared terms of the indices.

Looking for molecular structural descriptors which would be able to quantify specific properties of solutes, that is, the properties resulting from electron distribution over individual molecules, we turned our attention to topological indices. Recently we proposed (25) the quantum chemically based topological-electronic index T^E. The topological indices available until now contained only limited information relating to the interatomic distances. They did not utilize any significant information concerning the electronic structure of the solutes. The procedure applied to determine the T^E index is illustrated in Fig. 14.12. First the electronic charge distribution in the solute molecule is calculated. (At present this may be accomplished routinely by using computers and commercially available quantum chemical calculation programs.) To each individual vertex atom a number q is assigned, which is equal to its electronic excess charge. Then the distances $r_{i,j}$ between each pair of vertices are calculated. For every pair of vertices the absolute value of the excess charge difference is divided by the square of the respective interatomic distance. The resulting numbers are summed for all possible atomic pairs in the molecule. The index T^E has proved useful in QSRR analysis of gas chromatographic retention indices and should find application in the chemometrical analysis of HPLC capacity factors.

14.3.4.   Requirements of Statistically and Physically Meaningful QSRR

As long as one deals with homologous or closely congeneric series of solutes, usually simple linear regression analysis yields relationships between retention data and an individual structure descriptor of significant predictive value. With several independent structural variables affecting retention, the respective QSRR relationships are derived statistically. Advanced statistical techniques are now in routine use since efficient personal computers have become common research tools. The method most often used for deriving QSRR is multiparameter regression analysis, but the nonparameter de novo methods and factor analyses are also applied occasionally. Interestingly, some of the procedures were originally elab-

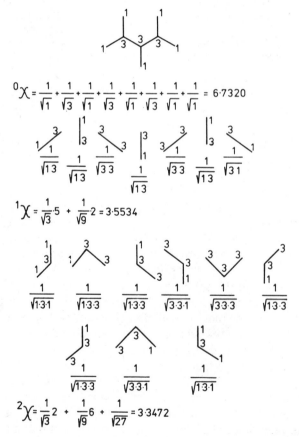

Figure 14.11. Calculation of molecular connectivity indices for 2,3,4-trimethylpentane.

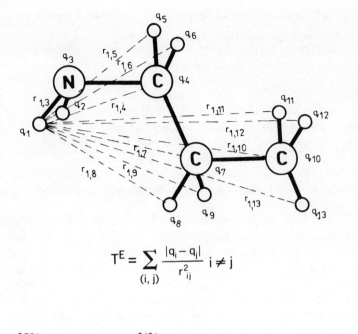

$$T^E = \sum_{(i,\, j)} \frac{|q_i - q_j|}{r_{ij}^2} \quad i \neq j$$

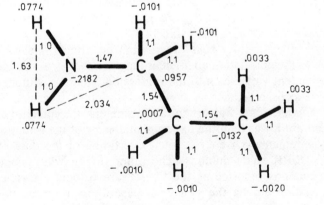

$$T^E = \frac{[.0774 - .0774]}{1.63^2} + \frac{[.0774 - (-.2182)]}{1.0^2} + \frac{[.0774 - .0957]}{2.034^2} + \cdots\cdots\cdots = 1.532$$

Figure 14.12. Procedure for determining the topological electronic index.

Table 14.3. Statistics Parameters Required for Evaluation of QSRR

---

log (retention parameter) $= f(aX_1, \ldots, zX_i)$

R: correlation coefficient

s: standard error of the estimate

F-test value $= \dfrac{(n - k - 1)R^2}{k(1 - R^2)}$

p: significance level

Intercorrelation matrix between variables considered

Confidence intervals (usually 95%) for regression coefficients

---

orated to solve problems in econometrics, psychology, and weather forecasting.

There is certainly no need at present to write individual programs for standard statistical methods of QSRR calculations. However, the programs available are sometimes used rather mechanically, and the results obtained can be statistically invalid if sufficient care is not taken. Among the more than 200 QSRR publications many are controversial from the statistical point of view. As far as the most often used QSRR procedure, multiparameter regression analysis, is concerned, the recommendations for reporting the results of correlation analysis in chemistry (26) should be observed. Thus the statistics analyzed and reported should include at least the data summarized in Table 14.3.

In searching for the "best" regression equation some workers pay attention to the correlation coefficient only. Thus they introduce individual independent variables into the regression equation without checking their statistical significance as retention descriptors. Occasionally, due to the introduction of several structural parameters, the standard deviations calculated are lower than the experimental error of the retention parameter measurements. Of course, such overfitting of the data makes the obtained QSRR value dubious. The other reason why sometimes the QSRR equations reported as the "best" are without relevance is high intercorrelation among the structural descriptors used as independent variables in the same regression equation. Also, there is a potential for chance correlation when too many variables are surveyed to correlate too few data. It is a rule of thumb that in a reliable regression equation there should be at least four to five data per independent variable. As a final remark, the simplicity principle should be observed, that is, if several correlation equations are equivalent, the simplest one should be chosen (27).

Factor analysis is applied in chemistry to determine the "intrinsic dimensionality" of certain experimentally determined chemical properties, that is, the number of "fundamental factors" required to account for the variance (28). Usually attempts are undertaken to give a physicochemical meaning to the individual abstract factors derived by the prescribed methods of linear algebra. The lack of physical identification of the factors does not limit the predictive power of the analysis, however. The example studies on the application of factor analysis to HPLC retention data may be the reports by Huber et al. (29) and Chretien and coworkers (30). Readers interested in the application of the de novo nonparameter method to the analysis of HPLC data are referred to Chen and Horvath (31).

14.4. MULTIPARAMETER QSRR

Since the early work by Martin (32) many reports have been published in which substituent increments to retention were determined. The contributions of the substituents to retention have often been found constant and additive for a given chromatographic system. Based on this observation, reliable predictions of retention behavior have been reported, especially in the case of RP HPLC or gas chromatography on nonpolar stationary phases (3). The most instructive report dealing with the functional group contribution to RP HPLC retention seems to be by Tomlinson et al. (33), in which these contributions were determined at conditions where enthalpy–entropy compensation was found. Recently Jandera (34) presented a consistent approach to the prediction of retention in HPLC based on the experimental determination of lipophilic and polar contributions to selectivity.

Another type of multiparameter QSRR is based on using substituent or fragmental structural constants as retention predictors. For example, Jinno and Kawasaki (35) used the following equation to describe log k' of substituted phenols chromatographed on octadecylsilica with acetonitrile–water (65:35) as the mobile phase, in terms of the Hammett σ and Hansch π substituent constants,

$$\log k' = 0.177\pi + 0.182\pi\cdot\sigma(1 - \pi) - 0.367 \qquad (9)$$

Equation 9 is characterized by a correlation coefficient of 0.954. It is not easy, however, to assign physical meaning to the product of π and σ.

From a chemometrics point of view multiparameter QSRR with nonempirical solute structure descriptors is the most interesting. Extensive studies were probably first reported by Spanish researchers in mid-1970 in the area of gas chromatography (36). Unlike gas chromatographic separation, where the differences in solute–carrier gas interactions can be neglected, there are three main variables determining the distribution of a solute between a mobile and a stationary liquid chromatographic zone. At a constant temperature of separation, the three variables are the chemical structure of the solute, the physicochemical properties of the mobile phase, and the physicochemical properties of the stationary phase. Thus if one gets numerical measures of the properties of the solutes, of the mobile phases, and of the stationary phases, one can attempt to derive a general relationship linking the appropriate quantities and retention parameters together. Recently we attempted to derive a QSRR equation that will account for changes in all three chromatographic variables (17, 18).

A set of 12 substituted benzene derivatives with various functional groups were selected (Table 14.4). The structural diversity among the solutes is evident from the point of view of nonspecific and polar properties. Yet the solutes selected form a related family of aromatic compounds. Changes in composition of the mobile phase methanol–water were also limited to the range of 35–65% v/v of methanol. The lower and upper concentrations limited the linearity of the dependence of log k'_{ij} for the ith solute determined on the jth stationary phase, on the mole fraction X of water in the binary solvent,

$$\log k'_{ij} = a_{ij}X + b_{ij} \tag{10}$$

where a_{ij} and b_{ij} are constants for a given solute i chromatographed on

Table 14.4. CNDO/2MO Parameters of Benzene Derivatives

Compound	Total Energy E_T (A.U.)	Maximum Excess Charge Difference Δ (electrons)	Dipole Moment μ (debyes)
1. Phenol	−65.5548	0.4328	1.7492
2. Acetophenone	−81.1756	0.5088	3.0417
3. Nitrobenzene	−94.8446	0.7774	5.0589
4. Methylbenzoate	−99.6184	0.6836	2.0376
5. p-Cresol	−74.2375	0.4265	1.7379
6. p-Ethylphenol	−82.6639	0.4275	2.3324
7. p-Propylphenol	−91.3479	0.4270	2.1472
8. 4-sec-Butylphenol	−100.1758	0.4246	1.8457
9. Aniline	−59.5473	0.3737	1.5206
10. N-Methylaniline	−68.2305	0.3564	1.1504
11. 4-Chloroacetophenone	−96.6241	0.5028	2.2894
12. 3,4-Dichloroacetophenone	−112.3142	0.4922	1.2809

an individual phase j. Also, for a given solute i chromatographed at a fixed mobile phase composition X, linearity has been found between $\log k'_{i,X}$ and the stationary phase composition determined as dimethyloctadecyl-silane coverage C_j,

$$\log k'_{i,X} = AC_j + B \tag{11}$$

where A and B are regression coefficients.

The coefficients of Eq. 10 were described by two-parameter regression equations in terms of the quantum chemically calculated total energy E_T and the polarity parameter Δ already discussed. Again, the submolecular polarity parameter Δ appeared statistically much more significant for retention description than the overall molecular dipole moment.

Eventually a general QSRR equation was derived describing the RP HPLC retention data as a function of the solute structural parameters E_T and Δ, the mole fraction of water in binary solvent with methanol X, and the density of the hydrocarbon coverage on the stationary phase C_j,

$$
\begin{aligned}
\log k'_{ij,X} = \; & [0.0454(\pm 0.0071)E_T(i) + 2.6493(\pm 0.9187)\Delta(i) \\
& - 0.1053(\pm 0.0672)C_j - 0.4946(\pm 0.5828)]X \\
& + [-0.0381(\pm 0.0039)E_T(i) + 2.1659(\pm 0.4919)\Delta(i) \\
& + 0.1696(\pm 0.0359)C_j + 1.2963(\pm 0.3120)]
\end{aligned}
\tag{12}
$$

The correlation between the retention data observed experimentally and those calculated by Eq. 12 is illustrated in Fig. 14.13. For 144 data points the correlation coefficient is $R = 0.986$.

14.5.   DETERMINATION OF HYDROPHOBICITY BY HPLC

Since the turn of the century it has been recognized, due to the works of Overton, Meyer, and Baum, that the lipophilic properties of drugs are of importance in determining their pharmacological activity (3). The measure of lipophilicity of chemical compounds is the logarithm of the partition coefficient $\log P$, determined in an n-octanol–water partitioning system.

The n-octanol–water system is the common reference system because it provides a single, continuous scale for measurements of the hydrophobicity. However, measurement of $\log P$ by the conventional "shake-flask" method is tedious and time consuming. It is difficult to determine $\log P$ for compounds that are poorly soluble in water or that cannot be

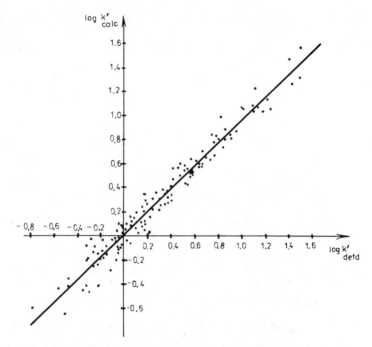

Figure 14.13. Correlation of calculated and observed logarithms of capacity factors for 12 test solutes chromatographed at different mobile and stationary phase compositions. [After R. Kaliszan, K. Osmialowski, S. A. Tomellini, S.-H. Hsu, S. D. Fazio, and R. A. Hartwick, Quantitative retention relationships as a function of mobile and C_{18} stationary phase composition for non-cogeneric solutes, *J. Chromatogr., 352* 141 (1986). With permission of Elsevier Science Publishers B.V., Amsterdam, The Netherlands.]

detected by routine analytical procedures. The shake-flask technique is also not applicable to surface-active and organometallic substances. It is difficult to determine reliable log P data for ionic substances, volatile compounds, and solutes for which association and dissociation processes are observed. Additional problems are caused by impurities, phase separation, and the formation of emulsions.

Assuming the extrathermodynamic LFER, one can expect the standard log P data to be linearly related to the partition chromatographic parameters, such as log k' from RP HPLC,

$$\log P = m \log k' + n \tag{13}$$

where m and n are constants. To date a great many papers have been published dealing with the chromatographic characterization of hydrophobicity and the application of chromatographically determined partitioning data in medicinal chemistry.

The advantages of the HPLC method of hydrophobicity evaluation over the classical shake-flask method are numerous (3, 6, 37–39). The HPLC method is fast; it is suitable for substances containing impurities and for substance mixtures. The method requires no quantitative determination and is applicable to volatile substances. It is highly reproducible and is suitable for compounds with a wide range of hydrophobicity. In addition, it provides precise control of pH and ionic strength during the separation process.

Much effort has been directed toward developing HPLC systems that mimic the conventional n-octanol–water partition system. Several procedures have been reported based on dynamically coating a stationary phase with n-octanol and using n-octanol–saturated aqueous eluent (40–42). Other coating agents, such as phosphatidylcholine (43), were also tried. There are serious disadvantages of the n-octanol-like chromatographic systems, however. There is some doubt as to whether the column characteristics change during use and, as practically no organic modifier can be added to the aqueous mobile phase (because of column stability), the chromatography of more hydrophobic solutes is impossible.

The stable RP HPLC systems most often used for hydrophobicity assessment employ alkyl (usually octadecyl) ligands chemically bonded to the silica support surface. Individual authors recommend a specific treatment of stationary phases before the use and the addition of various chemicals to the mobile phase (3, 44–46). Nonetheless, the correlations of such determined HPLC data with log P are good only as long as the solutes analyzed are more or less closely mutually related (congeneric). The decrease in the log P versus log k' corrrelation with an increasing structural diversity of solutes results mainly from specific interactions of the compounds chromatographed with residual silanol groups on the stationary phase material. It should be mentioned here that in spite of additional silylation ("capping") more than 50% of silanol groups on the silica support surface of the octadecylsilica phase remain unreacted. Due to the interaction with free silanol sites some solutes are strongly retained whereas others can be excluded from the pores of stationary phase material (14). Another disadvantage of alkyl-bonded silicas is their chemical instability at pH above 8. Thus in case of organic bases hydrophobicity cannot be determined directly. To overcome the problem Kraak et al. (47) used sodium dodecyl sulfate (SDS) as the pairing agent and determined the retention of bases as ion pairs with SDS at pH 4.

Having in mind the disadvantages of hydrocarbonaceous silica stationary phases, de Biasi et al. (48) studied the relationship between log

P and liquid chromatographic data derived on the styrene-divinylbenzene copolymeric stationary phase for a series of nonionized organic bases. The correlation of the parameters considered was only moderate (R = 0.906). The system proposed by de Biasi et al. (48) is not highly efficient and can only be operated under low pressure. The other stationary phase studied by the same authors was graphitized microparticulate carbon material recently introduced by Knox et al. (49). The retention data determined on that phase did not correlate with log P at all (R = 0.2).

Looking for an HPLC stationary phase that would be devoid of all the disadvantages of the materials used so far, we turned out attention (50) to a new polymer-coated reversed-phase material proposed by Schomburg and coworkers (51). To obtain a material stable over a wide pH range (especially at high pH conditions) and possessing no interfering silanol groups, poly(butadiene)-coated alumina was investigated. The poly(butadiene) was immobilized on the alumina support surface with the help of cross-linking reactions that involved radical formation. The material can be obtained from ES Industries (8 S. Maple Ave., Marlton, NJ 08053).

To balance reasonable retention times for the most strongly retained test solutes and reliable pH to control ionization, the methanol–buffer 50:50 v/v eluent was used. In the case of basic and neutral solutes the $0.05M$ $Na_2HPO_4/0.1M$ NaOH buffer was used with pH 10.7, with the ionic strength adjusted to 0.1. For acidic solutes a KCl/HCl buffer was applied with pH 1.65 and ionic strength 0.1.

In order that the HPLC capacity factor k' can be free-energy related and the chromatographic data compared between laboratories, the column dead volume has to be defined and determined unequivocally. We determined the dead volume, using a refraction index detector, as the elution volume of deuterated methanol with 100% CH_3OH as the mobile phase. All retention volumes were corrected for the extracolumn volume. The log k' values were correlated with the log P data taken as the mean of the values reported by Hansch and Leo (20) for unionized forms of test solutes. The test solutes selected are commercially available diverse structures—mostly organic bases, but some acids and neutrals included—of a wide range of hydrophobicity (Table 14.5).

The relationship between log P and log k' is illustrated in Fig. 14.14 and is characterized by the correlation coefficient R = 0.96 and a significance level of 3.14×10^{-12}. Of 24 test solutes studied, two (compounds 1 and 12 in Table 13.5) could not be included when deriving the correlation equation because their apparent k' data were negative. This fact illustrates limitations of the method when dealing with very hydrophilic solutes. Also other hydrophilic compounds (compounds 9 and 17

Table 14.5. Logarithms of HPLC Capacity Factors (log k') of Unionized Forms of Basic, Neutral, and Acidic Solutes and the Corresponding n-Octanol–Water Partition Coefficient (log P)

Compound	log P*	log k'†
1. Antipyrine	0.23	—
2. 4-Chloropyridine	1.28	−0.69
3. Acridine	3.40	0.41
4. Procaine	1.90	−0.64
5. (+)Ephedrine	1.02	−0.66
6. 4-Chlorobenzoic acid	2.65	0.11
7. 4-Chloroaniline	1.64	−0.10
8. p-Toluidine	1.42	−0.51
9. Benzamide	0.65	−1.55
10. Atropine	1.81	−0.33
11. 9-Aminoacridine	2.74	0.04
12. Sulfanilamide	−0.72‡	—
13. Benzoic acid	1.95	−0.40
14. Acetanilide	1.16	−0.99
15. Phenol	1.48	−0.58
16. Chlorobenzene	2.83	0.63
17. Caffeine	−0.07	−1.14
18. Aniline	1.08	−0.86
19. Diphenylamine	3.44	0.93
20. N-Phenylanthranilic acid	4.36	0.96
21. Acetophenone	1.66	−0.38
22. Phenothiazine	3.78	1.22
23. Biphenyl	4.06	1.32
24. Chlorpromazine	5.35	1.64

* Arithmetic mean of data compiled by Hansch and Leo (20).
† From HPLC at pH 1.65 for compounds 1, 6, 12, 13, 15, 17, and 20 and at pH 10.7 for the remaining compounds.
‡ Determined at "acidic pH" (20).

in Table 13.5) deviate most significantly from the correlation line. The relative error of k' values for solutes which elute close to the column dead volume makes the evaluation of hydrophobicity less reliable. This is the case for compounds of log P below 1, when a column of 150-mm length is used. For more hydrophilic substances a longer column can be applied or the poly(butadiene) coating of alumina increased.

There is a scattering of data in Fig. 14.14. This results partially from errors in log P data and is comparable to the scattering of data points reported for the log P versus log k' relationships derived for other limited

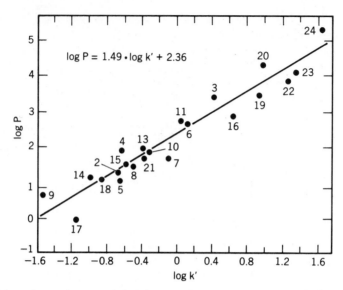

Figure 14.14. Correlation between RP HPLC derived capacity factors log k' and n-octanol–water partition coefficients log P for a set of basic, neutral, and acidic test solutes. Compounds are numbered as in Table 14.5.

sets of congeneric solutes. The question arises now whether log k' is a more reliable and reproducible measure of hydrophobicity than log P.

The most important feature of the method described is that it provides convenient hydrophobicity evaluation for basic compounds. Theoretically, for the stationary phase applied there is no upper limit of pH, and going down as far as pH 1.65 we did not observe a deterioration of the column performance. Thus with the poly(butadiene)-coated alumina phase the basic, neutral, and acidic compounds can be analyzed in the same chromatographic system. In other words, a continuous hydrophobicity scale may be obtained in a much easier, faster, and more reproducible manner than is the case with n-octanol–water systems.

14.6. HPLC PARAMETERS AND BIOACTIVITY

Retention parameters or chromatographically derived structural parameters are widely used in studies of quantitative structure–biological activity relationships (QSAR) aimed at a rational design of new drugs, pesticides, and other biologically active agents (3). Most often the chromatographic data are used in QSAR analysis as the descriptors of the lipophilicities of the bioactive agents, related more or less directly to the n-octanol–water partition system. There is no reason, however, to assume that the n-octanol–water system is the best for modeling complex processes in a living organism. On the other hand, one can imagine a specific chromatographic system that would mimic the biological transport and the receptor-binding phenomena. In such a chromatographic model system, the hydrophobic, electronic, and steric interactions would certainly be effective. Multidimensional chromatography could provide the means to obtain the systems required. At present, however, the state of the knowledge does not allow rational design of a proper multidimensional chromatographic system mimicking biological processes, and the trial and error approach offers little probability of success.

Besides the numerous QSAR publications reporting the use of HPLC data as lipophilicity descriptors there are a few studies that employ chromatographic parameters directly for bioactivity prediction. For example, Baker et al. (52) claimed the superiority of their HPLC retention index (defined by an analogy to the gas chromatographic Kovats index) over the n-octanol–water partition data for the quantitative description of the biological activity of propranolol and barbiturate analogs. However, in the case of the anti-inflammatory activity of the anthranilic acid derivatives, the same authors found a better description of pharmacological data when using the log P values. The Baker retention index was successfully applied in QSAR studies of 4-hydroxyquinoline-3-carboxylic acids as inhibitors of cell respiration (53) and for prediction of the antihypertensive activity of quinazolinamine derivatives (54).

Valko et al. (55) correlated bioactivity data of a group of new analgesic azidomorphines with RP HPLC parameters. The best linear correlation with the bioactivity measure provided log k' determined at 40% of acetonitrile in aqueous eluent. The attempt to rationalize that observation is described by the authors in the original paper (55).

The results obtained in QSAR studies on cardiac glycosides by Davydov (56) (Figure 14.15) seem especially impressive. The author related the toxicity LD of 22 glycosides to their retention parameter ln V_S, derived from HPLC on silica with bonded diphenylsilyl groups. The retention parameter V_S was defined as V_a/S, where V_a was the retention volume per gram of adsorbent and S was the specific surface area.

14.7. CONCLUDING REMARKS

Studies of quantitative relationships between descriptors of the structure of solutes and their chromatographic indices began only about 10 years

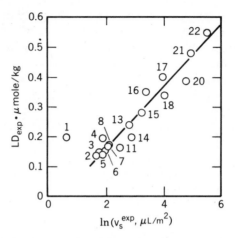

Figure 14.15. Correlation between biological activity of cardiac glycosides *LD* and their RP HPLC retention parameters. [After V. Ya. Davydov, Correlation between the molecular structure of cardiac glycosides, steroid hormones and carbohydrates and their retention in high performance liquid chromatography, *J. Chromatogr., 365* 123 (1986). With permission of Elsevier Science Publishers B.V., Amsterdam, The Netherlands.]

ago. In that short time, a significant amount of valuable information has already been gathered. Most significantly, the validity of the hope that QSRR could prove useful for practical medicinal chemistry applications has been amply demonstrated. Due to the QSRR approach, reliable HPLC methods have been elaborated, permitting a fast and convenient characterization of hydrophobicities of drugs and environmental pollutants.

The QSRR methodology has provided a new observation point for studies of the mechanism of chromatographic separations. Thanks to this approach some molecular structural factors affecting the retention of individual solutes have been found and quantified. For a designed set of test solutes, a comparison of the derived QSRR with the retention data generated at various mobile and stationary phase compositions allows a quantitative numerical characterization of the HPLC systems studied. Then a rational approach to optimize the separation conditions in specific analytical cases is possible instead of the trial-and-error method. Up till now, however, the QSRR have not been exploited enough by chromatographers. The reason is probably that because of the difficulties in the quantitative characterization of structural properties of chemical compounds the derived QSRR is valid only as long as one deals with a more or less congeneric group of compounds. It is a general chemistry problem, however, to relate structure and property, and chromatography has a good chance to solve it first.

In conclusion, it should be stressed that when deciding to do or apply QSRR, one must be cognizant of the limitations of the approach. It is essential to realize that the relationships of QSRR are derived statistically and thus their significance must be carefully checked. Next one has to realize that the reliability of numerical measures of structural properties of the solutes is often of limited precision. To a lesser extent this same limitation applies to the chromatographic data as well. For example, systematic errors in the determination of column dead volume can make the intercorrelation of various studies subject to error. Finally, one of the most important principles is that the relationships derived must be related to known chemical principles, and thus require a good working knowledge of chromatography and structural chemistry in order to avoid the folly of developing statistically valid correlations to chemically invalid principles.

REFERENCES

1. J. M. Prausnitz, *Science, 205,* 759 (1979).
2. R. P. W. Scott, in P. R. Brown and R. A. Hartwick, Eds., *High Performance Liquid Chromatography*, Wiley, New York, 1988 (this volume).
3. R. Kaliszan, *Quantitative Structure–Chromatographic Retention Relationships*, Wiley, New York, 1987.
4. E. Tomlinson, *British Pharmaceutical Conference Science Award Lecture*, Brighton, UK, 1981.
5. C. Reichardt, *Solvent Effects in Organic Chemistry*, Verlag Chemie, Weinheim, Federal Republic of Germany, 1979, p. 227.
6. R. Kaliszan, *CRC Crit. Rev. Anal. Chem., 16,* 323 (1986).
7. C. Horvath, W. Melander, and J. Molnar, *J. Chromatogr., 125,* 129 (1976).
8. A. Ben-Naim, *Hydrophobic Interactions*, Plenum, New York, 1980.
9. G. Vernin and M. Chanon, Eds., *Computers Aids to Chemistry*, Ellis Horwood Ltd., New York, 1986.
10. J. E. Leffler and E. Grunwald, *Rates and Equilibria of Organic Reactions*, Wiley, New York, 1963.
11. J. Shorter, *Correlation Analysis in Organic Chemistry: An Introduction to Linear Free-Energy Relationships*, Clarendon, Oxford, 1973, p. 64.
12. W. Melander, D. E. Campbell, and C. Horvath, *J. Chromatogr., 158,* 213 (1978).
13. J. H. Knox, R. Kaliszan, and G. J. Kennedy, *J. Chem. Soc., Faraday Symp., 15,* 113 (1980).
14. J. H. Knox and R. Kaliszan, *J. Chromatogr., 349,* 211 (1985).
15. E. Soczewinski and C. A. Wachtmeister, *J. Chromatogr., 7,* 311 (1962).

16. B. L. Karger, L. R. Snyder, and C. Eon, *J. Chromatogr.*, *125*, 71 (1976).
17. R. Kaliszan, K. Osmialowski, S. A. Tomellini, S.-H. Hsu, S. D. Fazio, and R. A. Hartwick, *Chromatographia, 20*, 705 (1985).
18. R. Kaliszan, K. Osmialowski, S. A. Tomellini, S.-H. Hsu, S. D. Fazio, and R. A. Hartwick, *J. Chromatogr.*, *352*, 141 (1986).
19. R. P. W. Scott, *J. Chromatogr.*, *122*, 35 (1976).
20. C. Hansch and A. Leo, *Substituent Constants for Correlation Analysis in Chemistry and Biology*, Wiley, New York, 1979.
21. R. Kaliszan, H. Lamparczyk, and A. Radecki, *Biochem. Pharmacol.*, *28*, 123 (1979).
22. S. A. Wise, W. J. Bonnett, F. R. Guenther, and W. E. May, *J. Chromatogr. Sci.*, *19*, 457 (1981).
23. M. Randic, *J. Am. Chem. Soc.*, *97*, 6609 (1975).
24. M. J. M. Wells, C. R. Clark, and R. M. Patterson, *J. Chromatogr. Sci.*, *19*, 573 (1981).
25. K. Osmialowski, J. Halkiewicz, and R. Kaliszan, *J. Chromatogr.*, *361*, 63 (1986).
26. M. Charton, S. Clementi, S. Ehrenson, O. Exner, J. Shorter, and S. Wold, *Quant. Struct. Act. Relat.*, *4*, 29 (1985).
27. J. K. Seydel and K.-J. Schapter, *Chemische Struktur und biologische Aktivität von Wirkstoffen. Methoden der quantitativen Struktur-Wirkung-Analyse*, Verlag Chemie, Weinheim, Federal Republic of Germany, 1979.
28. E. R. Malinowski and G. D. Howery, *Factor Analysis in Chemistry*, Wiley, New York, 1986.
29. J. F. K. Huber, C. A. M. Meijers, and J. A. R. J. Hulsman, *Anal. Chem.*, *44*, 111 (1982).
30. B. Walczak, L. Morin-Allory, M. Lafosse, M. Dreux, and J. R. Chretien, *J. Chromatogr.*, *395*, 183 (1987).
31. B.-K. Chen and C. Horvath, *J. Chromatogr.*, *171*, 15 (1979).
32. A. J. P. Martin, *Biochem. Soc. Symp.*, *3*, 4 (1950).
33. E. Tomlinson, H. Poppe, and J. C. Kraak, *Int. J. Pharm.*, *7*, 225 (1981).
34. P. Jandera, *J. Chromatogr.*, *352*, 11 (1986).
35. K. Jinno and K. Kawasaki, *Chromatographia, 18*, 90 (1984).
36. M. Gassiot-Matas and G. Firpo-Pamies, *J. Chromatogr.*, *187*, 1 (1980).
37. M. Harnisch, H. J. Mockel, and G. Schulze, *J. Chromatogr.*, *282*, 315 (1983).
38. T. Braumann, *J. Chromatogr.*, *373*, 191 (1986).
39. T. Hafkenscheid and E. Tomlinson, *Adv. Chromatogr.*, *25*, 1 (1986).
40. D. Henry, J. H. Block, J. L. Anderson, and G. R. Carlson, *J. Med. Chem.*, *19*, 619 (1976).
41. S. H. Unger and G. H. Chiang, *J. Med. Chem.*, *24*, 262 (1981).

42. S. J. Lewis, M. S. Mirrlees, and P. J. Taylor, *Quant. Struct. Act. Relat., 2,* 1 (1983).

43. K. Miyake, F. Kitaura, N. Mizuno, and H. Terada, *J. Chromatogr., 389,* 47 (1987).

44. J. J. Sabatka, D. J. Minick, T. K. Shumaker, G. L. Hodgson, Jr., and D. A. Brent, *J. Chromatogr., 384,* 349 (1987).

45. F. Gago, J. Alvarez-Builla, and J. Elguero, *J. Liq. Chromatogr., 10,* 1031 (1987).

46. M. C. Pietrogrande, F. Dondi, G. Blo, P. A. Borea, and C. Bighi, *J. Liq. Chromatogr., 10,* 1065 (1987).

47. J. C. Kraak, H. H. van Rooij, and J. L. G. Thus, *J. Chromatogr., 352,* 455 (1986).

48. V. de Biasi, W. J. Lough, and M. B. Evans, *J. Chromatogr., 353,* 279 (1986).

49. J. H. Knox, B. Kaur, and G. R. Millward, *J. Chromatogr., 352,* 3 (1986).

50. R. Kaliszan, R. W. Blain, and R. A. Hartwick, *Chromatographia, 25,* 5 (1988).

51. U. Bien-Vogelsang, A. Deege, H. Figge, J. Kohler, and G. Schomburg, *Chromatographia, 19,* 170 (1984).

52. J. K. Baker, D. O. Rauls, and R. F. Borne, *J. Med. Chem., 22,* 1302 (1979).

53. E. A. Coats, K. J. Shah, S. R. Milstein, C. S. Genther, D. M. Nene, J. Roesener, J. Schmidt, M. Pleiss, E. Wagner, and J. K. Baker, *J. Med. Chem., 25,* 57 (1982).

54. T. Sekiya, S. Yamada, S. Hata, and S.-I. Yamada, *Chem. Pharm. Bull., 31,* 2779 (1983).

55. K. Valko, T. Friedmann, J. Bati, and A. Nagykaldi, *J. Liq. Chromatogr., 7,* 2073 (1984).

56. V. Ya. Davydov, *J. Chromatogr., 365,* 123 (1986).

57. P. J. Schoenmakers, H. A. H. Billiet, and L. de Galan, *J. Chromatogr., 282,* 107 (1983).

58. A. Radecki, H. Lamparczyk, and R. Kaliszan, *Chromatographia, 12,* 595 (1979).

CHAPTER

15

FIELD-FLOW FRACTIONATION: AN HPLC ANALOGUE

LAYA F. KESNER and J. CALVIN GIDDINGS

Department of Chemistry
University of Utah
Salt Lake City, UTAH

15.1. INTRODUCTION

High performance liquid chromatography (HPLC), as shown by the contents of this book, is a powerful multifaceted collection of techniques. The similarly broad family of methods known as field-flow fractionation (FFF), according to strict definition, lies outside the HPLC category of techniques (1, 2). Yet there are many common threads joining HPLC and FFF (3): they are both flow-based elution separation techniques, their retention mechanisms are basically similar, band broadening has much the same origins, and they share common instrumental components and

601

procedures. However, there is one fundamental difference: FFF does not require a stationary phase to generate retention. According to most accepted definitions, including one promulgated by IUPAC (4), the distribution of species between a mobile and a stationary phase is an integral part of chromatography. However, FFF has no stationary phase into which molecules can partition. Nonetheless, FFF normally involves a special kind of partitioning, unusual because it occurs within a single phase. Therefore, although it creates a contradiction of terms, we can think of FFF as one-phase chromatography (5).

In the title we implied that FFF and HPLC are analogues. Dictionaries define analogues as similar or comparable in certain respects; they correspond in construction, function, qualities, and so on. We dwell for a moment on the analogous features of these two separation systems, an exercise that helps those unfamiliar with FFF to understand better how it works.

First of all, a cursory inspection of an FFF apparatus and the attendant experimental procedures leads one to conclude that FFF is closely related to liquid chromatography. A liquid phase is introduced into a separation column or channel; samples are injected, separated, and eluted; the column effluent is fed into a detector; a recorder or computer system displays peaks representing the sample fractions; finally, the fractions may be collected for further study (6).

The mechanisms of HPLC and FFF are also broadly similar. In HPLC, species are partitioned between moving regions (the mobile phase) and nonmoving regions (the stationary phase). The species are carried along at different velocities, depending on how strongly they partition into one or the other of the two phases. In FFF partitioning also occurs, but it is between regions having different velocities within a single flowing phase, inside the thin FFF channel (see Fig. 15.1). The displacement velocity of a component along the flow axis, as in HPLC, depends on whether the component is partitioned relatively more in the slow moving regions or in the high-velocity regions of the stream.

Since in FFF there are no phase differences or affinities to drive molecules into desired regions, other forces must be sought. Suitable forces are found to be those generated by externally applied fields and gradients acting in a direction perpendicular to the flow axis.

The regions of different flow velocity (into which components are to be partitioned by external fields) are always present in flow streams confined in a narrow tube or channel. Laminar flow in such conduits is characterized by a relatively high velocity near the channel center and a systematic decrease in velocity upon approaching the walls.

In practice, FFF is usually carried out in a ribbonlike channel of sub-

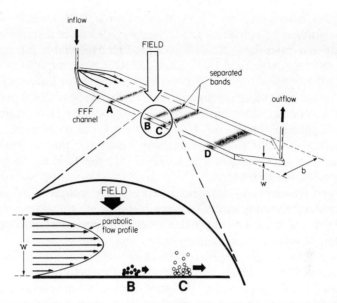

Figure 15.1. Illustration of FFF principle. Magnified channel edge view (bottom) illustrates normal operating mode.

millimeter thickness, as shown in Fig. 15.1. Across this thin dimension a typical bullet-shaped (parabolic) flow profile develops with the characteristics noted above: high flow in the center, zero at the walls (see lower panel of Fig. 15.1). The field is applied across the face of this channel and creates a driving force acting over the thin dimension of the channel. By driving different species to different degrees against the wall where the flow velocity is slowest, it is possible to differentially retain the components of a mixture.

We see, indeed, that the FFF process is normally one of selective partitioning, but it is confined to one phase and is driven by an external field. Thus we reemphasize the fundamental similarities of chromatography and FFF: they both generate separation within a confined flow stream, they accomplish this by differential partitioning, they end with sample elution, and they require similar equipment and procedures. As a consequence of these analogous features, the chromatographic techniques, including HPLC, are much more closely related to FFF than they are to other prominent separation methods such as electrophoresis and sedimentation (3).

Just as liquid chromatography can be subdivided into techniques based on the nature of the stationary phase, FFF possesses a variety of subtechniques determined by the type of external field applied. To date, thermal gradient, electric, magnetic, sedimentation (centrifugal or gravitational), and cross flow "fields" have produced the viable subtechniques of thermal FFF, electrical FFF, magnetic FFF, sedimentation FFF, and flow FFF (7). However, FFF is much broader than indicated by this choice: most of these subtechniques can, in theory, be realized in different operating modes, including normal FFF, steric FFF, hyperlayer FFF, cyclical-field FFF, and so on (7–9). One then has important choices concerning field strength, field programming, flow rates, flow splitters, channel dimensions, and so on. The variability of the FFF family is consequently immense (10, 11).

For practitioners and devotees of HPLC it is natural to ask why one would pursue the alternative strategy of replacing the stationary phase by an external field. By way of an answer, experience shows that this strategy enhances the capabilities of the separation system to work with high molecular weight species, both macromolecular and particulate in form. Where HPLC and FFF overlap, the two are largely complementary, generating separation based on different component properties and therefore producing different kinds of information concerning the samples in question (12).

Although HPLC has been widely used for high molecular weight species, there are intrinsic properties of chromatographic systems that make separation increasingly difficult as molecular weight goes up (2). For example, low molecular weight species will generally partition rapidly and reversibly between mobile and stationary phases. For high molecular weight species the mass transport processes underlying partitioning are more sluggish, but more importantly, they are increasingly difficult to maintain as reversible processes. This is because the component's net energy of interaction with a condensed phase or interface tends to increase with molecular size; when such interactions increase to a level significantly greater than mean thermal energy kT, the process is rendered virtually irreversible. It is necessary under such circumstances to seek special conditions and/or carefully generated gradients in order to restore reversibility. The basic difficulty of establishing and maintaining reversibility with macromolecules was recognized in the 1960s (13).

We note also that chromatographic retention forces are highly abrupt because they originate at the interface between two phases (1). The discontinuities can be associated with adsorption as well as structural alterations (such as denaturation) in complex high molecular weight species.

For fragile macromolecular species an additional problem with HPLC is the strong irregular shear forces generated by the liquid phase flowing under high pressure through a bed of tightly packed particles. Shear degradation and denaturation can result from such flow (1).

These HPLC limitations are largely absent in FFF. The flow in an FFF channel is smooth and laminar and has little tendency to disrupt susceptible species. Also, the forces leading to FFF retention are continuous and essentially uniform from one channel wall to the other; these forces are gentle to fragile species. In addition, adsorption phenomena, whose effects are much harsher, are greatly reduced in FFF because species encounter minimal surface area (1), consisting only of the channel walls (primarily just one channel wall). Furthermore, because the channel walls need only fulfill the requirements of containing the species and transmitting the field, the wall material is generally subject to replacement by more inert materials.

Finally, the reversibility of partitioning between two points in an FFF channel is readily achieved because no phase boundaries must be crossed and the driving forces are fully adjustable through changes in field strength. In fact, one of the major advantages of FFF is the presence of such tunable retention forces; these forces can be quickly and precisely adjusted to desired levels. Thus the driving forces can be readily altered in proceeding from one run to another in order to accommodate different types of samples without the need for a new channel system. The driving forces can also be varied continuously during the run in a procedure called field programming.

Field programming fully exploits the precise retention control possible in FFF (14). Species of widely divergent properties (such as a wide range of molecular weights) may be characterized within a single run. A high field is applied initially and is reduced systematically as some function of time. This programming permits highly retained species to emerge in a reasonable time without sacrificing the resolution of the less-retained species emerging at the beginning of the run. The field can be easily reduced to zero to permit flushing the channel of any residue material before subsequent runs.

An appealing aspect of FFF is its theoretical tractability (7). The response of a species to the applied field is generally described by a relatively simple proportionality containing some physicochemical property or properties of the species. Similarly the well-defined geometry of the FFF channel allows the formulation of an exact mathematical relationship for the channel flow profile. The field response equation can be combined with the flow equation to yield an expression for retention in terms of the physicochemical properties of the species and the applied field strength. The resulting retention equation permits (1) the determination of optimal field strength conditions for a required analysis and (2) the characterization of a complex sample without a calibration curve.

Another advantage of FFF is its sheer diversity. As pointed out, one has the choice of many different fields, different operating modes, variable programming conditions, and many other factors that, in combination, provide great latitude in adapting the FFF system to the material requiring analysis.

The variability of FFF, the gentleness and controllability of FFF forces, and the minimization of adsorption phenomena and shear degradation permit the FFF method to be used for species of virtually all types and origins ranging in effective molecular weight from under 10^3 to 10^{18}. This not only encompasses the macromolecular and colloidal range, but extends to a size covering most biological cells and particles up to 100-μm diameter.

We direct our attention now to the theory of FFF.

15.2.   THEORY

A feature shared by most forms of FFF is the ribbon-shaped separation channel shown in Fig. 15.1. The ratio of channel thickness w to channel breadth b or length L is small enough that the channel can be considered effectively as the space between two infinite parallel plates (15). Laminar flow between infinite parallel plates gives rise to a parabolic flow profile in which the flow rate close to either plate (channel wall) is low compared to the flow rate at the channel center (see Fig. 15.1). The perpendicular field of FFF differentially directs sample species to relatively faster or slower flow streamlines where they undergo differential migration. Thus zones are formed and separation is realized.

Below we develop the theoretical background for the normal operating mode of FFF; other modes such as steric FFF are touched on later.

15.2.1.   Steady-State Sample Layer

We now examine the action of the field or gradient on a sample species. The field is applied normal to the flow direction. A species which interacts with the field moves toward one wall of the channel (the accumulation wall) with a velocity governed by the degree of field–particle interaction, which is proportional to the field strength. The motion toward the accumulation wall is counteracted by a diffusive flux in response to the species' developing concentration gradient. Motion toward or motion away from the accumulation wall is opposed by the frictional resistance of carrier fluid; however, the friction effects in opposite directions generally cancel. At the steady state, where the field-induced flux balances the diffusional flux, the sample assumes a concentration profile of exponential form

$$c = c_0 \exp\left(\frac{-x}{l}\right) \tag{1}$$

where c is the sample concentration at a distance x from the accumulation wall, c_0 is the concentration at that wall, and l is the effective mean thickness of the sample layer (16). The parameter l, or the related dimensionless parameter $\lambda = l/w$, which is the ratio of the sample layer thickness to the channel thickness, defines the interaction between the species and the applied field. In general, we have

$$l = \frac{D}{U} \tag{2a}$$

and

$$\lambda = \frac{D}{Uw} = \frac{RT}{Fw} \tag{2b}$$

where D is the diffusion coefficient of the sample in the carrier liquid, U the field-induced velocity of the sample, R the gas constant, T the absolute temperature, and F the field-induced force per mole of sample component (7). The direct proportionality of λ and D illustrates the tendency of diffusion to increase the thickness of the sample layer; the inverse proportionality of λ and U indicates the tendency of the field to compress the sample layer. As mentioned, both U and D are inversely proportional to the friction coefficient which, consequently, does not appear in the final expression for λ. An exception is flow FFF, where U is independent of frictional considerations.

The various subtechniques of FFF differ not only in the type of applied field but also in the sample parameters responding to the field. Table 15.1 (5, 7) delineates the relationship of λ to the physicochemical constants of the sample components for the different FFF techniques. Electrical FFF, for example, exhibits a simple relationship: the product of electric field strength E and electrophoretic mobility μ is shown as inversely proportional to λ. In sedimentation FFF, λ is inversely proportional to the product of centrifugal acceleration G and sedimentation coefficient s. In thermal FFF, λ varies inversely as the product of the coefficient D_T of thermal

Table 15.1. Form of Retention Parameter λ for Different Subtechniques of Normal FFF*

Subtechnique	λ Expression	Component Parameters Controlling Retention
General	$\lambda = \dfrac{D}{Uw} = \dfrac{RT}{Fw}$	
Sedimentation FFF	$\lambda_s = \dfrac{D}{sGw} = \dfrac{RT}{GM(1 - \rho/\rho_s)w} = \dfrac{6kT}{\pi w G d^3 \Delta\rho}$	d, ρ_s, D, M
Flow FFF	$\lambda_F = \dfrac{DV^0}{\dot{V}_c w^2} = \dfrac{RTV^0}{3\pi\eta N\dot{V}_c w^2 d}$	D, d
Electrical FFF	$\lambda_E = \dfrac{D}{\mu Ew}$	μ, D
Thermal FFF	$\lambda_T = \dfrac{D}{D_T w(dT/dx)} = \dfrac{T}{\alpha w(dT/dx)}$	D_T, α

General symbols:

w = Channel thickness	*Particle characteristics*
k = Boltzmann's constant	D = Diffusion coefficient
U = Mean field-induced velocity	M = Molecular weight
F = Field-induced force per mole	d = Stokes diameter
R = Gas constant	μ = Electrophoretic mobility
T = Absolute temperature	D_T = Coefficient of thermal diffusion
G = Gravitational acceleration	α = Thermal diffusion factor
ρ = Solvent density	ρ_s = Density
V^0 = Void volume of column	$\Delta\rho = \rho_s - \rho$
$\dot{V}_c$ = Volumetric cross flow rate	s = Sedimentation coefficient
N = Avogadro's number	
η = Viscosity of solvent	
E = Electric field strength	

diffusion and the temperature drop ΔT across the channel. Flow FFF is unique in that the drift velocity U induced by the cross flow is uniform for all species and the differences in λ derive from differences in the sample diffusion coefficients which depend, in turn, on effective particle diameters.

It is worthwhile to note that in all these forms of FFF, λ can be plotted against the reciprocal of the applied field strength to obtain a straight line intersecting the origin. This plot directly provides a value of λ at any arbitrary field setting.

A review of the physicochemical constants influencing λ leads to the

FIELD

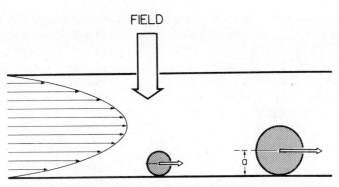

Figure 15.2. Schematic illustration of steric FFF operation. Reprinted from reference 52, by courtesy of The Humana Press, Inc.

conclusion that large, massive, and highly charged particles experience the greatest retention. This is a consequence of the fact that such particles have the greatest force exerted on them by the field; Eq. 2*b* shows that λ is inversely proportional to force F. However, we note that this conclusion and the others described are valid only for "normal" FFF. For other modes of operation, component retention is determined by other component properties or by other combinations of such properties. An example of such a changeover of properties is illustrated by steric FFF.

For particles greater than 1 μm in diameter, the steric mechanism takes over (17, 18). Although such a large particle is driven normally to the accumulation wall by the field, the bulk of the particle extends into the flow streamlines by virtue of its very size, even when the particle is touching the wall. The larger the particle, the greater will be the extension into the flow profile (see Fig. 15.2). Retention will therefore depend on size, more specifically on particle radius a, rather than on l, which is generally negligible for such large particles. Thus as illustrated in Fig. 15.2, the largest particles will elute first in steric FFF. This is an inverse size order relative to normal FFF. The transition from normal to steric FFF with the predicted inversion in elution order has been observed experimentally for sedimentation FFF (19).

15.2.2. Retention

An experimental parameter characterizing sample retention in both FFF and HPLC is the retention ratio R, which can be expressed in several forms,

$$R = \frac{\mathbf{V}}{\langle v \rangle} = \frac{V^0}{V_r} = \frac{t^0}{t_r} \tag{3}$$

where $\mathbf{V}$ is the zone migration velocity, $\langle v \rangle$ the mean velocity of the carrier stream, V^0 the dead- volume (in FFF the channel void volume), V_r the elution volume of the sample, t^0 the void time, and t_r the retention time of the sample. Combining Eq. 3 with the equation for the parabolic velocity profile and the expression for the sample concentration distribution (Eq. 1) yields an expression for R in terms of the dimensionless layer thickness λ applicable to normal FFF (20),

$$R = 6\lambda \left[\coth\left(\frac{1}{2\lambda}\right) - 2\lambda \right] \tag{4}$$

For $R < 0.5$ (corresponding to $\lambda < 0.1$) Eq. 4 is approximated quite well by the simple expression

$$R = 6\lambda \tag{5}$$

The $R < 0.5$ requirement is easily met under most experimental conditions.

We can use the above equations as a simple and direct way of determining λ and the physicochemical parameters on which λ depends (see Table 15.1) from measured retention data.

In the steric region λ in Eq. 5 is replaced by a/w, the ratio of particle radius to channel thickness. A factor γ is also introduced to account for nonideal effects, including viscous drag and hydrodynamic lift effects, characteristic of steric migration. We have (21)

$$R = 6\gamma \frac{a}{w} \tag{6}$$

We note that, in the steric region, R varies directly as the particle radius, whereas in normal FFF, R varies directly as λ, which varies inversely with particle size.

The selectivity S is a useful measure of the intrinsic resolving power of a separation technique. If samples are separated by virtue of differences in some parameter, such as mass m or molecular weight M in sedimentation FFF, the selectivity can be expressed as (2, 18)

$$S = \left| \frac{d \ln V_r}{d \ln M} \right| \tag{7}$$

Selectivity represents the fractional change in elution volume relative to a given fractional change in the parameter (in this case M) controlling the separation. Clearly a high value of S, represented by a large change in

V_r with a small change in M, is desirable for separation. At high retention levels, for which Eq. 5 is valid, we get

$$S_{max} = \left| \frac{d \ln \lambda}{d \ln M} \right| \tag{8}$$

Maximum S values based on mass for the presently practiced subtechniques of FFF range from 0.33 to 1; sedimentation FFF has the greatest selectivity, which approaches unity. Size exclusion chromatography, by contrast, possesses a maximum selectivity range of 0.05–0.22 (2).

15.2.3. Plate Height

Resolution in FFF, as in HPLC, is a function of zone broadening as well as of differential retention. Among the factors contributing to the plate height in both systems are longitudinal diffusion and nonequilibrium. In addition, various nonideal effects, which often can be minimized by judicious experimental design, are common to both methods. In FFF these arise as a consequence of relaxation, dead volume, finite injection volume and time, and irregularities in the channel surface. We examine briefly both the ideal and the nonideal FFF effects.

Longitudinal diffusion arises from the diffusion of individual particles along the flow (longitudinal) axis. As in HPLC, the longitudinal diffusion contribution to plate height H_D increases in proportion to the time the zone spends in the channel, that is, H_D is inversely proportional to flow velocity $\langle v \rangle$ (22). The magnitude of H_D, expressed as $2D/R\langle v \rangle$, is negligible at normal flow rates for macromolecules because of their small diffusion coefficients.

The nonequilibrium effect constitutes the major contribution to zone broadening in a well-designed FFF system. The nonequilibrium dispersion is caused by the random excursions of particles from their average distance from the wall given by l. The excursions temporarily place particles further from the wall where the flow velocity is greater than that at l or closer to the wall where the flow velocity is smaller. The fluctuations in migration velocity lead to zone broadening. This nonequilibrium effect is described by a plate height term proportional to velocity $\langle v \rangle$ (15, 16),

$$H_n = \frac{\chi w^2 \langle v \rangle}{D} \tag{9}$$

χ being a function of λ, which approaches $24\lambda^3$ for very small values of λ, where substantial retention is realized.

A possible nonideal source of zone broadening is relaxation, the finite amount of time required for the initial formation of the component's steady-state layer. Relaxation times vary from a few seconds to several minutes. The zone broadening H_r, which occurs if there is significant flow displacement before the steady state is established, is approximately (23)

$$H_r = \frac{17}{140} \frac{1}{L} \left(\frac{\lambda w^2 \langle v \rangle}{D} \right)^2 \tag{10}$$

where L is the channel length. The relaxation effect can be made negligible by the stop-flow technique, which consists in halting the longitudinal flow during the relaxation period.

An unavoidable source of zone broadening is the polydispersity of the sample. Most macromolecular samples contain a distribution of molecular weights or sizes. The FFF process fractionates the adjacent molecular weights or sizes; the narrow fraction thus appears as a broadened peak. The polydispersity contribution to plate height H_p in terms of the selectivity S is approximately (24)

$$H_p = LS^2(\mu - 1) \tag{11}$$

where the polydispersity μ is the ratio of the weight average molecular weight $\overline{M}_w$ to the number average molecular weight $\overline{M}_n$. Note that since H_p represents the fractionation of close lying species, a large H_p is symptomatic of high, not low, separation power.

Nonideal zone dispersion effects, which may be minimized in the same manner as in chromatography, include those arising in injection procedures and dead volumes. The open channel of FFF is subject to additional nonideal zone broadening if irregularities are present in the channel surfaces. The effects are negligible when channel walls are smooth and uniform.

The summary plate height expression for FFF is

$$H = \underbrace{\frac{2D}{R\langle v \rangle}}_{\substack{\text{longitudinal} \\ \text{diffusion}}} + \underbrace{\frac{xw^2\langle v \rangle}{D} + H_r}_{\text{nonequilibrium relaxation}} + \underbrace{H_p}_{\text{polydispersity}} + \sum H_j \tag{12}$$

where $\sum H_j$ represents miscellaneous nonideal contributions that can usually be made negligible by experimental manipulation.

15.2.4. Resolution

The resolution R_s of two-component peaks for FFF or HPLC is (22)

$$R_s = \frac{N^{1/2}}{4} \left(\frac{\Delta R}{\overline{R}} \right) \tag{13}$$

where N is the number of theoretical plates in the system, that is, the channel or column length divided by H. The selectivity is reflected in the $\Delta R/\overline{R}$ term. The $N^{1/2}$ term represents the system efficiency.

15.2.5. Programming

The programming of field strength addresses the problem encountered with wide ranging sample mixtures in which early peaks tend to be incompletely resolved and late peaks take too long to elute (25). The flow rate can also be programmed to advantage to solve this problem (26).

The theory of programming relates retention time t_r to channel length L by the integral equation (25)

$$L = \int_0^{t_r} R(t)\langle v \rangle(t)\, dt \tag{14}$$

In the case of field programming, R changes with time [as described by the function $R(t)$] as a consequence of its dependence on the varying field strength. Parabolic, linear, and abrupt changes in field strength with time have been used in sedimentation, flow, and thermal FFF (25, 27, 14, 28). Yau and Kirkland (29, 30) have developed a useful time-delayed exponential (TDE) field program for sedimentation FFF.

For flow programming (26), flow velocity varies with time as $\langle v \rangle(t)$. Theoretical analysis suggests substantial advantages for flow programming, but this approach has not been widely pursued. In addition, field and flow programming may, in theory, be used simultaneously (14). Finally, the programming of a carrier property (such as density) provides another option (25), closely related to gradient elution HPLC.

The first paper on FFF (31) identified programming as a potential asset. However, to evaluate fully the importance of programming, it is necessary to be aware of its limitations as well as its advantages (14). When a sample is relatively monodisperse, programming serves no useful function other than, possibly, to somewhat concentrate a hard-to-detect sample. Similarly, programming offers no advantage for the fractionation of a close binary sample. Programming, especially at an initially high field strength, may intensify some system nonidealities. Particle–surface interactions and particle–particle interactions are most pronounced at high field strength.

If the programming option is selected, the choice of initial field strength is crucial. The starting field must be strong enough to induce adequate retention ($R < 0.5$) in the smaller species without reducing R and λ for the larger species to such an extent that the layer thickness l is more compressed than the particle diameter. If the above conditions cannot be met, the smaller particles will undergo normal FFF while the larger particles experience steric FFF; separation then becomes complicated and/ or less effective.

Recent publications provide guidance concerning appropriate parameters for successful programming (32, 33). The parameter F_d, termed fractionating power, is a measure of the specific resolution of the system,

$$F_d = \frac{R_s}{\delta d/d} \qquad (15)$$

where R_s is the resolution between particles whose diameters differ by the small relative increment $\delta d/d$. For the isocratic (constant-field) sedimentation FFF of spherical particles, F_d is approximately proportional to the fourth power of the particle diameter. Alternately one may program an exponential field decay according to

$$S(t) = S_0 \exp\left[\frac{-(t - t_1)}{\tau'}\right] \qquad (16)$$

where the field strength (subsequent to primary relaxation) is held constant at S_0 for a set time t_1, $S(t)$ is the field strength at time t ($t \geq t_1$), and τ' is the exponential time constant. When $t_1 = \tau'$, this case becomes the TDE program of Yau and Kirkland (29). Equations have been developed (32) to yield suitable values for S_0 and τ'/t^0 to separate particles over a given diameter range (from d_1 to d_2) requiring separation at a specified F_d. Figure 15.3 illustrates conditions necessary to fractionate particles from diameter $d_1 = 0.1$ μm to $d_2 = 0.5$ μm requiring a minimum $F_d = 5$. (F_d corresponds to unit resolution between particles differing in size by 20%.) Thus the F_d curve must not drop below 5 anywhere in the d_1–d_2 range. However, the F_d curves with exponential programming peak well above 5. One would prefer F_d to be essentially constant over the particle size range studied; an overly large F_d requires an undue expenditure of time for the separation.

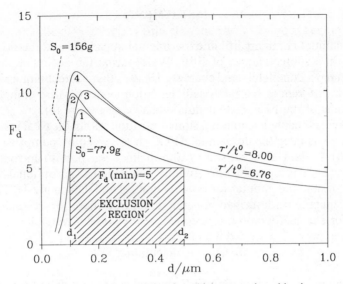

Figure 15.3. Illustration of exclusion region for which separation objectives are specified by d_1, d_2, and F_d(min). Various F_d curves chosen to avoid this region are also shown. These F_d curves are obtained by holding S_0 constant until the minimum F_d is reached. Other conditions for programmed sedimentation FFF system are $t^0 = 300$ s, $w = 0.0254$ cm, $\Delta\rho = 1.5$ g/cm^3, $\eta = 0.01$ P. Reprinted from reference 32, by courtesy of the American Chemical Society.

A power-law field-decay program has been found to generate a constant F_d over a substantial region (33). The field strength S is held constant at an initial level S_0 for a time-lag period t_1 and then decayed with time t according to the expression

$$S(t) = S_0 \left(\frac{t_1 - t_a}{t - t_a} \right)^p \tag{17}$$

where t_a and p are variable parameters. The F_d is constant for a wide range of particle diameter values if

$$p = 3n - 1 \tag{18}$$

where n expresses the power dependence of λ on d. For sedimentation FFF $n = 3$ and $p = 8$. For power programming, the constant F_d region is optimally extended toward smaller particle diameters if $t_a = -pt_1$. Thus the power program has essentially three independent parameters, namely, S_0, p, and either t_a/t^0 or t_1/t^0.

15.3.   INSTRUMENTATION

A substantial commonality in experimental apparatus and method exists among the subtechniques of FFF. We describe the sedimentation FFF system most completely and characterize the other subtechniques in terms of their differences from the sedimentation system. More complete descriptions of the FFF systems are available elsewhere (6).

The FFF channel occupies a similar position within the total instrument assembly as does the column in HPLC. It is served by pumps to deliver the carrier fluid at a constant rate, an injection port to introduce the sample, a detector to identify sample components, an optional fraction collector, and a computer for process control and data analysis. The auxiliary equipment, described in some detail elsewhere (6), resembles that used for chromatography; it is consequently widely available.

15.3.1.   Sedimentation FFF

The ribbonlike sedimentation FFF channel used in our laboratory is commonly assembled by sandwiching a stainless steel or Mylar spacer of desired thickness, from which the channel shape is cut and removed, between two stainless steel walls. The walls may be coated with Teflon or polyimide as necessary for inertness, or replaced by titanium or Hastelloy C for use with some physiological buffers. The channel is clamped or welded together and coiled to fit the inside circumference of a centrifuge basket, as shown in Fig. 15.4. A typical channel is 80–90 cm in tip-to-tip length, 2.0 cm in breadth, and 0.0127–0.0254 cm (0.005–0.01 in) in thickness. The centrifuge, which creates the sedimentation field, can be controlled by using a feedback loop to provide accurate field regulation and to permit speed variation for field programming.

The inlet and outlet flows are directed through the centrifuge shaft. Two outlets are used in an improved split outlet configuration (34). The split outlet design serves to collect selectively the laminae flowing near the accumulation wall (35). Because most of the solute becomes concentrated in these laminae, the split outlet configuration enhances detection sensitivity. The splitting system preferably consists of a thin rigid element extending across the breadth of the channel. The splitting is controlled by means of separate outlet flow streams regulated by needle valves. In an alternative configuration, two concentric stainless steel tubes, mounted such that the entrance to the inner tube is positioned very close to the accumulation wall while the outer tube is flush with the opposite wall, can be used (28).

Construction of the seal between the stationary and the spinning por-

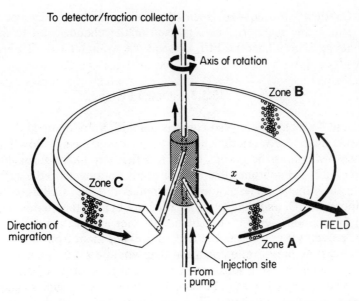

Figure 15.4. Sedimentation FFF channel coiled to fit within a centrifuge basket. Zones *A*, *B*, and *C* exhibit differential migration due to different centrifugal field interactions. Reprinted from reference 52, by courtesy of The Humana Press, Inc.

tions of the sedimentation FFF system is the primary experimental challenge. For aqueous systems, self-lubricating O rings and highly polished drive shafts have been most dependable. A spring loaded Teflon seal (Bal seal) has been used for nonaqueous systems (36).

Two commercial sedimentation FFF instruments are presently available. One of them (FFFractionation, Inc., Salt Lake City) closely parallels the instrumentation described above. The other (du Pont Instruments, Wilmington) has a face seal configuration which permits operation at high speeds. The du Pont instrument is strongly focused on the use of time-delayed exponential field programming.

15.3.2. Thermal FFF

The thermal FFF channel is formed by clamping a Mylar or Teflon spacer of desired thickness between two copper or copper alloy bars (6). The highly polished copper bars which serve as channel walls also accommodate heating elements or coolants to provide the temperature gradient, that is, the "field" in thermal FFF. The flat, planar channel is positioned horizontally with the cold wall as the bottom plate. If very high temperature gradients are desired, such that boiling of the carrier becomes a concern, a fine metering valve is used at the channel exit to increase channel pressure. Thermal FFF systems are available from FFFraction-ation, Inc.

15.3.3.   Flow FFF

The flow FFF channel is also constructed with a Teflon or Mylar spacer clamped tightly between the channel walls. In the case of flow FFF, the channel walls must be porous to allow a field-producing throughflow of carrier liquid. Membranes that are uniformly permeable to the carrier but not to the sample serve as ideal accumulation walls. Cellulose acetate, Millipore PTGC, and Amicon type YM5 are among the membrane types used for aqueous separation. The membranes are backed with stainless steel, polyethylene, or ceramic frits to prevent them from flexing under the cross flow pressure, a condition that would distort the parallel-plate channel geometry. In a substantially different form of flow FFF (37), the upper channel wall is made of a nonpermeable material such as glass. The channel thickness in flow FFF is typically 0.25–0.51 mm, somewhat greater than that used in other subtechniques to offset the difficulty of obtaining precisely uniform channel walls composed of membranes and frits.

The cross flow stream is delivered by a pump/unpump apparatus to ensure uniform cross flow. The channel flow is delivered by another pump. When a split channel outlet configuration is used, the two channel outlet flows are controlled by flow restrictors (28).

15.4.   SELECTED APPLICATIONS

The versatility of FFF as a method for the fractionation of macromolecular and particulate species is illustrated by a wide variety of materials studied successfully. This variety, originating in work at only a few laboratories, should expand rapidly as other groups become more active. Rather than survey the full range of existing applications, selected examples are chosen to illustrate the FFF approach.

15.4.1.   Sedimentation FFF

The separation power of sedimentation FFF is best illustrated using different populations of uniform latex spheres as model colloids. Figure 15.5

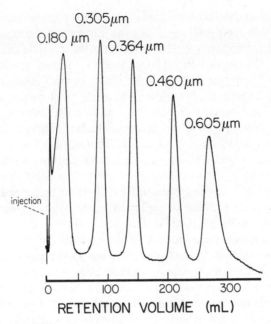

0.305μm
0.180 μm
0.364 μm
0.460 μm
0.605 μm

injection

0   100   200   300

RETENTION VOLUME (mL)

Figure 15.5. Separation of a mixture of five monodisperse polystyrene latex samples by field programmed sedimentation FFF.

shows the high level of resolution possible for various sizes (as indicated) of polystyrene latex beads using a programmed field run.

Stability of perfluorocarbon blood substitutes (38), particulate content in cataractous human lens (39), and physical characterization of the PBCV virus (40) are three examples of the broad range of important colloidal materials about which sedimentation FFF has provided useful analytical information. Recently the capabilities of sedimentation FFF have been expanded to permit the characterization of complex colloid populations having simultaneous variations in particle size and density (34, 41).

Recall that the sedimentation field forces particles toward the outer channel wall if the particles are denser than the carrier liquid ($\Delta \rho > 0$) and toward the inner wall if they are less dense than the carrier liquid ($\Delta \rho < 0$). Table 15.1 provides an expression for the dimensionless mean thickness λ of the particle layer in the region of the wall,

$$\lambda = \frac{6kT}{\pi w G d^3 \, \Delta \rho} \tag{19}$$

The symbols are defined in Table 15.1. For a given field strength G, particles which have the same density difference $\Delta\rho$ compared to the carrier fluid form layers based strictly on the particle diameter d. The mean layer thickness controls the species' access to the faster flow streamlines. The larger particles form more compact layers near the channel wall and experience the slow flow of the wall streamlines while the smaller particles form zones which extend into the faster streamlines near the channel center. The mean layer thickness is directly related to the retention volume V_r by Eqs. 3 and 4. Therefore particles of increasing diameters will exhibit a pattern of increasing elution volumes (or elution times).

The capability of sedimentation FFF to yield high-resolution particle (droplet) diameter profiles is exploited in the study of perfluorocarbon emulsion blood substitutes (38). Particle size is the characteristic of primary importance for such blood substitutes. Fine particles produce more stable emulsions than coarse particles, but larger particles are cleared from the circulation more rapidly than smaller ones. Sedimentation FFF was used to study Fluosol-43 and Fluosol-DA 20%, perfluorocarbon blood substitutes of known density, manufactured by the Green Cross Corporation of Osaka, Japan. Figure 15.6 shows fractograms for a fresh and a 56-day-old sample of Fluosol-DA 20%. The droplet diameter scale along the top axis has been calculated from Eqs. 3, 4, and 19. The blood substitute was studied in a $0.1M$ phosphate buffer solution of pH 7; the figure caption provides other relevant experimental conditions. The droplet di-

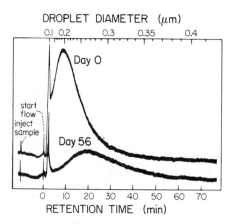

Figure 15.6. Time variation of fractograms for Fluosol-DA 20% studied by sedimentation FFF. The ordinate is detector response. Field strength—21.5 g; flow rate—60 mL/h. Particle diameter d scale is shown at top of figure. Reprinted from reference 38, by courtesy of Academic Press, Inc.

ameter profile for Fluosol-DA 20% is seen to shift to noticeably higher diameter values in less than 2 months. In this way sedimentation FFF provides a useful method of tracing aging effects in terms of droplet size distribution for these fluorocarbon emulsions. Clearly, this approach can be extended to all types of colloidal materials.

Particle content and particle size distribution are prominent characteristics which differentiate between a normal and a cataractous lens of the human eye. While the soluble high molecular weight species contained in a cataractous lens have been well characterized by such methods as gel permeation chromatography, gel electrophoresis, or light scattering, the urea-insoluble protein clusters mainly responsible for visual distortion have received little attention. With judicious use of field programming, sedimentation FFF can provide a fingerprint of the concentration and particle size distribution (and thus the light-scattering-induced vision impairment) for the urea-insoluble material in a period of less than 1 hour (39). Figure 15.7 shows a fractogram of the urea-insoluble material from the nucleus of a cataractous human lens. In these experiments the centrifuge rotational speed was controlled by computer according to the ex-

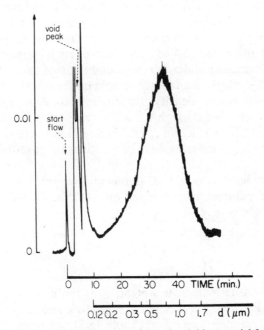

Figure 15.7. Sedimentation FFF fractogram of urea-insoluble material from nucleus of cataractous human lens. Reprinted from reference 39, by courtesy of Mount Sinai School of Medicine.

ponential function of Eq. 16. With the general term for field strength S replaced by the specific term for sedimentation FFF, acceleration G, and with $t_1 = \tau'$, we have

$$G(t) = G_0 e^{-(t-\tau')/\tau} \tag{20}$$

where $G(t)$ is the field strength at time t. In the present study G_0 was 86 g (1000 r/min), τ' was 5 min. The particle diameter scale superimposed on the fractogram is derived from the relationship of retention time to λ (Eqs. 3 and 4), the relationship between λ and the particle diameter for sedimentation FFF (Eq. 19), and the selected field programming expression (Eq. 20). The particle diameter scale is based on a typical protein density for the particles, 1.36 g/mL. Such a size analysis of urea-insoluble lens particulates may aid in understanding the mechanism of cataract formation.

The use of sedimentation FFF for multifaceted particle characterization is illustrated by the study of the PBCV virus (40). The well-developed theory of FFF accords the potential for determining physical parameters of a sample from the fractionation data. Retention volumes or times, experimentally measured quantities, are related to λ, the dimensionless retention parameter, by Eqs. 3 and 4. The λ equation, in turn, is comprised of the physical parameters of the sample, that is, volume, size, molecular weight, and density. In order to derive maximum information, the PBCV virus was investigated under varying conditions of carrier density, flow rate, and field strength in a tris-acetic acid buffer of pH 7.5; sucrose was used to vary the carrier density. Infectivity tests demonstrated that the virus particles migrated through the sedimentation FFF channel and seals without losing biological activity. The finding supports the contention that FFF is a gentle, nondestructive procedure for macromolecules and particles.

Effective molecular weight M' (molecular weight × buoyancy factor) was the initial parameter determined for the PBCV virus. The effective molecular weight is given by

$$M' = \frac{M|\Delta\rho|}{\rho_s} \tag{21}$$

where M is the true molecular weight, ρ_s is the density of the particle, and $\Delta\rho$ is the density difference between particle and carrier. Substituting from the M-containing λ equation (Table 15.1), we have

$$M' = \frac{RT}{\lambda wG} \tag{22}$$

where the symbols are defined in Table 15.1. A value for λ obtained from the experimental retention ratio (by means of Eq. 4) translates directly into an effective molecular weight through Eq. 22.

The PBCV experiments were performed at a variety of G values as well as at varying flow rates. Each series of experiments yielded a high precision for M', confirming the experimental consistency of sedimentation FFF. The effective molecular weight was found to be $253 \pm 12 \times 10^6$.

When the sample density is available from independent sources, Eq. 21 may be used to provide the true molecular weight M. Alternatively, sedimentation FFF can be used to approximate the sample density ρ_s as well as M. Equation 21 may be rearranged to yield

$$\rho = \rho_s \pm \left(\frac{\rho_s}{M}\right) M' \tag{23}$$

The carrier density is systematically changed in a series of runs by the addition of sucrose. The linear plot of ρ versus the corresponding M' values (from Eq.. 22) yields the particle density as the intercept and the true molecular weight from the slope.

Similarly, sedimentation FFF retention data can provide information about particle diameter by means of Eq. 19, contingent on knowledge of particle density either from independent sources or from the sedimentation FFF procedure described. Sedimentation FFF also provides a means for estimating the particle size independent of particle density. The contribution to zone broadening caused by nonequilibrium processes is velocity dependent (see Eq. 9). A plot of plate height H versus mean velocity $\langle v \rangle$ yields a value for the coefficient of the nonequilibrium term $C = \chi w^2/D$. Substitution of the Stokes–Einstein equation for D (42) yields an effective diameter for globular particles of

$$d = \frac{CkT}{3\pi\chi w^2 \eta} \tag{24}$$

where η is the viscosity of the carrier. The constant C together with the value for χ, a complicated but known function of λ, permit calculation of d from Eq. 24.

The values for the molecular weight, density, and diameter of the PBCV virus determined from the preceding methods compare reasonably well with values determined by conventional methods, although the latter are far from consistent. Clearly, sedimentation FFF serves as a rapid primary or complementary technique for such colloid characterization.

We have seen that the action of sedimentation FFF on a sample particle is a function of both the particle density and the particle diameter. The simultaneous dependence of the retention time on two physicochemical parameters of the sample may complicate the fractionation as well as the determination of the parameters. However, recent developments in sedimentation FFF turn this adversity into advantage (34). In particular, if the sample consists of two subpopulations of fairly different density, such as organic and inorganic particulates in a natural water sample, the carrier for sedimentation FFF could be modified to assume an intermediate density. In the case of the higher density material, the density difference $\Delta\rho$ of Eq. 19 is positive and the sample would accumulate at the outer wall of the sedimentation FFF channel. At the same time, $\Delta\rho$ is negative for the less dense material which would move toward the inner channel wall. Thus the sample would divide into two subpopulations, one at each wall, each undergoing a normal sedimentation FFF process. By splitting the flow at the outlet and collecting the inner wall and outer wall flow substreams individually, we would simultaneously characterize the two subpopulations and achieve sample enrichment at each outlet as well.

To test this concept, a mixture of two subpopulations was composed by mixing 0.620-μm polystyrene (PS, $\rho_s = 1.05$ g/mL) latex beads and 0.371-μm polymethylmethacrylate (PMMA, $\rho_s = 1.21$ g/mL) latex beads. The outlet flows were adjusted such that the inner outlet collected 45% of the channel flow from the laminae near the inner wall while the outer outlet collected the remaining 55% of channel flow from the laminae near the outer wall. Figure 15.8 shows the two detector signals obtained as the density of the carrier was varied from run to run. At the normal aqueous carrier density of 0.997 g/mL, both PS and PMMA particles accumulate at the outer wall where their FFF migration produces a single peak (Fig. 15.8a). In Fig. 15.8b the carrier density is increased to 1.018 g/mL by the addition of sucrose. Both components still undergo FFF at the outer wall, but the PS peak has shifted its position considerably due to its greater particle size and larger relative buoyancy effect. At the carrier density ρ = 1.054 g/mL of Fig. 15.8c, the PS is almost neutrally buoyant and consequently appears as a void peak distributed between the two outlets. Note that the denser PMMA continues to emerge from the outer wall outlet, but at slightly smaller elution volumes because of its higher density.

Figure 15.8d and e illustrates the benefits gained by running a split outlet system at an intermediate carrier density (ρ = 1.073 g/mL in Fig. 15.8d, ρ = 1.108 g/mL in Fig. 15.8e) at which the denser PMMA undergoes sedimentation FFF at the outer wall and the less dense PS undergoes sedimentation FFF at the inner wall. Each latex generates is own fractogram under these conditions. From these the two particle size distributions can be obtained free of interference from one another. There will be no overlap even for broad distributions.

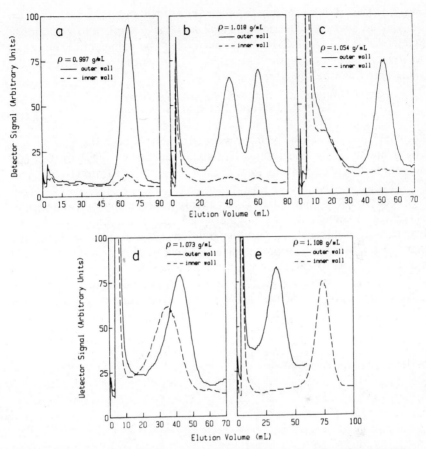

Figure 15.8. Fractograms illustrating separation of a mixture of 0.620-μm PS and 0.371-μm PMMA latex beads achieved in split outlet sedimentation FFF system by incremental increase in carrier density. Field strength—27.4 g. Reprinted from reference 34, by courtesy of Academic Press, Inc.

The physical characterization of the subpopulations can be pursued further. Equation 23 can be written in terms of the particle diameter d,

$$\rho = \rho_s \pm \left(\frac{6kT}{\pi d^3 wG} \right) \frac{1}{\lambda} \qquad (25)$$

A plot of carrier density ρ versus $1/\lambda$ yields a straight line with the intercept equal to the particle density ρ_s and the slope related to the particle diameter d. The line resulting from the outer wall PMMA data of Fig. 15.8 indicated a particle density of 1.2 g/mL and a particle diameter of 0.353

μm. The pair of lines arising from the outer wall and inner wall data for PS yielded particle densities of 1.047 and 1.046 g/mL and particle diameters of 0.564 and 0.551 μm, respectively. The values obtained are in reasonable agreement with those yielded by other methods as well as the values provided by the supplier (34).

15.4.2. Thermal FFF

Polymer samples, generally in organic carriers, have been the main targets of thermal FFF investigations. The ability of thermal FFF to separate polymeric species according to molecular weight is illustrated in Fig. 15.9 (7).

Recent work in thermal FFF has involved extension of the scope to very high molecular weight polymers (43) and measurement of the polydispersity of ultranarrow polymer fractions (24). Studies involving the competitive (44) and complementary (12) relationship between FFF and size-exclusion chromatography have been reported.

Ultrahigh molecular weight polymers present a challenge for separation and characterization. They are susceptible to shear degradation and, when

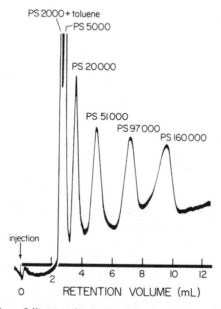

Figure 15.9. Separation of linear polystyrene polymer fractions of indicated molecular weights in toluene/ethylbenzene by thermal FFF using a temperature drop of 60°C. Reprinted from reference 7, by courtesy of Marcel Dekker, Inc.

present with lower molecular weight material, demand a separation technique of very broad range. Thermal FFF answers both challenges. (1) In a recent study (43) a polystyrene sample of nominal molecular weight 20.6 $\times$ 10^6 dissolved in tetrahydrofuran (THF) carrier was eluted from a thermal FFF column. A fraction was collected and reinjected. The fraction was observed to elute as a narrow peak near the original position, demonstrating both successful fractionation and a lack of shear degradation. (2) A thermal FFF column possesses no upper limit to the elution range; field programming can be instituted if necessary to accommodate samples of both low and high molecular weight as first demonstrated in 1976 (27).

The relationship of polymer molecular weight and thermal FFF retention is found in the λ expression of Table 15.1. The rigorous calculation of the retention parameter λ from the experimental retention volume is more involved for thermal than for sedimentation FFF because the departure from isoviscous flow in the former must be taken into account (45). The molecular weight dependence of λ arises from the molecular weight dependence of the ratio D/D_T, which is described by a simple power law (46, 47). A convenient linear relationship between λ and M becomes

$$\log (\lambda \Delta T) = -b' \log M + \log A' \qquad (26)$$

where b' and A' are the constants of the power law expression (43). Figure 15.10 shows a log–log plot of Eq. 26 for a broad range of molecular weights of linear polystyrenes in THF as carrier. The temperature drop ΔT for the different runs varied from 8 to 81°C. The linearity of Eq. 26 is confirmed for samples of average molecular weight as high as 20.6 $\times$ 10^6 (43).

Thermal FFF presents several advantages for the determination of polymer molecular weight distributions (MWD) or polymer polydispersities. Not only is selectivity high, but as mentioned, there is little limitation on the elution volume range for a given FFF system nor are high molecular weight polymers subjected to large shear stresses. However, the overriding advantage is the capability of isolating MWD (polydispersity) from other sources of zone broadening.

Consider the plate height equation for FFF (Eq. 12). For large molecules with small diffusion coefficients, the longitudinal diffusion term is negligible. The nonequilibrium term, having a first-order dependence on carrier velocity, can be determined by studies of plate height as a function of carrier velocity. The relaxation effects, having an approximate second-order velocity dependence, can be avoided by using a stop flow procedure. Finally, nonideal zone dispersion effects can be minimized by care-

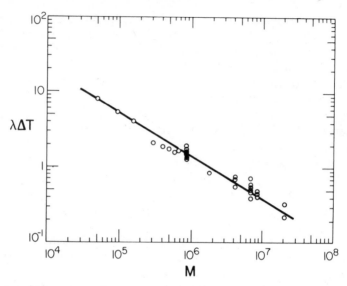

Figure 15.10. Thermal FFF retention data obtained for a broad range of molecular weights of linear polystyrenes in THF with temperature drops ΔT ranging from 8 to 81°C and a cold wall temperature of 15°C. Reprinted from reference 43, by courtesy of the American Chemical Society.

ful experimental design. Therefore for narrow MWD, linear plots of experimental plate height H as a function of carrier velocity $\langle v \rangle$ will yield H_p, the polydispersity contribution to plate height, as an intercept. From H_p one gets polydispersity μ as shown by Eq. 11. In order to maximize the ratio of the polydispersity contribution to the nonequilibrium contribution, that is, to obtain the greatest accuracy in the MWD or polydispersity measurement, it is advantageous to minimize the channel thickness w and maximize the temperature gradient dT/dx (24).

Plate height–velocity studies were conducted on four narrow polystyrene standards using ethylbenzene as carrier (24). For each molecular weight standard two ΔT's, for which selectivity values in the range of $S = 0.5–0.6$ had been determined, were used. For a given set of conditions, plate height measurements were made using carrier velocities $\langle v \rangle$ of 2–12 cm/min. Figure 15.11 is a typical plot of plate height versus flow velocity. The intercept representing H_p is converted to a polydispersity value by means of Eq. 11. The position of the H_p value equivalent to a polydispersity of 1.04 is shown for comparison. The polydispersities determined by the above method are listed in Table 15.2. In all cases the values are well below the ceiling values (1.06) provided by the manufacturers.

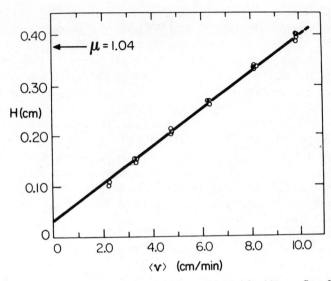

Figure 15.11. Plate height H versus carrier solvent velocity $\langle v \rangle$ with stop flow for a linear polystyrene with $\overline{M}_w = 170{,}000$ in a thermal FFF channel. Temperature drop—30°C; cold wall temperature—21°C. Reprinted from reference 24, by courtesy of John Wiley and Sons.

Polymer molecular weight distributions and polydispersity measurements are commonly performed using size-exclusion chromatography (SEC). It is instructive to compare the relative advantages of SEC and thermal FFF in comparable experimental situations (44). For such a comparison, experiments were done in which the SEC column used was a commercial Ultrastyragel column, 30 cm long with an inner diameter of 7.8 mm. The column is applicable over a molecular weight range of 50 × 10^2 to 4000 × 10^3. The thermal FFF channel was 0.0762 mm thick with a volume of 0.75 mL. The latter was pressurized to 100 lb/in² to permit

Table 15.2. Results of Linear Fit to Plate Height versus Flow Velocity Data for Linear Polystyrene Using Stop Flow Technique

		Stop Flow	
$\overline{M}_w$	ΔT (K)	H_p (cm)	μ
170,000	30	0.0311	1.0034
295,000	24	0.0433	1.0055
350,000	24	0.0468	1.0059

a hot wall temperature of 72°C, above the normal boiling point of the carrier, tetrahydrofuran. The ΔT was 50°C. Five polystyrene samples with the manufacturers' nominal molecular weights of 50, 100, 233, 411, and 600×10^3 daltons were studied.

Figure 15.12 illustrates the fractionation of 100 and 233×10^3 dalton samples by SEC and thermal FFF (44). Resolution values were $R_s = 1.40$ for SEC and $R_s = 1.72$ for thermal FFF. It is useful to note that the elution order is opposite for the two methods. The high molecular weight material emerges first in SEC. In thermal FFF, the high molecular weight material is most affected by the field, is more highly retained, and emerges last. While the SEC peaks are narrower (the retention time scale in the figure is expanded), thermal FFF provides a greater differential displacement and higher resolution. Understanding of the resolution anomaly requires a discussion of the two separate components of resolution: selectivity and column efficiency. An approximate equation for resolution is (44)

$$R_s = \underbrace{\frac{N^{1/2}}{4}}_{\text{efficiency}} \underbrace{\frac{d \ln V_r}{d \ln M}}_{\text{selectivity}} \frac{\Delta M}{M} \tag{27}$$

where N is the average number of plates, $(d \ln V_r)/(d \ln M)$ is the selectivity, ΔM the difference in molecular weight of the two components, and M the average molecular weight.

The selectivity, or fractional change in elution volume relative to the fractional increment in molecular weight, is a measure of how far apart various polymer samples are distributed along the elution time or volume axis. Recall that the volume range can be many times the channel volume in FFF. The selectivity is determined by measuring the tangent to the experimentally determined V_r versus M calibration curve at various molecular weights. The thermal FFF system has about five times the maximum selectivity of the SEC system. It should also be noted that the selectivity for components falling below the maximum can be altered by a change in ΔT for the thermal FFF channel.

The second component of resolution, column efficiency, is more difficult to evaluate for polymers. The polydispersity of the standard samples in the form of H_p contributes to the measured zone broadening H, which is the measure of column efficiency. However, thermal FFF provides a means of determining polydispersity as described, or thermal FFF retention data can be used in conjunction with Eqs. 9–12. Table 15.3 lists the selectivity S, the corrected (for polydispersity) plate count N_c, and the corrected resolution ratios R_s for the polystyrene mixtures. SEC currently

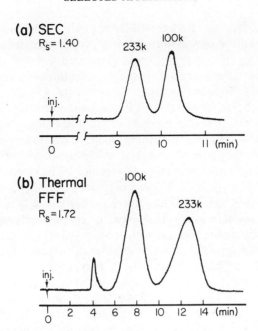

(a) SEC
$R_s = 1.40$

233k

100k

inj.

0   9   10   11 (min)

(b) Thermal FFF
$R_s = 1.72$

100k

233k

inj.

0   2   4   6   8   10   12   14 (min)

Figure 15.12. Separation of polystyrene samples of indicated molecular weights by SEC and thermal FFF. SEC flow rate–1.0 mL/min; thermal FFF flow rate—0.2 mL/min; temperature drop for thermal FFF—50°C; cold wall temperature—22°C. Reprinted from reference 44, by courtesy of Elsevier Science.

possesses a considerably greater efficiency than thermal FFF. For the mixture of Fig. 15.12, SEC provides a corrected number of plates of 5971, as compared to 302 for thermal FFF.

When the selectivity and the efficiency are combined to give the resolution (Eq. 27), thermal FFF exhibits greater resolution for all except the lowest molecular weight mixture, a mixture outside of the normal range of thermal FFF separation. This superiority is largely due to the

Table 15.3. Selectivity S, Corrected Plate Count N_c, and Corrected Resolution Ratios R_s for Polystyrene Mixtures

Molecular Weight (10^3 daltons)	$\dfrac{S(\text{FFF})}{S(\text{SEC})}$	$\dfrac{N_c(\text{FFF})}{N_c(\text{SEC})}$	$\dfrac{R_s(\text{FFF})}{R_s(\text{SEC})}$
50/100	4.61	0.0283	0.78
100/233	5.45	0.0506	1.23
233/350	5.04	0.1077	1.65
350/600	5.25	0.3780	3.23

fact that selectivity, the strength of thermal FFF, appears as a first-order term in the resolution expression of Eq. 27, while efficiency, the strength of SEC, appears as a $\frac{1}{2}$-order term. Further improvements in thermal FFF should be achievable based on increasing column efficiency, a process largely directed by existing theory.

The combined use of thermal FFF and SEC promises the characterization of polymers based on both molecular size and chemical composition (12). Retention and separation in SEC are based primarily on molecular size; differences in chemical constitution are of consequence only to the extent that they affect size. In thermal FFF, retention λ is based on the ratio of the ordinary diffusion coefficient D to the thermal diffusion coefficient D_T (Table 15.1). The diffusion coefficient D is inversely dependent on the molecular size. However, the D_T parameter is independent of molecular weight (as well as branching structure), but is sensitive to the chemical nature of the polymer samples (12). Therefore the D determined by SEC can be used with the λ corresponding to experimental thermal FFF retention to calculate D_T, which reflects chemical composition (12). Figure 15.13 illustrates this effect by showing SEC and thermal FFF runs for a mixture of 2.4×10^5 molecular weight polymethylmethacrylate and 2.0×10^5 molecular weight polystyrene (12). The diffusion coefficients for the components, 3.67×10^{-7} cm²/s and $3.85 \times$

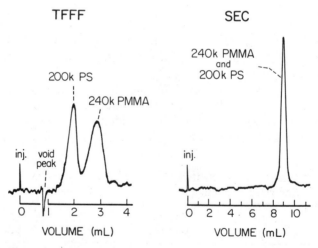

Figure 15.13. Illustration of chemical composition effects in thermal FFF as shown by attempted separation of 2.4×10^5 MW polymethylmethacrylate and 2.0×10^5 MW polystyrene by both thermal FFF and SEC. Thermal FFF flow rate—0.13 mL/min; SEC flow rate—1.0 mL/min; temperature drop for thermal FFF—41°C; cold wall temperature—21°C. Reprinted from reference 12, by courtesy of the American Chemical Society.

10^{-7} cm^2/s, differ little. Consequently, fractionation by SEC is not achieved. However, the D_T values of 1.31 cm^2/sK and 0.92 cm^2/sK differ sufficiently for successful thermal FFF separations. Such data suggest that compositional variations, such as those found in copolymers and polymer blends, can be evaluated by thermal FFF once the molecular size has been determined by SEC.

15.4.3. Flow FFF

Flow FFF is the most universally applicable of the FFF subtechniques. The field is effectively a cross flow, perpendicular to the channel axis, which will drive virtually any type of suspended sample species toward the accumulation wall. The λ equation of Table 15.1 indicates that D is the particle parameter that determines the retention. The diffusion coefficient is, in turn, a function of molecular weight.

The broad applicability of flow FFF permits its usage for materials of wide-ranging properties, such as the fulvic and humic acids of natural waters and soils (48). Recent improvements in flow FFF have dealt with flow rate and programming as well as with outlet stream splitting (28). An interesting recent modification involves an asymmetrical channel having only one permeable wall (37).

Humic substances, the colored organic substances from soils and natural waters, have generated much interest because of their effect on water pollution. They have been implicated in the formation of chlorinated hydrocarbons in treatment plants, the binding and transport of heavy metals, and the solubilization of nonpolar organic substances (48). Molecular weight estimates for the humic substances have varied from 500 to 200,000, but 800–3000 are probably more realistic values (49, 50). Flow FFF has been used to determine molecular weight distributions of the humic substances using a series of polystyrenesulfonate (PSS) standards of molecular weights 4000–100,000 for a calibration curve (48). Again refer to the λ equation of Table 15.1. The diffusion coefficient D can be determined from the retention data. Parameter D is generally a power law function of molecular weight (42). Figure 15.14 shows the log diffusion coefficient versus log molecular weight for the PSS standards as well as for several fulvate and humate standards. The values for the humic substances fall on the same line as the PSS standards, a confirmation of the validity of using the PSS samples as standards for the humic substances. Average molecular weight determinations by flow FFF for reference humic substances were in agreement with those obtained by other methods. However, the principal challenge in this field is the acquisition of MWDs. Estimated MWDs from digitized flow FFF data suggest the fol-

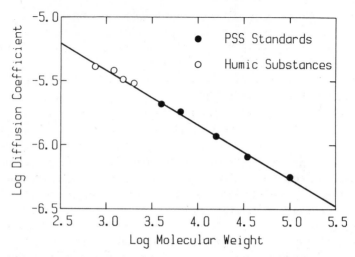

Figure 15.14. Molecular weight calibration line for humic materials, constructed using measured diffusion coefficients of polystyrenesulfonate molecular weight standards. Data points for reference humic substances arise from diffusion coefficients measured at peak maxima of flow FFF fractograms and from molecular weights estimated by independent methods. Reprinted from reference 48, by courtesy of the American Chemical Society.

lowing trends. Fulvic acids are in general smaller than humic acids. The humates from different sources, in the order of increasing size and polydispersity, were aquatic < soil < peat bog < lignite coal. Two highly colored natural waters directly injected without pretreatment yielded molecular weight distributions closest to that of the soil humate (48).

Improved resolution, decreased separation time, and increased sample detectability are among the goals of a separation method optimization. The theory of FFF provides guidance to achieve these goals. If we begin with the resolution equation (Eq. 13), substitute the nonequilibrium plate height (Eq. 9) for high retention, insert the expression for λ from Table 15.1, and write $\langle v \rangle$ as $\dot{V}L/V^0$, with $\dot{V}$ being the volumetric flow rate along the axis of the channel, then the resolution equation becomes

$$R_s = 0.0510 \frac{\Delta R}{\overline{R}} \frac{w^2}{\overline{D}V^0} \left(\frac{\dot{V}_c^3}{\dot{V}} \right)^{1/2} \tag{28}$$

where $\overline{R}$ and $\overline{D}$ are effective averages for the two components. Equation 28 states clearly that resolution improves significantly with increased cross flow rates. Similarly, solving for retention time from Eq. 3, substituting $\langle v \rangle$ as above, and using the λ expression in the case of high retention, gives

$$t_r = \frac{w^2}{6D} \frac{\dot{V}_c}{\dot{V}}$$ (29)

A reduction of separation time requires that the axial flow rate be increased more than the cross flow rate. The decreased t_r resulting from increased $\dot{V}$ and the increased resolution from increased $\dot{V}_c$ are offset somewhat by zone dilution and overloading. Outlet stream splitting, in which the streamlines near the accumulation wall are collected separately from the remainder of the channel streamlines, promises to work effectively against these drawbacks (28). Stream splitting, by concentrating the sample material at the outlet, counteracts the excessive dilution normally arising from very high levels of retention.

High retention is also associated with excessive retention time. As mentioned, field programming can obviate the difficulty.

In addition to the variation in flow rate, programming, and outlet stream splitting, an asymmetrical flow was developed as a modification to flow FFF (37). Only the accumulation wall of this modification is permeable to carrier flow. The opposite wall is a glass plate, which permits observation of the channel contents. The channel has one incoming flow stream (at the inlet) and two outgoing streams, one the diffuse stream through the membrane. The advantage of the new design is its technical simplicity. The traditional design requires a uniform cross flow of carrier through both channel walls, a condition often difficult to achieve. However, new theoretical expressions are necessary because the new design creates both longitudinal and transverse velocity gradients (37).

Figure 15.15 illustrates the fractionation potential of the new design. Cytochrome-C (molecular weight 12,384), albumin (molecular weight 68,000), and thyroglobulin (molecular weight 669,000) were separated using a pH 7.3 tris buffer as carrier. Diffusion coefficients calculated from retention data were in good agreement with accepted values. The resolution of Fig. 15.15 appears to be comparable to traditional flow FFF (51), but the fractionation time is shorter.

15.4.4. Steric FFF

At the limit of large particles (> 1 μm), all normal forms of FFF converge to the "steric" mechanism. Therefore the task of fractionating many large particles of interest, such as biological cells, chromatographic supports, fly ash, latices, glass beads, crushed coal, and coal liquefaction residues,

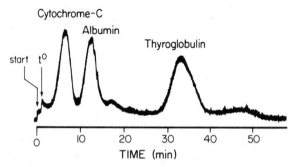

Figure 15.15. Separation of cytochrome-C, albumin, and thyroglobulin in asymmetrical flow FFF system. Reprinted from reference 37, by courtesy of the American Chemical Society.

falls to steric FFF and its modifications. Recent work has included red blood cell fractionation (52) and a systematic study of the effects of flow velocity and field strength on fractionation (53). Steric FFF as the limiting mechanism of flow FFF has been investigated (10).

Figure 15.16 illustrates the separation of a mixture of fixed avian and human red blood cells in ≈20 min under the influence of a 3.6-g sedimentation field (52). Recall that in the steric region larger particles elute earlier because their volume extends into the faster flow streamlines. The carrier flow rate, 80.4 mL/h for the separation in Fig. 15.16, led to the

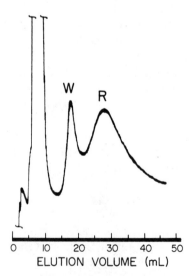

Figure 15.16. Steric FFF fractogram of a mixture of fixed avian (W) and human (R) red blood cells. Reprinted from reference 52, by courtesy of The Humana Press, Inc.

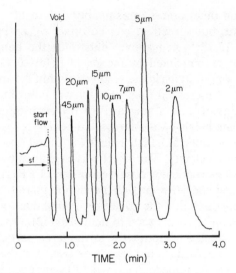

Figure 15.17. Steric FFF fractogram of seven latex bead diameters at 38-mL/min flow rate and 343 g (1400 r/min) field strength. Reprinted from reference 53, by courtesy of the American Chemical Society.

reasonably rapid elution of the cell components. However, much higher flow rates (and higher separation speeds) can be realized under appropriate field strength conditions. For Fig. 15.17 (53) the flow rate was increased to 2280 mL/h, and the separation time (of latex standards) was accordingly reduced to 3.5 min. This run required a field strength of 343 g.

The degree to which increased flow rate generates increased separation speed is influenced by the presence of hydrodynamic lift forces, which increase in magnitude with flow velocity. The lift forces elevate the particles above the wall contact level. The lift forces, if made excessive by high flow rates, must be counteracted by increasing the applied field.

A systematic study (53) of the effects of flow velocity and field strength on fractionation in the steric region of sedimentation FFF was carried out using a set of different sizes of latex particles (2–45-μm diameter). At a fixed sedimentation field of 7.0 g, the initially satisfactory resolution of six sizes of latex particles deteriorated significantly as the flow rate increased from 2 mL/min to 8 mL/min and to 21 mL/min. In a second series, the flow rate was held constant at 21 mL/min, while the sedimentation field was increased from 7.0 g (as in the first experiment) to 142 g, leading to a final fractionation of the latex particles with a better resolution (and much higher speed) than the initial resolution of the first experiment (53).

Clearly, resolution deteriorates (presumably due to lift forces) with increasing flow rate, but the effect can be offset by an increase in field strength. Figure 15.17, noted above, illustrates the benefits of the increased flow rate accompanied by increased field; seven latex particle diameters were essentially baseline resolved within 3.5 min (53).

The concern that increased flow velocity will contribute to increased zone broadening as occurs in normal FFF (through the nonequilibrium contribution to plate height) was addressed in an H versus $\langle v \rangle$ experiment using fixed red blood cells (52). Plate height was found to actually decrease with increasing flow velocity. The effect was explained by the hypothesis that most of the zone broadening observed in steric FFF is a polydispersity effect. Closely sized species approach a common migration velocity with increased flow rate such that the zone broadening due to polydispersity, actually a separation effect, is seen to decrease (52).

Steric FFF has been studied mainly as the large-particle limiting case of sedimentation FFF. Were it not for the existence of the hydrodynamic lift forces, the steric FFF mechanism would be independent of the type

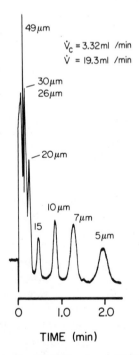

Figure 15.18. Fractogram of latex beads by flow/steric FFF. Reprinted from reference 10, by courtesy of the American Chemical Society.

of field employed. The particle would be driven to the channel wall, remain there during migration, and elute at a time that is a function of the particle diameter alone. However, the necessity of counteracting the lift forces with an increased driving force introduces a dependence of retention on the particular type of driving force. (The lift forces and driving force both influence the γ of Eq. 6.) For example, in sedimentation/steric FFF, the driving force is proportional to the particle diameter to the third power while for flow/steric FFF the driving force is directly proportional to the particle diameter. The γ and R values, consequently, are expected to vary considerably more with particle diameter in flow/steric FFF than in sedimentation/steric FFF (10). Furthermore, the selectivity based on particle diameter is expected to be greater for flow/steric FFF than for sedimentation/steric FFF. The first flow/steric FFF study confirms these predictions (10). The elution spectrum of 5-, 7-, 10-, 15-, 20-, 26-, 30-, and 49-μm spherical latex beads in an aqueous suspension is shown in Fig. 15.18. The volumetric channel flow rate was high, 1158 mL/h, or 19.3 mL/min, to complete the fractionation in approximately 2 min after relaxation. The cross-flow rate was 3.32 mL/min. The larger particles emerge in rapid order, after which the smaller particles, less affected by lift forces, emerge very well resolved. The diameter-based selectivity approximated using the elution times of the 5- and 10-μm particles is greater than unity; sedimentation/steric FFF yields selectivity values of only about 0.5 (10).

ACKNOWLEDGMENT

This work was supported by grant GM10851-30 from the National Institutes of Health.

REFERENCES

1. J. C. Giddings, *J. Chromatogr., 125,* 3 (1976).
2. J. C. Giddings, *Pure Appl. Chem., 51,* 1459 (1979).
3. J. C. Giddings, *J. Chromatogr., 395,* 19 (1987).
4. Analytical Chemistry Division Commission on Analytical Nomenclature, *Pure Appl. Chem., 37,* 447–462 (1974).
5. J. C. Giddings, S. R. Fisher, and M. N. Myers, *Am. Lab., 10,* 15 (1978).
6. J. C. Giddings, M. N. Myers, K. D. Caldwell, and S. R. Fisher, in D. Glick, Ed., *Methods of Biochemical Analysis,* vol. 26, Wiley, New York, 1980, p. 79.

7. J. C. Giddings, *Sep. Sci. Technol., 19*, 831 (1984).

8. J. C. Giddings, *Sep. Sci. Technol., 18*, 765 (1983).

9. J. C. Giddings, *Anal. Chem., 58*, 2052 (1986).

10. J. C. Giddings, X. Chen, K.-G. Wahlund, and M. N. Myers, *Anal. Chem., 59*, 1957 (1987).

11. J. C. Giddings, in J. D. Navratil and C. J. King, Eds., *Chemical Separations*, vol. 1, Litarvan, Denver, 1986, p. 3.

12. J. J. Gunderson and J. C. Giddings, *Macromolecules, 19*, 2618 (1986).

13. J. C. Giddings, *J. Gas Chromatogr., 5*, 143 (1967).

14. J. C. Giddings and K. D. Caldwell, *Anal. Chem., 56*, 2093 (1984).

15. J. C. Giddings, *J. Chem. Phys., 49*, 81 (1968).

16. J. C. Giddings, *J. Chem. Ed., 50*, 667 (1973).

17. J. C. Giddings and M. N. Myers, *Sep. Sci. Technol., 13*, 637 (1978).

18. M. N. Myers and J. C. Giddings, *Anal. Chem., 54*, 2284 (1982).

19. S. Lee and J. C. Giddings, *Anal. Chem.,* submitted.

20. M. E. Hovingh, G. H. Thompson, and J. C. Giddings, *Anal. Chem., 42*, 195 (1970).

21. R. E. Peterson II, M. N. Myers, and J. C. Giddings, *Sep. Sci. Technol., 19*, 307 (1984).

22. J. C. Giddings, *Dynamics of Chromatography*, pt. I, Dekker, New York, 1965.

23. L. K. Smith, M. N. Myers, and J. C. Giddings, *Anal. Chem., 49*, 1750 (1977).

24. M. E. Schimpf, M. N. Myers, and J. C. Giddings, *J. Appl. Polym. Sci., 33*, 117 (1987).

25. F. J. F. Yang, M. N. Myers, and J. C. Giddings, *Anal. Chem., 46*, 1924 (1974).

26. J. C. Giddings, K. D. Caldwell, J. F. Moellmer, T. H. Dickinson, M. N. Myers, and M. Martin, *Anal. Chem., 51*, 30 (1979).

27. J. C. Giddings, L. K. Smith, and M. N. Myers, *Anal. Chem., 48*, 1587 (1976).

28. K.-G. Wahlund, H. S. Winegarner, K. D. Caldwell, and J. C. Giddings, *Anal. Chem., 58*, 573 (1986).

29. W. W. Yau and J. J. Kirkland, *Sep. Sci. Technol., 16*, 577 (1981).

30. J. J. Kirkland, C. H. Dilks, and W. W. Yau, *J. Chromatogr., 255*, 255 (1983).

31. J. C. Giddings, *Sep. Sci., 1*, 123 (1966).

32. J. C. Giddings, P. S. Williams, and R. Beckett, *Anal. Chem., 59*, 28 (1987).

33. P. S. Williams and J. C. Giddings, *Anal. Chem., 59*, 2038 (1987).

34. H. K. Jones, K. Phelan, M. N. Myers, and J. C. Giddings, *J. Colloid Interface Sci., 120*, 140 (1987).

35. J. C. Giddings, H. C. Lin, K. D. Caldwell, and M. N. Myers, *Sep. Sci. Technol., 18*, 293 (1983).

36. K. D. Caldwell, G. Karaiskakis, M. N. Myers, and J. C. Giddings, *J. Pharm. Sci., 70*, 1350 (1981).

37. K.-G. Wahlund and J. C. Giddings, *Anal. Chem.*, *59*, 1332 (1987).

38. F.-S. Yang, K. D. Caldwell, J. C. Giddings, and L. Astle, *Anal. Biochem.*, *138*, 488 (1984).

39. K. D. Caldwell, B. J. Compton, J. C. Giddings, and R. J. Olson, *Invest. Ophthalm. Vis. Sci.*, *25*, 153 (1984).

40. C. R. Yonker, K. D. Caldwell, J. C. Giddings, and J. L. van Etten, *J. Virol. Meth.*, *11*, 145 (1985).

41. H. K. Jones and J. C. Giddings, *Anal. Chem.*, submitted.

42. C. Tanford, *Physical Chemistry of Macromolecules*, Wiley, New York, 1961, chap. 6.

43. Y. S. Gao, K. D. Caldwell, M. N. Myers, and J. C. Giddings, *Macromolecules, 18*, 1272 (1985).

44. J. J. Gunderson and J. C. Giddings, *Anal. Chim. Acta, 189*, 1 (1986).

45. J. J. Gunderson, K. D. Caldwell, and J. C. Giddings, *Sep. Sci. Technol., 19*, 667 (1984).

46. M. Martin and R. Reynaud, *Anal. Chem., 52*, 2293 (1980).

47. M. N. Myers, K. D. Caldwell, and J. C. Giddings, *Sep. Sci., 9*, 47 (1974).

48. R. Beckett, Z. Jue, and J. C. Giddings, *Environ. Sci. Technol., 21*, 289 (1987).

49. R. L. Malcolm, in C. R. Aiken, D. M. McKnight, R. L. Wershaw, and P. MacCarthy, Eds., *Humic Substances in Soil Sediment and Water*, Wiley, New York, 1985, chap. 7.

50. E. M. Thurman, *Organic Geochemistry of Natural Waters*, Martinus Nijhoff/ Junk, The Hague, The Netherlands, 1985, pp. 304–312.

51. J. C. Giddings, *Anal. Chem., 57*, 945 (1985).

52. K. D. Caldwell, Z.-Q. Cheng, P. Hradecky, and J. C. Giddings, *Cell Biophys., 6*, 233 (1984).

53. T. Koch and J. C. Giddings, *Anal. Chem., 58*, 994 (1986).

CHAPTER

16

MULTIDIMENSIONAL TECHNIQUES IN CHROMATOGRAPHY

NICHOLAS SAGLIANO, Jr., THOMAS V. RAGLIONE, and
RICHARD A. HARTWICK

Department of Chemistry
Rutgers University
Piscataway, New Jersey

16.1. INTRODUCTION: PERFORMANCE DESCRIPTORS FOR SEPARATION SYSTEMS

16.1.1. Peak Capacity as a Performance Criterion

Chemical separations may be defined as "the art and science of maximizing separative transport relative to dispersive transport" (1). Since the advent of chromatography, researchers have strived to minimize the dispersive transport that occurs during chromatographic separations

643

while maximizing the differential transport component, or selectivity, which is the driving force of the separation. The introduction, first of pellicular supports and chemically bonded phases, followed shortly thereafter by porous microparticulate materials, elevated high performance liquid chromatography (HPLC) to the status of a modern analytical tool. Increases in the efficiency and selectivity of modern capillary gas chromatography (GC) have improved even more dramatically, and its interfacing to mass spectrometry, has created, arguably, one of the most powerful analytical tools in modern chemistry.

One measure of the separation power of a chromatographic system is the column peak capacity n_c, originally introduced in Davis and Giddings (2). Peak capacity is an artificial number, which represents the maximum number of separated zones possible within some arbitrary coordinate distance x of the chromatogram. For the gradient elution technique in LC or for temperature programming in GC, where peak widths tend to be approximately constant, Davis and Giddings have shown (2) that the maximum number of peaks that can fit between retention volumes V_1 and V_2 at some minimum spacing x_0 is

$$n_c = \frac{V_2 - V_1}{x_0} \tag{1}$$

The spacing factor x_0 can be expressed in terms of resolution R_s,

$$x_0 = 4\sigma R_s \tag{2}$$

where σ is the average standard deviation of the adjacent zones. To the extent that the chromatogram is homogeneous over the entire elution range, that is, from V_m through V_{max}, the peak capacity of the column can be given by

$$n_c = \frac{V_{max} - V_m}{x_0} \tag{3}$$

In the isocratic mode, where peak widths are smoothly increasing, the column peak capacity for a homogeneously distributed chromatogram is

$$n_c = \frac{\sqrt{N}}{4R_s} \ln \frac{V_{max}}{V_m} \tag{4}$$

For the typical case where $V_{max}/V_m \approx 7$ and a minimum acceptable R_s

≈ 1, then (2)

$$n_c \approx \tfrac{1}{2}\sqrt{N} \qquad (5)$$

In the typical HPLC column, producing at best perhaps 20,000 theoretical plates, peak capacities on the order of 70–100 might be expected. In capillary GC, with plate counts approaching 500,000, peak capacities about three times higher could be anticipated.

The limitations of single-mode separations become evident in Eq. 5. Sixteenfold efficiency increases are required to merely double the peak capacity of a chromatographic system. Advances in both theory and technique have been such that modern chromatographic separations are commonly operated at levels not far removed from their theoretical limits, so that incremental rather than quantum advances in single-mode chromatographic performance are to be anticipated in the near future. It will remain for the introduction of entirely new column technologies, such as capillary LC, for substantial performance increases to be realized.

16.1.2. Statistical Theory of Component Overlap

Before addressing the peak capacity increases possible via multidimensional techniques, it is necessary to examine the statistical theory of peak overlap, as originally conceived and developed by Davis and Giddings (2). In moderately complex mixtures, the frequency of zone overlap can be estimated by statistical means (2–4). The fundamental assumption is that the solute zones will distribute themselves randomly along some x axis, representing volume, distance, time, and so on. Specifically, the assumption is made (2) that the probability that a peak maximum falls within some small interval δx is constant, and given by $\lambda\,\delta x$. This probability, taken over some region of the x axis large enough to hold a statistically significant number of zones, leads to the Poisson distribution. Accordingly, the probability that the distance δx between peaks will exceed some minimum value x_0 is

$$P_{\delta x \to x_0} = e^{-\{\lambda x_0\}} \qquad (6)$$

For $\overline{m}$ components emerging in sequence, the number of apparent peaks p is given by

$$p = \overline{m}\, e^{-\{\lambda x_0\}} \qquad (7)$$

The probability of isolating the peak on both sides is the product of the two probabilities, which when multiplied by the estimated number of components $\overline{m}$, yields the number of expected singlets s,

$$s = \overline{m} \; e^{-\{2\lambda x_0\}} \tag{8}$$

Obtaining the value of $\overline{m}$ represents a special problem, which has been addressed in detail by Davis and Giddings (2–4). The reader is referred to these original works for more details. In general, however, $\overline{m}$ can be estimated either by varying the separation distance x_0, by running separations on columns of varying efficiency, or, more realistically, by applying statistical theory to a single chromatogram. In this latter case the number of gaps between peaks on a given chromatogram p', which exceed some arbitrary distance x_0', is expected to follow

$$\ln p' = \ln \overline{m} - \left(\overline{m} \, \frac{x_0'}{x} \right) \tag{9}$$

The value of p' is obtained for a series of x_0' values, which are plotted in a $\ln p'$ versus x_0'/X graph, the slope or intercept of which yields an estimate of $\overline{m}$.

Having estimated the number of components in a mixture, the value of λ can be expressed as

$$\lambda = \frac{\overline{m}}{X} \tag{10}$$

Thus λ is the component peak density, or the number of components distributed over coordinate distance X. It can further be shown (4) that

$$\alpha = \frac{\overline{m}}{n_c} \tag{11}$$

where α is defined as the saturation factor of the separation. In addition, the number of apparent peaks p expected in a separation can be expressed as

$$p = \overline{m} \; e^{-\alpha} \tag{12}$$

with a predicted number of singlet, or pure, peaks being given by

$$s = \overline{m} \; e^{-\{2\alpha\}} \tag{13}$$

Since the success of a separation is normally judged by the number of singlets observed, Davis and Giddings have further defined what they refer to as a success ratio SR, which is simply

$$SR = \frac{s}{m} = e^{-\{2\alpha\}} \tag{14}$$

16.1.3. Multidimensional Systems for Complex Separations

16.1.3.1. Success Ratios and Their Implications for Multidimensional Separations

The implications of the preceding theory to realistic separations are quite astounding. Figure 16.1 shows the number of theoretical plates required to achieve a peak singlet with a given level of probability from a mixture containing $\overline{m}$ compounds. For example, even with a moderately complex mixture of 20 compounds, upward of 200,000 theoretical plates are necessary to achieve 90% probability for a singlet, with nearly 800,000 plates being required for 95% probability. When one considers that modern HPLC columns typically produce about 20,000 plates, from a statistical viewpoint, the probabilities of achieving clean separations on even moderately complex samples can be quite low. In capillary GC the situation is somewhat better, but clearly, it is quite possible to saturate the separating power of a system with only moderately complex mixtures.

The values of the success ratios expected at various levels of system saturation are given in Table 16.1. Systems operated at saturation levels of unity produce success ratios of only 0.14, while saturation values of 50% yield success ratios of only 0.37. For success ratios of 90% to be achieved, the chromatographic system must operate at saturation levels of less than 0.053, while 99% ratios require saturation levels of only 0.005. It should be pointed out that the theoretical treatment of the probability of peak overlap assumes that compounds elute randomly across the separation axis. In practice, the components in a sample generally are related in some way (polarity, molecular weight, hydrophobicity). They may tend to cluster into regions, or to elute nonuniformly for one reason or another. Thus the actual success ratio may be different from what theory predicts. However, the broad trends of the statistical theories described remain valid.

These numbers are quite astonishing, especially in light of the fact that chromatographic systems, especially HPLC, are commonly operated at relatively high levels of saturation. For clean separations of even moderately complex mixtures it is clear that chromatographic systems exhib-

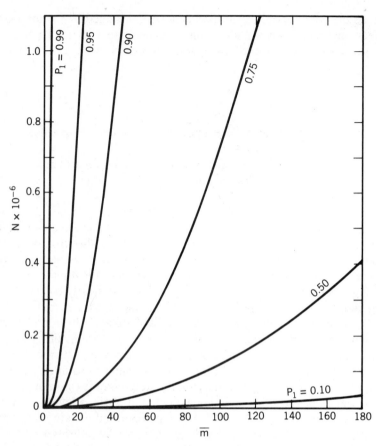

Figure 16.1. Number of theoretical plates required versus estimated number of components in a mixture for different probability levels. [From Ref. (2).]

Table 16.1. Loss in Efficiency for Total Zone Transfer*

k'	Eluted Volume (μL)	Loss in N (%)
1	82.3	22
2	123.4	50
3	164.5	89

* Loss in efficiency measured for peak eluting with $k' = 2$ on second column. Column 1: 4.6 mm × 250 mm, 5000 theoretical plates; column 2: 4.6 mm × 250 mm, 10,000 theoretical plates.

648

iting peak capacities orders of magnitude greater than those in use today are required. Since unimodal chromatographic systems are already operating at such high levels of performance, it is clear that multidimensional systems offer one of the most feasible routes to achieving the necessary peak capacities.

16.1.3.2. Peak Capacities of Multidimensional Systems

Freeman (5) has shown that there is a maximum limit for a separation scheme of j' stages, which is given by the relationship (derived from Eq. 4)

$$n_T = n_{c_1} \times n_{c_2} \times n_{c_3} \times \cdots \times n_{c_i} = \prod_{c_i=1}^{j} n_{c_i} \tag{15}$$

This relationship predicts that the maximum peak number n_T is the product of the individual n_c for each stage i of the separation. This will be true only if the separation mechanism for each stage is completely independent (nonredundant) from that of any other stage in the separation scheme. If each separation mode has approximately the same peak capacity, as is often the case in GC–GC and LC–LC separations, then

$$n_T = n_c^j \tag{16}$$

The mode can be considered the separation mechanism that provides for the retention selectivity observed. Therefore Eq. 16 holds only for a completely nonredundant separation scheme. Clearly, such a relationship is very difficult to achieve in practice since there are always secondary effects involved in any chromatographic separation. In the case of complete redundancy throughout the separation scheme, for j identical modes, Eq. 4 collapses to

$$n_T = \frac{\sqrt{jN}}{4R_s} \ln \left(j \frac{V_{max}}{V_m} \right) \tag{17}$$

This relationship can approximated by

$$n_T \approx j^{\frac{1}{2}} n_c \tag{18}$$

which is analogous to the relationship illustrated in Eq. 5 for a single column, which in effect is what the coupling of two or more chemically

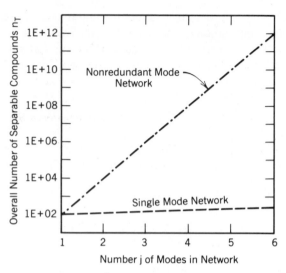

Figure 16.2. Effect of system redundancy on total peak capacity for n coupled columns, plotted assuming equal peak capacities for each column. [Adapted from Ref. (5).]

similar modes produces. Figure 16.2, taken from Freeman's work (5), compares the theoretical peak capacities for separation networks constructed with either single-mode separation schemes (Eq. 18) or nonredundant multidimensional schemes (Eq. 16) with the number of modes employed. From this graph the power relationship of nonredundant modes is clearly seen. These two curves represent mathematical limits between which practical separations can be expected to fall, since most solutes are not retained exclusively by a single separation mode (retention mechanism). This redundancy reduces the specificity of the combined systems and, consequently, the total peak capacity.

16.1.3.3. Peak Capacities in Elution–Elution or Elution–Displacement Systems

Chromatographic systems may be operated in either the elution or the development mode. This gives rise to four combinations for two-dimensional chromatographic designs—elution–elution, elution–displacement, displacement–displacement, and displacement–elution. Of these, only displacement–displacement is capable of sequentially generating all of the separation information available within a given mixture without recourse to complicated valving and peak trapping. In the other combinations, temporal constraints prevent the second separation from operating in real time with the first, except in a few isolated instances of a very high-speed separation coupled with a very slow-speed primary separation.

This is in sharp contrast to GC–MS, which may be considered as one of the most powerful multidimensional separation devices. The reasons for this power are that (1) the two modes of separation are essentially nonredundant and (2) the time frames of the two techniques are orders of magnitude different, such that a real-time combination of the tools is possible.

The operational requirement that a secondary separation be able to complete its function within the time frame of a single eluted band from the primary dimension reduces the effective peak capacity that a real-time elution–elution system can generate. Since under the best conditions, perhaps two or three zones from the first stage might be operated upon by the second stage, the value of n_t in terms of actual isolated peaks would approach $(\approx 2-3)n_c$ for a two-stage system.

However, it is critical to realize that the "quality" of such a separation is in no way diminished over the total theoretical peak capacities as derived previously. Only the total "quantity of information" is reduced, that is, the overall success ratio of a separation, as expressed in Eqs. 11–14, would be calculated on the basis of Eq. 15, even though the actual number of observable bands would be significantly lower. In practice this is simply the sum of the peak capacities of the two columns used. This is an important point that is often overlooked in manuscripts in the field.

16.2. MULTIDIMENSIONAL INSTRUMENTATION CONSIDERATIONS

16.2.1. Introduction: Design Principles for Multichromatographic Systems

The ideal multidimensional chromatographic separation might be visualized as one in which the first column separates perhaps ionic from nonionic compounds. The nonionic solutes would then be shunted to a second system, which separated saturated from unsaturated compounds. The unsaturated solutes could then be diverted to a third phase system capable of separating the aromatics from the remaining unsaturated compounds. Such a system would be difficult to implement since class-specific phases are needed for each stage, and solvent compatibility between stages can be a problem. Workable systems can be achieved, however, with work progressing on coupled separations such as size exclusion coupled to GC (6) and ring size fractionation of polyaromatic hydrocarbons (PAH) prior to GC (7).

The design problem for elution–elution systems, the subject with which the remainder of this chapter will deal almost exclusively, can be generally divided into two main areas, (1) physical considerations and (2) chemical compatibilities.

16.2.1.1. Physical Zone Transfer

Physical considerations are related to the problem of physically transferring the packet of zones separated from one dimension to the next. This must occur within acceptable limits of dispersion, and must be capable of reproducible, quantitative transfer. In some techniques, such as LC–GC, the problem of a phase change compounds the volumetric dissimilarities between the two instruments. In both LC–LC and LC–GC, these have been overcome, as is shown later, either by eluent splitting, by LC miniaturization, by on-column concentration (Grob's retention gap), or by a combination of the above.

16.2.1.2. Chemical Compatibilities

The second general area of concern in multidimensional designs is that of chemical compatibility of the systems. In LC–LC the coupling of normal and reversed-phase systems can be limited by the solvent incompatibilities of the systems. This is unfortunate, since the least redundancy of mechanisms is likely to occur in the most dissimilar modes. However, even normal and reversed-phase systems can be coupled if the volume of the primary system is made small enough such that the solvent introduced into the secondary column is insignificantly small. Another example of this problem is that of LC–GC, where polar samples eluted from LC must be derivatized before introduction to GC. Solutions to these and other linkage problems are reviewed in later sections.

Smoothly integrated coupled chromatographic systems have been researched and designed for virtually all of the chromatographic modes, and solutions to many specific design problems have been found. Despite some unavoidable constraints and limitations, the orders of magnitude increases in peak capacity available from multidimensional systems make their application to many complex mixtures imperative.

16.3. LC–LC IMPLEMENTATIONS

16.3.1. Zone Transfer

Information must not be lost or masked by the movement of a zone from one system to the next. Therefore the method of zone transfer is critical to the success of a multicolumn system. In coupling an elution method to another method (by either displacement or elution) one must wait for the compound of interest to elute from the first column before the second separation may be performed. If the zones of interest are separated in the first few centimeters of the column, then the last several centimeters merely serve to dilute the band by increasing its width. An ideal column design would be one in which such solutes could be shunted laterally from the column within their separation "window," allowing other bands not yet separated to continue their development. While such column designs have not yet been widely investigated (but perhaps should be), in practice this implies that the column length should be carefully tuned to match the zones of interest, although by doing so, information in other zones could be compromised.

Three basic techniques have evolved for column–column transfers, (1) transferring the entire zone, (2) heart-cutting, and (3) scaling the zone volume of the first to the second system. The first of these is the easiest to implement and is routinely used in the simple coupled column experiment. However, due to the size of the transfer volume (see Table 16.1) the loss in efficiency of the final column can be quite substantial.

The second approach, heart-cutting, is a poor solution due to the inevitable discrimination resulting from errors in the sampling of the zone. In Fig. 16.3 a hypothetical transfer zone for a 4.6 mm i.d. column is illustrated where the thin line represents a typical transfer volume, 10 μL. It is clear that unless all of the compounds within the zone perfectly overlap each other, sampling the front and the back of the zone will result

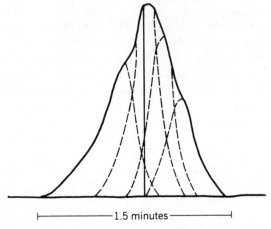

├──────────1.5 minutes──────────┤

Figure 16.3. Hypothetical zone transfer of 10 μL (vertical line) for conventional column. Note change in composition of transferred zone that would result if peak were sampled in front, middle, and tail of peak.

in two entirely different answers. It is also obvious that the use of simple heart-cutting under the conditions illustrated is a poor choice for quantitative work.

The last technique, scaling the elution volume of the first system to meet the injection volume requirements of the second system, has been explored in our laboratory (8, 9) for both an LC–LC and an LC–GC (7, 10) system. By properly matching the elution and injection volumes, the entire zone may be transferred from one system to the other, thus eliminating the discrimination errors inherent in heart-cutting. The relation that governs this approach for LC–LC is given by

$$\frac{6d_{c_1}^2\epsilon_1 L_1(1 + k_1')}{\sqrt{N_1}} \geq \frac{\theta d_{c_2}^2\epsilon_2 L_2(1 + k_2')}{\sqrt{N_2}}$$

where subscripts 1 and 2 refer to columns 1 and 2; N is theoretical plate count; ϵ the column porosity; L the column length, d_c the column diameter, and θ the resolution reduction factor. Substituting reasonable values for the porosity, length and plate count and setting a maximum loss of resolution to 10% reduces this relationship to

$$d_{c_1} \leq \frac{0.40d_{c_2}}{\sqrt{1 + k_1'}}$$

Thus for peaks eluting with $k_1' < 5$, the 1.0:4.6-mm column-diameter ratio is ideally suited for the quantitative zone transfer with less than 10% loss in the resolution in the second stage. Also, since the volume being transferred to the second system is acceptably small, the need to perform on-column concentration is eliminated. This eliminates the requirement for reequilibration of the second stage between runs.

In LC the available polarity range and other interactions that govern selectivity are greater compared to those available for GC separations. This permits greater differences in the separation modes and allows for more powerful multidimensional separations. Thus LC has an advantage over GC in the establishment of systems that are truly multimodal.

The concept of an on-line LC–LC system assumes that a band of solutes from the first LC system is transferred to a second LC system to be further separated. Thus the zone transfer methods in which the solute bands are to be brought from the first system to the next system, as well as the mixture being analyzed, dictate the overall system requirements, such as mobile phases, column parameters (diameters, lengths, minimum efficiencies), and column packings (stationary phase bondings, particle sizes, and porosities). LC–LC systems, therefore, are problem-specific;

there is no one standard system description. At presently all of the transfer methods mentioned have been employed for LC–LC systems.

16.3.2. Review of LC–LC Separations

The first reported use of an on-line LC–LC system was by Huber et al. (11) in 1973. A low-dispersion switching valve was used to pass early eluting peaks from a short low-capacity column to a longer high-capacity column for further separation. The later eluting peaks bypassed the longer and more highly retentive column and were detected as they exited the first column (Fig. 16.4).

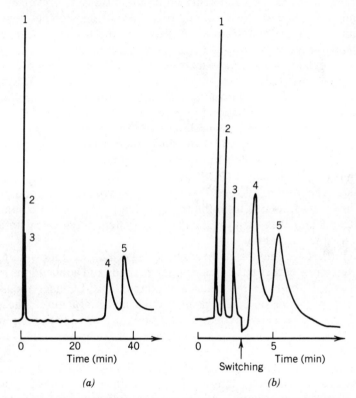

Figure 16.4. Column stripping. Sample: 1 = toluene; 2 = azobenzene; 3 = 2-nitrobenzene; 4 = p-cresol; 5 = phenol; injection volume = 10 μL. Column: packing, 10% (w/w) 1,2,3-tris(cyanoethoxy)propane on silica support, 4–8 μm; eluent, 2,2,4-trimethylpentane. Detector: UV, 270 ± 10 nm. (a) Columns 1–2 for total sample. (b) Columns 1 + 2 for components 1–3; column 1 alone for components 4 and 5. Column 1, 30 × 3 mm; column 2, 290 × 3 mm. Flow rate: column 1, 2.27 mL/min; columns 1 + 2, 1.33 mL/min. [Reprinted with permission from Ref. (11).]

The next significant advances appeared in 1978 from three different groups. An on-column concentration technique for a gel permeation chromatography/reversed-phase chromatography (GPC–RPC) system was performed by Erni and Frei (12). This combination was used for the analysis of senna glycosides in plant extracts. When attempting the transfer of a large volume band, the zone should be concentrated on the head of the second column to minimize the deleterious effects on efficiency that the large transfer volume can have. Thus it is essential that the mobile phase of the first dimension be made much weaker than that of the second dimension. Erni and Frei used an aqueous GPC system buffered at pH 7.0 for the molecular weight size fractionation. The GPC fraction of interest was preferentially concentrated on the head of the reversed-phase column and then eluted with a mild organic-aqueous step gradient.

Johnson, Gloor, and Majors (13) reported the use of a heart-cutting technique for the separation of various low molecular weight solutes from high molecular weight solute matrices. A high performance exclusion chromatography (HPEC) system was coupled to a reversed-phase system through a six-port sampling valve. The 10-μL loop allowed them to transfer slices of the HPEC fraction to the reversed-phase system. Figure 16.5 shows the separation of a disulfide mixture by HPEC/RP.

Willmott, Mackenzie, and Dolphin (14) incorporated an automated three-way valve controlled by a microcomputer to study a mixture of polychlorinated biphenyls (PCB) and polychlorinated dibenzo-p-dioxins (PCDD). The system allowed them either to bypass the second column or to heart-cut fractions from the first column to the second column. An alumina packing was used for the first LC system and a silica stationary phase for the second column.

From these techniques many applications of LC–LC followed. Majors reviewed many earlier multidimensional methods and applications in 1980 (15). Some of the more recent uses of coupled LC–LC have focused on analytical problems of a biological or environmental nature. These biological analyses usually involve a solute mixture of wide-ranging polarities. The necessity for a separation system that can accommodate this range of solute polarities is easily achieved by the coupling of two dissimilar retention modes. Thus Benjamin et al. (16) employed a reversed-phase/cation exchange system to separate an antiviral agent of significant polarity and the relatively nonpolar ester and diester adducts.

Another, more novel, approach was that of Kemper et al. (17) for the analysis of catecholamines in human urine samples. Their system consisted of a phenylboronic acid bonded silica HPLAC/cation exchange HPLC combination. The catecholamines were preferentially retarded by the HPLAC phase and then transferred to the cation exchange column for further separation and detection.

contribution to peak variance by the first-stage components was mea-
~red for the thymidine peak. The exponentially modified Gaussian
(EMG) approach of Foley and Dorsey (19) was employed to measure the
~eak variance. The additional variance introduced by the first separation
~tage was found to be ≈5%, which agrees well with theoretical
considerations.

16.4. LC–GC INSTRUMENTAL DESIGNS

16.4.1. History of Early Experiments

The LC–GC experiment furnishes some very interesting problems to the
chromatographer. The problems encountered in interfacing LC and GC
run parallel to those encountered in the marriage of LC and MS. First,
since the two individual techniques employ mobile phases of different
physical states, the transferred zone must undergo a phase transformation. In addition the GC separation requires injection volumes that are
usually only a fraction of the elution volumes of the first dimension. Third,
many of the solvents employed in the first (LC) system are incompatible
with the seocnd dimension.

The first reported use of on-line LC–GC was in 1979 by Majors (20)
and the first published results appeared in 1983 by Apffel and McNair
(21). Both groups employed a conventional HPLC system coupled to a
capillary GC. Apffel and McNair analyzed the various fractions of gasoline; saturates, unsaturates, aromatics, and polars. However, the key to
the success of multimodal separations is the method of zone transfer.
Since the above authors employed a conventional LC system, they were
only able to heart-cut a minute slice of the zone of interest to GC for
further analysis. Thus although both groups clearly demonstrated the
power of LC–GC, they were able to generate only qualitative information
about their samples.

16.4.2. GC Retention Gaps

The next approach to on-line LC–GC was by Grob and coworkers (22),
who employed a new type of on-column concentration technique coined

Figure 16.7. LC–GC separation of diethylstilbestrol (DES) in bovine urine. (*a*) LC trace
after derivatization of DES to dipentafluorobenzyl ether. Transferred peak contains DES
as well as other compounds. (*b*) FID trace of LC fraction. Position and height of main peak
correspond to 10 ppb of DES in urine. [Reprinted with permission from Ref. (22).]

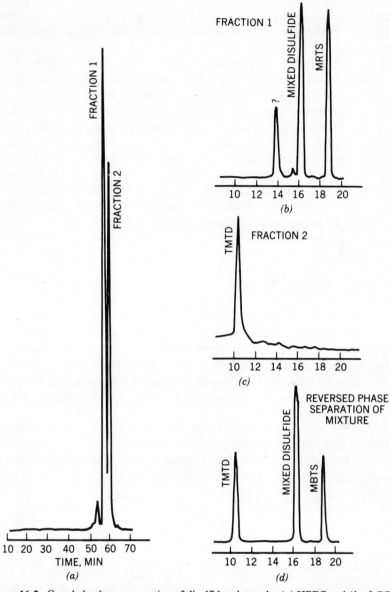

Figure 16.5. Coupled column separation of disulfide mixture by (*a*) HPEC and (*b–d*) RPC.
(*a*) HPEC columns—50 cm 3000H, 50 cm 2000H, and 80 cm 1000H MicroPak TSK (8 mm
id); THF: flow rate—1 mL/min; detection—215 nm (1.0 AUFS), injection volume—200 μL.
(*b*) RPC: 25 cm × 2.2 mm MicroPak MCH; flow rate—0.5 mL/min; injection volume—10
μL; gradient acetonitrile–water (20:80, v/v) to 100% acetonitrile at 3% acetonitrile/min;
detection—254 nm (0.05 AUFS). [Reprinted with the permission from Ref. (13).]

A powerful example employing novel stationary phases to further enhance the selectivity of multimodal LC–LC is given by the work of Linder
et al. (18). In the study of nitrogenous PAHs in diesel exhaust particulate
extracts, a pyrene–butyric acid–amide (PBA) phase was prepared for use
with a reversed-phase column. Fractions from the PBA system were preconcentrated on the reversed-phase column before further analysis to
accommodate the large transfer volume from the PBA system.

More recently a third transfer technique was established where the
volume of the transferred band is made compatible with subsequent stages
by reducing the column dimensions of the initial stages, that is, a microbore (1.0-mm id) column for the first stage and an analytical (4.6-mm id)
column for the second separation stage. It has been shown (9) that coupling a 1-mm microbore column to a 4.6-mm conventional column should
allow total transfer of the zone with a minimal loss in efficiency (<5%).

The coupling of a small-diameter column to a larger column has been
shown (9) to be sample mass conservative and also to have a negligible
effect on the second system's efficiency N. Table 16.2 compares the variance measured for each peak of the nucleosides separated by the second
stage of the multidimensional separation, as seen in Fig. 16.6. The results
compare the average peak areas for each component under two conditions. The first column indicates the average peak areas observed where
the peaks are eluted through the microbore stage, are collected in a sampling loop, and then are transferred to the analytical stage. The second
column indicates the average peak area for the same solute sample injected directly with the microbore injection valve into the secondary
stage, that is, the microbore column, detector, and sampling valve are
removed from the separation scheme. Thus any significant difference between these areas would indicate discrimination losses. No significant
differences at the 99% confidence level were observed.

Using the same bypass approach as followed in the previous paragraph,

Table 16.2. Effect of Microbore System on Peak Area

Nucleoside	Coupled-System Microbore Peak Area (10^3 V·min)	System Bypassed Peak Area (10^3 V·min)
d-Cytidine	143.4 ± 6.4	152.1 ± 3.0
d-Guanosine	59.7 ± 6.4	61.7 ± 3.0
Thymidine	133.1 ± 6.4	127.3 ± 4.8
d-Adenosine	51.6 ± 4.8	52.6 ± 4.8

Source: Reprinted with permission from Ref. (18).

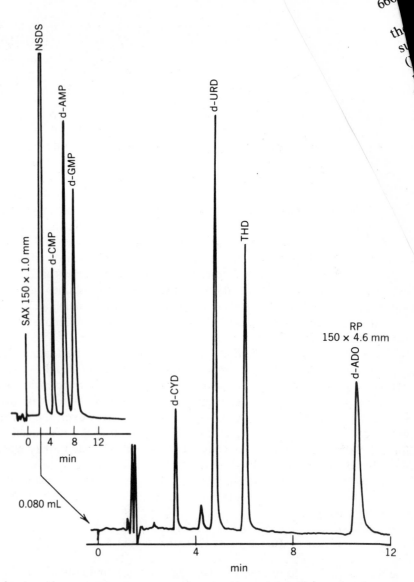

Figure 16.6. Separation of major deoxyribonucleosides and their 5′-monophosphate deox-
ynucleotides. Columns: 25 cm × 1.0 mm SAX column interfaced with 15 cm × 4.6 mm
RPC column; mobile phases—(SAX) 0.025M KH$_2$PO$_4$ at pH 3.9; and (RPC) 90:10 (v/v)
0.025M KH$_2$PO$_4$ at pH 3.9/methanol; detection—UV, 254 nm (0.1 AUFS). [Reprinted with
permission from Ref. (9).]

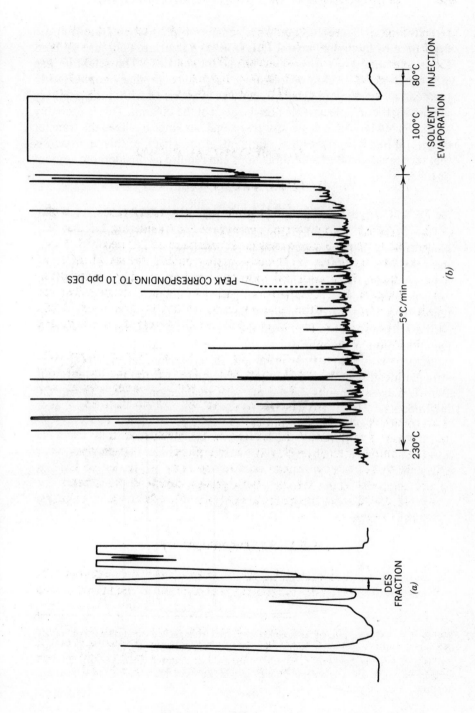

retention gap. Figure 16.7 shows a separation of the heptafluorobutyro derivative of diethylstilbestrol (DES) in bovine urine. (The use of DES for accelerating growth is illegal in many countries.) The retention gap (23) consisted of a 50-m deactivated fused silica capillary connected to the head of the column. The solvent was evaporated within the capillary depositing the sample as a plug on the head of the column. This is probably the preferred interface design for trace analysis since it allows the transfer of several hundred microliters of eluent. A drawback of this approach is that the transfer and evaporation steps can require a considerable amount of time.

16.4.3. Packed Fused-Silica HPLC

Advances in LC miniaturization allowed Cortes and coworkers (24) to apply Grob's technique to an on-line system by more closely matching the LC elution to the GC injection volume. Their system consisted of a packed capillary LC system coupled to a capillary GC system. For total zone transfer (approximately 25–40 μL) they employed a short 4-m retention gap, differing only in size from Grob's original work. Figure 16.8 shows a separation of coal tar on the capillary LC–capillary GC system. One limitation to the miniaturization of the LC is that highly specialized equipment is required (that is, pump, injector, and detector). Also the stringent dead volume requirements of capillary LC (on the order of 50–100 nL places demands on the physical installation of the LC and GC instruments.

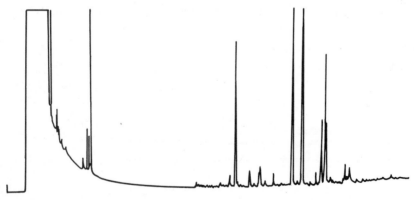

Figure 16.8. GC trace of coal tar sample transferred from packed capillary LC system. GC conditions: 30 m × 250 μm id Suplecowax 10, 0.25-μm film; overn: 115°C, 7 min 5°C/min to 240°C. Retention gap: 4 m × 250 μm id fused silica. Transfer volume: 40 μL. LC conditions: 96 cm × 250 μm id capillary packed with Zorbax ODS; d_p = 7 μm; eluent: acetonitrile. Injection volume 60: nL.

16.4.4. Isotachic Eluent Splitting

The third approach taken in interfacing LC and GC is that of eluent splitting. When splitting the eluent, the LC elution volume is no longer restricted by the GC injection volume requirements. The discrimination
errors that are associated with heart-cutting are minimized since the splitter allows one to skim a low volume image from the entire zone of interest
(25, 26). Figure 16.9 shows the GC trace of the three-ring fractionation
of a coal tar sample on the LC–GC interface employing microbore HPLC
and eluent splitting. The major advantage of this interface is that it allows
the use of commercially available instrumentation, removing the necessity
of matching LC and GC volume requirements. Conversely, its major limitation is loss of sensitivity, although it should be mentioned that the total
mass transferred woud be equivalent to LC miniaturization, since the
transferred zone could be considered as simply a small-diameter column
feeding into the GC system.

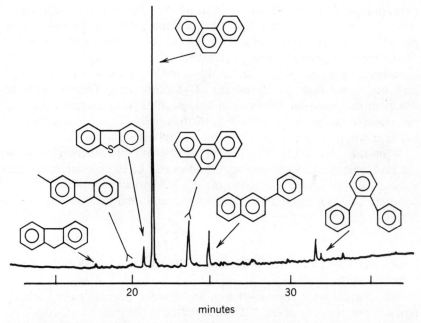

Figure 16.9. GC trace of three-ring fraction of solvent-refined coal sample. LC conditions:
two propyl amine columns (25 cm × 1 mm each) mobile phase, hexane; injection volume,
1 μL. Split ratio 2:1; transferred volume, 10 μL. GC conditions, 50:1 split; 20-m SE-30
column; FID detection, 50°C hold 3 min, 10°C/min to 150°C, then 5°C/min to 265°C hold 2
min.

In summary, the problem of physically matching the inlet volume requirements of a GC to the elution volumes of an LC have been addressed by three major routes, namely, the use of a retention gap on the GC, miniaturization of the LC, and eluent splitting. In practice, the approaches are not mutually exclusive. For example, despite miniaturization, Cortes used the retention gap technique of Grob to increase sensitivity. Raglione had used eluent splitters to essentially simulate a micro-LC in terms of elution volumes, so that existing LC instrumentation could be applied. In practice, all three techniques are used on a "mix and match" basis to meet the demands of a particular analysis. If sensitivity is the prime concern, then the retention gap technique is preferred. For ease of use, simple splitting might be desired.

16.4.5. Postcolumn Reactors

Another issue, mentioned earlier, is that of chemical incompatibilities between LC and GC. Most applications to date have focused on normal phase separations, typically of fuels, coal tars, or similar species. However, in many situations it would be desirable to derivatize a solute prior to its introduction into the GC. This is done routinely for many compounds; when someone collects an LC or thin-layer chromatography (TLC) fraction, then he or she derivatizes it prior to GC.

Recently Raglione et al. (27, 28) have made progress in introducing a postcolumn reaction system into the LC–GC interface. They have shown the quantitative on-line analysis of lipid profiles from bacterial cultures. An example of this is shown in Fig. 16.10, where a lipid fraction from an LC separation is sent to the GC after methyl ester formation.

With the successful introduction of postcolumn reaction systems, the problems of chemical incompatibilities can often be overcome. This substantially broadens the scope of LC–GC instrumentation to include pharmaceuticals, metabolites, pesticides, and many other classes of compounds.

16.5. FUTURE DIRECTIONS FOR MULTIDIMENSIONAL SEPARATIONS

In this chapter only elution–elution multimodal systems have been examined, and the area of GC–GC was omitted in the interest of space. The area of displacement–displacement separations, such as two-dimensional TLC, is probably one of the most widely used multidimensional techniques. As recently pointed out by Giddings (29), two-dimensional continuous displacement systems are theoretically superior to elution–elution

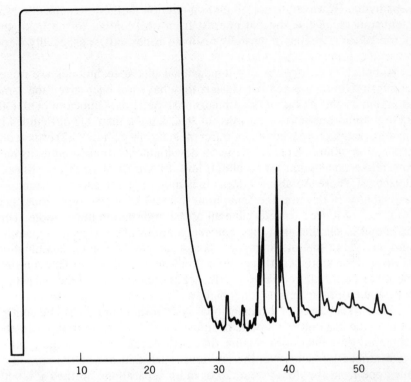

Figure 16.10. On-line LC–PCR–GC of biological lipids. Separation by HPLC, followed by methylation, and GC of lipid fraction from staphylococcus aureus. 50-m retention gap was used for GC. [From Refs. (28).]

methods (noncontinuous). The displacement experiment has several advantages, (1) zone transfer to other dimensions becomes unnecessary and (2) there is no loss in the quantity of information since the entire sample is subjected to the second separation (rather than only one band). The limitations of two-dimensional planar chromatographic separations include the problems of detection and sample collection. Only in relatively isolated situations would it be superior to detect the solute band in the presence of the adsorbent, such as aflatoxin separations. In most other instances it is far easier to detect the solute in the free fluid.

The instrumental problems of two-dimensional planar separations employing free-zone detection have been investigated by Guiochon and coworkers (30) with promising, but limited success. Elution–elution methods, such as LC–LC and LC–GC as well as GC–GC, SFC–SVC, and SFC–GC, offer the advantages of readily available instrumentation of high sensitivity. However, since all elution–elution methods require the efficient transfer of the zone of interest from one column to the next, the potential for remixing of partially resolved zones exists, especially if too large of a primary cut is taken.

Although in theory the two-dimensional planar techniques are richer in information and potentially faster, they have not been given the same attention as elution-based two-dimensional methods. Guiochon et al. (30) have pointed out several reasons for this, among them (1) difficulties in making plates composed of two different stationary phases, (2) cross contamination of mobile phases between developments (that is, aqueous mobile phase deactivates silica plate), and (3) a lack of good quantitative detectors. There are also (4) difficulties in automating such a system, (5) an inability to develop the separation in more than two directions, and (6) a lack of a need to determine all of the information in a sample. This last point is significant in that many separations are designed to quantify only several of the total components in a sample. However, examination of the general statistical theory presented earlier will show that even simple separations can require powerful separation systems if the necessary statistical confidence interval is high.

The ultimate two- or n-dimensional system should probably be continuous if the maximum amount of information is to be extracted in the minimum time. This will require that each successive separation either be significantly faster than the previous, such that during the elution of one zone from the nth system, 20 or more separations on the $(n + 1)$th system may be completed, or all information from the complementary separative transport mechanisms be generated simultaneously in the true n-dimensional system. This requirement virtually eliminates the possibility of real-time continuous interfacing elution chromatographic systems, except in the trivial instances of very high speed, low efficiency $(n + 1)$th separations.

An interesting compromise between elution and displacement two-dimensional systems has been explored by Shelley and French (31). Their multimodal system consisted of a packed capillary HPLC coupled to an HPTLC. Some of the advantages of such a design are (1) increased throughput, since the second system is not "recycled" as in an elution method and (2) no loss of information, since the entire LC separation is placed on the plate. With such a system the $(n + 1)$th dimension no longer has a time dependence, this having been largely removed by trapping the solute on the relatively slowly diffusing TLC plate.

Finally, for all of the various arguments discussed, multidimensional separations may well find their strongest applications in the domain of preparative LC. The design of preparative HPLC would benefit if the

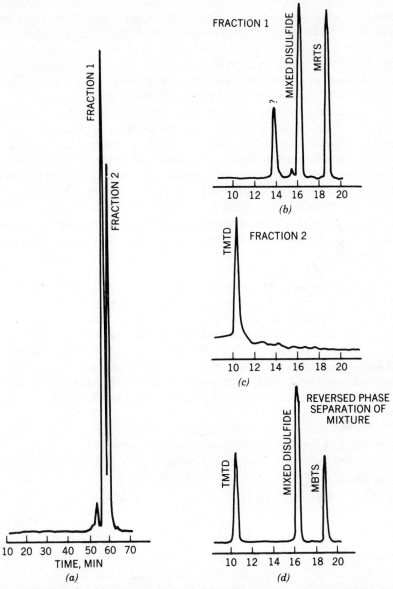

Figure 16.5. Coupled column separation of disulfide mixture by (a) HPEC and (b–d) RPC. (a) HPEC columns—50 cm 3000H, 50 cm 2000H, and 80 cm 1000H MicroPak TSK (8 mm id); THF: flow rate—1 mL/min; detection—215 nm (1.0 AUFS), injection volume—200 μL. (b) RPC: 25 cm × 2.2 mm MicroPak MCH; flow rate—0.5 mL/min; injection volume—10 μL; gradient acetonitrile–water (20:80, v/v) to 100% acetonitrile at 3% acetonitrile/min; detection—254 nm (0.05 AUFS). [Reprinted with the permission from Ref. (13).]

657

A powerful example employing novel stationary phases to further enhance the selectivity of multimodal LC–LC is given by the work of Linder et al. (18). In the study of nitrogenous PAHs in diesel exhaust particulate extracts, a pyrene–butyric acid–amide (PBA) phase was prepared for use with a reversed-phase column. Fractions from the PBA system were preconcentrated on the reversed-phase column before further analysis to accommodate the large transfer volume from the PBA system.

More recently a third transfer technique was established where the volume of the transferred band is made compatible with subsequent stages by reducing the column dimensions of the initial stages, that is, a microbore (1.0-mm id) column for the first stage and an analytical (4.6-mm id) column for the second separation stage. It has been shown (9) that coupling a 1-mm microbore column to a 4.6-mm conventional column should allow total transfer of the zone with a minimal loss in efficiency (<5%).

The coupling of a small-diameter column to a larger column has been shown (9) to be sample mass conservative and also to have a negligible effect on the second system's efficiency N. Table 16.2 compares the variance measured for each peak of the nucleosides separated by the second stage of the multidimensional separation, as seen in Fig. 16.6. The results compare the average peak areas for each component under two conditions. The first column indicates the average peak areas observed where the peaks are eluted through the microbore stage, are collected in a sampling loop, and then are transferred to the analytical stage. The second column indicates the average peak area for the same solute sample injected directly with the microbore injection valve into the secondary stage, that is, the microbore column, detector, and sampling valve are removed from the separation scheme. Thus any significant difference between these areas would indicate discrimination losses. No significant differences at the 99% confidence level were observed.

Using the same bypass approach as followed in the previous paragraph,

Table 16.2. Effect of Microbore System on Peak Area

Nucleoside	Coupled-System Microbore Peak Area (10^3 V·min)	System Bypassed Peak Area (10^3 V·min)
d-Cytidine	143.4 ± 6.4	152.1 ± 3.0
d-Guanosine	59.7 ± 6.4	61.7 ± 3.0
Thymidine	133.1 ± 6.4	127.3 ± 4.8
d-Adenosine	51.6 ± 4.8	52.6 ± 4.8

Source: Reprinted with permission from Ref. (18).

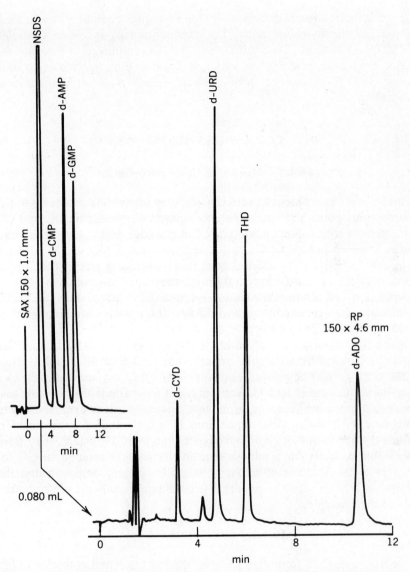

Figure 16.6. Separation of major deoxyribonucleosides and their 5'-monophosphate deoxyribonucleotides. Columns: 25 cm × 1.0 mm SAX column interfaced with 15 cm × 4.6 mm RPC column; mobile phases—(SAX) 0.025*M* KH₂PO₄ at pH 3.9; and (RPC) 90:10 (v/v) 0.025*M* KH₂PO₄ at pH 3.9/methanol; detection—UV, 254 nm (0.1 AUFS). [Reprinted with permission from Ref. (9).]

the contribution to peak variance by the first-stage components was measured for the thymidine peak. The exponentionally modified Gaussian (EMG) approach of Foley and Dorsey (19) was employed to measure the peak variance. The additional variance introduced by the first separation stage was found to be ≈5%, which agrees well with theoretical considerations.

16.4. LC–GC INSTRUMENTAL DESIGNS

16.4.1. History of Early Experiments

The LC–GC experiment furnishes some very interesting problems to the chromatographer. The problems encountered in interfacing LC and GC run parallel to those encountered in the marriage of LC and MS. First, since the two individual techniques employ mobile phases of different physical states, the transferred zone must undergo a phase transformation. In addition the GC separation requires injection volumes that are usually only a fraction of the elution volumes of the first dimension. Third, many of the solvents employed in the first (LC) system are incompatible with the seocnd dimension.

The first reported use of on-line LC–GC was in 1979 by Majors (20) and the first published results appeared in 1983 by Apffel and McNair (21). Both groups employed a conventional HPLC system coupled to a capillary GC. Apffel and McNair analyzed the various fractions of gasoline; saturates, unsaturates, aromatics, and polars. However, the key to the success of multimodal separations is the method of zone transfer. Since the above authors employed a conventional LC system, they were only able to heart-cut a minute slice of the zone of interest to GC for further analysis. Thus although both groups clearly demonstrated the power of LC–GC, they were able to generate only qualitative information about their samples.

16.4.2. GC Retention Gaps

The next approach to on-line LC–GC was by Grob and coworkers (22), who employed a new type of on-column concentration technique coined

→

Figure 16.7. LC–GC separation of diethylstilbestrol (DES) in bovine urine. (*a*) LC trace after derivatization of DES to dipentafluorobenzyl ether. Transferred peak contains DES as well as other compounds. (*b*) FID trace of LC fraction. Position and height of main peak correspond to 10 ppb of DES in urine. [Reprinted with permission from Ref. (22).]

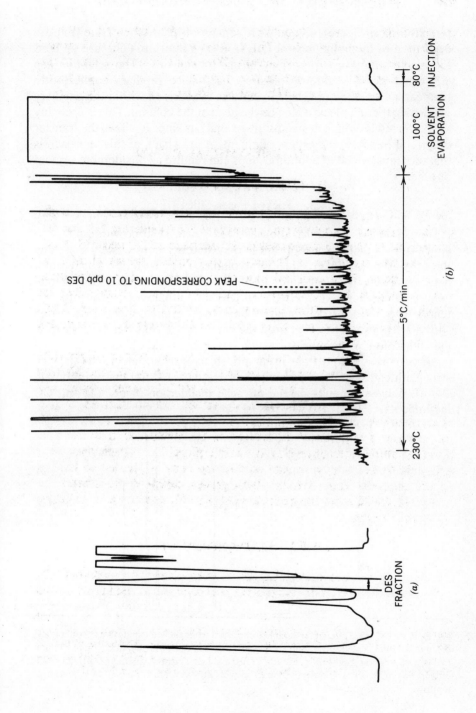

retention gap. Figure 16.7 shows a separation of the heptafluorobutyro derivative of diethylstilbestrol (DES) in bovine urine. (The use of DES for accelerating growth is illegal in many countries.) The retention gap (23) consisted of a 50-m deactivated fused silica capillary connected to the head of the column. The solvent was evaporated within the capillary depositing the sample as a plug on the head of the column. This is probably the preferred interface design for trace analysis since it allows the transfer of several hundred microliters of eluent. A drawback of this approach is that the transfer and evaporation steps can require a considerable amount of time.

16.4.3. Packed Fused-Silica HPLC

Advances in LC miniaturization allowed Cortes and coworkers (24) to apply Grob's technique to an on-line system by more closely matching the LC elution to the GC injection volume. Their system consisted of a packed capillary LC system coupled to a capillary GC system. For total zone transfer (approximately 25–40 μL) they employed a short 4-m retention gap, differing only in size from Grob's original work. Figure 16.8 shows a separation of coal tar on the capillary LC–capillary GC system. One limitation to the miniaturization of the LC is that highly specialized equipment is required (that is, pump, injector, and detector). Also the stringent dead volume requirements of capillary LC (on the order of 50–100 nL places demands on the physical installation of the LC and GC instruments.

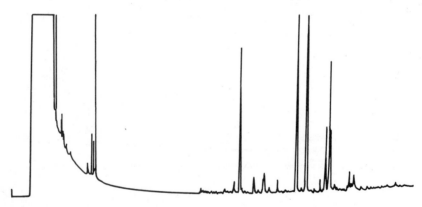

Figure 16.8. GC trace of coal tar sample transferred from packed capillary LC system. GC conditions: 30 m × 250 μm id Suplecowax 10, 0.25-μm film; overn: 115°C, 7 min 5°C/min to 240°C. Retention gap: 4 m × 250 μm id fused silica. Transfer volume: 40 μL. LC conditions: 96 cm × 250 μm id capillary packed with Zorbax ODS; d_p = 7 μm; eluent: acetonitrile. Injection volume 60: nL.

16.4.4.   Isotachic Eluent Splitting

The third approach taken in interfacing LC and GC is that of eluent splitting. When splitting the eluent, the LC elution volume is no longer restricted by the GC injection volume requirements. The discrimination errors that are associated with heart-cutting are minimized since the splitter allows one to skim a low volume image from the entire zone of interest (25, 26). Figure 16.9 shows the GC trace of the three-ring fractionation of a coal tar sample on the LC–GC interface employing microbore HPLC and eluent splitting. The major advantage of this interface is that it allows the use of commercially available instrumentation, removing the necessity of matching LC and GC volume requirements. Conversely, its major limitation is loss of sensitivity, although it should be mentioned that the total mass transferred woud be equivalent to LC miniaturization, since the transferred zone could be considered as simply a small-diameter column feeding into the GC system.

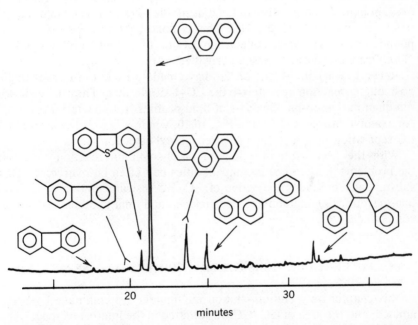

minutes

Figure 16.9. GC trace of three-ring fraction of solvent-refined coal sample. LC conditions: two propyl amine columns (25 cm × 1 mm each) mobile phase, hexane; injection volume, 1 μL. Split ratio 2:1; transferred volume, 10 μL. GC conditions, 50:1 split; 20-m SE-30 column; FID detection, 50°C hold 3 min, 10°C/min to 150°C, then 5°C/min to 265°C hold 2 min.

In summary, the problem of physically matching the inlet volume requirements of a GC to the elution volumes of an LC have been addressed by three major routes, namely, the use of a retention gap on the GC, miniaturization of the LC, and eluent splitting. In practice, the approaches are not mutually exclusive. For example, despite miniaturization, Cortes used the retention gap technique of Grob to increase sensitivity. Raglione had used eluent splitters to essentially simulate a micro-LC in terms of elution volumes, so that existing LC instrumentation could be applied. In practice, all three techniques are used on a "mix and match" basis to meet the demands of a particular analysis. If sensitivity is the prime concern, then the retention gap technique is preferred. For ease of use, simple splitting might be desired.

16.4.5. Postcolumn Reactors

Another issue, mentioned earlier, is that of chemical incompatibilities between LC and GC. Most applications to date have focused on normal phase separations, typically of fuels, coal tars, or similar species. However, in many situations it would be desirable to derivatize a solute prior to its introduction into the GC. This is done routinely for many compounds; when someone collects an LC or thin-layer chromatography (TLC) fraction, then he or she derivatizes it prior to GC.

Recently Raglione et al. (27, 28) have made progress in introducing a postcolumn reaction system into the LC–GC interface. They have shown the quantitative on-line analysis of lipid profiles from bacterial cultures. An example of this is shown in Fig. 16.10, where a lipid fraction from an LC separation is sent to the GC after methyl ester formation.

With the successful introduction of postcolumn reaction systems, the problems of chemical incompatibilities can often be overcome. This substantially broadens the scope of LC–GC instrumentation to include pharmaceuticals, metabolites, pesticides, and many other classes of compounds.

16.5. FUTURE DIRECTIONS FOR MULTIDIMENSIONAL SEPARATIONS

In this chapter only elution–elution multimodal systems have been examined, and the area of GC–GC was omitted in the interest of space. The area of displacement–displacement separations, such as two-dimensional TLC, is probably one of the most widely used multidimensional techniques. As recently pointed out by Giddings (29), two-dimensional continuous displacement systems are theoretically superior to elution–elution

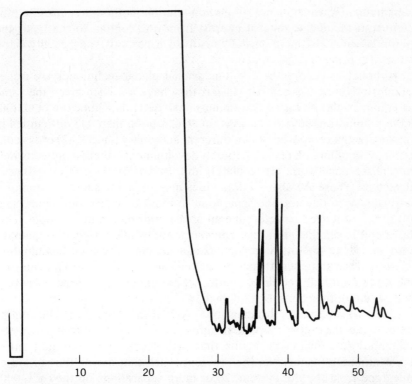

Figure 16.10. On-line LC–PCR–GC of biological lipids. Separation by HPLC, followed by methylation, and GC of lipid fraction from staphylococcus aureus. 50-m retention gap was used for GC. [From Refs. (28).]

methods (noncontinuous). The displacement experiment has several advantages, (1) zone transfer to other dimensions becomes unnecessary and (2) there is no loss in the quantity of information since the entire sample is subjected to the second separation (rather than only one band). The limitations of two-dimensional planar chromatographic separations include the problems of detection and sample collection. Only in relatively isolated situations would it be superior to detect the solute band in the presence of the adsorbent, such as aflatoxin separations. In most other instances it is far easier to detect the solute in the free fluid.

The instrumental problems of two-dimensional planar separations employing free-zone detection have been investigated by Guiochon and coworkers (30) with promising, but limited success. Elution–elution methods, such as LC–LC and LC–GC as well as GC–GC, SFC–SVC, and SFC–GC, offer the advantages of readily available instrumentation of high sensitivity. However, since all elution–elution methods require the efficient transfer of the zone of interest from one column to the next, the potential for remixing of partially resolved zones exists, especially if too large of a primary cut is taken.

Although in theory the two-dimensional planar techniques are richer in information and potentially faster, they have not been given the same attention as elution-based two-dimensional methods. Guiochon et al. (30) have pointed out several reasons for this, among them (1) difficulties in making plates composed of two different stationary phases, (2) cross contamination of mobile phases between developments (that is, aqueous mobile phase deactivates silica plate), and (3) a lack of good quantitative detectors. There are also (4) difficulties in automating such a system, (5) an inability to develop the separation in more than two directions, and (6) a lack of a need to determine all of the information in a sample. This last point is significant in that many separations are designed to quantify only several of the total components in a sample. However, examination of the general statistical theory presented earlier will show that even simple separations can require powerful separation systems if the necessary statistical confidence interval is high.

The ultimate two- or n-dimensional system should probably be continuous if the maximum amount of information is to be extracted in the minimum time. This will require that each successive separation either be significantly faster than the previous, such that during the elution of one zone from the nth system, 20 or more separations on the $(n + 1)$th system may be completed, or all information from the complementary separative transport mechanisms be generated simultaneously in the true n-dimensional system. This requirement virtually eliminates the possibility of real-time continuous interfacing elution chromatographic systems, except in the trivial instances of very high speed, low eficiency ($n + 1$)th separations.

An interesting compromise between elution and displacement two-dimensional systems has been explored by Shelley and French (31). Their multimodal system consisted of a packed capillary HPLC coupled to an HPTLC. Some of the advantages of such a design are (1) increased throughput, since the second system is not "recycled" as in an elution method and (2) no loss of information, since the entire LC separation is placed on the plate. With such a system the $(n + 1)$th dimension no longer has a time dependence, this having been largely removed by trapping the solute on the relatively slowly diffusing TLC plate.

Finally, for all of the various arguments discussed, multidimensional separations may well find their strongest applications in the domain of preparative LC. The design of preparative HPLC would benefit if the column were divided into numerous short coupled segments. Discrete solute zones would then be shunted into and out of the nearest column segment, while the rest of the solute bands could continue migration for further separation. Likewise, fractions ready for collection could be directed into still other phases whereby solvent removal and zone concentration (32) could be effected.

Multidimensional separations offer a rich area of research, both in coupling seemingly disparate separative transport mechanisms and in instrumental design. As routine, sensitive instruments are developed, it seems inevitable that the application of multidimensional separations will increase, offering new tools to the working chromatographer.

REFERENCES

1. J. C. Giddings, *Dynamics of Chromatography*, vol. 1, Dekker, New York, 1965.

2. J. M. Davis and J. C. Giddings, *Anal. Chem., 55*, 418–423 (1983).

3. J. M. Davis and J. C. Giddings, *Anal. Chem., 57*, 2168–2177 (1985).

4. J. M. Davis and J. C. Giddings, *Anal. Chem., 57*, 2178–2182 (1985).

5. D. H. Freeman, *Anal. Chem., 53*, 2 (1981).

6. C. V. Philip and R. G. Anthony, *J. Chromatogr. Sci., 24*, 438 (1986).

7. T. V. Raglione and R. A. Hartwick, *Anal. Chem., 58*, 2680–2683 (1986).

8. N. Sagliano, Jr., and R. A. Hartwick, presented at the 1986 Eastern Analytical Symposium, New York, paper 364.

9. N. Sagliano, Jr., S. H. Hsu, T. V. Floyd, T. V. Raglione, and R. A. Hartwick, *J. Chromatogr. Sci., 23*, 238 (1985).

10. T. V. Raglione, N. Sagliano, Jr., and R. A. Hartwick, *LC–GC Mag., 4*, 328–338 (1986).

11. J. F. K. Huber, R. Van der Linden, E. Ecker, and M. Oreans, *J. Chromatogr., 83*, 267–277 (1973).

12. F. Erni and F. W. Frei, *J. Chromatogr., 149*, 561–569 (1978).

13. B. L. Johnson, R. Gloor, and R. E. Majors, *J. Chromatogr., 149*, 571–585 (1978).

14. F. W. Willmott, I. Mackenzie, and R. J. Dolphin, *J. Chromatogr., 167*, 31–39 (1978).

15. R. E. Majors, *J. Chromatogr. Sci., 18*, 571–579 (1980).

16. E. J. Benjamin, B. A. Firestone, and J. A. Schneider, *J. Chromatogr. Sci., 23*, 168–170 (1985).

17. K. Kemper, E. Hagemeier, D. Ahrens, K. S. Boos, and E. Schlimme, *Chromatographia, 19*, 288–291 (1985).

18. W. Linder, W. Posch, O. S. Wolfbeis, and P. Trithart, *Chromatographia, 20*, 213–218 (1985).

19. J. P. Foley and J. G. Dorsey, *Anal. Chem., 55*, 730 (1983).

20. R. E. Majors, presented at the Pittsburgh Conference on Analytical Chemistry and Applied Spectroscopy, 1979, paper 116.

21. J. A. Apffel and H. McNair, *J. Chromatogr., 279*, 139 (1983).

22. K. Grob, Jr., H. P. Neukom, and R. Etter, *J. Chromatogr., 357*, 416–422 (1986).

23. K. Grob, Jr., *J. Chromatogr., 237*, 15 (1982).

24. H. J. Cortes, C. D. Pfeiffer, and B. E. Richter, *J. High Resol. Chromatogr. Chromatogr. Commun., 8*, 469 (1985).

25. T. V. Raglione, J. A. Troskosky, and R. A. Hartwick, *J. Chromatogr., 409*, 205–212 (1987).

26. T. V. Raglione, J. A. Troskosky, and R. A. Hartwick, *J. Chromatogr., 409*, 213 (1987).

27. T. V. Raglione and R. A. Hartwick, presented at the Pittsburgh Conference on Analytical Chemistry and Applied Spectroscopy, 1987, paper 144.

28. T. V. Raglione and R. A. Hartwick, *J. Chromatogr.,* in press.

29. J. C. Giddings, *J. High Resol. Chromatogr. Chromatogr. Commun., 10*, 312–323 (1987).

30. G. Guiochon, M. F. Gonnord, M. Zakaria, L. A. Beaver, and A. M. Siouffi, *Chromatographia, 17*, 121 (1983).

31. D. Shelley and M. French, presented at the Pittsburgh Conference on Analytical Chemistry and Applied Spectroscopy, 1987, paper 84.

32. S. H. Hsu, T. V. Raglione, S. A. Tomellini, T. V. Floyd, N. Sagliano, Jr., and R. A. Hartwick, *J. Chromatogr., 367*, 293 (1986).

INDEX

669

(*continued from front*)